HANDBOOK ON
DATA ENVELOPMENT ANALYSIS

Recent titles in the
INTERNATIONAL SERIES IN
OPERATIONS RESEARCH & MANAGEMENT SCIENCE
Frederick S. Hillier, Series Editor, *Stanford University*

Ramík, J. & Vlach, M. / *GENERALIZED CONCAVITY IN FUZZY OPTIMIZATION AND DECISION ANALYSIS*

Song, J. & Yao, D. / *SUPPLY CHAIN STRUCTURES: Coordination, Information and Optimization*

Kozan, E. & Ohuchi, A. / *OPERATIONS RESEARCH/ MANAGEMENT SCIENCE AT WORK*

Bouyssou et al. / *AIDING DECISIONS WITH MULTIPLE CRITERIA: Essays in Honor of Bernard Roy*

Cox, Louis Anthony, Jr. / *RISK ANALYSIS: Foundations, Models and Methods*

Dror, M., L'Ecuyer, P. & Szidarovszky, F. / *MODELING UNCERTAINTY: An Examination of Stochastic Theory, Methods, and Applications*

Dokuchaev, N. / *DYNAMIC PORTFOLIO STRATEGIES: Quantitative Methods and Empirical Rules for Incomplete Information*

Sarker, R., Mohammadian, M. & Yao, X. / *EVOLUTIONARY OPTIMIZATION*

Demeulemeester, R. & Herroelen, W. / *PROJECT SCHEDULING: A Research Handbook*

Gazis, D.C. / *TRAFFIC THEORY*

Zhu, J. / *QUANTITATIVE MODELS FOR PERFORMANCE EVALUATION AND BENCHMARKING*

Ehrgott, M. & Gandibleux, X. / *MULTIPLE CRITERIA OPTIMIZATION: State of the Art Annotated Bibliographical Surveys*

Bienstock, D. / *Potential Function Methods for Approx. Solving Linear Programming Problems*

Matsatsinis, N.F. & Siskos, Y. / *INTELLIGENT SUPPORT SYSTEMS FOR MARKETING DECISIONS*

Alpern, S. & Gal, S. / *THE THEORY OF SEARCH GAMES AND RENDEZVOUS*

Hall, R.W./*HANDBOOK OF TRANSPORTATION SCIENCE - 2nd Ed.*

Glover, F. & Kochenberger, G.A. / *HANDBOOK OF METAHEURISTICS*

Graves, S.B. & Ringuest, J.L. / *MODELS AND METHODS FOR PROJECT SELECTION: Concepts from Management Science, Finance and Information Technology*

Hassin, R. & Haviv, M./ *TO QUEUE OR NOT TO QUEUE: Equilibrium Behavior in Queueing Systems*

Gershwin, S.B. et al/ *ANALYSIS & MODELING OF MANUFACTURING SYSTEMS*

Maros, I./ *COMPUTATIONAL TECHNIQUES OF THE SIMPLEX METHOD*

Harrison, Lee & Neale/ *THE PRACTICE OF SUPPLY CHAIN MANAGEMENT: Where Theory And Application Converge*

Shanthikumar, Yao & Zijm/ *STOCHASTIC MODELING AND OPTIMIZATION OF MANUFACTURING SYSTEMS AND SUPPLY CHAINS*

Nabrzyski, J., Schopf, J.M., Węglarz, J./ *GRID RESOURCE MANAGEMENT: State of the Art and Future Trends*

Thissen, W.A.H. & Herder, P.M./ *CRITICAL INFRASTRUCTURES: State of the Art in Research and Application*

Carlsson, C., Fedrizzi, M., & Fullér, R./ *FUZZY LOGIC IN MANAGEMENT*

Soyer, R., Mazzuchi, T.A., & Singpurwalla, N.D./ *MATHEMATICAL RELIABILITY: An Expository Perspective*

Talluri, K. & van Ryzin, G./ *THE THEORY AND PRACTICE OF REVENUE MANAGEMENT*

Kavadias, S. & Loch, C.H./*PROJECT SELECTION UNDER UNCERTAINTY: Dynamically Allocating Resources to Maximize Value*

Sainfort, F., Brandeau, M.L., Pierskalla, W.P./ *HANDBOOK OF OPERATIONS RESEARCH AND HEALTH CARE: Methods and Applications*

* A list of the early publications in the series is at the end of the book *

HANDBOOK ON
DATA
ENVELOPMENT
ANALYSIS

edited by

William W. Cooper
University of Texas at Austin, U.S.A.

Lawrence M. Seiford
University of Michigan at Ann Arbor, U.S.A.

Joe Zhu
Worcester Polytechnic Institute, U.S.A.

KLUWER ACADEMIC PUBLISHERS
Boston / Dordrecht / London

Distributors for North, Central and South America:
Kluwer Academic Publishers
101 Philip Drive
Assinippi Park
Norwell, Massachusetts 02061 USA
Telephone (781) 871-6600
Fax (781) 871-9045
E-Mail: kluwer@wkap.com

Distributors for all other countries:
Kluwer Academic Publishers Group
Post Office Box 322
3300 AH Dordrecht, THE NETHERLANDS
Telephone 31 786 576 000
Fax 31 786 576 254
E-mail: services@wkap.nl

 Electronic Services <http://www.wkap.nl>

Library of Congress Cataloging-in-Publication Data

A C.I.P. Catalogue record for this book is available from the Library of Congress.

HANDBOOK ON DATA ENVELOPMENT ANALYSIS, edited by William W. Cooper,
Lawrence M. Seiford, and Joe Zhu

ISBN 1-4020-7797-1

Printed on acid-free paper.

Printed in the United States of America.

To Alec
J.Z.

CONTENTS

Preface   xi
William W. Cooper, Lawrence M. Seiford and Joe Zhu

1   **Data Envelopment Analysis:**   1
     History, Models and Interpretations
     William W. Cooper, Lawrence M. Seiford and Joe Zhu

2   **Returns to Scale in DEA**   41
     Rajiv D. Banker, William W. Cooper,
     Lawrence M. Seiford and Joe Zhu

3   **Sensitivity Analysis in DEA**   75
     William W. Cooper, Shanling Li, Lawrence M. Seiford
     and Joe Zhu

4   **Incorporating Value Judgments in DEA**   99
     Emmanuel Thanassoulis, Maria Conceição Portela, and
     Rachel Allen

5   **Distance Functions with Applications to DEA**   139
     Rolf Färe, Shawna Grosskopf and Gerald Whittaker

6   **Qualitative Data in DEA**   153
     Wade D. Cook

7   **Congestion: Its Identification and Management**   177
     with DEA
     William W. Cooper, Honghui Deng,
     Lawrence M. Seiford and Joe Zhu

8   **Malmquist Productivity Index:**   203
     Efficiency Change Over Time
     Kaoru Tone

9   **Chance Constrained DEA**   **229**
     William W. Cooper, Zhimin Huang and Susan X. Li

10   **Performance of the Bootstrap for DEA**   **265**
     Estimators and Iterating the Principle
     Léopold Simar and Paul W.Wilson

11   **Statistical Tests Based on DEA Efficiency Scores**   **299**
     Rajiv D. Banker and Ram Natarajan

12   **Performance Evaluation in Education:**   **323**
     Modeling Educational Production
     John Ruggiero

13   **Assessing Bank and Bank Branch Performance:**   **349**
     Modeling Considerations and Approaches
     Joseph C. Paradi, Sandra Vela and Zijiang Yang

14   **Engineering Applications of Data Envelopment**   **401**
     Analysis: Issues and Opportunities
     Konstantinos P. Triantis

15   **Benchmarking in Sports: Bonds or Ruth:**   **443**
     Determining the Most Dominant Baseball Batter
     Using DEA
     Timothy R. Anderson

16   **Assessing the Selling Function in Retailing:**   **455**
     Insights from Banking, Sales forces, Restaurants
     & Betting shops
     Antreas D. Athanassopoulos

17   **Health Care Applications: From Hospitals to**   **481**
     Physicians, From Productive Efficiency to
     Quality Frontiers
     Jon A. Chilingerian and H. David Sherman

18   **DEA Software Tools and Technology:**   **539**
         A State-of-the-Art Survey
         Richard Barr

         Notes about Authors   **567**

         Author Index   **579**

         Subject Index   **587**

Preface

Data Envelopment Analysis (DEA) is a relatively new "data-oriented" approach for evaluating the performances of a set of entities called Decision-Making Units (DMUs) which convert multiple inputs into multiple outputs. DEA has been used in evaluating the performances of many different kinds of entities engaged in many different kinds of activities in many different contexts. It has opened up possibilities for use in cases which have been resistant to other approaches because of the complex and often unknown nature of the relations between the multiple inputs and outputs involved in many of these activities, which are often reported in non-commeasurable units. DEA has also been used to supply new insights into activities and entities that have previously been evaluated by other methods.

This handbook is intended to represent a milestone in the progression of DEA. Written by experts, who are often major contributors to the topics to be covered, it includes a comprehensive review and discussion of basic DEA models, extensions to the basic DEA methods, and a collection of DEA applications in the areas of banking, education, sports, retail, health care, and a review of current DEA software technology.

This handbook's chapters are organized into three categories: (i) basic DEA models, concepts, and their extensions; (ii) DEA applications; and (iii)

DEA software packages. The first category consists of eleven chapters. Chapter 1, by Cooper, Seiford and Zhu, covers the various models and methods for treating "technical" and "allocative" efficiency. It includes a new "additive" model for treating "allocative" and "overall" efficiency that can be used when the usual "ratio" form of the efficiency measure gives unsatisfactory or misleading results. Chapter 2, by Banker, Cooper, Seiford and Zhu, deals with returns to scale (RTS) and the ways in which this topic is treated with different models and methods. The emphasis in this chapter is on relationships between models and methods and the RTS characterizations that they produce. This chapter also introduces a new method for determining "exact" elasticities of scale in place of previous approaches, which are limited because they can only establish "bounds" on the elasticities. Chapter 3, by Cooper, Li, Seiford and Zhu, describes ways to determine the "stability" and "sensitivity" of DEA efficiency evaluations in the presence of stipulated variations in the data. The sensitivity analyses covered in this chapter extend from variations in *one* "data point' and include determining the sensitivity of DEA efficiency evaluations when *all* data points are varied simultaneously.

In Chapter 4, Thanassoulis, Portela and Allen treat ways for analysts or decision-makers to incorporate value judgments into DEA analyses, including *a priori* information in both the "dual" (or "multiplier") and "direct" (or "envelopment") models. The authors then describe the effects of incorporating such information on other parts of a DEA such as possible effects on RTS characterizations. Chapter 5, by Färe, Grosskopf and Whittaker, applies distance functions to DEA. The treatment in this chapter covers distance functions and their duality relations, as defined in the relations between distance and cost functions established by R.W. Shephard — which are here extended to "directed distance" functions and the profit function duals that can be obtained from them. Chapter 6, by Cook, discusses how to treat qualitative data in DEA. The emphasis is on cases in which the data are ordinal and not cardinal. This extends DEA so it can treat problems in which the data can be ordered but the numbers utilized to represent the ordering do not otherwise lend themselves to the usual arithmetic operations such as addition, multiplication, etc.

Chapter 7, by Cooper, Deng, Seiford and Zhu, treats "congestion" and discusses modeling to identify the amounts and sources of this particularly severe form of "technical" inefficiency. Chapter 8, by Tone, provides a comprehensive study of Malmquist productivity-index-number calculations for use in identifying and evaluating technology and efficiency changes that may occur between different periods. This chapter also includes a new "slacks-based" index formulation that reflects *all* inefficiencies that the model can identify. This approach removes an inadequacy that may be present in the usual calculations, which do not reflect inefficiencies associated with non-zero slack.

The final three chapters in this category are directed to probabilistic and statistical characterizations of the efficiency evaluation models discussed in Chapter 1. Chapter 9, by Cooper, Huang and Li, turns to probabilistic formulations as in "chance-constrained programming." "Joint" chance constraints, as well as the more customary types, are covered. All of this is accompanied by discussions of uses of both types of constraints in some of the applications of these chance-constrained-programming formulations of DEA. Chapter 10 by Simar and Wilson utilizes the relatively recently developed methods associated with "bootstrapping" and shows how these methods may be used to obtain statistical tests and estimates of DEA results. Chapter 11 by Banker and Natarajan is directed to the more classical methods of "statistically consistent estimates". Hence both classical and more recently developed approaches are brought to bear on statistical characterizations that are now available for use with DEA.

The second category of the topics covered in this handbook involves six DEA applications chapters. Chapter 12, by Ruggiero, deals with DEA applications in education, along with a discussion of the treatment of non-discretionary variables. Chapter 13, by Paradi, Vela and Yang, provides a detailed discussion of DEA applications to banking with an emphasis on factors, circumstance and formulations that need to be considered in actual applications. It also includes a comprehensive list of DEA bank branch models in the literature. Chapter 14, by Triantis, discusses DEA applications in engineering and includes a comprehensive bibliography of published DEA engineering applications. As this chapter shows, engineering uses of DEA have been relatively few but this is a field that is rich with potential applications of DEA ranging from engineering designs to uses of DEA to evaluate performances and to locate deficiencies in already functioning systems. Chapter 15, by Anderson, provides an application to the sport of baseball and uses the concept of "super-efficiency" in DEA to supply insights into the debate on who had the most dominant baseball batting season: Babe Ruth or Barry Bonds? Chapter 16, by Athanassopoulos, deals with DEA applications in the retail trades with reports on case studies based on research undertaken by Athanassopoulos in commercial banking, restaurants, brewing markets and betting shops in the U.K. Chapter 17, by Chilingerian and Sherman, offers a succinct history of health care applications of DEA and discusses the models and the motivations behind the applications with an eight-step application procedure and some "do's and don'ts" in DEA health care applications with an emphasis on the need for including "quality" measures of the services provided.

The final category consists of Chapter 18, by Barr, which provides a detailed and comprehensive review of currently available commercial and non-commercial DEA software packages and related technologies.

We hope this DEA handbook can serve as a comprehensive reference for researchers and practitioners and as a guide for further developments and uses of DEA. We welcome your comments, criticisms, and suggestions.

William W. Cooper
University of Texas at Austin
Austin, TX University Station 1 B6500
cooperw@mail.utexas.edu

Lawrence M. Seiford
University of Michigan at Ann Arbor
Ann Arbor, MI 48109-2117
seiford@umich.edu

Joe Zhu
Worcester Polytechnic Institute
Worcester, MA 01609
jzhu@wpi.edu

Chapter 1

DATA ENVELOPMENT ANALYSIS
History, Models and Interpretations

William W. Cooper[1], Lawrence M. Seiford[2] and Joe Zhu[3]
*[1] Red McCombs School of Business, University of Texas at Austin, Austin, TX 78712 USA
email: cooperw@mail.utexas.edu*

*[2] Department of Industrial and Operations Engineering, University of Michigan at Ann Arbor,
Ann Arbor, MI 48102 USA   email: seiford@umich.edu*

*[3] Department of Management, Worcester Polytechnic Institute, Worcester, MA 01609   USA
email: jzhu@wpi.edu*

Abstract: In a relatively short period of time Data Envelopment Analysis (DEA) has grown into a powerful quantitative, analytical tool for measuring and evaluating performance. DEA has been successfully applied to a host of different types of entities engaged in a wide variety of activities in many contexts worldwide. This chapter discusses the fundamental DEA models and some of their extensions.

Key words: Data envelopment analysis (DEA); Efficiency; Performance

1.   INTRODUCTION

Data Envelopment Analysis (DEA) is a relatively new "data oriented" approach for evaluating the performance of a set of peer entities called Decision Making Units (DMUs) which convert multiple inputs into multiple outputs. The definition of a DMU is generic and flexible. Recent years have seen a great variety of applications of DEA for use in evaluating the performances of many different kinds of entities engaged in many different activities in many different contexts in many different countries. These DEA applications have used DMUs of various forms to evaluate the performance of entities, such as hospitals, US Air Force wings, universities, cities, courts, business firms, and others, including the performance of countries, regions, etc. Because it requires very few assumptions, DEA has also opened up possibilities for use in cases which have been resistant to other approaches because of the complex (often unknown) nature of the relations between the multiple inputs and multiple outputs involved in DMUs.

As pointed out in Cooper, Seiford and Tone (2000), DEA has also been used to supply new insights into activities (and entities) that have previously been evaluated by other methods. For instance, studies of benchmarking practices with DEA have identified numerous sources of inefficiency in some of the most profitable firms - firms that had served as benchmarks by reference to this (profitability) criterion – and this has provided a vehicle for identifying better benchmarks in many applied studies. Because of these possibilities, DEA studies of the efficiency of different legal organization forms such as "stock" vs. "mutual" insurance companies have shown that previous studies have fallen short in their attempts to evaluate the potentials of these different forms of organizations. Similarly, a use of DEA has suggested reconsideration of previous studies of the efficiency with which pre- and post-merger activities have been conducted in banks that were studied by DEA.

Since DEA in its present form was first introduced in 1978, researchers in a number of fields have quickly recognized that it is an excellent and easily used methodology for modeling operational processes for performance evaluations. This has been accompanied by other developments. For instance, Zhu (2002) provides a number of DEA spreadsheet models that can be used in performance evaluation and benchmarking. DEA's empirical orientation and the absence of a need for the numerous *a priori* assumptions that accompany other approaches (such as standard forms of statistical regression analysis) have resulted in its use in a number of studies involving efficient frontier estimation in the governmental and nonprofit sector, in the regulated sector, and in the private sector. See, for instance, the use of DEA to guide removal of the Diet and other government agencies from Tokyo to locate a new capital in Japan, as described in Takamura and Tone (2003).

In their originating study, Charnes, Cooper, and Rhodes (1978) described DEA as a 'mathematical programming model applied to observational data [that] provides a new way of obtaining empirical estimates of relations - such as the production functions and/or efficient production possibility surfaces – that are cornerstones of modern economics'.

Formally, DEA is a methodology directed to frontiers rather than central tendencies. Instead of trying to fit a regression plane through the *center* of the data as in statistical regression, for example, one 'floats' a piecewise linear surface to rest on top of the observations. Because of this perspective, DEA proves particularly adept at uncovering relationships that remain hidden from other methodologies. For instance, consider what one wants to mean by "efficiency", or more generally, what one wants to mean by saying that one DMU is more efficient than another DMU. This is accomplished in a straightforward manner by DEA without requiring explicitly formulated assumptions and variations with various types of models such as in linear and nonlinear regression models.

Relative efficiency in DEA accords with the following definition, which has the advantage of avoiding the need for assigning a priori measures of relative importance to any input or output,

Definition 1.1 (Efficiency – Extended Pareto-Koopmans Definition): Full (100%) efficiency is attained by any DMU if and only if none of its inputs or outputs can be improved without worsening some of its other inputs or outputs.

In most management or social science applications the theoretically possible levels of efficiency will not be known. The preceding definition is therefore replaced by emphasizing its uses with only the information that is empirically available as in the following definition:

Definition 1.2 (Relative Efficiency): A DMU is to be rated as fully (100%) efficient on the basis of available evidence if and only if the performances of other DMUs does not show that some of its inputs or outputs can be improved without worsening some of its other inputs or outputs.

Notice that this definition avoids the need for recourse to prices or other assumptions of weights which are supposed to reflect the relative importance of the different inputs or outputs. It also avoids the need for explicitly specifying the formal relations that are supposed to exist between inputs and outputs. This basic kind of efficiency, referred to as "technical efficiency" in economics can, however, be extended to other kinds of efficiency when data such as prices, unit costs, etc., are available for use in DEA.

In this chapter we discuss the mathematical programming approach of DEA that implements the above efficiency definition. Section 2 of this chapter provides a historical perspective on the origins of DEA. Section 3 provides a description of the original "CCR ratio model" of Charnes, Cooper, and Rhodes (1978) which relates the above efficiency definition to other definitions of efficiency such as the ones used in engineering and science, as well as in business and economics. Section 4 describes some methodological extensions that have been proposed. Section 5 expands the development to concepts like "allocative" (or price) efficiency which can add additional power to DEA when unit prices and costs are available. This is done in section 5 and extended to profit efficiency in section 6 after which a conclusion section 7 is supplied.

2.   BACKGROUND AND HISTORY

In an article which represents the inception of DEA, Farrell (1957) was motivated by the need for developing better methods and models for evaluating productivity. He argued that while attempts to solve the problem usually produced careful measurements, they were also very restrictive because they failed to combine the measurements of multiple inputs into any satisfactory overall measure of efficiency. Responding to these inadequacies of separate indices of labor productivity, capital productivity, etc., Farrell proposed an activity analysis approach that could more adequately deal with the problem. His measures were intended to be applicable to any productive organization; in his words, '... from a workshop to a whole economy'. In the process, he extended the concept of "productivity" to the more general concept of "efficiency".

Our focus in this chapter is on basic DEA models for measuring the efficiency of a DMU *relative* to similar DMUs in order to estimate a 'best practice' frontier. The initial DEA model, as originally presented in Charnes, Cooper, and Rhodes (CCR) (1978), built on the earlier work of Farrell (1957).

This work by Charnes, Cooper and Rhodes originated in the early 1970s in response to the thesis efforts of Edwardo Rhodes at Carnegie Mellon University's School of Urban & Public Affairs - now the H.J. Heinz III School of Public Policy and Management. Under the supervision of W.W. Cooper, this thesis was to be directed to evaluating educational programs for disadvantaged students (mainly black or Hispanic) in a series of large scale studies undertaken in U.S. public schools with support from the Federal government. Attention was finally centered on Program Follow Through - a huge attempt by the U.S. Office (now Department) of Education to apply principles from the statistical design of experiments to a set of matched schools in a nation-wide study. Rhodes secured access to the data being processed for that study by Abt Associates, a Boston based consulting film, under contract with the US Office of Education. The data base was sufficiently large so that issues of degrees of freedom, etc., were not a serious problem despite the numerous input and output variables used in the study. Nevertheless, unsatisfactory and even absurd results were secured from all of the statistical-econometric approaches that Rhodes attempted to use.

While trying to respond to this situation, Rhodes called Cooper's attention to M.J. Farrell's seminal article "The Measurement of Productive Efficiency," in the 1957 Journal of the Royal Statistical Society. In this article Farrell used "activity analysis concepts" to correct what he believed were deficiencies in commonly used index number approaches to productivity (and like) measurements.

Cooper had previously worked with A. Charnes in order to give computationally implementable form to Tjalling Koopmans' "activity analysis concepts." So, taking Farrell's statements at face value, Cooper and Rhodes formalized what was involved in the definitions that were given in section 1 of this chapter. These definitions then provided the guides that were used for their subsequent research.

The name of Pareto is assigned to the first of these two definitions for the following reasons. In his Manual of Political Economy (1906) the Swiss-Italian economist, Vilfredo Pareto, established the basis of modern "welfare economics", i.e., the part of economics concerned with evaluating public policies, by noting that a social policy could be justified if it made some persons better off without making others worse off. In this way the need for making comparisons between the value of the gains to some and the losses to others could be avoided. This avoids the necessity of ascertaining the "utility functions" of the affected individuals and/or to "weight" the relative importance of each individual's gains and losses.

This property, known as the "Pareto criterion" as used in welfare economics, was carried over, or adapted, in Activity Analysis of Production and Allocation, a book edited by Koopmans (1951). In this context, it was "final goods" which were accorded this property, in that they were all constrained so that no final good was allowed to be improved if this improvement resulted in worsening one or more other final goods. These final goods (=outputs) were to be satisfied in stipulated amounts while inputs were to be optimally determined in response to the prices and amounts exogenously fixed for each output (=final good). Special attention was then directed by Koopmans to "efficiency prices" which are the prices associated with efficient allocation of resources (=inputs) to satisfy the pre-assigned demands for final goods. For a succinct summary of the mechanisms involved in this "activity analysis" approach, see p. 299 in Charnes and Cooper (1961).

Pareto and Koopmans were concerned with analyses of entire economies. In such a context it is reasonable to allow input prices and quantities to be determined by reference to their ability to satisfy final demands. Farrell, however, extended the Pareto-Koopmans property to inputs as well as outputs and explicitly eschewed any use of prices and/or related "exchange mechanisms." Even more importantly, he used the performance of other DMUs to evaluate the behavior of each DMU relative to the outputs and the inputs they all used. This made it possible to proceed empirically to determine their relative efficiencies.

The resulting measure which is referred to as the "Farrell measure of efficiency," was regarded by Farrell as restricted to meaning "technical efficiency" or the amount of "waste" that can be eliminated without worsening any input or output. This was then distinguished by Farrell from "allocative" and "scale" efficiencies as adapted from the literature of economics. These additional efficiencies will be discussed later in this chapter where the extensions needed to deal with problems that were encountered in DEA attempts to use these concepts in actual applications will also be discussed. Here we want to note that Farrell's approach to efficiency evaluations, as embodied in the "Farrell measure," carries with it an assumption of equal access to inputs by all DMUs. This does not mean that all DMUs use the same input amounts, however, and, indeed, part of their efficiency evaluations will depend on the input amounts used by each DMU as well as the outputs which they produce.

This "equal access assumption" is a mild one, at least as far as data availability is concerned. It is less demanding than the data and other requirements needed to deal with aspects of performance such as "allocative" or "scope" and "scale efficiencies." Furthermore, as discussed below, this assumption can now be relaxed. For instance, one can introduce "non-discretionary variables and constraints" to deal with conditions beyond the control of a DMU's management--in the form of "exogenously" fixed resources which may differ for each DMU. One can also introduce "categorical variables" to insure that evaluations are effected by reference to DMUs which have similar characteristics, and still other extensions and relaxations are possible, as will be covered in the discussions that follow.

To be sure, the definition of efficiency that we have referred to as "Extended Pareto-Koopmans Efficiency" and "Relative Efficiency" were formalized by Charnes, Cooper and Rhodes rather than Farrell. However, these definitions conform both to Farrell's models and the way Farrell used them. In any case, these were the definitions that Charnes, Cooper and Rhodes used to guide the developments that we next describe.

The Program Follow Through data with which Rhodes was concerned in his thesis recorded "outputs" like "increased self esteem in a disadvantaged child" and "inputs" like "time spent by a mother in reading with her child," as measured by psychological tests and prescribed record keeping and reporting practices. Farrell's elimination of the need for information on prices proved attractive for dealing with outputs and inputs like these--as reported for each of the schools included in the Program Follow Through experiment.

Farrell's empirical work had been confined to single-output cases and his sketch of extensions to multiple outputs did not supply what was required for applications to large data sets like those involved in Program Follow Through. To obtain what was needed in computationally implementable form, Charnes, Cooper and Rhodes developed the dual pair of linear programming problems that are modeled in the next section, section 3. It was then noticed that Farrell's measure failed to account for the non-zero slacks, which is where the changes in proportions connected with mix inefficiencies are located (in both outputs and inputs). The possible presence of non-zero slack as a source of these mix inefficiencies also requires attention even when restricted to "technical efficiency."

We now emphasize the problems involved in dealing with these slacks because a considerable part of the DEA (and related) literatures continues to be deficient in its treatment of non-zero slack even today. A significant part of the problem to be dealt with, as we noted above, involves the possible presence of alternate optima in which the same value of the Farrell measure could be associated with zero slack in some optima but not in others. Farrell introduced "points at infinity" in what appears to have been an attempt to deal with this problem but was unable to give operationally implementable form to this concept. Help in dealing with this problem was also not available from the earlier work of Sidney Afriat (1972), Ronald Shephard (1970) or Gerhard Debreu (1951). To address this problem, Charnes, Cooper and Rhodes introduced mathematical concepts that are built around the "non-Archimedean" elements associated with $\varepsilon > 0$ which handles the problem by insuring that slacks are always maximized without altering the value of the Farrell measure.

The dual problems devised by Cooper and Rhodes readily extended the above ideas to multiple outputs and multiple inputs in ways that could locate inefficiencies in each input and each output for every DMU. Something more was nevertheless desired in the way of summary measures. At this point, Cooper invited A. Charnes to join him and Rhodes in what promised to be a very productive line of research. Utilizing the earlier work of Charnes and Cooper (1962), which had established the field of "fractional programming," Charnes was able to put the dual linear programming problems devised by Cooper and Rhodes into the equivalent ratio form represented in (1.1) below and this provided a basis for unifying what had been done in DEA with long standing approaches to efficiency evaluation and analysis used in other fields, such as engineering and economics.

Since the initial study by Charnes, Cooper, and Rhodes some 2000 articles have appeared in the literature. See Cooper, Seiford and Tone

(2000). See also G. Tavares (2003). Such rapid growth and widespread (and almost immediate) acceptance of the methodology of DEA is testimony to its strengths and applicability. Researchers in a number of fields have quickly recognized that DEA is an excellent methodology for modeling operational processes, and its empirical orientation and minimization of a *priori* assumptions has resulted in its use in a number of studies involving efficient frontier estimation in the nonprofit sector, in the regulated sector, and in the private sector.

At present, DEA actually encompasses a variety of alternate (but related) approaches to evaluating performance. Extensions to the original CCR work have resulted in a deeper analysis of both the "multiplier side" from the dual model and the "envelopment side" from the primal model of the mathematical duality structure. Properties such as isotonicity, nonconcavity, economies of scale, piecewise linearity, Cobb-Douglas loglinear forms, discretionary and nondiscretionary inputs, categorical variables, and ordinal relationships can also be treated through DEA. Actually the concept of a frontier is more general than the concept of a "production function" which has been regarded as fundamental in economics in that the frontier concept admits the possibility of multiple production functions, one for each DMU, with the frontier boundaries consisting of "supports" which are "tangential" to the more efficient members of the set of such frontiers.

3.   CCR DEA MODEL

To allow for applications to a wide variety of activities, we use the term Decision Making Unit (=DMU) to refer to any entity that is to be evaluated in terms of its abilities to convert inputs into outputs. These evaluations can involve governmental agencies and not-for-profit organizations as well as business firms. The evaluation can also be directed to educational institutions and hospitals as well as police forces (or subdivision thereof) or army units for which comparative evaluations of their performance are to be made.

We assume that there are n DMUs to be evaluated. Each DMU consumes varying amounts of m different inputs to produce s different outputs. Specifically, DMU_j consumes amount x_{ij} of input i and produces amount y_{rj} of output r. We assume that $x_{ij} \geq 0$ and $y_{rj} \geq 0$ and further assume that each DMU has at least one positive input and one positive output value.

We now turn to the "ratio-form" of DEA. In this form, as introduced by Charnes, Cooper, and Rhodes, the ratio of outputs to inputs is used to measure the relative efficiency of the $DMU_j = DMU_o$ to be evaluated relative to the ratios of all of the $j = 1, 2, ..., n$ DMU_j. We can interpret the

CCR construction as the reduction of the multiple-output /multiple-input situation (for each DMU) to that of a single 'virtual' output and 'virtual' input. For a particular DMU the ratio of this single virtual output to single virtual input provides a measure of efficiency that is a function of the multipliers. In mathematical programming parlance, this ratio, which is to be maximized, forms the objective function for the particular DMU being evaluated, so that symbolically

$$\max h_o(u,v) = \sum_r u_r y_{ro} / \sum_i v_i x_{io} \tag{1.1}$$

where it should be noted that the variables are the u_r's and the v_i's and the y_{ro}'s and x_{io}'s are the observed output and input values, respectively, of DMU_o, the DMU to be evaluated. Of course, without further additional constraints (developed below) (1.1) is unbounded.

A set of normalizing constraints (one for each DMU) reflects the condition that the virtual output to virtual input ratio of every DMU, including $DMU_j = DMU_o$, must be less than or equal to unity. The mathematical programming problem may thus be stated as

$$\max h_o(u,v) = \sum_r u_r y_{ro} / \sum_i v_i x_{io} \tag{1.2}$$
subject to
$$\sum_r u_r y_{rj} / \sum_i v_i x_{ij} \le 1 \text{ for } j = 1, \dots, n,$$
$$u_r, v_i \ge 0 \text{ for all } i \text{ and } r.$$

<u>Remark:</u> A fully rigorous development would replace $u_r, v_i \ge 0$ with

$$\frac{u_r}{\sum_{i=1}^{m} v_i x_{io}}, \frac{u_r}{\sum_{i=1}^{m} v_i x_{io}} \ge \varepsilon > 0 \text{ where } \varepsilon \text{ is a non-Archimedean element smaller than}$$

any positive real number. See Arnold et al. (1998). This condition guarantees that solutions will be positive in these variables. It also leads to the $\varepsilon > 0$ in (1.6) which, in turn, leads to the 2nd stage optimization of the slacks as in (1.10).

The above ratio form yields an infinite number of solutions; if (u^*, v^*) is optimal, then $(\alpha u^*, \alpha v^*)$ is also optimal for $\alpha > 0$. However, the transformation developed by Charnes and Cooper (1962) for linear fractional programming selects a representative solution [i.e., the solution (u, v) for which $\sum_{i=1}^{m} v_i x_{io} = 1$] and yields the equivalent linear programming problem in which the change of variables from (u, v) to (μ, v) is a result of the Charnes-Cooper transformation,

$$\max z = \sum_{r=1}^{s} \mu_r y_{ro}$$

subject to

$$\sum_{r=1}^{s} \mu_r y_{rj} - \sum_{i=1}^{m} v_i x_{ij} \le 0 \qquad (1.3)$$

$$\sum_{i=1}^{m} v_i x_{io} = 1$$

$$\mu_r, v_i \ge 0$$

for which the LP dual problem is

$$\theta^* = \min \theta$$

subject to

$$\sum_{j=1}^{n} x_{ij} \lambda_j \le \theta x_{io} \qquad i = 1,2,...,m; \qquad (1.4)$$

$$\sum_{j=1}^{n} y_{rj} \lambda_j \ge y_{ro} \qquad r = 1,2,...,s;$$

$$\lambda_j \ge 0 \qquad j = 1,2,...,n.$$

This last model, (1.4), is sometimes referred to as the "Farrell model" because it is the one used in Farrell (1957). In the economics portion of the DEA literature it is said to conform to the assumption of "strong disposal" because it ignores the presence of non-zero slacks. In the operations research portion of the DEA literature this is referred to as "weak efficiency."

Possibly because he used the literature of "activity analysis"[1] for reference, Farrell also failed to exploit the very powerful dual theorem of linear programming which we have used to relate the preceding problems to each other. This also caused computational difficulties for Farrell because he did not take advantage of the fact that activity analysis models can be converted to linear programming equivalent that provide immediate access to the simplex and other methods for efficiently solving such problems. See, e.g., Charnes and Cooper (1961, Ch. IX). We therefore now begin to bring these features of linear programming into play.

By virtue of the dual theorem of linear programming we have $z^* = \theta^*$. Hence either problem may be used. One can solve say (1.4), to obtain an efficiency score. Because we can set $\theta = 1$ and $\lambda_k^* = 1$ with $\lambda_k^* = \lambda_o^*$ and all other $\lambda_j^* = 0$, a solution of (1.4) always exists. Moreover this solution implies $\theta^* \le 1$. The optimal solution, θ^*, yields an efficiency score for a particular DMU. The process is repeated for each DMU_j i.e., solve (1.4), with $(X_o, Y_o) = (X_k, Y_k)$, where (X_k, Y_k) represent vectors with components x_{ik}, y_{rk} and, similarly (X_o, Y_o) has components x_{ok}, y_{ok}. DMUs for which $\theta^* < 1$ are inefficient, while DMUs for which $\theta^* = 1$ are boundary points.

Some boundary points may be "weakly efficient" because we have non-zero slacks. This may appear to be worrisome because alternate optima may

---

[1] See T.C. Koopmans (1951).

have non-zero slacks in some solutions, but not in others. However, we can avoid being worried even in such cases by invoking the following linear program in which the slacks are taken to their maximal values.

$$\max \sum_{i=1}^{m} s_i^- + \sum_{r=1}^{s} s_r^+$$

subject to

$$\sum_{j=1}^{n} x_{ij}\lambda_j + s_i^- = \theta^* x_{io} \quad i = 1,2,...,m; \quad (1.5)$$

$$\sum_{j=1}^{n} y_{rj}\lambda_j - s_r^+ = y_{ro} \quad r = 1,2,...,s;$$

$$\lambda_j, s_i^-, s_r^+ \geq 0 \ \forall i, j, r$$

where we note the choices of s_i^- and s_r^+ do not affect the optimal θ^* which is determined from model (1.4).

These developments now lead to the following definition based upon the "relative efficiency" definition 1.2 which was given in section 1 above.

Definition 1.3 (DEA Efficiency): The performance of DMU_o is fully (100%) efficient if and only if both (i) $\theta^* = 1$ and (ii) all slacks $s_i^{-*} = s_r^{+*} = 0$.

Definition 1.4 (Weakly DEA Efficient): The performance of DMU_o is weakly efficient if and only if both (i) $\theta^* = 1$ and (ii) $s_i^{-*} \neq 0$ and/or $s_r^{+*} \neq 0$ for some i and r in some alternate optima.

It is to be noted that the preceding development amounts to solving the following problem in two steps:

$$\min \theta - \varepsilon(\sum_{i=1}^{m} s_i^- + \sum_{r=1}^{s} s_r^+)$$

subject to

$$\sum_{j=1}^{n} x_{ij}\lambda_j + s_i^- = \theta x_{io} \quad i = 1,2,...,m; \quad (1.6)$$

$$\sum_{j=1}^{n} y_{rj}\lambda_j - s_i^+ = y_{ro} \quad r = 1,2,...,s;$$

$$\lambda_j, s_i^-, s_r^+ \geq 0 \ \forall i, j, r$$

where the s_i^- and s_r^+ are slack variables used to convert the inequalities in (1.4) to equivalent equations. Here $\varepsilon > 0$ is a so-called non-Archimedean element defined to be smaller than <u>any</u> positive real number. This is equivalent to solving (1.4) in two stages by first minimizing θ, then fixing $\theta = \theta^*$ as in (1.2), where the slacks are to be maximized without altering the previously determined value of $\theta = \theta^*$. Formally, this is equivalent to granting "preemptive priority" to the determination of θ^* in (1.3). In this manner, the fact that the non-Archimedean element ε is defined to be smaller than any positive real number is accommodated without having to specify the value of ε.

Alternately, one could have started with the output side and considered instead the ratio of virtual input to output. This would reorient the objective from max to min, as in (1.2), to obtain

$$\text{Min } \sum_i v_i x_{io} / \sum_r u_r y_{ro}$$
$$\text{Subject to}$$
$$\sum_i v_i x_{ij} / \sum_r u_r y_{rj} \geq 1 \text{ for } j = 1, \ldots, n, \tag{1.7}$$
$$u_r, v_i \geq \varepsilon > 0 \text{ for all } i \text{ and } r.$$

where $\varepsilon > 0$ is the previously defined non-Archimedean element.

Again, the Charnes-Cooper (1962) transformation for linear fractional programming yields model (1.8) (multiplier model) below, with associated dual problem, (1.9) (envelopment model), as in the following pair,

$$\min q = \sum_{i=1}^{m} v_i x_{io}$$
$$\text{subject to}$$
$$\sum_{i=1}^{m} v_i x_{ij} - \sum_{r=1}^{s} \mu_r y_{rj} \geq 0 \tag{1.8}$$
$$\sum_{r=1}^{s} \mu_r y_{ro} = 1$$
$$\mu_r, v_i \geq \varepsilon, \quad \forall r, i$$

$$\max \phi + \varepsilon \left(\sum_{i=1}^{m} s_i^- + \sum_{r=1}^{s} s_r^+ \right)$$
$$\text{subject to}$$
$$\sum_{j=1}^{n} x_{ij} \lambda_j + s_i^- = x_{io} \quad i = 1,2,\ldots,m; \tag{1.9}$$
$$\sum_{j=1}^{n} y_{rj} \lambda_j - s_r^+ = \phi y_{ro} \quad r = 1,2,\ldots,s;$$
$$\lambda_j \geq 0 \qquad\qquad j = 1,2,\ldots,n.$$

See pp 75-76 in Cooper, Seiford and Tone (2000) for a formal development of this transformation and modification of the expression for $\varepsilon > 0$. See also the remark following (1.2).

Here we are using a model with an output oriented objective as contrasted with the input orientation in (1.6). However, as before, model (1.9) is calculated in a two-stage process. First, we calculate ϕ^* by ignoring the slacks. Then we optimize the slacks by fixing ϕ^* in the following linear programming problem,

$$\max \sum_{i=1}^{m} s_i^- + \sum_{r=1}^{s} s_r^+$$

subject to

$$\sum_{j=1}^{n} x_{ij}\lambda_j + s_i^- = x_{io} \qquad i = 1,2,...,m; \qquad (1.10)$$

$$\sum_{j=1}^{n} y_{rj}\lambda_j - s_r^+ = \phi^* y_{ro} \qquad r = 1,2,...,s;$$

$$\lambda_j \geq 0 \qquad j = 1,2,...,n.$$

We then modify the previous input-oriented definition of DEA efficiency to the following output-oriented version.

Definition 1.5: DMU_o is efficient if and only if $\phi^* = 1$ and $s_i^{-*} = s_r^{+*} = 0$ for all i and r. DMU_o is weakly efficient if $\phi^* = 1$ and $s_i^{-*} \neq 0$ and (or) $s_r^{+*} \neq 0$ for some i and r in some alternate optima.

Table 1-1 presents the CCR model in input- and output-oriented versions, each in the form of a pair of dual linear programs.

Table 1-1. CCR DEA Model

Input-oriented	
Envelopment model	Multiplier model
$\min \theta - \varepsilon(\sum_{i=1}^{m} s_i^- + \sum_{r=1}^{s} s_r^+)$ subject to $\sum_{j=1}^{n} x_{ij}\lambda_j + s_i^- = \theta x_{io} \quad i = 1,2,...,m;$ $\sum_{j=1}^{n} y_{rj}\lambda_j - s_r^+ = y_{ro} \quad r = 1,2,...,s;$ $\lambda_j \geq 0 \qquad j = 1,2,...,n.$	$\max z = \sum_{r=1}^{s} \mu_r y_{ro}$ subject to $\sum_{r=1}^{s} \mu_r y_{rj} - \sum_{i=1}^{m} v_i x_{ij} \leq 0$ $\sum_{i=1}^{m} v_i x_{io} = 1$ $\mu_r, v_i \geq \varepsilon > 0$
Output-oriented	
Envelopment model	Multiplier model
$\max \phi + \varepsilon(\sum_{i=1}^{m} s_i^- + \sum_{r=1}^{s} s_r^+)$ subject to $\sum_{j=1}^{n} x_{ij}\lambda_j + s_i^- = x_{io} \quad i = 1,2,...,m;$ $\sum_{j=1}^{n} y_{rj}\lambda_j - s_r^+ = \phi y_{ro} \quad r = 1,2,...,s;$ $\lambda_j \geq 0 \qquad j = 1,2,...,n.$	$\min q = \sum_{i=1}^{m} v_i x_{io}$ subject to $\sum_{i=1}^{m} v_i x_{ij} - \sum_{r=1}^{s} \mu_r y_{rj} \geq 0$ $\sum_{r=1}^{s} \mu_r y_{ro} = 1$ $\mu_r, v_i \geq \varepsilon > 0$

These are known as CCR (Charnes, Cooper, Rhodes, 1978) models. If the constraint $\sum_{j=1}^{n} \lambda_j = 1$ is adjoined, they are known as BCC (Banker, Charnes, Cooper, 1984) models. This added constraint introduces an additional variable, μ_0, into the (dual) multiplier problems. As will be seen in the next chapter, this extra variable makes it possible to effect returns-to-scale evaluations (increasing, constant and decreasing). So the BCC model is also referred to as the VRS (Variable Returns to scale) model and distinguished form the CCR model which is referred to as the CRS (Constant Returns to Scale) model.

We now proceed to compare and contrast the input and output orientations of the CCR model. To illustrate the discussion to follow we will employ the example presented in Figure 1-1 consisting of five DMUs, labeled P1, …, P5, each consuming a single input to produce a single output.

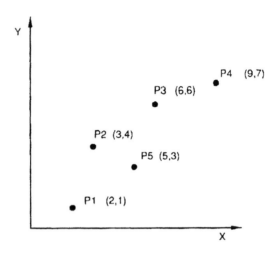

Figure 1-1. Example DMUs

The numbers in parentheses in Figure 1-1 are interpreted as coordinate values which correspond to the input and output of a *DMU_j* represented as Pj, j = 1, …, 5. In each case the value on the left in the parentheses is the input and the value on the right is the output for the Pj alongside which these values are listed.

To assist the reader in verifying the model interpretations which follow, Table 1-2 contains optimal solution values for the five example DMUs for both of the dual LP problems of the CCR model. For example, to evaluate the efficiency of P5 (*DMU₅* in Table 1-2), we can solve the following input-oriented envelopment CCR model:

$$\min \theta$$
subject to
$$2\lambda_1 + 3\lambda_2 + 6\lambda_3 + 9\lambda_4 + 5\lambda_5 \le 5\theta \quad (input)$$
$$1\lambda_1 + 4\lambda_2 + 6\lambda_3 + 7\lambda_4 + 3\lambda_5 \ge 3 \quad (output)$$
$$\lambda_1, \lambda_2, \lambda_3, \lambda_4, \lambda_5 \ge 0$$

which yields the values of $\theta^* = 9/20$, $\lambda_2^* = ¾$ and $\lambda_j^* = 0$ ($j \neq 2$) (see the last two columns in the last row of the upper portion of Table 1-2).

Alternatively, we can solve the input-oriented multiplier CCR model,

$$\max z = 3\mu$$
$$subject \quad to$$
$$1\mu - 2v \leq 0 \quad (P1)$$
$$4\mu - 3v \leq 0 \quad (P2)$$
$$6\mu - 6v \leq 0 \quad (P3)$$
$$7\mu - 9v \leq 0 \quad (P4)$$
$$3\mu - 5v \leq 0 \quad (P5)$$
$$5v = 1$$
$$\mu, v \geq 0$$

which yields $z^* = 9/20$, $\mu^* = 3/20$ and $v^* = 1/5$. Hence we have $\theta^* = z^*$. Moreover, with $\mu^* = 3/20$ and $v^* = 1/5$, this is also the value of $h_o(u^*, v^*) = 3 \cdot \frac{3}{20} / 5 \cdot \frac{1}{5} = \frac{9}{20}$ for the corresponding ratio model obtained from (1.2).

Table 1-2. Optimal solution values for the CCR model

	DMU	z^*	μ^*	v^*	θ^*	λ^*
	1	3/8	3/8	1/2	3/8	$\lambda_2 = 1/4$
	2	1	¼	1/3	1	$\lambda_2 = 1$
Input-oriented	3	¾	1/8	1/6	¾	$\lambda_2 = 3/2$
	4	7/12	1/12	1/9	7/12	$\lambda_2 = 7/4$
	5	9/20	3/20	1/5	9/20	$\lambda_2 = 3/4$
	1	8/3	1	4/3	8/3	$\lambda_2 = 2/3$
	2	1	¼	1/3	1	$\lambda_2 = 1$
Output-oriented	3	4/3	1/6	2/9	4/3	$\lambda_2 = 2$
	4	12/7	1/7	4/21	12/7	$\lambda_2 = 3$
	5	20/9	1/3	4/9	20/9	$\lambda_2 = 5/3$

A DMU is inefficient if the efficiency score given by the optimal value for the LP problem is less than one ($\theta^* < 1$ or $z^* < 1$). If the optimal value is equal to one *and* if there exist positive optimal multipliers ($\mu_r > 0$, $v_i > 0$), then the DMU is efficient. Thus, all efficient points lie on the frontier. However, a DMU can be a boundary point ($\theta^* = 1$) and be inefficient. Note that the complementary slackness condition of linear programming yields a condition for efficiency which is equivalent to the above; the constraints involving X_0 and Y_0, must hold with equality, i.e., $X_0 = X\lambda^*$ and $Y_0 = Y\lambda^*$ for all optimal λ^*, where X_0 and Y_0 are vectors and X and Y are matrices.

An inefficient DMU can be made more efficient by projection onto the frontier. In an input orientation one improves efficiency through proportional reduction of inputs, whereas an output orientation requires proportional augmentation of outputs. However, it is necessary to distinguish between a boundary point and an efficient boundary point. Moreover, the efficiency of a boundary point can be dependent upon the model orientation.

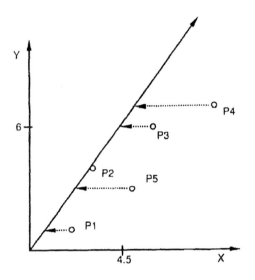

Figure 1-2. Projection to frontier for the input-oriented CCR model

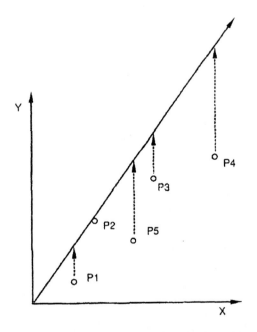

Figure 1-3. Projection to frontier for the output-oriented CCR model

The efficient frontier and DEA projections are provided in Figures 1-2 and 1-3 for the input-oriented and output-oriented CCR models, respectively. In both cases, the efficient frontier obtained from the CCR model is the ray $\{\alpha(x_2, y_2)| \ \alpha \geq 0\}$, where x_2 and y_2 are the coordinates of P2.

As can be seen from the points designated by the arrow head, an inefficient DMU may be projected to different points on the frontier under the two orientations. However, the following theorem provides a correspondence between solutions for the two models.

Theorem 1.1: Let (θ^*, λ^*) be an optimal solution for the input oriented model in (1.9). Then $(1/\theta^*, \lambda^*/\theta^*) = (\phi^*, \hat{\lambda}^*)$ is optimal for the corresponding output oriented model. Similarly if $(\phi^*, \hat{\lambda}^*)$ is optimal for the output oriented model then $(1/\phi^*, \hat{\lambda}^*/\phi^*) = (\theta^*, \lambda^*)$ is optimal for the input oriented model. The correspondence need not be 1-1, however, because of the possible presence of alternate optima.

For an input orientation the projection $(X_0,Y_0) \rightarrow (\theta^* X_0, Y_0)$ always yields a boundary point. But technical efficiency is achieved only if also all slacks are zero in all alternate optima so that $\theta^* X_0 = X \lambda^*$ and $Y_0 = Y \lambda^*$ for all optimal λ^*. Similarly, the output-oriented projection $(X_0,Y_0) \rightarrow (X_0, \phi^* Y_0)$ yields a boundary point which is efficient (technically) only if $\phi^* Y_0 = Y \lambda^*$ and $X_0 = X \lambda^*$ for all optimal λ^*. That is, the constraints are satisfied as equalities in all alternate optima for (1.4). To achieve technical efficiency the appropriate set of constraints in the CCR model must hold with equality.

To illustrate this, we consider a simple numerical example used in Zhu (2002) as shown in Table 1-3 where we have five DMUs representing five supply chain operations. Within a week, each DMU generates the same profit of $2,000 with a different combination of supply chain cost and response time.

Table 1-3. Supply Chain Operations Within a Week

| DMU | Inputs | | Output |
	Cost ($100)	Response time (days)	Profit ($1,000)
1	1	5	2
2	2	2	2
3	4	1	2
4	6	1	2
5	4	4	2

Source: Zhu (2002).

We now turn to the BCC model for which Figure 1-4 presents the five DMUs and the piecewise linear DEA frontier. DMUs 1, 2, 3, and 4 are on the frontier. If we adjoin the constraint $\sum_{j=1}^{n} \lambda_j = 1$ to model (1.4) for DMU5, we get from the data of Table 1-3,

$$\text{Min } \theta$$

Subject to
$$1\,\lambda_1 + 2\lambda_2 + 4\lambda_3 + 6\lambda_4 + 4\lambda_5 \le 4\theta$$
$$5\,\lambda_1 + 2\lambda_2 + 1\lambda_3 + 1\lambda_4 + 4\lambda_5 \le 4\theta$$
$$2\,\lambda_1 + 2\lambda_2 + 2\lambda_3 + 2\lambda_4 + 2\lambda_5 \ge 2$$
$$\lambda_1 + \lambda_2 + \lambda_3 + \lambda_4 + \lambda_5 = 1$$
$$\lambda_1, \lambda_2, \lambda_3, \lambda_4, \lambda_5 \ge 0$$

This model has the unique optimal solution of $\theta^* = 0.5$, $\lambda_2^* = 1$, and $\lambda_j^* = 0$ ($j \neq 2$), indicating that DMU5 needs to reduce its cost and response time to the amounts used by DMU2 if it is to be efficient This example indicates that technical efficiency for DMU5 is achieved at DMU_2 on the boundary.

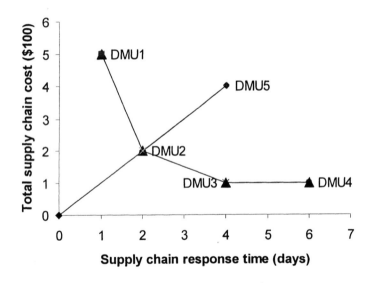

Figure 1-4. Five Supply Chain Operations

Source: Zhu (2002).

Now, if we similarly use model (1.4) with $\sum_{j=1}^{n} \lambda_j = 1$ for DMU4, we obtain $\theta^* = 1$, $\lambda_4^* = 1$, and $\lambda_j^* = 0$ ($j \neq 4$), indicating that DMU4 is on the frontier and is a boundary point. However, Figure 1-4 indicates that DMU4 can still reduce its response time by 2 days to achieve coincidence with DMU3 on the efficiency frontier. This input reduction is the input slack and the constraint with which it is associated is satisfied as a strict inequality in this solution. Hence, DMU4 is weakly efficient.

The nonzero slack can be found by using model (1.5). With the constraint $\sum_{j=1}^{n} \lambda_j = 1$ adjoined and setting $\theta^* = 1$ yields the following model,

$$\text{Max } s_1^- + s_2^- + s_1^+$$
Subject to
$$1\,\lambda_1 + 2\lambda_2 + 4\lambda_3 + 6\lambda_4 + 4\lambda_5 + s_1^- = 6\,\theta^* = 6$$
$$5\,\lambda_1 + 2\lambda_2 + 1\lambda_3 + 1\lambda_4 + 4\lambda_5 + s_2^- = 1\,\theta^* = 1$$
$$2\,\lambda_1 + 2\lambda_2 + 2\lambda_3 + 2\lambda_4 + 2\lambda_5 - s_1^+ = 2$$
$$\lambda_1 + \lambda_2 + \lambda_3 + \lambda_4 + \lambda_5 = 1$$
$$\lambda_1,\, \lambda_2,\, \lambda_3, \lambda_4,\, \lambda_5,\, s_1^-,\, s_2^-,\, s_1^+ \geq 0$$

The optimal slacks are $s_1^{-*} = 2$, $s_2^{-*} = s_1^{+*} = 0$, with $\lambda_3^* = 1$ and all other $\lambda_j^* = 0$.

4. EXTENSIONS TO THE CCR MODEL

A number of useful enhancements have appeared in the literature. We here limit our coverage to five of the extensions that illustrate the adaptability of the basic DEA methodology. The extensions discussed below allow an analyst to treat both nondiscretionary and categorical inputs and outputs and to incorporate judgment or ancillary managerial information. They are also easily extended to investigate efficiency changes over multiple time periods, and to measure congestion.

4.1 Nondiscretionary Inputs and Outputs

The above model formulations implicitly assume that all inputs and outputs are discretionary, i.e., can be controlled by the management of each DMU and varied at its discretion. Thus, failure of a DMU to produce maximal output levels with minimal input consumption results in a worsened efficiency score. However, there may exist exogenously fixed (or nondiscretionary) inputs or outputs that are beyond the control of a DMU's management. Instances from the DEA literature include snowfall or weather in evaluating the efficiency of maintenance units, soil characteristics and topography in different farms, number of competitors in the branches of a restaurant chain, local unemployment rates which affect the ability to attract recruits by different U.S. Army recruitment stations, age of facilities in different universities, and number of transactions (for a purely gratis service) in library performance.

For example, Banker and Morey (1986a), whose formulations we use, illustrate the impact of exogenously determined inputs that are not controllable in an analysis of a network of fast food restaurants. In their study, each of the 60 restaurants in the fast food chain consumes six inputs to produce three outputs. The three outputs (all controllable) correspond to breakfast, lunch, and dinner sales. Only two of the six inputs, expenditures for supplies and expenditures for labor, are discretionary. The other four inputs (age of store, advertising level as determined by national headquarters, urban/rural location, and drive-in capability) are beyond the control of the individual restaurant manager in this chain.

The key to the proper mathematical treatment of a nondiscretionary variable lies in the observation that information about the extent to which a nondiscretionary input variable may be reduced is beyond the discretion of the individual DMU managers and thus cannot be used by them.

Suppose that the input and output variables may each be partitioned into subsets of discretionary (D) and nondiscretionary (N) variables. Thus,

$$I = \{1, 2, ..., m\} = I_D \cup I_N \text{ with } I_D \cap I_N = \varnothing$$

and

$$O = \{1, 2, ..., s\} = O_D \cup O_N \text{ with } O_D \cap O_N = \varnothing$$

where I_D, O_D and I_N, O_N refer to discretionary (D) and nondiscretionary (N) input, I, and output, O, variables, respectively and $\varnothing$ is the empty set.

To evaluate managerial performance in a relevant fashion we may need to distinguish between discretionary and non-discretionary inputs as is done in the following modified version of a CCR model.

$$\min \theta - \varepsilon \left(\sum_{i \in I_D} s_i^- + \sum_{r=1}^{s} s_r^+ \right)$$

subject to

$$\sum_{j=1}^{n} x_{ij} \lambda_j + s_i^- = \theta x_{io} \qquad i \in I_D;$$

$$\sum_{j=1}^{n} x_{ij} \lambda_j + s_i^- = x_{io} \qquad i \in I_N \tag{1.11}$$

$$\sum_{j=1}^{n} y_{rj} \lambda_j - s_r^+ = y_{ro} \qquad r = 1,2,...,s;$$

$$\lambda_j \geq 0 \qquad j = 1,2,...,n.$$

It is to be noted that the θ to be minimized appears only in the constraints for which $i \in I_D$, whereas the constraints for which $i \in I_N$ operate only indirectly (as they should) because the input levels x_{io} for $i \in I_N$, are not subject to managerial control. It is also to be noted that the slack variables associated with I_N, the non-discretionary inputs, are not included in the objective of (1.11) and hence the non-zero slacks for these inputs do not enter directly into the efficiency scores to which the objective is oriented.

The necessary modifications to incorporate nondiscretionary variables for the output-oriented CCR model is given by

$$\max \phi + \varepsilon(\sum_{i=1}^{m} s_i^- + \sum_{r \in O_D} s_r^+)$$

subject to

$$\sum_{j=1}^{n} x_{ij}\lambda_j + s_i^- = x_{io} \qquad i = 1,2,...,m;$$

$$\sum_{j=1}^{n} y_{rj}\lambda_j - s_r^+ = \phi y_{ro} \qquad r \in O_D;$$ (1.12)

$$\sum_{j=1}^{n} y_{rj}\lambda_j - s_r^+ = y_{ro} \qquad r \in O_N;$$

$$\lambda_j \geq 0 \qquad\qquad j = 1,2,...,n.$$

We should point out that there can be subtle issues associated with the concept of controllable outputs that may be obscured by the symmetry of the input/output model formulations. Specifically, switching from an input to an output orientation is not always as straightforward as it may appear. Interpretational difficulties for outputs not directly controllable may be involved as in the case of outputs influenced through associated input factors. An example of such an output would be sales that are influenced by advertising from the company headquarters, but are not directly controllable by district mangers. Finally, chapter 12 by Ruggiero provides some treatments for nondiscretionary variables.

4.2   Categorical Inputs and Outputs

Our previous development assumed that all inputs and outputs were in the same category. However this need not be the case as when some restaurants in a fast food chain have a dive-in facility and some do not. See Banker and Morey (1986b) for a detailed discussion.

To see how this can be handled, suppose that an input variable can assume one of L levels $(1, 2, ..., L)$. These L values effectively partition the set of DMUs into categories. Specifically, the set of DMUs $K = \{1, 2, ..., n\}$ $= K_1 \cup K_2 \cup ... \cup K_L$, where $K_f = \{j \mid j \in K$ and input value is $f\}$ and $K_i \cap K_j = \emptyset$, $i \neq j$. We wish to evaluate a DMU with respect to the envelopment surface determined for the units contained in it and all preceding categories. The following model specification allows $DMU_o \in K_f$.

$$\min \theta$$
subject to
$$\sum_{j \in \bigcup_{f=1}^{K_f} K_f} x_{ij} \lambda_j + s_i^- = \theta x_{io} \qquad i = 1,...,m; \qquad (1.13)$$

$$\sum_{j \in \bigcup_{f=1}^{K_f} K_f} y_{rj} \lambda_j - s_r^+ = y_{ro} \qquad r = 1,2,...,s;$$

$$\lambda_j \geq 0 \qquad\qquad\qquad j = 1,2,...,n.$$

Thus, the above specification allows one to evaluate all DMUs $l \in D_1$ with respect to the units in K_1, all DMUs $l \in K_2$ with respect to the units in $K_1 \cup K_2$, ..., all DMUs $l \in K_C$ with respect to the units in $\bigcup_{f=1}^{K_C} K_f$, etc. Although our presentation is for the input-oriented CCR model, it should be obvious that categorical variables can also be incorporated in this manner for any DEA model. In addition, the above formulation is easily implemented in the underlying LP solution algorithm via a candidate list.

The preceding development rests on the assumption that there is a natural nesting or hierarchy of the categories. Each DMU should be compared only with DMUs in its own and more disadvantaged categories, i.e., those operating under the same or worse conditions. If the categories are not comparable (e.g., public universities vs. private universities), then a separate analysis should be performed for each category.

4.3   Incorporating Judgment or A Priori Knowledge

Perhaps the most significant of the proposed extensions to DEA is the concept of restricting the possible range for the multipliers. In the CCR model, the only explicit restriction on multipliers is positivity, as noted for the $\varepsilon > 0$ in (1.8). This flexibility is often presented as advantageous in applications of the DEA methodology, since a priori specification of the multipliers is not required. and each DMU is evaluated in its best possible light.

In some situations, however, this complete flexibility may give rise to undesirable consequences, since it can allow a DMU to appear efficient in ways that are difficult to justify. Specifically, the model can assign unreasonably low or excessively high values to the multipliers in an attempt to drive the efficiency rating for a particular DMU as high as possible.

Three situations for which it has proven beneficial to impose various levels of control are the following:

1. the analysis would otherwise ignore additional information that cannot be directly incorporated into the model that is used, e.g., the envelopment model;

2.   management has strong preferences about the relative importance of different factors and what determines best practice; and

3.   for a small sample of DMUs, the method fails to discriminate, and all are efficient.

Using the multiplier models that duality theory makes available for introducing restrictions on the multipliers can affect the solutions that can be obtained from the corresponding envelopment model. Proposed techniques for enforcing these additional restrictions include imposing upper and lower bounds on individual multipliers (Dyson and Thanassoulis, 1988; Roll, Cook, and Golany, 1991); imposing bounds on ratios of multipliers (Thompson et al., 1986); appending multiplier inequalities (Wong and Beasley, 1990); and requiring multipliers to belong to given closed cones (Charnes et al., 1989).

To illustrate the general approach, suppose we wish to incorporate additional inequality constraints of the following form into (1.3) or, more generally, into the multiplier model in Table 1-1:

$$\alpha_i \le \frac{v_i}{v_{i_o}} \le \beta_i, \qquad i = 1,...,m$$

$$\delta_r \le \frac{\mu_r}{\mu_{r_o}} \le \gamma_r, \qquad r = 1,...,s \tag{1.14}$$

Here, v_{i_o} and μ_{r_o} represent multipliers which serve as "numeraires" in establishing the upper and lower bounds represented here by α_i, β_i, and by δ_r, γ_r for the multipliers associated with inputs i =1, ..., m and outputs r = 1, ..., s where $\alpha_{i_o} = \beta_{i_o} = \delta_{r_o} = \gamma_{r_o} = 1$. The above constraints are called Assurance Region (AR) constraints as developed by Thompson et al. (1986) and defined more precisely in Thompson et al. (1990).

Uses of such bounds are not restricted to prices. They may extend to "utils" or any other units that are regarded as pertinent. For example, Zhu (1996a) uses an assurance region approach to establish bounds on the weights obtained from uses of Analytic Hierarchy Processes in Chinese textile manufacturing in order to include bounds on these weights that better reflect local government preferences in measuring textile manufacturing performances.

There is another approach called the "cone-ratio envelopment approach" which can also be used for this purpose. See Charnes et al. (1990). We do not examine this approach in detail, but rather only note that the assurance region approach can also be given an interpretation in terms of cones. See Cooper et al. (1996).

The generality of these AR constraints provides flexibility in use. Prices, utils and other measures may be accommodated and so can mixtures of such concepts. Moreover, one can first examine provisional solutions and then tighten or loosen the bounds until one or more solutions is attained that appears to be reasonably satisfactory to decision makers who cannot state the values for their preferences in an a priori manner.

The assurance region approach also greatly relaxes the conditions and widens the scope for use of a priori conditions. In some cases, the conditions to be comprehended may be too complex for explicit articulation, in which case additional possibilities are available from other recent advances. For instance, *instead* of imposing bounds on allowable variable values, the cone-ratio envelopment approach *transforms* the data. Brockett et al. (1997) provide an example in which a bank regulatory agency wanted to evaluate "risk coverage" as well as the "efficiency" of the banks under its jurisdiction. Bounds could not be provided on possible tradeoffs between risk coverage and efficiency, so this was accomplished by using a set of banks identified as "excellent" (even when they were not members of the original (regulatory) set). Then, employing data from these excellent banks, a cone-ratio envelopment was used to transform the data into improved values that could be used to evaluate each of the regulated banks operating under widely varying conditions. This avoided the need for jointly specifying what was meant by "adequate" risk coverage and efficiency not only in each detail, but also in all of the complex interplays between risk and efficiency that are possible in bank performances. The non-negativity imposed on the slacks in standard CCR model was also relaxed. This then made it possible to identify deficiencies which were to be repaired by increasing expense items such as "bad loan allowances" (as needed for risk coverage) even though this worsened efficiency as evaluated by the transformed data.

There are in fact a number of ways in which we can incorporate a priori knowledge or requirements as conditions into DEA models. For examples see the "Prioritization Models" of Cook et al. (1992) and the "Preference Structure Model" of Zhu (1996b). See Chapter 4 for a detailed discussion on incorporation of value judgments.

4.4   Window Analysis

In the examples of the previous sections, each DMU was observed only once, i.e., each example was a cross-sectional analysis of data. In actual studies, observations for DMUs are frequently available over multiple time periods (time series data), and it is often important to perform an analysis where interest focuses on changes in efficiency over time. In such a setting, it is possible to perform DEA over time by using a moving average analogue, where a DMU in each different period is treated as if it were a "different" DMU. Specifically, a DMU's performance in a particular period is contrasted with its performance in other periods in addition to the performance of the other DMUs.

The *window analysis* technique that operationalizes the above procedure can be illustrated with the study of aircraft maintenance operations, as described in Charnes et al. (1985). In this study, data were obtained for 14 *(n* = 14) tactical fighter wings in the U.S. Air Force over seven (p = 7) monthly periods. To perform the analysis using a three-month (w = 3) window, one proceeds as follows.

Each DMU is represented as if it were a different DMU for each of the three successive months in the first window (Ml, M2, M3) consisting of the months at the top of Table 1-4. An analysis of the 42 (= *nw* = 3 × 14) DMUs can then be performed. The window is then shifted one period by replacing M1 with M4, and an analysis is performed on the second three-month set (M2, M3, M4) of these 42 DMUs. The process continues in this manner, shifting the window forward one period each time and concluding with the final (fifth) analysis of 42 DMUs for the last three months (M5, M6, M7). (In general, one performs p - w + 1 separate analyses, where each analysis examines *nw* DMUs).

Table 1-4 illustrates the results of this analysis in the form of efficiency scores for the performance of the airforce wings as taken from Charnes et al. (1985). The structure of this table portrays the underlying framework of the analysis. For the first "window," wing A is represented in the constraints of the DEA model as though it were a different DMU in months 1, 2, and 3. Hence, when wing 1 is evaluated for its month-1 efficiency, its own performance data for months 2 and 3 are included in the constraint sets along with similar performance data of the other wings for months 1, 2, and 3. Thus the results of the "first window" analysis consist of the 42 scores under the column headings for Month 1-Month 3 in the first row for each wing. For example, wing A had efficiency ratings of 97.89, 97.31, and 98.14 for its performance in months 1, 2, and 3, respectively, as shown in the first row for Wing A in Table 1-4. The second row of data for each wing is the result of analyzing the second window of 42 DMUs, which result from dropping the month-1 data and appending the month-4 data.

The arrangement of the results of a window analysis as given in Table 1-4 facilitates the identification of trends in performance, the stability of reference sets, and other possible insights. For illustration, "row views" clarify performance trends for wings E and M. Wing E improved its performance in month 5 relative to prior performance in months 3 and 4 in the third window, while wing M's performance appears to deteriorate in months 6 and 7. Similar "column views" allow comparison of wings (DMUs) across different reference sets and hence provide information on the stability of these scores as the reference sets change.

Table 1-4. Window Analysis with Three-Month Window

Wing	Month 1	Month 2	Month 3	Month 4	Month 5	Month 6	Month 7
Wing-A	97.89	97.31	98.14				
		97.36	97.53	97.04			
			96.21	95.92	94.54		
				95.79	94.63	97.64	
					94.33	97.24	97.24
Wing-B	93.90	95.67	96.14				
		96.72	96.42	94.63			
			95.75	94.14	93.26		
				94.54	93.46	96.02	
					93.02	96.02	94.49
Wing-C	93.77	91.53	95.26				
		91.77	95.55	94.29			
			93.21	95.04	94.83		
				93.20	93.09	92.21	
					93.59	92.32	92.83
Wing-D	99.72	96.15	95.06				
		97.91	95.70	100.0			
			94.79	100.0	94.51		
				99.71	94.39	94.76	
					94.95	94.67	89.37
Wing-E	100.0	100.0	100.0				
		100.0	100.0	100.0			
			98.97	99.05	100.0		
				99.37	100.0	100.0	
					100.0	100.0	100.0
Wing-F	97.42	93.48	96.07				
		93.60	96.24	93.56			
			94.46	91.75	92.49		
				91.73	92.32	92.35	
					92.68	91.98	99.64
Wing-G	90.98	92.80	95.96				
		93.67	96.80	99.52			
			93.34	94.48	91.73		
				91.94	89.79	95.58	
					89.35	95.14	96.38
Wing-H	100.0	100.0	100.0				
		100.0	100.0	100.0			
			100.0	100.0	100.0		
				100.0	100.0	100.0	
					100.0	100.0	100.0
Wing-I	99.11	95.94	99.76				
		96.04	100.0	100.0			
			98.16	98.99	94.59		
				98.97	94.62	99.16	
					94.68	98.92	97.28
Wing-J	92.85	90.90	91.62				
		91.50	92.12	94.75			
			90.26	93.39	93.83		
				92.92	93.84	95.33	
					94.52	96.07	94.43
Wing-K	86.25	84.42	84.03				
		84.98	84.47	93.74			
			83.37	82.54	80.26		
				82.39	80.14	79.58	
					80.96	78.66	79.75
Wing-L	100.0	100.0	100.0				
		100.0	100.0	99.55			
			100.0	99.39	97.39		
				100.0	96.85	100.0	
					96.66	100.0	100.0
Wing-M	100.0	100.0	100.0				
		100.0	100.0	100.0			
			100.0	100.0	100.0		
				100.0	100.0	98.75	
					100.0	98.51	99.59
Wing-N	100.0	100.0	98.63				
		100.0	100.0	100.0			
			99.45	100.0	100.0		
				100.0	100.0	100.0	
					100.0	100.0	100.0

The utility of Table 1-4 can be further extended by appending columns of summary statistics (mean, median, variance, range, etc.) for each wing to reveal the relative stability of each wing's results. See, for instance, the drop in the efficiency from 93.74 to 82.54 in Month 4 for Wing K.

The window analysis technique represents one area for further research extending DEA. For example, the problem of choosing the width for a window (and the sensitivity of DEA solutions to window width) is currently determined by trial and error. Similarly, the theoretical implications of representing each DMU as if it were a different DMU for each period in the window remain to be worked out in full detail.

5.   ALLOCATIVE AND OVERALL EFFICIENCY

To this point we have confined attention to "technical efficiency" which, as explained immediately after definition 1.2, does not require a use of prices or other "weights." Now we extend the analysis to situations in which unit prices and unit costs are available. This allows us to introduce the concepts of "allocative" and "overall" efficiency and relate them to "technical efficiency" in a manner first introduced by M.J. Farrell (1957).

For this introduction we utilize Figure 1-5 in which the solid line segments connecting points ABCD constitute an "isoquant" or "level line' that represents the different amounts of two inputs (x_1, x_2) which can be used to produce the same amount (usually one unit) of a given output. This line represents the "efficiency frontier" of the "production possibility set" because it is not possible to reduce the value of one of the inputs without increasing the other input if one is to stay on this isoquant.

The dashed line represents an isocost (=budget) line for which (x_1, x_2) pairs on this line yield the same total cost, when the unit costs are c_1 and c_2, respectively. When positioned on C the total cost is k. However, shifting this budget line upward in parallel fashion until it reaches a point of intersection with R would increase the cost to $k' > k$. In fact, as this Figure shows, k is the minimum total cost needed to produce the specified output since any parallel shift downward below C would yield a line that fails to intersect the production possibility set. Thus, the intersection at C gives an input pair (x_1, x_2) that minimizes the total cost of producing the specified output amount and the point C is therefore said to be "allocatively" as well as "technically" efficient.

Now let R represent an observation that produced this same output amount. The ratio $0 \le OQ/OR \le 1$ is said to provide a "radial" measure of technical efficiency,[2] with $0 \leq 1 - (OQ/OR) \leq 1$ yielding a measure of technical inefficiency.

Now consider the point P which is at the intersection of this cost line through C with the ray from the origin to R. We can also obtain a radial measure of "overall efficiency" from the ratio $0 \leq OP/OR \leq 1$. In addition, we can form the ratio $0 \leq OP/OQ \leq 1$ to obtain a measure of what Farrell (1957) referred to as "price efficiency" but is now more commonly called "allocative efficiency." Finally, we can relate these three measures to each other by noticing that

$$\frac{OP}{OQ} \frac{OQ}{OR} = \frac{OP}{OR} \tag{1.15}$$

which we can verbalize by saying that the product of allocative and technical efficiency eqauls overall efficiency in these radial measures.

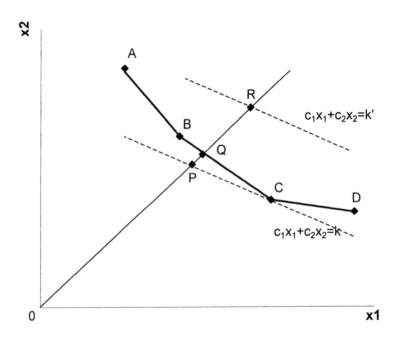

Figure 1-5. Allocative and Overall Efficiency

---

[2] Actually this "radial measure" is a ratio of two measures of distance, one of which measures the distance from the origin to Q and the other which measures the distance from the origin to R to obtain $0 \leq d(O,Q)/d(O,R) \leq 1$ where $d(....)$ means "distance." See I. Bardhan et al. (1996) for a proof in terms of the Euclidean measure of distance--although other measures of distance may also be used.

To implement these ideas we use the following model, as taken from Cooper, Seiford and Tone (2000, p. 236),

$$\min \sum_{i=1}^{m} c_{io} x_i$$

subject to

$$\sum_{j=1}^{n} x_{ij}\lambda_j \le x_i, \quad i = 1,...,m \qquad (1.16)$$

$$\sum_{j=1}^{n} y_{rj}\lambda_j \ge y_{ro}, \quad r = 1,...,s$$

$$L \le \sum_{j=1}^{n} \lambda_j \le U$$

where the objective is to choose the x_i and λ_j values to minimize the total cost of satisfying the output constraints. The c_{io} in the objective represent unit costs. This formulation differs from standard models, as in Färe, Grosskopf and Lovell (1985, 1994), in that these unit costs are allowed to vary from one DMU$_o$ to another in (1.16). In addition the values of $\sum_{j=1}^{n} \lambda_j$ are limited above by U and below by L--according to the returns-to-scale conditions that are imposed. See next chapter. Here we only note that the choice L=U=1 makes this a BCC model, whereas $L = 0$, $U = \infty$ converts it to a CCR model. See the discussion following Table 1-1. Finally, using the standard approach, we can obtain a measure of relative cost (=overall) efficiency by utilizing the ratio

$$0 \le \frac{\sum_{i=1}^{m} c_{io} x_i^*}{\sum_{i=1}^{m} c_{io} x_{io}} \le 1 \qquad (1.17)$$

where the x_i^* are the optimal values obtained from (1.16) and the x_{io} are the observed values for DMU$_o$.

The use of a ratio like (1.17) is standard and yields an easily understood measure. It has shortcomings, however, as witness the following example from Tone and Sahoo (2003): Let γ_a and γ_b represent cost efficiency, as determined from (1.17), for DMUs a and b. Now suppose $x_{ia}^* = x_{ib}^*$ and $x_{ia} = x_{ib}$, $\forall i$, but $c_{ia} = 2c_{ib}$ so that the unit costs for a are twice as high as for b in every input. We then have

$$\gamma_a = \frac{\sum_{i=1}^{m} c_{ia} x_{ia}^*}{\sum_{i=1}^{m} c_{ia} x_{ia}} = \frac{\sum_{i=1}^{m} 2c_{ib} x_{ib}^*}{\sum_{i=1}^{m} 2c_{ib} x_{ib}} = \frac{\sum_{i=1}^{m} c_{ib} x_{ib}^*}{\sum_{i=1}^{m} c_{ib} x_{ib}} = \gamma_b \qquad (1.18)$$

Thus, as might be expected with a use of ratios, important information may be lost since $\gamma_a = \gamma_b$ conceals the fact that a is twice as costly as b.

6. PROFIT EFFICIENCY

We now introduce another type of model called the "additive model" to evaluate technical inefficiency. First introduced in Charnes et al. (1985) this model has the form

$$
\begin{aligned}
&\max \; \sum_{r=1}^{s} s_r^+ + \sum_{i=1}^{m} s_i^- \\
&\text{subject to} \\
&y_{ro} = \sum_{j=1}^{n} y_{rj}\lambda_j - s_r^+, \quad r = 1,2,\ldots,s \\
&x_{io} = \sum_{j=1}^{n} x_{ij}\lambda_j + s_i^-, \quad i = 1,2,\ldots,m \\
&1 = \sum_{j=1}^{n} \lambda_j \\
&0 \le \lambda_j, s_r^+ \; s_i^-; \forall i,j,r.
\end{aligned}
\tag{1.19}
$$

This model uses a metric that differs from the one used in the "radial measure" model.[3] It also dispenses with the need for distinguishing between an "output" and an "input" orientation as was done in the discussion leading up to (1.10) because the objective in (1.19) simultaneously maximizes outputs and minimizes inputs--in the sense of vector optimizations. This can be seen by utilizing the solution to (1.19) to introduce new variables $\hat{y}_{ro}$, $\hat{x}_{io}$ defined as follows,

$$
\begin{aligned}
\hat{y}_{ro} &= y_{ro} + s_r^{+*} \ge y_{ro}, \quad r = 1,\ldots,s, \\
\hat{x}_{io} &= x_{io} - s_i^{-*} \le x_{io}, \quad i = 1,2,\ldots,m.
\end{aligned}
\tag{1.20}
$$

Now note that the slacks are all independent of each other. Hence an optimum is not reached until it is not possible to increase an output $\hat{y}_{ro}$ or reduce an input $\hat{x}_{io}$ without decreasing some other output or increasing some other input. The following theorem, which follows immediately, is proved in Cooper, Seiford and Tone (2000),

Theorem 1.2: DMUo is efficient if and only if all slacks are zero in an optimum solution.

---

[3] The additive model uses what is called the " ℓ_1 metric" in mathematics, and the "city block metric" in operations research. See Appendix A in Charnes and Cooper (1961).

As proved in Ahn, Charnes and Cooper (1988) one can also relate solutions in the additive model to those in the radial measure model via

Theorem 1.3: DMUo is efficient for an additive model if and only if it is efficient for the corresponding radial model.

Here the term "corresponding" means that the constraint sets are the same so that $\sum_{j=1}^{n}\lambda_j = 1$ appears as a constraint in the additive model if and only if it also appears in the radial measure model to which it is being compared.

We now use the class of additive models to develop a different route to treating technical, allocative and overall inefficiencies and their relations to each other. This can help to avoid difficulties in treating possibilities like "negative" or "zero" profits which are not easily treated by the ratio approaches, as in (1.17), which are commonly used in the DEA literature. See the discussion in the appendix to Cooper, Park and Pastor (1999) from which the following development is taken. See also Chapter 8 in Cooper, Seiford and Tone (2000).

First we observe that we can multiply the output slacks by unit prices and the input slacks by unit costs after we have solved (1.19) and thereby accord a monetary value to this solution. Then we can utilize (1.20) to write

$$
\sum_{r=1}^{s}p_{ro}s_r^{+*} + \sum_{i=1}^{m}c_{io}s_i^{-*} = \left(\sum_{r=1}^{s}p_{ro}\hat{y}_{ro} - \sum_{i=1}^{m}p_{ro}y_{ro}\right) + \left(\sum_{i=1}^{m}c_{io}x_{io} - \sum_{i=1}^{m}c_{io}\hat{x}_{io}\right)
$$
$$
= \left(\sum_{r=1}^{s}p_{ro}\hat{y}_{ro} - \sum_{i=1}^{m}c_{io}\hat{x}_{io}\right) - \left(\sum_{r=1}^{s}p_{ro}y_{ro} - \sum_{i=1}^{m}c_{io}x_{io}\right). \tag{1.21}
$$

From the last pair of parenthesized expressions we find that, at an optimum, the objective in (1.19) after multiplication by unit prices and costs is equal to the profit available when production is technically efficient minus the profit obtained from the observed performance. Hence when multiplied by unit prices and costs the solution to (1.19) provides a measure in the form of the amount of the profits lost by not performing in a technically efficient manner term by term if desired.

Remark: We can, if we wish restate this measure in ratio form because, by definition,

$$
\sum_{r=1}^{s}p_{ro}\hat{y}_{ro} - \sum_{i=1}^{m}c_{io}\hat{x}_{io} \geq \sum_{r=1}^{s}p_{ro}y_{ro} - \sum_{i=1}^{m}c_{io}x_{io}
$$

Therefore,

$$
1 \geq \frac{\sum_{r=1}^{s}p_{ro}y_{ro} - \sum_{i=1}^{m}c_{io}x_{io}}{\sum_{r=1}^{s}p_{ro}\hat{y}_{ro} - \sum_{i=1}^{m}c_{io}\hat{x}_{io}} \geq 0
$$

and the upper bound is attained if and only if performance is efficient.

We can similarly, develop a measure of allocative efficiency by means of the following additive model,

$$\max \sum_{r=1}^{s} p_{ro}\hat{s}_r^+ + \sum_{i=1}^{m} c_{io}\hat{s}_i^-$$

subject to

$$\hat{y}_{ro} = \sum_{j=1}^{n} y_{rj}\hat{\lambda}_j - \hat{s}_r^+, \quad r = 1,2,...,s$$

$$\hat{x}_{io} = \sum_{j=1}^{n} x_{ij}\hat{\lambda}_j - \hat{s}_i^-, \quad i = 1,2,...,m \qquad (1.22)$$

$$1 = \sum_{j=1}^{n} \hat{\lambda}_j$$

$$0 \le \hat{\lambda}_j \;\; \forall j; \; \hat{s}_i^-, \; \hat{s}_r^+ \text{ free } \forall i,r.$$

Comparison with (1.19) reveals the following differences: (1) the objective in (1.19) is replaced by one which is monetized (2) the y_{ro} and x_{io} in (1.19) are replaced by $\hat{y}_{ro}$ and $\hat{x}_{io}$ in (1.22) as obtained from (1.20) and, finally, (3) the slack values in (1.22) are not constrained in sign as is the case in (1.19). This last relaxation, we might note, is needed to allow for substitutions between the different output and the different input amounts, as may be needed to achieve the proportions required for allocative efficiency. See Cooper, Park and Pastor (2000).

Finally, we use the following additive model to evaluate overall (profit) efficiency--called "graph efficiency" in Färe, Grosskopf and Lovell (1985, 1994),

$$\max \sum_{r=1}^{s} p_{ro}s_r^+ + \sum_{i=1}^{m} c_{io}s_i^-$$

subject to

$$y_{ro} = \sum_{j=1}^{n} y_{rj}\lambda_j - s_r^+, \quad r = 1,2,...,s$$

$$x_{io} = \sum_{j=1}^{n} x_{ij}\lambda_j - \hat{s}_i^-, \quad i = 1,2,...,m \qquad (1.23)$$

$$1 = \sum_{j=1}^{n} \lambda_j$$

$$0 \le \lambda_j \;\; \forall j; \; s_i^-, \; s_r^+ \text{ free } \forall i,r.$$

We then have the relation set forth in the following

Theorem 1.4: The value (=total profit foregone) of overall inefficiency for DMUo as obtained from (1.23) is equal to the value of technical inefficiency as obtained from (1.19) plus the value of allocative inefficiency as obtained from (1.22). i.e.,

$$\max \left(\sum_{r=1}^{s} p_{ro}s_r^+ + \sum_{i=1}^{m} c_{io}s_i^- \right) = \left(\sum_{r=1}^{s} p_{ro}s_r^{+*} + \sum_{i=1}^{m} c_{io}s_i^{-*} \right) + \left(\sum_{r=1}^{s} p_{ro}\hat{s}_r^{+*} + \sum_{i=1}^{m} c_{io}\hat{s}_i^{-*} \right).$$

Table 1-5, as adapted from Cooper, Seiford and Tone (2000), can be used to construct an example to illustrate this theorem. The body of the table records the performances of 3 DMUs in terms of the amount of the one output they all produce, y, and the amount of the two inputs x_1, x_2 they all use. For simplicity it is assumed that they all receive the same unit price p=6 as recorded on the right and incur the same unit costs $c_1 = \$4$ and $c_2 = \$2$ for the inputs as shown in the rows with which they are associated. The bottom of each column records the profit, π, made by each DMU in the periods of interest.

Table 1-5. Price-Cost-Profit Data

	DMU$_1$	DMU$_2$	DMU$_3$	\$
y	4	4	2	6
x_1	4	2	4	4
x_2	2	4	6	2
π	4	8	-16	

DMU$_3$, as shown at the bottom of its column is, by far, the worst performer having experienced a loss of \$16. This loss, however, does not account for all of the lost profit possibilities. To discover this value we turn to (1.23) and apply it to the data in Table 1.5. This produces the following model to evaluate the overall inefficiency of DMU$_3$.

$$\max \ 6s^+ + 4s_1^- + 2s_2^-$$
$$\text{subject to}$$
$$2 = 4\lambda_1 + 4\lambda_2 + 2\lambda_3 - s^+$$
$$4 = 4\lambda_1 + 2\lambda_2 + 4\lambda_3 \qquad + s_1^- \qquad\qquad (1.24)$$
$$6 = 2\lambda_1 + 4\lambda_2 + 6\lambda_3 \qquad\qquad - s_2^-$$
$$1 = \lambda_1 + \lambda_2 + \lambda_3$$
$$0 \le \lambda_1, \ \lambda_2, \ \lambda_3 ; s^+, \ s_1^-, \ s_2^- \ \text{free}$$

An optimum solution to this problem is $\lambda_2^* = 1$, $s^{+*} = 2$, $s_1^{-*} = 2$, $s_2^{-*} = 2$ and all after variables zero. Utilizing the unit price and costs exhibited in the objective of (1.24) we therefore find

$$6s^{+*} + 4s_1^{-*} + 2s_1^{-*} = \$6 \times 2 + \$4 \times 2 + \$2 \times 2 = \$24 \ .$$

This is the value of the foregone profits arising from inefficiencies in the performance of DMU$_3$. Eliminating these inefficiencies would have wiped out the \$16 loss and replaced it with an \$8 profit. This is the same amount of profit as DMU$_2$, which is the efficient performer used to effect this evaluation of DMU$_3$ via $\lambda_2^* = 1$ in the above solution.

We now utilize theorem 1.4 to further identify the sources of this lost profit. For this purpose we first apply (1.19) to the data of Table 1-5 in order to determine the lost profits from the technical inefficiency of DMU$_3$ via

$$\begin{aligned}
&\max \ s^+ + s_1^- + s_1^- \\
&\text{subject to} \\
&\quad 2 = 4\lambda_1 + 4\lambda_2 + 2\lambda_3 - s^+ \\
&\quad 4 = 4\lambda_1 + 2\lambda_2 + 4\lambda_3 \quad\quad + s_1^- \\
&\quad 6 = 2\lambda_1 + 4\lambda_2 + 6\lambda_3 \quad\quad\quad\quad - s_2^- \\
&\quad 1 = \lambda_1 + \lambda_2 + \lambda_3 \\
&\quad 0 \le \lambda_1, \ \lambda_2, \ \lambda_3, \ s^+, \ s_1^-, \ s_2^-.
\end{aligned} \tag{1.25}$$

This has an optimum with $\lambda_1^* = 1$, $s^{+*} = 2$, $s_1^{-*} = 0$, $s_2^{-*} = 4$ and all other variables zero so that multiplying these values by their unit price and unit costs we find that the lost profits due to technical inefficiencies are

$$\$6 \times 2 + \$4 \times 0 + \$2 \times 4 = \$20 .$$

For allocative inefficiency we apply (1.20) and (1.22) to the data in Table 1-5 and get.

$$\begin{aligned}
&\max \ 6\hat{s}^+ + 4\hat{s}_1^- + 2\hat{s}_2^- \\
&\text{subject to} \\
&\quad 4 = 4\hat{\lambda}_1 + 4\hat{\lambda}_2 + 2\hat{\lambda}_3 - \hat{s}^+ \\
&\quad 4 = 4\hat{\lambda}_1 + 2\hat{\lambda}_2 + 4\hat{\lambda}_3 \quad\quad + \hat{s}_1^- \\
&\quad 2 = 2\hat{\lambda}_1 + 4\hat{\lambda}_2 + 6\hat{\lambda}_3 \quad\quad\quad\quad + \hat{s}_2^- \\
&\quad 1 = \hat{\lambda}_1 + \hat{\lambda}_2 + \hat{\lambda}_3 \\
&\quad 0 \le \hat{\lambda}_1, \ \hat{\lambda}_2, \ \hat{\lambda}_3; \ \hat{s}^+, \ \hat{s}_1^-, \ \hat{s}_2^- \text{ free.}
\end{aligned}$$

An optimum is $\hat{\lambda}_2^* = 1$, $\hat{s}^{+*} = 0$, $\hat{s}_1^{-*} = 2$, $\hat{s}_2^{-*} = -2$ with all other variables zero so the profit lost from allocative efficiency is

$$\$4 \times 2 + \$2(-2) = \$4 ,$$

which accounts for the remaining $4 of the $24 lost profit obtained from overall inefficiency via (1.24). Here we might note that an increase in $\hat{x}_2$ -- see the second expression in (1.20)--is more than compensated for by the offsetting decrease in $\hat{x}_1$ en route to the proportions needed to achieve allocative efficiency.

Finally we supply a tabulation obtained from these solutions in Table 1-6.

Table 1-6. Solution Detail

Model Variable	Overall	Technical	Allocative
s^+	2	2	0
s_1^-	2	0	2
s_2^-	2	4	-2
π	24	20	4

Remark 1: Adding the figures in each row of the last two columns yields the corresponding value in the column under overall efficiency. This will always be true for the dollar value in the final row, by virtue of theorem 1.3, but it need not be true for the other rows because of the possible presence of alternate optima.

Remark 2: The solutions need not be "units invariant." That is, the optimum solutions for the above models may differ if the unit prices and unit costs used are stated in different units. See pp.228 ff. in Cooper, Park and Pastor (2000) for a more detailed discussion and methods for making the solutions units invariant.

7. CONCLUSIONS

This chapter has provided an introduction to DEA and some of its uses. However it is far from having exhausted the possibilities that DEA offers. For instance we have here focused on what is referred to in the DEA (and economics) literature as technical efficiency. For perspective we concluded with discussions of allocative and overall efficiency when costs or profits are of interest and the data are available. This does not exhaust the possibilities. There are still other types of efficiency that can be addressed with DEA. For instance, returns-to-scale inefficiencies, as covered in the next chapter, can offer additional possibilities which identify where additional shortcomings can appear when unit prices or unit costs are available which are of interest to potential users. We have not even fully exploited uses of our technical inefficiency models, as developed in this chapter. For instance, uses of DEA identify DMUs that enter into the optimal evaluations and hence can serve as "benchmark DMUs" to determine how best to eliminate these inefficiencies.

Topics like these will be discussed in some of the chapters that follow. However, the concept of technical inefficiency provides a needed start which will turn out to be basic for all of the other types of efficiency that may be of interest. Technical efficiency is also the most elemental of the various efficiencies that might be considered in that it requires only minimal information and minimal assumptions for its use. It is also fundamental because other types of efficiency such as allocative efficiency and returns to scale efficiency require technical efficiency to be attained before these can be achieved. This will all be made clearer in the chapters that follow.

REFERENCES

1. Afriat, S., 1972, Efficiency estimation of production functions, *International Economic Review* 13, 568-598.
2. Ahn, T., A. Charnes and W.W. Cooper, 1988, "Efficiency Characterizations in Different DEA Models," *Socio-Economic Planning Sciences* 22, 253-257.
3. Arnold, V., I. Bardhan, W.W. Cooper and A. Gallegos, 1998, "Primal and Dual Optimality in Computer Codes Using Two-Stage Solution Procedures in DEA" in J. Aronson and S. Zionts, eds., *Operations Research Methods, Models and Applications* (Westpost, Conn: Quorum Books).
4. Banker, R., A. Charnes and W.W. Cooper, 1984, Some models for estimating technical and scale inefficiencies in data envelopment analysis, Management Science 30, 1078-1092.
5. Banker, R.D. and R.C. Morey, 1986a, Efficiency analysis for exogenously fixed inputs and outputs, Operations Research 34, No. 4, 513-521.
6. Banker, R.D. and R.C. Morey, 1986b, The use of categorical variables in data envelopment analysis, Management Science 32, No. 12, 1613-1627.
7. Bardhan, I., W.F. Bowlin, W.W. Cooper and T. Sueyoshi, 1996, "Models and Measures for Efficiency Dominance in DEA, Part I: Additive Models and MED Measures," *Journal of the Operational Research Society of Japan* 39, 322-332.
8. Brockett, P.L., A. Charnes, W.W. Cooper, Z.M. Huang, and D.B. Sun, 1997, Data transformations in DEA cone ratio envelopment approaches for monitoring bank performances, European Journal of Operational Research 98, No. 2, 250-268.
9. Brockett, P.L., W.W. Cooper, H.C. Shin and Y. Wang, 1998, Inefficiency and congestion in Chinese production before and after the 1978 economic reforms. Socio-Econ Plann Sci 32, 1-20.
10. Charnes, A. and W.W. Cooper, 1962, Programming with linear fractional functionals, Naval Research Logistics Quarterly 9, 181-185.
11. Charnes, A., W.W. Cooper, and E. Rhodes, 1978, Measuring the efficiency of decision making units, European Journal of Operational Research 2, 429-444.

12. Charnes A, C.T. Clark, W.W. Cooper, B. Golany, 1985, A developmental study of data envelopment analysis in measuring the efficiency of maintenance units in the US air forces. In R.G. Thompson and R.M. Thrall, Eds., Annals Of Operation Research 2:95-112.

13. Charnes, A., W.W. Cooper, D.B. Sun, and Z.M. Huang, 1990, Polyhedral cone-ratio DEA models with an illustrative application to large commercial banks, Journal of econometrics 46, 73-91.

14. Charnes, A., W.W. Cooper, Q.L. Wei, and Z.M. Huang, 1989, Cone ratio data envelopment analysis and multi-objective programming, International Journal of Systems Science 20, 1099-1118.

15. Charnes, A. and W.W. Cooper, 1961, *Management Models and Industrial Applications of Linear Programming,* 2 vols., with A. Charnes (New York: John Wiley and Sons, Inc.).

16. Charnes, A., W.W. Cooper, B. Golany, L. Seiford and J. Stutz, (1985) "Foundations of Data Envelopment Analysis for Pareto-Koopmans Efficient Empirical Production Functions," *Journal of Econometrics* (1985), 30, pp. 91-107.

17. Cook, W.D., M. Kress, and L.M. Seiford, 1992, Prioritization models for frontier decision making units in DEA, European Journal of Operational Research 59, No. 2, 319-323.

18. Cooper, W.W., R.G. Thompson, and R.M. Thrall, 1996, Extensions and new developments in data envelopment analysis, Annals of Operations Research 66, 3-45.

19. Cooper, W.W., Seiford, L.M. and Tone, K., 2000, Data Envelopment Analysis: A Comprehensive Text with Models, Applications, References and DEA-Solver Software, Kluwer Academic Publishers, Boston.

20. Cooper, W.W., Seiford, L.M. and Zhu, Joe, A unified additive model approach for evaluating inefficiency and congestion with associated measures in DEA. Socio-Economic Planning Sciences, Vol. 34, No. 1 (2000), 1-25.

21. Cooper, W.W., H. Deng, Z. M. Huang and S. X. Li, 2002, A one-model approach to congestion in data envelopment analysis, Socio-Economic Planning Sciences 36, 231-238.

22. Cooper, W.W., K.S. Park and J.T. Pastor, 2000, "Marginal Rates and Elasticities of Substitution in DEA." *Journal of Productivity Analysis* 13, 2000, pp. 105-123.

23. Cooper, W.W., K.S. Park and J.T. Pastor, 1999, "RAM: A Range Adjusted Measure of Inefficiency for Use with Additive Models and Relations to Other Models and Measures in DEA", *Journal of Productivity Analysis* 11, 5-42.

24. Debreu, G., 1951, The coefficient of resource utilization, Econometrica 19, 273-292.
25. Dyson, R.G. and E. Thanassoulis, 1988, Reducing weight flexibility in data envelopment analysis, Journal of the Operational Research Society 39, No. 6, 563-576.
26. Färe, R., S. Grosskopf and C.A.K. Lovell, 1985, The Measurement of Efficiency of Production. Boston:Kluwer Nijhoff Publishing Co.
27. Färe, R., S. Grosskopf and CAK Lovell, 1994, *Production Frontiers* (Cambridge: Cambridge University Press).
28. Farrell, M.J., 1957, The measurement of productive efficiency, Journal of Royal Statistical Society A 120, 253-281.
29. Koopmans, T.C., 1951, Analysis of Production as an efficient combination of Activities, in T.C. Koopmans, ed. Wiley, New York.
30. Roll, Y., W.D. Cook, and B. Golany, 1991, Controlling factor weights in data envelopment analysis, IIE Transactions, 23, 2-9.
31. Shephard, R.W., 1970, Theory of Cost and Production Functions, Princeton University Press, Princeton, NJ.
32. Takamura, T. and K. Tone, 2003, "A Comparative Site Evaluation Study for Relocating Japanese Government Agencies Out of Tokyo," *Socio-Economic Planning Sciences* 37, 85-102.
33. Tavares, G., 2003, "A Bibliography of Data Envelopment Analysis (1978-2001)," *Socio-Economic Planning Sciences*(to appear).
34. Thompson, R.G., F.D.Jr. Singleton, R.M. Thrall, and B.A. Smith, 1986, Comparative site evaluation for locating a high-energy physics lab in Texas, Interfaces 16, 35-49.
35. Thompson, R.G., L. Langemeier, C. Lee, E. Lee, and R. Thrall, 1990, The role of multiplier bounds in efficiency analysis with application to Kansas farming, Journal of Econometrics 46, 93-108.
36. Tone, K. and B.K. Sahoo, 2003, "A Reexamination of Cost Efficiency and Cost Elasticity in DEA," *Management Science* (forthcoming).
37. Wong, Y.-H.B. and J.E. Beasley, 1990, Restricting weight flexibility in data envelopment analysis, Journal of the Operational Research Society 41, 829-835.
38. Zhu, J., 1996a, DEA/AR analysis of the 1988-1989 performance of the Nanjing Textiles Corporation, Annals of Operations Research 66, 311-335.
39. Zhu, J., 1996b, Data envelopment analysis with preference structure, Journal of the Operational Research Society 47, No. 1, 136-150.
40. Zhu, J., 2000, Multi-factor performance measure model with an application to Fortune 500 companies. European Journal of Operational Research 123, No. 1, 105-124.

41. Zhu, J. 2002, Quantitative Models for Performance Evaluation and Benchmarking: Data Envelopment Analysis with Spreadsheets and DEA Excel Solver, Kluwer Academic Publishers, Boston.

Part of the material in this chapter is adapted from the Journal of Econometrics, Vol. 46, Seiford, L.M. and Thrall, R.M., Recent developments in DEA: The mathematical programming approach to frontier analysis, 7-38, 1990, with permission from Elsevier Science.

Chapter 2

RETURNS TO SCALE IN DEA

Rajiv D. Banker[1], William W. Cooper[2], Lawrence M. Seiford[3] and Joe Zhu[4]
[1] *School of Management, University of Texas at Dallas, Richardson, TX 75088-0688 USA*
email: rbanker@utdallas.edu

[2] *Red McCombs School of Business, University of Texas at Austin, Austin, TX 78712 USA*
email: cooperw@mail.utexas.edu

[3] *Department of Industrial and Operations Engineering, University of Michigan at Ann Arbor,*
Ann Arbor, MI 48102 USA   email: seiford@umich.edu

[4] *Department of Management, Worcester Polytechnic Institute, Worcester, MA 01609   USA*
email: jzhu@wpi.edu

Abstract: This chapter discusses returns to scale (RTS) in data envelopment analysis (DEA). The BCC and CCR models are treated in input oriented forms while the multiplicative model is treated in output oriented form. (This distinction is not pertinent for the additive model which simultaneously maximizes outputs and minimizes inputs in the sense of a vector optimization.) Quantitative estimates in the form of scale elasticities are treated in the context of multiplicative models, but the bulk of the discussion is confined to qualitative characterizations such as whether RTS is identified as increasing, decreasing or constant. This is discussed for each type of model and relations between the results for the different models are established. The opening section describes and delimits approaches to be examined. The concluding section outlines further opportunities for research and an Appendix discusses other approaches in DEA treatment of RTS.

Key words: Data envelopment analysis (DEA), Efficiency, Returns to scale (RTS)

1.   INTRODUCTION

It has long been recognized that Data Envelopment Analysis (DEA) by its use of mathematical programming is particularly adept at estimating inefficiencies in multiple input and multiple output production correspondences. Following Charnes, Cooper and Rhodes (CCR, 1978), a number of different DEA models have now appeared in the literature (see Cooper, Seiford and Tone, 2002). During this period of model development, the economic concept of returns to scale (RTS) has also been widely studied within the different frameworks provided by these methods and this is the topic to which this chapter is devoted.

In the literature of classical economics, returns to scale (RTS) have typically been defined only for single output situations. RTS are considered to be increasing if a proportional increase in all the inputs results in a more than proportional increase in the single output. Let α represent the proportional input increase and β represent the resulting proportional increase of the single output. Increasing returns to scale prevail if $\beta > \alpha$ and decreasing returns to scale prevail if $\beta < \alpha$. Banker (1984), Banker, Charnes and Cooper (1984) and Banker and Thrall (1992) extend the RTS concept from the single output case to multiple output cases using DEA.

Two paths may be followed in treating returns to scale (RTS) in DEA. The first path, developed by Färe, Grosskopf and Lovell (FGL, 1985, 1994), determines RTS by a use of ratios of radial measures. These ratios are developed from model pairs which differ only in whether conditions of convexity and sub-convexity are satisfied. The second path stems from work by Banker (1984), Banker, Charnes and Cooper (1984) and Banker and Thrall (1992). This path, which is the one we follow, includes, but is not restricted to, radial measure models. It extends to additive and multiplicative models as well, and does so in ways that provide opportunities for added insight into the nature of RTS and its treatment by the methods and concepts of DEA.

The FGL approach has now achieved a considerable degree of uniformity that has long been available -- as in FGL (1985), for instance. See also FGL (1994). We therefore treat their approach in the Appendix to this chapter. This allows us to center this chapter on more recently developed methods for treating returns to scale with each of the different models. These treatments have therefore been available only in widely scattered literatures. We also delineate relations that have been established between these different treatments and extend this to relations that have also been established with the FGL approach. See Banker, Chang and Cooper (1996), Zhu and Shen (1995), Seiford and Zhu (1999) and Färe and Grosskopf (1994).

The plan of development in this chapter starts with a recapitulation of results from the very important paper by Banker and Thrall (1992). Although developed in the context of radial measure models, we also use the Banker and Thrall (1992) results to unify the treatment of all of the models we cover. This is done after we first cover the radial measure models that are treated by Banker and Thrall (1992). Proofs of their theorems are not supplied because these are already available in Banker and Thrall (1992). Instead refinements from Banker, Bardhan and Cooper (1996) and from Banker, Chang and Cooper (1996) are introduced which are directed to (a) providing simpler forms for implementing the Banker-Thrall theorems and (b) eliminating some of the assumptions underlying these theorems.

We then turn to concepts such as the MPSS (Most Productive Scale Size) introduced by Banker (1984) to treat multiple output - multiple input cases in DEA to extend returns-to-scale concepts built around the single output case in classical economics. Additive and multiplicative models are then examined and the latter are used to introduce (and prove) new theorems for determining scale elasticities.

The former (i.e., the additive case) is joined with a "goal vector" approach introduced by Thrall (1996a) in order to make contact with "invariance" and "balance" ideas that play prominent roles in the "dimensional analysis" used to guide the measurements used in the natural sciences (like physics). We next turn to the class of multiplicative models where, as shown by Charnes et al. (1982, 1983) and Banker and Maindiratta (1986), the piecewise linear frontiers usually employed in DEA are replaced by a frontier that is piecewise Cobb - Douglas (= log linear). Scale elasticity estimates are then obtained from the exponents of these "Cobb-Douglas like" functions for the different segments that form a frontier, which need not be concave. A concluding section points up issues for further research.

The Appendix of this chapter presents the FGL approach. We then present a simple RTS approach developed by Zhu and Shen (1995) and Seiford and Zhu (1999) to avoid the need for checking the multiple optimal solutions. This approach will substantially reduce the computational burden, because it relies on the standard CCR and BCC computational codes.

2.   RTS APPROACHES WITH BCC MODELS

For ease of reference, we here present the BCC models. Suppose, that we have n DMUs (Decision Making Units) where every $DMU_j, j = 1, 2, ..., n$, produces the same s outputs in (possibly) different amounts, y_{rj} ($r = 1, 2, ..., s$), using the same m inputs, x_{ij} ($i = 1, 2, ..., m$), also in (possibly) different amounts. The efficiency of a specific DMU_o can be evaluated by the "BCC model" of DEA in "envelopment form" as follows,

$$\min \quad \theta_o - \varepsilon(\sum_{i=1}^{m} s_i^- + \sum_{r=1}^{s} s_r^+)$$

subject to

$$\theta_o x_{io} = \sum_{j=1}^{n} x_{ij}\lambda_j + s_i^- \quad i = 1,2,...,m;$$

$$y_{ro} = \sum_{j=1}^{n} y_{rj}\lambda_j - s_r^+ \quad r = 1,2,...,s; \qquad (2.1)$$

$$1 = \sum_{j=1}^{n} \lambda_j$$

$$0 \leq \lambda_j, s_i^-, s_r^+ \qquad \forall i, r, j.$$

where, as discussed for the expression (1.6) in chapter 1, $\varepsilon > 0$ is a non-Archimedean element defined to be smaller than any positive real number.

As noted in the abstract we confine attention to input oriented versions of these radial measure models and delay discussion of changes in mix, as distinct from changes in scale, until we come to the class of additive models where input and output orientations are treated simultaneously. Finally, we do use output orientations in the case of multiplicative models because, as will be seen, the formulations in that case do not create problems in distinguishing between scale and mix changes.

Remark: We should however note that input and output oriented models may give different results in their returns to scale findings. See Figure 2-1 and related discussion below. Thus the result secured may depend on the orientation used. Increasing returns to scale may result from an input oriented model, for example, while an application of an output oriented model may produce a decreasing returns to scale characterization from the same data. See Golany and Yu (1994) for treatments of this problem.

The dual (multiplier) form of the BCC model represented in (2.1) is obtained from the same data which are then used in the following form,

$$\max \quad z = \sum_{r=1}^{s} u_r y_{ro} - u_o$$

subject to

$$\sum_{r=1}^{s} u_r y_{rj} - \sum_{i=1}^{m} v_i x_{ij} - u_o \leq 0 \qquad j = 1,...,n$$

$$\sum_{i=1}^{m} v_i x_{io} = 1 \qquad (2.2)$$

$$v_i \geq \varepsilon, \quad u_r \geq \varepsilon, \quad u_o \text{ free in sign}$$

The above formulations assume that $x_{ij}, y_{rj} \geq 0 \quad \forall i, r, j$. All variables in (2.2) are also constrained to be non-negative – except for u_o which may be positive, negative or zero with consequences that make it possible to use optimal values of this variable to identify RTS.

When a DMU_o is efficient in accordance with the Definition 1.3 in chapter 1, the optimal value of u_o, i.e., u_o^*, in (2.2), can be used to characterize the situation for Returns to Scale (RTS).

RTS generally has an unambiguous meaning only if DMU_o is on the efficiency frontier -- since it is only in this state that a tradeoff between inputs and outputs is <u>required</u> to improve one or the other of these elements. However, there is no need to be concerned about the efficiency status in our analyses because efficiency can always be achieved as follows. If a DMU_o is not BCC efficient, we can use optimal values from (2.1) to project this DMU onto the BCC efficiency frontier via the following formulas,

$$\begin{cases} \hat{x}_{io} = \theta_o^* x_{io} - s_i^{-*} = \sum_{j=1}^{n} x_{ij} \lambda_j^*, & i = 1, \ldots, m \\ \hat{y}_{ro} = y_{ro} + s_r^{+*} = \sum_{j=1}^{n} y_{rj} \lambda_j^*, & r = 1, \ldots, s \end{cases} \qquad (2.3)$$

where the symbol "*" denotes an optimal value. These are sometimes referred to as the "CCR Projection Formulas" because Charnes, Cooper and Rhodes (1978) showed that the resulting $\hat{x}_{io} \leq x_{io}$ and $\hat{y}_{ro} \geq y_{ro}$ correspond to the coordinates of a point on the efficiency frontier. They are, in fact, coordinates of the point used to evaluate DMU_o when (2.1) is employed.

Suppose we have five DMUs, A, B, C, D, and H as shown in Figure 2-1. Ray OBC is the constant returns to scale (CRS) frontier. AB, BC and CD constitute the BCC frontier, and exhibit increasing, constant and decreasing returns to scale, respectively. B and C exhibit CRS. On the line segment AB, increasing returns to scale (IRS) prevail to the left of B for the BCC model and on the line segment CD, decreasing (DRS) prevail to the right of C. By applying (2.3) to point H, we have a frontier point H' on the line segment AB where IRS prevail. However, if we use the output-oriented BCC model, the projection is on to H" where DRS prevail. This is due to the fact that the input-oriented and the output-oriented BCC models yield different projection points on the BCC frontier and it is on the frontier that returns to scale is determined. See Zhu (2002) for discussion on "returns to scale regions".

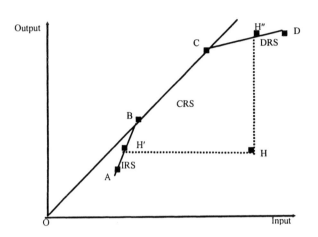

IRS: Increasing RTS, CRS: Constant RTS, DRS: Decreasing RTS
Figure 2-1. Returns to Scale

We now present our theorem for returns to scale (RTS) as obtained from Banker and Thrall (1992, p. 79) who identify RTS with the sign of u_o^* in (2.2) as follows:

Theorem 2.1
The following conditions identify the situation for RTS for the BCC model given in (2.2),
(i) Increasing RTS prevail at $(\hat{x}_o, \hat{y}_o)$ if and only if $u_o^* < 0$ for all optimal solutions.
(ii) Decreasing RTS prevail at $(\hat{x}_o, \hat{y}_o)$ if and only if $u_o^* > 0$ for all optimal solutions.
(iii) Constant RTS prevail at $(\hat{x}_o, \hat{y}_o)$ if and only if $u_o^* = 0$ for at least one optimal solution.

Here, it may be noted, $(\hat{x}_o, \hat{y}_o)$ are the coordinates of the point on the efficiency frontier which is obtained from (2.3) in the evaluation of DMU$_o$ via the solution to (2.1). Note, therefore, that a use of the projection makes it unnecessary to <u>assume</u> that the points to be analyzed are all on the BCC efficient frontier – as was assumed in Banker and Thrall (1992).

An examination of all optimal solutions can be onerous. Therefore, Banker and Thrall (1992) provide one way of avoiding a need for examining <u>all</u> optimal solutions. However, Banker, Bardhan and Cooper (1996) is the approach which will be used here because it avoids the possibility of infinite solutions which are present in the Banker-Thrall approach. In addition, the Banker, Bardhan, Cooper (1996) approach insures that the returns-to-scale analyses are conducted on the efficiency frontier. This is accomplished as follows.

Suppose an optimum has been achieved with $u_o^* < 0$. As suggested by Banker, Bardhan and Cooper (1996), the following model may then be employed to avoid having to explore all alternate optima,

$$
\begin{aligned}
&\text{maximize} \quad \hat{u}_o \\
&\text{subject to} \\
&\sum_{r=1}^{s} u_r y_{rj} - \sum_{i=1}^{m} v_i x_{ij} - \hat{u}_o \leq 0, \quad j = 1,...,n; j \neq o, \\
&\sum_{r=1}^{s} u_r \hat{y}_{ro} - \sum_{i=1}^{m} v_i \hat{x}_{io} - \hat{u}_o \leq 0, \quad j = o, \\
&\qquad\qquad \sum_{i=1}^{m} v_i \hat{x}_{io} = 1, \\
&\sum_{r=1}^{s} u_r \hat{y}_{ro} \qquad\qquad - \hat{u}_o = 1, \\
&\qquad v_i, u_r \geq 0 \text{ and } \hat{u}_o \leq 0,
\end{aligned}
\tag{2.4}
$$

where the $\hat{x}_{io}$ and $\hat{y}_{ro}$ are obtained from (2.3).

With these changes of data the constraints for (2.4) are in the same form as (2.2) except for the added conditions $\sum_{r=1}^{s} u_r \hat{y}_{ro} - \hat{u}_o = 1$ and $\hat{u}_o \leq 0$. The first of these conditions helps to ensure that we will be confined to the efficiency frontier. The second condition allows us to determine whether an optimal value can be achieved with max $\hat{u}_o = 0$. If $\hat{u}_o^* = 0$ can be obtained then condition (iii) of Theorem 2.1 is satisfied and returns to scale are constant. If, however, max $\hat{u}_o = \hat{u}_o^* < 0$ then, as set forth in (i) of Theorem 2.1, returns to scale are increasing. In either case, the problem is resolved and the need for examining all alternate optima is avoided in this way of implementing Theorem 2.1.

We can deal in a similar manner with the case when $u_o^* > 0$ by (a) reorienting the objective in (2.4) to "minimize" $\hat{u}_o$ and (b) replacing the constraint $\hat{u}_o \leq 0$ with $\hat{u}_o \geq 0$. All other elements of (2.4) remain the same and if min $\hat{u}_o = \hat{u}_o^* > 0$ then condition (ii) of Theorem 2.1 is applicable while if min $\hat{u}_o = \hat{u}_o^* = 0$ then condition (iii) is applicable.

Reference to Figure 2-2 can help us to interpret these results. This Figure portrays the case of one input, x, and one output, y. The coordinates of each point are listed in the order (x,y). Now, consider the data for A which has the coordinates (x=1, y=1), as shown at the bottom of Figure 2-2. The "supports" at A form a family which starts at the vertical line (indicated by the dotted line) and continues through rotations about A until coincidence is achieved with the line connecting A to B. All of these supports will have negative intercepts so $u_o^* < 0$ and the situation for A is one of increasing returns to scale as stated in (i) of Theorem 2.1.

The reverse situation applies at D. Starting with the horizontal line indicated by the dots, supports can be rotated around D until coincidence is achieved with the line connecting D and C. In all cases the intercept is positive so $u_o^* > 0$ and returns to scale are decreasing as stated in (ii) of Theorem 2.1.

Rotations at C or B involve a family of supports in which at least one member will achieve coincidence with the broken line going through the origin, so that, in at least this one case, we will have $u_o^* = 0$, in conformance with the condition for constant returns to scale in (iii) of Theorem 2.1.

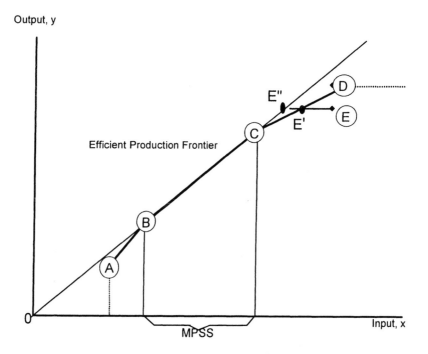

$$A = (1, 1), \ B=(3/2, 2), \ C=(3, 4), \ D = (4,5), \ E = (4, 9/2)$$

Figure 2-2. Most Productive Scale Size

Finally, we turn to E, the only point which is BCC inefficient in Figure 2-2. Application of (2.3), however, projects E into E' -- a point on the line between C and D -- and, therefore, gives the case of decreasing returns to scale with a unique solution of $\hat{u}_o^* > 0$. Hence all possibilities are comprehended by Theorem 2.1 for the qualitative returns to scale characterizations which are of concern here. Thus, for these characterizations only the signs of the non-zero values of $\hat{u}_o^*$ suffice.

3.   RTS APPROACHES WITH CCR MODELS

We now turn to the CCR models which, as discussed in chapter 1 take the following form,

$$\text{minimize } \theta - \varepsilon \left(\sum_{i=1}^{m} s_i^- + \sum_{r=1}^{s} s_r^+ \right)$$

subject to

$$\theta x_{io} = \sum_{j=1}^{n} x_{ij} \lambda_j + s_i^-$$

$$y_{ro} = \sum_{j=1}^{n} y_{rj} \lambda_j - s_r^+, \qquad (2.5)$$

$$0 \le \lambda_j, s_i^-, s_r^+ \forall i, j, r.$$

As can be seen, this model is the same as the "envelopment form" of the BCC model in (2.1) except for the fact that the condition $\sum_{j=1}^{n} \lambda_j = 1$ is omitted. In consequence, the variable u_o, which appears in the "multiplier form" for the BCC model in (2.2), is omitted from the dual (multiplier) form of this CCR model. The projection formulas expressed in (2.3) are the same for both models. We can therefore use these same projections to move all points onto the efficient frontier for (2.5) and proceed directly to returns to scale characterizations for (2.5) which are supplied by the following theorem from Banker and Thrall (1992).

Theorem 2.2
The following conditions identify the situation for RTS for the CCR model given in (2.5)
(i) Constant returns to scale prevail at $(\hat{x}_o, \hat{y}_o)$ if $\sum \lambda_j^* = 1$ in any alternate optimum.
(ii) Decreasing returns to scale prevail at $(\hat{x}_o, \hat{y}_o)$ if $\sum \lambda_j^* > 1$ for all alternate optima.
(iii) Increasing returns to scale prevail at $(\hat{x}_o, \hat{y}_o)$ if $\sum \lambda_j^* < 1$ for all alternate optima.

Following Banker, Chang and Cooper (1996), we can avoid the need for examining all alternate optima. This is done as follows. Suppose an optimum has been obtained for (2.5) with $\sum \lambda_j^* < 1$. We then replace (2.5) with

$$\text{maximize } \sum_{j=1}^{n} \hat{\lambda}_j + \varepsilon \left(\sum_{i=1}^{m} \hat{s}_i^- + \sum_{r=1}^{s} \hat{s}_r^+ \right)$$

subject to

$$\theta^* x_{io} = \sum_{j=1}^{n} x_{ij} \hat{\lambda}_j + \hat{s}_i^-, \quad \text{for } i=1,\dots, m \qquad (2.6)$$

$$y_{ro} = \sum_{j=1}^{n} y_{rj} \hat{\lambda}_j - \hat{s}_r^+, \quad \text{for } r=1,\dots, s$$

$$1 \ge \sum_{j=1}^{n} \hat{\lambda}_j$$

with $0 \le \hat{\lambda}_j, \hat{s}_i^-, \hat{s}_r^+ \forall i, j, r,$

where θ^* is the optimal value of θ secured from (2.5).

Remark: This model can also be used for setting scale-efficient targets when multiple optimal solutions in model (2.5) are present. See Zhu (2000, 2002).

We note for (2.5) that we may omit the 2 stage process described for the CCR model in chapter 1 -- i.e., the process in which the sum of the slacks are maximized in stage 2 after θ^* has been determined. This is replaced with a similar two-stage process for (2.6) because only the optimal value of θ is needed from (2.5) to implement the analysis now being described. The optimal solution to (2.6) then yields values of $\hat{\lambda}_j^*, j = 1, \ldots, n$, for which the following theorem is immediate,

Theorem 2.3
Given the existence of an optimal solution with $\sum \lambda_j^* < 1$ in (2.5), the returns to scale at $(\hat{x}_o, \hat{y}_o)$ are constant if and only if $\sum \hat{\lambda}_j^* = 1$ and returns to scale are increasing if and only if $\sum \hat{\lambda}_j^* < 1$ in (2.6).

Consider A = (1, 1) as shown at the bottom of Figure 2-2. Because we are only interested in θ^*, we apply (1.4) in chapter 1 to obtain

minimize θ
subject to

$$1\theta \geq 1\lambda_A + \frac{3}{2}\lambda_B + 3\lambda_C + 4\lambda_D + 4\lambda_E$$

$$1 \leq 1\lambda_A + 2\lambda_B + 4\lambda_C + 5\lambda_D + \frac{9}{2}\lambda_E \qquad (2.7)$$

$$0 \leq \lambda_A, \lambda_B, \lambda_C, \lambda_D, \lambda_E.$$

This problem has $\theta^* = 3/4$ and hence A is found to be inefficient. Next, we observe that this problem has alternate optima because this same $\theta^* = 3/4$ can be obtained from either $\lambda_B^* = 1/2$ or from $\lambda_C^* = 1/4$ with all other $\lambda^* = 0$. For each of these optima, we have $\sum \lambda_j^* < 1$, so we utilize (2.6) and write

maximize $\hat{\lambda}_A + \hat{\lambda}_B + \hat{\lambda}_C + \hat{\lambda}_D + \hat{\lambda}_E + \varepsilon(\hat{s}^- + s^+)$
subject to

$$\frac{3}{4} = 1\hat{\lambda}_A + \frac{3}{2}\hat{\lambda}_B + 3\hat{\lambda}_C + 4\hat{\lambda}_D + 4\hat{\lambda}_E + \hat{s}^-$$
$$1 = 1\hat{\lambda}_A + 2\hat{\lambda}_B + 4\hat{\lambda}_C + 5\hat{\lambda}_D + \frac{9}{2}\hat{\lambda}_E - \hat{s}^+$$
$$1 \geq \hat{\lambda}_A + \hat{\lambda}_B + \hat{\lambda}_C + \hat{\lambda}_D + \hat{\lambda}_E \qquad (2.8)$$
$$0 \leq \hat{\lambda}_A, \hat{\lambda}_B, \hat{\lambda}_C, \hat{\lambda}_D, \hat{\lambda}_E.$$

so that $\sum \hat{\lambda}_j^* \equiv \hat{\lambda}_A^* + \hat{\lambda}_B^* + \hat{\lambda}_C^* + \hat{\lambda}_D^* + \hat{\lambda}_E^*$ with all $\hat{\lambda}$ non-negative. An optimal solution is $\hat{\lambda}_B^* = 1/2$ and all other $\hat{\lambda}^* = 0$. Hence $\sum \hat{\lambda}_j^* < 1$, so from Theorem 2.3, increasing returns to scale prevail at A.

We are here restricting attention to solutions of (2.6) with $\sum_{j=1}^n \hat{\lambda}_j \leq 1$, as in the constraint of (2.6), but the examples we provide below show how to treat situations in which θ^* is associated with solutions of (2.5) that have values $\sum \lambda_j^* > 1$.

Consider E = (4, 9/2) for (2.6), as a point which is not on either (i) the BCC efficiency frontier represented by the solid lines in Figure 2-2 or (ii) the CCR efficiency frontier represented by the broken line from the origin. Hence both the BCC and CCR models find E to be inefficient. Proceeding via the CCR envelopment model in (2.5) with the slacks omitted from the objective, we get

$$\text{minimize } \theta$$
$$\text{subject to}$$
$$4\theta \geq 1\lambda_A + \frac{3}{2}\lambda_B + 3\lambda_C + 4\lambda_D + 4\lambda_E,$$
$$\frac{9}{2} \leq 1\lambda_A + 2\lambda_B + 4\lambda_C + 5\lambda_D + \frac{9}{2}\lambda_E, \qquad (2.9)$$
$$0 \leq \lambda_A, \lambda_B, \lambda_C, \lambda_D, \lambda_E.$$

Again we have alternate optima with, now, $\theta^* = 27/32$ for either $\lambda_B^* = 9/4$ or $\lambda_C^* = 9/8$ and all other $\lambda^* = 0$. Hence, in both cases we have $\sum \hat{\lambda}_j^* > 1$. Continuing in an obvious way, we next reorient the last constraint and the objective in (2.6) to obtain

$$\text{minimize } (\hat{\lambda}_A + \hat{\lambda}_B + \hat{\lambda}_C + \hat{\lambda}_D + \hat{\lambda}_E) - \varepsilon(\hat{s}^- + s^+)$$
$$\text{subject to}$$
$$\tfrac{27}{8} = 1\hat{\lambda}_A + \tfrac{3}{2}\hat{\lambda}_B + 3\hat{\lambda}_C + 4\hat{\lambda}_D + 4\hat{\lambda}_E + \hat{s}^-$$
$$\tfrac{9}{2} = 1\hat{\lambda}_A + 2\hat{\lambda}_B + 4\hat{\lambda}_C + 5\hat{\lambda}_D + \tfrac{9}{2}\hat{\lambda}_E - \hat{s}^+ \qquad (2.10)$$
$$1 \leq \hat{\lambda}_A + \hat{\lambda}_B + \hat{\lambda}_C + \hat{\lambda}_D + \hat{\lambda}_E$$
$$0 \leq \hat{\lambda}_A, \hat{\lambda}_B, \hat{\lambda}_C, \hat{\lambda}_D, \hat{\lambda}_E.$$

This has its optimum at $\hat{\lambda}_C^* = 9/8$ with all other $\hat{\lambda}^* = 0$. So, in conformance with Theorem 2.3, as given for (2.6), we associate E with decreasing returns to scale.

There is confusion in the literature on the returns-to-scale characterizations obtained from Theorems 2.1 and 2.2 and the BCC and the CCR models with which they are associated. Hence, we proceed further as follows.

As noted earlier, returns to scale generally has an unambiguous meaning only for points on the efficiency frontier. When the BCC model as given in (2.1) is used on the data in Figure 2-2, the primal model projects E into E' with coordinates (7/2, 9/2) on the segment of the line y = 1 + x which connects C to D on the BCC efficiency frontier. Comparing this with E = (4, 9/2) identifies E as having an inefficiency in the amount of 1/2 unit in its input. This is a technical inefficiency, in the terminology of DEA. Turning to the dual for E formed from the BCC model, as given in (2.2), we obtain $u_o^* = 1/4$. Via Theorem 2.1 this positive value of u_o^* suggests that returns to scale are either decreasing or constant at E'= (28/8, 9/2) --the point to which E is projected in order to obtain access to model (2.4). Substitution in the latter model yields a value of $\hat{u}_o^* \doteq 2/7$, which is also positive, thereby identifying E' with the decreasing returns to scale that prevail for the BCC model on this portion of the efficiency frontier in Figure 2-2.

Next we turn to the conditions specified in Theorem 2.2 which are identified with the CCR envelopment model (2.5). Here we find that the projection is to a new point E" = (27/8, 9/2) which is on the line y=4/3x corresponding to the broken line from the origin that coincides with the segment from B to C in Figure 2-2. This ray from the origin constitutes the efficiency frontier for the CCR model which, when used in the manner we have previously indicated, simultaneously evaluates the technical and returns-to-scale performances of E. In fact, as can be seen from the solution to (2.9), this evaluation is effected by either $\hat{\lambda}_B^* = 9/4$ or $\hat{\lambda}_C^* = 9/8$ -- which are variables associated with vectors in a "constant returns-to-scale region" that we will shortly associate with "most productive scale size" (MPSS) for the BCC model. The additional 1/8 unit input reduction effected in going from E' to E" is needed to adjust to the efficient mix that prevails in this MPSS region which the CCR model is using to evaluate E.

Thus, the CCR model as given in (2.5) simultaneously evaluates scale and purely technical inefficiencies, while the BCC model, as given in (2.1), separates out the scale inefficiencies for evaluation in its associated dual (=multiplier) form as given in (2.2). Finally, as is well known, a simplex method solution to (2.1) automatically supplies the solution to its dual in (2.2). Thus, no additional computations are required to separate the purely technical inefficiency characterizations obtained form (2.1) and the returns-to-scale characterizations obtained from (2.2). Both sets of values are obtainable from a solution to (2.1).

We now introduce the following theorem which will allow us to consider the relations between Theorems 2.1 and 2.2 in the returns to scale characterization.

Theorem 2.4
Suppose DMU_o is designated as efficient by the CCR model, DMU_o then it is also designated as efficient by the BCC model.

<u>Proof</u>: The CCR and BCC models differ only because the latter has the additional constraint $\sum_{j=1}^n \lambda_j = 1$. The following relation must therefore hold
$$\theta_{CCR}^* - \varepsilon\left(\sum_{i=1}^m s_i^{-*} + \sum_{r=1}^s s_r^{+*}\right) \le \theta_{BCC}^* - \varepsilon\left(\sum_{i=1}^m s_i^{-*} + \sum_{r=1}^s s_r^{+*}\right)$$

where the expressions on the left and right of the inequality respectively designate optimal values for objective of the CCR and BCC models.

Now, suppose DMU_o is found to be efficient with the CCR model. This implies $\theta^*_{CCR} = 1$ and all slacks are zero for the expression on the left. Hence, we will have

$$1 \leq \theta^*_{BCC} - \varepsilon(\sum_{i=1}^{m} s_i^{-*} + \sum_{r=1}^{s} s_r^{+*})$$

However, the x_{io} and y_{ro} values appear on both the left and right sides of the corresponding constraints in the DEA models. Hence, choosing $\lambda^*_o = \lambda^*_j = 1$, we can always achieve equality with θ^*_{BCC} which is the lower bound in this exact expression with all slacks zero. Thus, this DMU_o will also be characterized as efficient by the BCC model whenever it is designated as efficient by the CCR model. . ∎

We now note that the reverse of this theorem is not true. That is, a DMU_o may be designated as efficient by the BCC model but not by the CCR model. Even when both models designate a DMU_o as inefficient, moreover, the measures of inefficiency may differ. Application of the BCC model to point E in Figure 2-2, for example, will designate E' on the line connecting C and D to evaluate its efficiency. However, utilization of the CCR model will designate E'' with $\theta^*_{CCR} < \theta^*_{BCC}$ so that $1 - \theta^*_{CCR} > 1 - \theta^*_{BCC}$ which shows a greater inefficiency value for DMU_o when the CCR model is used.

Because DEA evaluates relative efficiency, it will always be the case that at least one DMU will be characterized as efficient by either model. However, there will always be at least one point of intersection between these two frontiers. Moreover, the region of the intersection will generally expand a DMU set to be efficient with the CCR model. The greatest spread between the envelopments will then constitute extreme points that define the boundaries of the intersection between the CCR and BCC models.

The way Theorem 2.2 effects its efficiency characterization is by models of linear programming algorithms that use "extreme point" methods. That is, the solutions are expressed in terms of "basis sets" consisting of extreme points. The extreme points B and C in Figure 2-2 can constitute active members of such an "optimal basis" where, by "active member", we refer to members of a basis which have non-zero coefficients in an optimal solution.

Now, as shown in Cooper, Seiford and Tone (2000), active members of an optimal basis are necessarily efficient. For instance, $\lambda^*_B = \frac{1}{2}$ in the solution to (2.8) designates B as an active member of an optimal basis and the same is true for $\lambda^*_C = \frac{1}{4}$ and both B and C are therefore efficient. In both cases, we have $\sum \lambda^*_j < 1$ and we have increasing returns to scale at point (3/4, 1) on the constant returns to scale ray which is used to evaluate A. In other words, $\sum \lambda^*_j < 1$ shows that B and C both lie below the region of intersection because its coordinates are smaller in value than the corresponding values of the active members in the optimal basis.

Turning to the evaluation of E, we have $\theta^* = 27/32 < 1$ showing that E is inefficient in (2.10). Either $\lambda_B^* = 9/4$ or $\lambda_C^* = 9/8$ can serve as active member in the basis. Thus, to express E" in terms of these bases, we have

$$\underset{E''}{(\frac{27}{8},\frac{9}{2})} = \underset{B}{\frac{9}{4}(\frac{3}{2},2)} = \underset{C}{\frac{9}{8}(3,4)}$$

because E" lies above the region of intersection, as shown by $\sum \lambda_j^* > 1$ for either of the optimal solutions.

As shown in the next section of this chapter, Banker (1984) refers to the region between B and C as the region of most productive scale size (MPSS). For justification, we might note that the slope of the ray from the origin through B and C is steeper than the slope of any other ray from the origin that intersects the production possibility set (PPS). This measure with the output per unit input is maximal relative to any other ray that intersects PPS.

Hence, Theorem 2.2 is using the values of $\sum \lambda_j^*$ to determine whether returns to scale efficiency has achieved MPSS and what needs to be done to express this relative to the region of MPSS. We can therefore conclude with a corollary to Theorem 2.4: DMU_o will be at MPSS if and only if $\sum \lambda_j^* = 1$ in an optimal solution when it is evaluated by a CCR model.

To see how this all comes about mathematically and how it relates to the RTS characterization, we note that the optimal solution for the CCR model consists of all points on the ray from the origin that intersect the MPSS region. If the point being evaluated is in MPSS, it can be expressed as a convex combination of the extreme points of MPSS so that $\sum \lambda_j^* = 1$. If the point is above the region, its coordinate values will all be larger than their corresponding coordinates in MPSS so that we will have $\sum \lambda_j^* > 1$. If the point is below the region, we will have $\sum \lambda_j^* < 1$. Because the efficient frontier, as defined by the BCC model, is strictly concave, the solution will designate this point as being in the region of constant, decreasing or increasing RTS, respectively.

Thus, the CCR model simultaneously evaluates RTS and technical inefficiency while the BCC model separately evaluates technical efficiency with θ_{BCC}^* from the envelopment model and RTS with u_o^* obtained from the multiplier model. As Figure 2-2 illustrates, at point E, the evaluation for the CCR model is global with returns to scale always evaluated relative to MPSS. The evaluation for the BCC model is local with u_o^* being determined by the facet of the efficient frontier in which the point used to evaluate DMU_o is located. As a consequence, it always be the case that $\theta_{CCR}^* < \theta_{BCC}^*$ unless the point used to evaluate DMU_o is in the region of MPSS, in which case $\theta_{CCR}^* = \theta_{BCC}^*$ will obtain.

4. MOST PRODUCTIVE SCALE SIZE

There is some ambiguity in dealing with points like B and C in Figure 2-2 because the condition that prevails depends on the direction in which movement is to be effected. As noted by Førsund (1996) this situation was dealt with by Ragnar Frisch -- who pioneered empirical studies of production and suggested that the orientation should be toward maximizing the output per unit input when dealing with technical conditions of efficiency. See Frisch (1964). However, Frisch (1964) dealt only with the case of single outputs. Extensions to multiple output-multiple input situations can be dealt with by the concept of Most Productive Scale Size (MPSS) as introduced into the DEA literature by Banker (1984). To see what this means consider

$$(X_o\alpha, Y_o\beta) \qquad (2.11)$$

with $\beta, \alpha \geq 0$ representing scalars and X_o and Y_o representing input and output vectors, respectively, for DMU$_o$. We can continue to move toward a possibly better (i.e., more productive) returns-to-scale situation as long as $\beta/\alpha \neq 1$. In other words, we are not at a point which is MPSS when either (a) all outputs can be increased in proportions that are at least as great as the corresponding proportional increases in all inputs needed to bring them about, or (b) all inputs can be decreased in proportions that are at least as great as the accompanying proportional reduction in all outputs. Only when $\beta/\alpha=1$, or $\alpha=\beta$, will returns to scale be constant, as occurs at MPSS.

One way to resolve problems involving returns to scale for multiple output-multiple input situations would use a recourse to prices, costs (or similar weights) to determine a "best" or "most economical" scale size. Here, however, we are using the concept of MPSS in a way that avoids the need for additional information on unit prices, costs, etc., by allowing all inputs and outputs to vary simultaneously in the proportions prescribed by α and β in (2.11). Hence, MPSS allows us to continue to confine attention to purely technical inefficiencies, as before, while allowing for other possible choices after scale changes and size possibilities have been identified and evaluated in our DEA analyses.

The interpretation we have just provided for (2.11) refers to returns to scale locally, as is customary – e.g., in economics. However, this does not exhaust the uses that can be made of Banker's (1984) MPSS. For instance, we can now replace our preceding local interpretation of (2.11) by one which is oriented globally. That is, we seek to characterize the returns to scale conditions for DMUo with respect to MPSS instead of restricting this evaluation to the neighborhood of the point (X_o, Y_o) where, say, a derivative is to be evaluated. See Varian (1984, p. 20) for economic interpretations of restrictions needed to justify uses of derivatives. We also do this in a way that enables us to relate Theorems 2.1 and 2.2 to each other and thereby provide further insight into how the BCC and CCR models relate to each other in scale size (and other) evaluations.

For these purposes, we introduce the following formulation,

$$
\begin{aligned}
&\text{maximize } \beta/\alpha \\
&\text{subject to} \\
&\beta Y_o \leq \sum_{j=1}^{n} Y_j \lambda_j, \\
&\alpha X_o \geq \sum_{j=1}^{n} X_j \lambda_j, \\
&1 = \sum_{j=1}^{n} \lambda_j, \\
&0 \leq \beta, \alpha, \lambda_j, j = 1,...,n.
\end{aligned} \tag{2.12}
$$

Now note that the condition $\sum \lambda_j = 1$ appears just as it does in (2.1). However, in contrast to (2.1), we are now moving to a global interpretation by jointly maximizing the proportional increase in outputs and minimizing the proportional decrease in inputs. We are also altering the characterizations so that these α and β values now yield new vectors $\hat{X}_o = \alpha X_o$ and $\hat{Y}_o = \beta Y_o$, which we can associate with points which are MPSS, as in the following

Theorem 2.5
A necessary condition for DMU_o, with output and input vectors Y_o and X_o, to be MPSS is max $\beta/\alpha=1$ in (2.12), in which case returns to scale will be constant.

Theorem 2.5 follows from the fact that $\beta = \alpha = 1$ with $\lambda_j = 0$, $\lambda_o = 1$ for $j \neq o$ is a solution of (2.12), so that, always, max $\beta/\alpha = \beta^*/\alpha^* \geq 1$. See the appendix in Cooper, Thompson and Thrall (1996) for a proof and a reduction of (2.12) to a linear programming equivalent.

We illustrate with D=(4,5) in Figure 2-2 for which we utilize (2.12) to obtain

$$
\begin{aligned}
&\text{Maximize } \beta/\alpha \\
&\text{subject to} \\
&5\beta \leq 1\lambda_A + 2\lambda_B + 4\lambda_C + 5\lambda_D + \tfrac{9}{2}\lambda_E \\
&4\alpha \geq 1\lambda_A + \tfrac{3}{2}\lambda_B + 3\lambda_C + 4\lambda_D + 4\lambda_E \\
&1 = \lambda_A + \lambda_B + \lambda_C + \lambda_D + \lambda_E \\
&0 \leq \lambda_A, \lambda_B, \lambda_C, \lambda_D, \lambda_E.
\end{aligned} \tag{2.13}
$$

This has an optimum at $\lambda_B^* = 1$ with $\alpha^* = 3/8$ and $\beta^* = 2/5$ to give $\beta^*/\alpha^* = 16/15 > 1$. Thus, MPSS is not achieved. Substituting in (2.13) with $\lambda_B^* = 1$, we can use this solution to obtain $4\alpha^* = 3/2$ and $5\beta^* = 2$ which are the coordinates of B in Figure 2-2. Thus, D=(4,5) is evaluated globally by reference to B=(3/2,2), which is in the region of constant returns to scale and hence is MPSS.

There is also an alternate optimum to (2.13) with $\lambda_C^* = 1$ and $\alpha^* = 3/4$, $\beta^* = 4/5$ so, again, $\beta^*/\alpha^* = 16/15$, and D is not at MPSS. Moreover, $4\alpha^* = 5$, $5\beta^* = 4$ gives the coordinates of C = (3, 4). Thus, D is again evaluated globally by a point in the region of MPSS. Indeed, any point in this region of MPSS would give the same value of $\beta^*/\alpha^* = 16/15$, since all such points are representable as convex combinations of B and C.

Theorem 2.6

Sign conditions for BCC and CCR models:
(i) The case of increasing returns to scale. $u_o^* < 0$ for all optimal solutions to (2.2) if and only if $(\sum \lambda_j^* - 1) < 0$ for all optimal solutions to (2.5).
(ii) The case of decreasing returns to scale. $u_o^* > 0$ for all optimal solutions to (2.2) if and only if $(\sum \lambda_j^* - 1) > 0$ for all optimal solutions to (2.5).
(iii) The case of constant returns to scale. $u_o^* = 0$ for some optimal solutions to (2.2) if and only if $(\sum \lambda_j^* - 1) = 0$ for some optimal solution to (2.5).

This theorem removes the possibility that uses of the CCR and BCC models might lead to different RTS characterizations. It is also remarkable because differences might be expected from the fact that (2.2) effects its evaluations locally with respect to a neighboring facet while (2.5) effects its evaluations globally with respect to a facet (or point) representing MPSS.

To see what this means we focus on active members of an optimal solution set as follows.

Turning to E in Figure 2-2 we see that it is evaluated by E' when (2.1) is used. This point, in turn, can be represented as a convex combination of C and D with both of the latter vectors constituting active members of the optimal basis. The associated support coincides with the line segment connecting C and D with a (unique) value $u_o^* > 0$ so returns to scale are decreasing, as determined from (2.2). This is a local evaluation. When (2.5) is used, the projection is to E", with alternate optima at B or C respectively serving as the only active member of the optimal basis. Hence the evaluation by the CCR model is effected globally. Nevertheless, the same decreasing returns to scale characterization is secured.

We now note that E" may be projected into the MPSS region by means of the following formulas,

$$\frac{\theta^* x_{io} - s_i^{-*}}{\sum_{j=1}^{n} \hat{\lambda}_j^*}$$
$$\frac{y_{ro} + s_i^{+*}}{\sum_{j=1}^{n} \hat{\lambda}_j^*} \tag{2.14}$$

where the denominators are secured from (2.6). This convexification of (2.3), which is due to Banker and Morey (1986), provides a different projection than (2.3). We illustrate for E" by using the solutions for (2.9) to obtain

$$\frac{4\theta^* - s_i^{-*}}{9/4} = \frac{27/8}{9/4} = 3/2$$

$$\frac{y_{ro} + s_i^{+*}}{9/4} = \frac{9/2}{9/4} = 2.$$

This gives the coordinates of B from one optimal solution. The other optimal solution yields the coordinates of C via

$$\frac{4\theta^* - s_i^{-*}}{9/8} = \frac{27/8}{9/8} = 3$$

$$\frac{y_{ro} + s_i^{+*}}{9/8} = \frac{9/2}{9/8} = 4.$$

This additional step brings us into coincidence with the results already described for the MPSS model given in (2.13). Consistency is again achieved even though the two models proceed by different routes. The MPSS model in (2.12) bypasses the issue of increasing vs. decreasing returns to scale and focuses on the issue of MPSS, but this same result can be achieved for (2.5) by using the additional step provided by the projection formula (2.14).

5.   ADDITIVE MODELS

The model (2.12), which we used for MPSS, avoids the problem of choosing between input and output orientations, but this is not the only type of model for which this is true. The additive models to be examined in this section also have this property. That is, these models simultaneously maximize outputs and minimize inputs, in the sense of vector optimizations.

The additive model we select is

$$\max \quad \sum_{i=1}^{m} g_i^- s_i^- + \sum_{r=1}^{s} g_r^+ s_r^+$$

subject to

$$\sum_{j=1}^{n} x_{ij} \lambda_j + s_i^- = x_{io}, \quad i = 1,2,...,m$$

$$\sum_{j=1}^{n} y_{rj} \lambda_j - s_r^+ = y_{ro}, \quad r = 1,2,...,s \qquad (2.15)$$

$$\sum_{j=1}^{n} \lambda_j = 1$$

$$\lambda_j, s_i^-, s_r^+ \geq 0.$$

This model utilizes the "goal vector" approach of Thrall (1996a) in which the slacks in the objective are accorded "goal weights" which may be subjective or objective in character. Here we want to use these "goal weights" to ensure that the units of measure associated with the slack variables do not affect the optimal solution choices.

Employing the language of "dimensional analysis," as in Thrall (1996a), we want these weights to be "contragredient" in order to insure that the resulting objective will be "dimensionless." That is, we want the solutions to be free of the dimensions in which the inputs and outputs are stated. An example is the use of the input and output ranges in Cooper, Park and Pastor (1999) to obtain $g_i = 1/R_i^-$, $g_r = 1/R_r^+$ where R_i^- is the range for the ith input and R_r^+ is the range for the rth output. This gives each term in the objective of (2.15) a contragredient weight. The resulting value of the objective is dimensionless, as follows from the fact that the s_i^- and s_r^+ in the numerators are measured in the same units as the R_i^- and R_r^+ in the denominators. Hence the units of measure cancel.

The condition for efficiency given in Definition 1.3 in chapter 1 for the CCR model is now replaced by the following simpler condition,

Definition 2.1: A DMU_o evaluated by (2.15) is efficient if and only if all slacks are zero.

Thus, in the case of additive models it suffices to consider only condition (ii) in Definition 1.3. Moreover this condition emerges from the second stage solution procedure associated with the non-Archimedean $\varepsilon > 0$ in (1.1). Hence we might expect that returns-to-scale characterizations will be related, as we will now see.

To start our returns-to-scale analyses for these additive models we first replace the CCR projections of (2.3) with

$$\hat{x}_{io} = x_{io} - s_i^{-*}, \quad i = 1,...,m$$

$$\hat{y}_{ro} = y_{ro} + s_r^{+*}, \quad r = 1,...,s \qquad (2.16)$$

where s_i^{-*} and s_r^{+*} are optimal slacks obtained from (2.15). Then we turn to the dual (multiplier) model associated with (2.15) which we write as follows,

$$\min \ \sum_{i=1}^{m} v_i x_{io} - \sum_{r=1}^{s} \mu_{ri} y_{ro} + u_o$$

subject to

$$\sum_{i=1}^{m} v_i x_{ij} - \sum_{r=1}^{s} \mu_{ri} y_{rj} + u_o \geq 0, \quad j = 1,..,m \qquad (2.17)$$

$$v_i \geq g_i^-, \ \mu_r \geq g_r^+; \ u_o \ \textit{free}.$$

We are thus in position to use Theorem 2.1 for "additive" as well as "radial measures" as reflected in the BCC and CCR models discussed in earlier parts of this chapter. Hence we again have recourse to this theorem where, however, we note the difference in objectives between (2.2) and (2.17), including the change from $-u_o$ to $+u_o$. As a consequence of these differences we also modify (2.4) to the following,

Maximize $\hat{u}_o$

subject to

$$\sum_{r=1}^{s} \mu_r y_{rj} - \sum_{i=1}^{m} v_i x_{ij} - \hat{u}_o \leq 0, j = 1,..,m; j \neq o$$

$$\sum_{r=1}^{s} \mu_r \hat{y}_{ro} - \sum_{i=1}^{m} v_i \hat{x}_{io} - \hat{u}_o = 0 \qquad (2.18)$$

$$\mu_r \geq g_r^+, v_i \geq g_i^-, \hat{u}_o \leq 0.$$

Here we have assumed that $u_o^* < 0$ was achieved in a first-stage use of (2.17). Hence, if $\hat{u}_o^* < 0$ is maximal in (2.18) then returns to scale are increasing at $(\hat{x}_o, \hat{y}_o)$ in accordance with (i) in Theorem 2.1 whereas if $\hat{u}_o^* = 0$ then (iii) applies and returns to scale are constant at this point $(\hat{x}_o, \hat{y}_o)$ on the efficiency frontier.

For $u_o^* > 0$ in stage one, the objective and the constraint on $\hat{u}_o$ are simply reoriented in the manner we now illustrate by using (2.15) to evaluate E in Figure 2-2 via

$$\max \quad s^- + s^+$$

subject to

$$\lambda_A + \tfrac{3}{2}\lambda_B + 3\lambda_C + 4\lambda_D + 4\lambda_E + s^- = 4$$

$$\lambda_A + 2\lambda_B + 4\lambda_C + 5\lambda_D + \tfrac{9}{2}\lambda_E - s^+ = \tfrac{9}{2}$$

$$\lambda_A + \lambda_B + \lambda_C + \lambda_D + \lambda_E = 1$$

$$s^-, s^+, \lambda_A, \lambda_B, {}_,\lambda_C, \lambda_D, \lambda_E \geq 0.$$

where we have used unit weights for the g_i^-, g_r^+, to obtain the usual additive model formulation. (See Thrall (1996b) for a discussion of the applicable condition for a choice of such "unity" weights.) This has an optimal solution with $\lambda_C^* = \lambda_D^* = s^{-*} = \frac{1}{2}$ and all other variables zero. To check that this is optimal we turn to the corresponding dual (multiplier) form for the above envelopment model which is

$$\min \quad 4v - \frac{9}{2}\mu + u_o$$

subject to

$$v - \mu + u_o \geq 0$$
$$\frac{3}{2}v - 2\mu + u_o \geq 0$$
$$3v - 4\mu + u_o \geq 0$$
$$4v - 5\mu + u_o \geq 0$$
$$4v - \frac{9}{2}\mu + u_o \geq 0$$
$$v, \mu \geq 1, u_o \text{ free}$$

The solution $v^* = \mu^* = u_o^* = 1$ satisfies all constraints and gives $4v^* - \frac{9}{2}\mu^* + u_o^* = \frac{1}{2}$. This is the same value as in the preceding problem so that, by the dual theorem of linear programming, both solutions are optimal.

To determine the conditions for returns to scale we use (2.16) to project E into E' with coordinates $(\hat{x}, \hat{y}) = (\frac{7}{2}, \frac{9}{2})$ in Figure 2-2. Then we utilize the following reorientation of (2.18),

$$\min \quad \hat{u}_o$$

subject to

$$v - \mu + \hat{u}_o \geq 0$$
$$\frac{3}{2}v - 2\mu + \hat{u}_o \geq 0$$
$$3v - 4\mu + \hat{u}_o \geq 0$$
$$4v - 5\mu + \hat{u}_o \geq 0$$
$$\frac{7}{2}v - \frac{9}{2}\mu + \hat{u}_o = 0$$
$$v, \mu \geq 1, \hat{u}_o \geq 0.$$

This also gives $v^* = u^* = \hat{u}_o^* = 1$ so the applicable condition is (ii) in Theorem 2.1. Thus returns to scale are decreasing at E', the point on the BCC efficiency frontier which is shown in Figure 2-2.

6. MULTIPLICATIVE MODELS

The treatments to this point have been confined to "qualitative" characterizations in the form of identifying whether RTS are "increasing," "decreasing," or "constant." There is a literature – albeit a relatively small one – which is directed to "quantitative" estimates of RTS in DEA. Examples are the treatment of scale elasticities in Banker, Charnes and Cooper (1984), Førsund (1996) and Banker and Thrall (1992). However, there are problems in using the standard DEA models, as is done in these studies, to obtain scale elasticity estimates. Førsund (1996), for instance, lists a number of such problems. Also the elasticity values in Banker and

Thrall (1992) are determined only within upper and lower bounds. This is an inherent limitation that arises from the piecewise linear character of the frontiers for these models. Finally, attempts to extend the Färe, Grosskopf and Lovell (1985, 1994) approaches to the determination of scale elasticities have not been successful. See the criticisms in Førsund (1996, p. 296) and Fukuyama (2000, p. 105). (Multiple output-multiple input production and cost functions which meet the sub- and super-additivity requirements in economics are dealt with in Panzar and Willig (1977). See also Baumol, Panzar and Willig (1982).)

This does not, however, exhaust the possibilities. There is yet another class of models referred to as "multiplicative models" which were introduced by this name into the DEA literature in Charnes et al. (1982) – see also Banker et al. (1981) -- and extended in Charnes et al. (1983) to accord these models non-dimensional (=units invariance) properties like those we have just discussed. Although not used very much in applications these multiplicative models can provide advantages for extending the range of potential uses for DEA. For instance, they are not confined to efficiency frontiers which are concave. They can be formulated to allow the efficiency frontiers to be concave in some regions and non-concave elsewhere. See Banker and Maindiratta (1986). They can also be used to obtain "exact" estimates of elasticities in manners that we now describe.

The models we use for this discussion are due to Banker and Maindiratta (1986) -- where analytical characterizations are supplied along with confirmation in controlled-experimentally designed simulation studies.

We depart from the preceding development and now use an output oriented model which has the advantage of placing this development in consonance with the one in Banker and Maindiratta (1986) -- <u>viz.</u>,

$$\max \ \gamma_o$$
$$\text{subject to}$$
$$\prod_{j=1}^{n} x_{ij}^{\lambda_j} \le x_{io}, \quad i=1,...,m$$
$$\prod_{j=1}^{n} y_{rj}^{\lambda_j} \ge \gamma_o y_{ro}, \quad r=1,...,s \qquad (2.19)$$
$$\sum_{j=1}^{n} \lambda_j = 1$$
$$\gamma_o, \lambda_j \ge 0.$$

To convert these inequalities to equations we use

$$e^{s_i^-} \prod_{j=1}^{n} x_{ij}^{\lambda_j} = x_{io}, \quad i = 1,...,m$$

and (2.20)

$$e^{-s_r^+} \prod_{j=1}^{n} y_{rj}^{\lambda_j} = \gamma_o y_{ro}, \quad r = 1,...,s$$

and replace the objective in (2.19) with $\gamma e^{\varepsilon\left(\sum_{r=1}^{s} s_r^+ + \sum_{i=1}^{m} s_i^-\right)}$, where $s_i^-, s_r^+ \geq 0$ represent slacks. Employing (2.20) and taking logarithms we replace (2.19) with

$$\min \quad -\tilde{\gamma}_o - \varepsilon(\sum_{r=1}^{s} s_r^+ + \sum_{i=1}^{m} s_i^-)$$

subject to

$$\tilde{x}_{io} = \sum_{j=1}^{n} \tilde{x}_{ij}\lambda_j + s_i^-, \quad i = 1,...,m$$

$$\tilde{\gamma}_o + \tilde{y}_{ro} = \sum_{j=1}^{n} \tilde{y}_{rj}\lambda_j - s_r^+, \quad r = 1,...,s \qquad (2.21)$$

$$1 = \sum_{j=1}^{n} \lambda_j$$

$$\lambda_j, s_r^+, s_i^- \geq 0, \forall j, r, i.$$

where "~" denotes "logarithm" so the $\tilde{x}_{ij}$, $\tilde{y}_{rj}$ and the $\tilde{\gamma}_o$, $\tilde{x}_{io}$, $\tilde{y}_{ro}$ are in logarithmic units.

The dual to (2.21) is

$$\max \quad \sum_{r=1}^{s} \beta_r \tilde{y}_{ro} - \sum_{i=1}^{m} \alpha_i \tilde{x}_{io} - \alpha_o$$

subject to

$$\sum_{r=1}^{s} \beta_r \tilde{y}_{rj} - \sum_{i=1}^{m} \alpha_i \tilde{x}_{ij} - \alpha_o \leq 0, \quad j = 1,...,n$$

$$\sum_{r=1}^{s} \beta_r = 1 \qquad (2.22)$$

$$\alpha_i \geq \varepsilon, \ \beta_r \geq \varepsilon; \ \alpha_o \ \textit{free in sign.}$$

Using α_i^*, β_r^* and α_o^* for optimal values, $\sum_{r=1}^{s} \beta_r^* \tilde{y}_{ro} - \sum_{i=1}^{m} \alpha_i^* \tilde{x}_{io} - \alpha_o^* = 0$ represents a supporting hyperplane (in logarithmic coordinates) for DMU$_o$, where efficiency is achieved. We may rewrite this log-linear supporting hyperplane in terms of the original input/output values:

$$\prod_{r=1}^{s} y_{ro}^{\beta_r^*} = e^{\alpha_o^*} \prod_{i=1}^{m} x_{io}^{\alpha_i^*} \qquad (2.23)$$

Then, in the spirit of Banker and Thrall (1992), we introduce

Theorem 2.7
Multiplicative Model RTS,
(i) RTS are increasing if and only if $\sum \alpha_i^* > 1$ for all optimal solutions to (2.23).
(ii) RTS are decreasing if and only if $\sum \alpha_i^* < 1$ for all optimal solutions to (2.23).
(iii) RTS are constant if and only if $\sum \alpha_i^* = 1$ for some optimal solutions to (2.23).

To see what this means we revert to the discussion of (2.11) and introduce scalars a, b in (aX_o, bY_o). In conformance with (2.23) this means

$$e^{\alpha_o^*} \prod_{i=1}^{m} (ax_{io})^{\alpha_i^*} = \prod_{r=1}^{s} (by_{ro})^{\beta_r^*} \qquad (2.24)$$

so that the thus altered inputs and outputs satisfy this extension of the usual Cobb-Douglas types of relations.

The problem now becomes: given an expansion a > 1, contraction a < 1, or neither, i.e., a = 1, for application to all inputs, what is the value of b that positions the solution in the supporting hyperplane at this point? The answer is given by the following

Theorem 2.8
If (aX_o, bY_o) lies in the supporting hyperplane then $b = a^{\sum_{i=1}^{m} \alpha_i^*}$.

Proof: This proof is adopted from Banker et al. (2003). Starting with the expression on the left in (2.25) we can write

$$e^{\alpha_o^*} \prod_{i=1}^{m} (ax_{io})^{\alpha_i^*} = a^{\sum_{i=1}^{m} \alpha_i^*} e^{\alpha_o^*} \prod_{i=1}^{m} x_{io}^{\alpha_i^*} = \frac{a^{\sum_{i=1}^{m} \alpha_i^*}}{b} \prod_{r=1}^{s} (by_{ro})^{\beta_r^*} \qquad (2.25)$$

by using the fact that $\sum_{r=1}^{s} \beta_r^* = 1$ in (2.22) and $e^{\alpha_o^*} \prod_{i=1}^{m} x_{io}^{\alpha_i^*} = \prod_{r=1}^{s} y_{ro}^{\beta_r^*}$ in (2.23).

Thus, to satisfy the relation (2.24) we must have $b = a^{\sum_{i=1}^{m} \alpha_i^*}$ as the theorem asserts. ∎

Via this Theorem, we have the promised insight into reasons why more than proportionate output increases are associated with $\sum_{i=1}^{m} \alpha_i^* > 1$, less than proportionate increases are associated with $\sum_{i=1}^{m}\alpha_i^* < 1$ and constant returns to scale is the applicable condition when $\sum_{i=1}^{m}\alpha_i^* = 1$.

There may be alternative optimal solutions for (2.22) so the values for the α_i^* components need not be unique. For dealing with alternate optima, we return to (2.19) and note that a necessary condition for efficiency is $\gamma_o^* = 1$. For full efficiency we must also have all slacks at zero in (2.20). An adaptation of (2.3) to the present problem therefore gives

$$\prod_{j=1}^{n} x_{ij}^{\lambda_j^*} = e^{-s_i^{-*}} x_{io} = x_{io}', \qquad i = 1,...,m$$
$$\prod_{j=1}^{n} y_{rj}^{\lambda_j^*} = e^{s_{ri}^{+*}} \gamma_o^* y_{ro} = y_{ro}', \qquad r = 1,...,s$$

(2.26)

and x_{io}', y_{ro}' are the coordinates of the point on the efficiency frontier used to evaluate DMU$_o$.

Thus, we can extend the preceding models in a manner that is now familiar. Suppose we have obtained an optimal solution for (2.22) with $\sum_{i=1}^{m}\alpha_i^* < 1$. We then utilize (2.26) to form the following problem

$$\max \quad \sum_{i=1}^{m}\alpha_i$$

subject to

$$\sum_{r=1}^{s}\beta_r \tilde{y}_{rj} - \sum_{i=1}^{m}\alpha_i \tilde{x}_{ij} - \alpha_o \leq 0, \quad j = 1,...,n; j \neq o$$
$$\sum_{r=1}^{s}\beta_r \tilde{y}_{ro}' - \sum_{i=1}^{m}\alpha_i \tilde{x}_{io}' - \alpha_o = 0$$
$$\sum_{r=1}^{s}\beta_r \qquad\qquad = 1 \qquad\qquad (2.27)$$
$$\sum_{i=1}^{m}\alpha_i \qquad\qquad \leq 1$$
$$\alpha_i \geq \varepsilon, \beta_r \geq \varepsilon; \; \alpha_o \text{ free in sign.}$$

If $\sum_{i=1}^{m}\alpha_i^* = 1$ in (2.27), then returns to scale are constant by (iii) of Theorem 2.8. If the maximum is achieved with $\sum_{i=1}^{m}\alpha_i^* < 1$, however, condition (ii) of Theorem 2.7 is applicable and returns to scale are decreasing at the point x_{io}', y_{ro}'; $i = 1, ..., m; r = 1, ..., s$.

If we initially have $\sum_{i=1}^{m}\alpha_i^* > 1$ in (2.22), we replace $\sum_{i=1}^{m}\alpha_i^* \leq 1$ with $\sum_{i=1}^{m}\alpha_i^* \geq 1$ in (2.27) and also change the objective to minimize $\sum_{i=1}^{m}\alpha_i^*$. If the optimal value is greater than one, then (i) of Theorem 2.7 is applicable and the RTS are increasing. On the other hand, if we attain $\sum_{i=1}^{m}\alpha_i^* = 1$ then condition (iii) applies and returns to scale are constant.

Theorem 2.8 also allows us to derive pertinent scale elasticities in a straightforward manner. Thus, using the standard logarithmic derivative formulas for elasticities we obtain

$$\frac{d\ln b}{d\ln a} = \frac{a}{b}\frac{db}{da} = \sum_{i=1}^{m}\alpha_i^*. \qquad (2.28)$$

Consisting of a sum of component elasticities, one for each input, this overall measure of elasticity is applicable to the value of the multiplicative expression with which DMU_o is associated.

The derivation in (2.28) holds only for points where this derivative exists. However, we can bypass this possible source of difficulty by noting that Theorem 2.8 allows us to obtain this elasticity estimate via

$$\frac{\ln b}{\ln a} = \sum_{i=1}^{m} \alpha_i^*. \tag{2.29}$$

Further, as discussed in Cooper, Thompson and Thrall (1996), it is possible to extend these concepts to the case in which all of the components of Y_o are allowed to increase by at least the factor b. However, we cannot similarly treat the constant, a, as providing an upper bound for the inputs since mix alterations are not permitted in the treatment of returns to scale in economics. See Varian (1984, p. 20) for requirements of RTS characterizations in economics.

In conclusion we turn to properties of units invariance for these multiplicative models. Thus we note that $\sum_{i=1}^{m} \alpha_i^*$ is units invariant by virtue of the relation expressed in (2.28). The property of units invariance is also exhibited in (2.29) since a and b are both dimension free. Finally, we also have

Theorem 2.9
The model given in (2.19) and (2.20) is dimension free. That is, changes in the units used to express the input quantities x_{ij} or the output quantities y_{rj} in (2.19) will not affect the solution set or alter the value of max $\gamma_o = \gamma_o^*$.

Proof: Let

$$\begin{aligned} x_{ij}' &= c_i x_{ij}, & x_{io}' &= c_i x_{io}, & i = 1,...,m \\ y_{rj}' &= k_r y_{rj}, & y_{ro}' &= k_r y_{ro}, & r = 1,...,s \end{aligned} \tag{2.30}$$

where the c_i and k_r are any collection of positive constants. By substitution in the constraints for (2.20) we then have

$$\begin{aligned} e^{s_i^-} \prod_{j=1}^{n} x_{ij}'^{\lambda_j} &= x_{io}', & i = 1,...,m \\ e^{s_r^+} \prod_{j=1}^{n} y_{rj}'^{\lambda_j} &= \gamma_o y_{ro}', & r = 1,...,s \\ \sum_{j=1}^{n} \lambda_j &= 1, \ \lambda_j \geq 0, & j = 1,...,n \end{aligned} \tag{2.31}$$

Utilization of (2.30) therefore gives

$$e^{s_i^-} c_i^{\frac{\sum_{j=1}^n \lambda_j}} \prod_{j=1}^n x_{ij}^{\lambda_j} = c_i x_{io}, \qquad i = 1,\ldots,m$$

$$e^{-s_r^+} k_r^{\frac{\sum_{j=1}^n \lambda_j}} \prod_{j=1}^n y_{rj}^{\lambda_j} = \gamma_o k_r y_{ro}, \quad r = 1,\ldots,s \qquad (2.32)$$

$$\sum_{j=1}^n \lambda_j = 1, \ \lambda_j \geq 0, \quad j = 1,\ldots,n$$

However, $\sum_{j=1}^n \lambda_j = 1$, so $c_i^{\sum_{j=1}^n \lambda_j} = c_i$ and $k_r^{\sum_{j=1}^n \lambda_j} = k_r \ \forall i, r$. Therefore,

these constants, which appear on the right and left of (2.32), all cancel. Thus, all solutions to (2.31) are also solutions to (2.20) and <u>vice versa</u>. It follows that the optimal value of one program is also optimal for the other. ∎

We now conclude our discussion of these multiplicative models with the following

Corollary to Theorem 2.9
The restatement of (2.20) in logarithmic form yields a model which is translation invariant.

<u>Proof:</u> Restating (2.31) in logarithmic form gives

$$s_i^- + \sum_{j=1}^n (\tilde{x}_{ij} + \tilde{c}_i)\lambda_j = \tilde{x}_{io} + \tilde{c}_i, \qquad i = 1,\ldots,m$$

$$-s_r^+ + \sum_{j=1}^n (\tilde{y}_{rj} + \tilde{k}_r)\lambda_j = \tilde{y}_{ro} + \tilde{k}_r + \tilde{\gamma}_o, \quad r = 1,\ldots,s \qquad (2.33)$$

$$\sum_{j=1}^n \lambda_j = 1, \lambda_j \geq 0, \qquad j = 1,\ldots,n.$$

Once more utilizing $\sum \lambda_j = 1$ we eliminate the $\tilde{c}_i$ and $\tilde{k}_r$ on both sides of these expressions and obtain the same constraints as in (2.21). Thus, as before, the solution sets are the same and an optimum solution for one program is also optimal for the other -- including the slacks. ∎

7.   SUMMARY AND CONCLUSION

Although we have now covered all of the presently available models, we have not covered all of the orientations in each case. Except for the multiplicative models we have not covered output oriented objectives for a variety of reasons. There are no real problems with the mathematical development but further attention must be devoted to how changes in input scale and input mix should be treated when all outputs are to be scaled up in the same proportions. See the discussion in Cooper, Thompson and Thrall (1996).

As also noted in Cooper, Thompson and Thrall (1996), the case of increasing returns to scale can be clarified by using Banker's most productive scale size to write $(X_o\alpha, Y_o\beta)$. The case $1 < \beta/\alpha$ means that all outputs are increased by at least the factor β and returns to scale are increasing as long as this condition holds. The case $1 > \beta/\alpha$ has the opposite meaning--viz., no output is increasing at a rate that exceeds the rate at which all inputs are increased. Only for constant returns to scale do we have $1 = \beta/\alpha$, in which case all outputs and all inputs are required to be increasing (or decreasing) at the same rate so no mix change is involved for the inputs.

The results in this chapter (as in the literature to date) are restricted to this class of cases. This leaves unattended a wide class of cases. One example involves the case where management interest is centered on only subsets of the outputs and inputs. A direct way to deal with this situation is to partition the inputs and outputs of interest and designate the conditions to be considered by $(X_o^I\alpha, X_o^N, Y_o^I\beta, Y_o^N)$ where I designates the inputs and outputs that are of interest to management and N designates those which are not of interest (for such scale returns studies). Proceeding as described in the present chapter and treating X_o^N and Y_o^N as "exogenously fixed," in the spirit of Banker and Morey (1986), would make it possible to determine the situation for returns to scale with respect to the thus designated subsets. Other cases involve treatments with unit costs and prices as in FGL (1994) and Sueyoshi (1999).

The developments covered in this chapter have been confined to technical aspects of production. Our discussions follow a long-standing tradition in economics which distinguishes scale from mix changes by not allowing the latter to vary when scale changes are being considered. This permits the latter (i.e., scale changes) to be represented by a single scalar -- hence the name. However, this can be far from actual practice, where scale and mix are likely to be varied simultaneously when determining the size and scope of an operation. See the comments by a steel industry consultant that are quoted in Cooper, Seiford and Tone (2000, p. 130) on the need for reformulating this separation between mix and scale changes in order to achieve results that more closely conform to needs and opportunities for use in actual practice.

There are, of course, many other aspects to be considered in treating returns to scale besides those attended to in the present chapter. Management efforts to maximize profits, even under conditions of certainty, require simultaneous determination of scale, scope and mix magnitudes with prices and costs known, as well as the achievement of the technical efficiency which is always to be achieved with <u>any</u> set of positive prices and costs. The topics treated in this chapter do not deal with such price-cost information. Moreover, the focus is on <u>ex</u>-<u>post</u> <u>facto</u> analysis of already effected decisions. This can have many uses, especially in the control aspects of management where evaluations of performance are required. Left unattended in this chapter, and in much of the DEA literature, is the <u>ex</u> <u>ante</u> (planning) problem of how to use this knowledge in order to determine how to blend scale and scope with mix and other efficiency considerations when effecting future-oriented decisions.

REFERENCES

1. Banker, R.D., 1984, Estimating most productive scale size using Data Envelopment Analysis, European Journal of Operational Research 17, 35-44.
2. Banker, R.D., I. Bardhan and W.W. Cooper, 1996, A note on returns to scale in DEA, European Journal of Operational Research 88, 583-585.
3. Banker, R.D., H. Chang and W.W. Cooper, 1996, Equivalence and implementation of alternative methods for determining returns to scale in Data Envelopment Analysis," European Journal of Operational Research 89, 473-481.
4. Banker, R., A. Charnes and W.W. Cooper, 1984, Some models for estimating technical and scale inefficiencies in data envelopment analysis, Management Science 30, 1078-1092.
5. Banker, R.D., A. Charnes, W.W. Cooper and A. Schinnar, 1981, A bi-extremal principle for Frontier Estimation and Efficiency Evaluation, Management Science 27, 1370-1382.
6. Banker, R.D., and A. Maindiratta, 1986, Piecewise loglinear estimation of efficient production surfaces, Management Science 32, 126-135.
7. Banker, R.D. and R. Morey, 1986, Efficiency analysis for exogenously fixed inputs and outputs, Operations Research 34, 513-521.
8. Banker, R.D. and R.M. Thrall, 1992, Estimation of returns to scale using Data Envelopment Analysis, European Journal of Operational Research 62, 74-84.
9. Banker, R.D., W.W. Cooper, L.M. Seiford, R.M. Thrall and J. Zhu, 2003, Returns to scale in different DEA models, European Journal of Operational Research (to appear).
10. Baumol, W.J., J.C. Panzar and R.D. Willig, 1982, Contestable Markets, New York: Harcourt Brace Jovanovich.

11. Charnes, A., W.W. Cooper, and E. Rhodes, 1978, Measuring the efficiency of decision making units, European Journal of Operational Research 2, 429-444.
12. Charnes, A, W.W. Cooper, L.M. Seiford and J. Stutz, 1982, A multiplicative model for efficiency analysis, Socio-Economic Planning Sciences 16, 213-224.
13. Charnes, A, W.W. Cooper, L.M. Seiford and J. Stutz, 1983, Invariant multiplicative efficiency and piecewise Cobb-Douglas envelopments, Operations Research Letters 2, 101-103.
14. Charnes, A., W.W. Cooper and R.M. Thrall, 1991, A structure for classifying efficiencies and inefficiencies in DEA, Journal of Productivity Analysis 2, 197-237.
15. Cooper, W.W., K.S. Park and J.T. Pastor, 1999, RAM: A range adjusted measure of efficiency, Journal of Productivity Analysis 11, 5-42.
16. Cooper, W.W., Seiford, L.M. and Tone, K., 2000, Data Envelopment Analysis: A Comprehensive Text with Models, Applications, References and DEA-Solver Software, Kluwer Academic Publishers, Boston.
17. Cooper, W.W., R.G. Thompson and R.M. Thrall, 1996, Extensions and new developments in DEA, The Annals of Operations Research 66, 3-45.
18. Färe, R. and S. Grosskopf, 1994, Estimation of returns to scale using data envelopment analysis: A comment, European Journal of Operational Research 79, 379-382.
19. Färe, R., S. Grosskopf and C.A.K. Lovell, 1985, The Measurement of Efficiency of Production. Boston: Kluwer .Nijhoff. Publishing Co.
20. Färe, R., S. Grosskopf and C.A.K. Lovell, 1994. Production Frontiers, Cambridge University Press.
21. Førsund, F.R., 1996, On the calculation of scale elasticities in DEA models, Journal of Productivity Analysis 7, 283-302.
22. Frisch, R.A., 1964, Theory of Production, Dordrecht: D. Rieoel.
23. Fukuyama, H., 2000, Returns to scale and scale elasticity in Data Envelopment Analysis, European Journal of Operational Research 125, 93-112.
24. Golany, B. and G. Yu, 1994, Estimating returns to scale in DEA, European Journal of Operational Research 103, 28-37.
25. Panzar, J.C. and R.D. Willig, 1977, Economies of scale in multi-output production, Quarterly Journal of Economics XLI. 481-493.
26. Seiford, L.M. and J. Zhu, 1999, An investigation of returns to scale under Data Envelopment Analysis, Omega 27, 1-11.

27. Sueyoshi, T., 1999, DEA duality on returns to scale (RTS) in production and cost analyses: an occurrence of multiple solutions and differences between production-based and cost-based RTS estimates, Management Science 45, 1593-1608.
28. Thrall, R.M., 1996a, Duality, classification and slacks in DEA, Annals of Operations Research 66, 109-138.
29. Thrall, R.M., 1996b, The lack of invariance of optimal dual solutions under translation invariance, Annals of Operations Research 66, 103-108.
30. Varian, H., 1984, Microeconomic Analysis, New York: W.W. Norton.
31. Zhu, J., 2000, Setting scale efficient targets in DEA via returns to scale estimation methods, Journal of the Operational Research Society 51, No. 3, 376-378.
32. Zhu, J., 2002, Quantitative Models for Performance Evaluation and Benchmarking: Data Envelopment Analysis with Spreadsheets and DEA Excel Solver, Kluwer Academic Publishers, Boston.
33. Zhu, J. and Z. Shen, 1995, A discussion of testing DMUs' returns to scale, European Journal of Operational Research 81, 590-596.

APPENDIX

In this Appendix, we first present the FGL approach. We then present a simple RTS approach without the need for checking the multiple optimal solutions as in Zhu and Shen (1995) and Seiford and Zhu (1999) where only the BCC and CCR models are involved. This approach will substantially reduce the computational burden, because it relies on the standard CCR and BCC computational codes (see Zhu (2002) for a detailed discussion).

To start, we add to the BCC and CCR models by the following DEA model whose frontier exhibits non-increasing returns to scale (NIRS), as in Färe, Grosskopf and Lovell (FGL, 1985, 1994)

$$\theta^*_{NIRS} = \min \quad \theta_{NIRS}$$

subject to

$$\theta_{NIRS} x_{io} = \sum_{j=1}^{n} x_{ij}\lambda_j + s_i^- \quad i = 1,2,...,m;$$

$$y_{ro} = \sum_{j=1}^{n} y_{rj}\lambda_j - s_r^+ \quad r = 1,2,...,s; \quad \text{(A.1)}$$

$$1 \geq \sum_{j=1}^{n}\lambda_j$$

$$0 \leq \lambda_j, s_i^-, s_r^+ \quad \forall i, r, j.$$

The development used by FGL (1985, 1994) rests on the following relation

$$\theta^*_{CCR} \leq \theta^*_{NIRS} \leq \theta^*_{BCC}$$

where "*" refers to an optimal value and θ^*_{NIRS} is defined in (A.1) while θ^*_{BCC} and θ^*_{CCR} refer to the BCC and CCR models as developed in Theorems 2.3 and 2.4.

FGL utilize this relation to form ratios that provide measures of RTS. However, we turn to the following tabulation which relates their RTS characterization to Theorems 2.3 and 2.4 (and accompanying discussion). See also Färe and Grosskopf (1994), Banker, Chang and Cooper (1996), and Seiford and Zhu (1999)

	FGL Model	RTS	CCR Model
Case 1	If $\theta^*_{CCR} = \theta^*_{BCC}$	Constant	$\sum \lambda^*_j = 1$
Case 2	If $\theta^*_{CCR} < \theta^*_{BCC}$ then		
Case 2a	If $\theta^*_{CCR} = \theta^*_{NIRS}$	Increasing	$\sum \lambda^*_j < 1$
Case 2b	If $\theta^*_{CCR} < \theta^*_{NIRS}$	Decreasing	$\sum \lambda^*_j > 1$

It should be noted that the problem of non-uniqueness of results in the presence of alternative optima is not encountered in the FGL approach (unless output-oriented as well as input-oriented models are used) whereas they do need to be coincided, as in Theorem 2.3. However, Zhu and Shen (1995) and Seiford and Zhu (1999) develop an alternative approach that is not troubled by the possibility of such alternative optima.

We here present their results with respect to Theorems 2.3 and 2.4 (and accompanying discussion). See also Zhu (2002).

	Seiford and Zhu (1999)	RTS	CCR Model
Case 1	If $\theta^*_{CCR} = \theta^*_{BCC}$	Constant	$\sum \lambda^*_j = 1$
Case 2	$\theta^*_{CCR} \neq \theta^*_{BCC}$		
Case 2a	If $\sum \lambda^*_j < 1$ in any CCR outcome	Increasing	$\sum \lambda^*_j < 1$
Case 2b	If $\sum \lambda^*_j > 1$ in any CCR outcome	Decreasing	$\sum \lambda^*_j > 1$

The significance of Seiford and Zhu's (1999) approach lies in the fact that the possible alternate optimal λ^*_j obtained from the CCR model only affect the estimation of RTS for those DMUs that truly exhibit constant returns to scale, and have nothing to do with the RTS estimation on those DMUs that truly exhibit increasing returns to scale or decreasing returns to scale. That is, if a DMU exhibits increasing returns to scale (or decreasing returns to scale), then $\sum^n_j \lambda^*_j$ must be less (or greater) than one, no matter whether there exist alternate optima of λ_j, because these DMUs do not lie in the MPSS region. This finding is also true for the u_o^* obtained from the BCC multiplier models.

Thus, in empirical applications, we can explore RTS in two steps. First, select all the DMUs that have the same CCR and BCC efficiency scores regardless of the value of $\sum_j^n \lambda_j^*$ obtained from model (2.5). These DMUs are constant returns to scale. Next, use the value of $\sum_j^n \lambda_j^*$ (in any CCR model outcome) to determine the RTS for the remaining DMUs. We observe that in this process we can safely ignore possible multiple optimal solutions of λ_j.

Part of the material in this chapter is adapted from European Journal of Operational Research, Vol 154, Banker, R.D., Cooper, W.W., Seiford, L.M., Thrall, R.M. and Zhu, J, Returns to scale in different DEA models, 345-362, 2004, with permission from Elsevier Science.

Chapter 3

SENSITIVITY ANALYSIS IN DEA

William W. Cooper[1], Shanling Li[2], Lawrence M. Seiford[3] and Joe Zhu[4]

[1] Red McCombs School of Business, University of Texas at Austin, Austin, TX 78712 USA
email: cooperw@mail.utexas.edu

[2] Faculty of Management, McGill University, 1001 Sherbrooke Street West, Montreal, Quebec
H3A 1G5, Canada   email: shanling.li@mcgill.ca

[3] Department of Industrial and Operations Engineering, University of Michigan at Ann Arbor,
Ann Arbor, MI 48102 USA   email: seiford@umich.edu

[4] Department of Management, Worcester Polytechnic Institute, Worcester, MA 01609   USA
email: jzhu@wpi.edu

Abstract: This chapter presents some of the recently developed analytical methods for
studying the sensitivity of DEA results to variations in the data. The focus is
on the stability of classification of DMUs (Decision Making Units) into
efficient and inefficient performers. Early work on this topic concentrated on
developing algorithms for conducting such analyses after it was noted that
standard approaches for conducting sensitivity analyses in linear programming
could not be used in DEA. However, recent work has bypassed the need for
such algorithms. It has also evolved from the early work that was confined to
studying data variations in one input or output for one DMU. The newer
methods described in this paper make it possible to analyze the sensitivity of
results when all data are varied simultaneously for all DMUs.

Key words: Data envelopment analysis (DEA), Efficiency, Stability, Sensitivity

1.   INTRODUCTION

This chapter surveys analytical approaches that have been developed to treat sensitivity and stability analyses in DEA. We may classify the approaches to this topic into two categories that we can characterize as (i) substantive and (ii) methodological. The term "substantive" refers to generalizations or characterizations directed to the properties of DEA. An early example is *Measuring Efficiency: An Assessment of Data Envelopment Analysis. New Directions for Program Evaluation.* Authored by Sexton, Silkman and Hogan (1986), this book concludes that DEA results are likely to be unstable because its evaluations are based on "outlier" observations. We may contrast this with the "methodological" approaches to sensitivity and stability analysis. As the term suggests, this category is concerned with the development of tools and concepts that can be used to determine the degree of sensitivity to data variations in any particular application of DEA.

Possibly because they were not formulated in a manner that could provide guidance as to where further research or adaptations for use might best be pointed, little progress has been made in such substantive approaches. By way of contrast, there has been a great deal of progress in the category that we have referred to as methodological approaches. The early work in this area, which focused on analyses of a single input or output for a single DMU, has now moved to sensitivity analyses directed to evaluating the stability of DEA results when <u>all</u> inputs and outputs are varied simultaneously in <u>all</u> DMUs. Unlike other, looser characterizations of stability (as in the substantive approaches) these analytical approaches have been precise as to the nature of the problems to be treated. They have focused almost exclusively on changes from efficient to inefficient status for the DMUs being analyzed.

The present chapter is devoted to these methodological approaches. Hence it, too, will focus on the sensitivity and stability of efficient vs. inefficient classifications of DMUs under different ranges of data variations.

The approaches to be examined are deterministic in the same sense that the models covered in the preceding chapters are deterministic and so we leave the use of stochastic approaches with accompanying tests of stability and significance to later chapters in this handbook, as in the chapter by Banker. Attention here is confined to studies involving only variations in data. Other topics such as sensitivity to model changes or diminution and augmentation in the number of DMUs (as in the statistical studies of sampling distributions) are covered in the chapter by Simar and Wilson.

2. SENSITIVITY ANALYSIS APPROACHES

The topic of sensitivity (= stability or robustness) analysis has taken a variety of forms in the DEA literature. One part of this literature studies responses with given data when DMUs are deleted or added to the set being considered. See Wilson (1995) and the discussion of "window analysis" in Chapter 1. Another part of this literature deals with increases or decreases in the number of inputs and outputs to be treated. Analytically oriented treatments of these topics are not lacking but most of the literature on this topic has taken the form of simulation studies, as in Banker, Chang and Cooper (1996). We also do not examine sensitivity studies related to choices of different DEA models as in Ahn and Seiford (1993). In any case we restrict the discussion in this chapter to analytically formulated (mathematical) methods for examining stability and sensitivity of results to data variations with given variables and given models.

As in statistics or other empirically oriented methodologies, there is a problem involving degrees of freedom, which is compounded in DEA because of its orientation to *relative* efficiency. In the envelopment model, the number of degrees of freedom will increase with the number of DMUs and decrease with the number of inputs and outputs. A rough rule of thumb which can provide guidance is to choose a value of n that satisfies $n \geq \max\{m \times s, 3(m + s)\}$, where n= number of DMUs, m = number of inputs and s = number of outputs. Hereafter we assume that this (or other) degrees of freedom conditions are satisfied and that there is no trouble from this quarter.

2.1 Algorithmic Approaches

Research on analytical approaches to sensitivity analysis in DEA was initiated in Charnes et al. (1985) where it was noted that the methods used for this purpose in linear programming were not appropriate to DEA. The discussion in Charnes et al. (1985) took the form of algorithmic developments which built on the earlier work in Charnes and Cooper (1968) -- after noting that variations in the data for the DMU_o being analyzed could alter the inverse matrix that is generally used in linear programming approaches to sensitivity analyses. Building on the earlier work of Charnes and Cooper (1968), this work by Charnes et al. (1985) was therefore directed to developing algorithms that would avoid the need for additional matrix inversions. Originally confined to treating a single input or output this line of work was extended and improved in a series of papers published by Charnes and Neralic. For a summary discussion see Charnes and Neralic (1992) and Neralic (1997).

2.2   Metric Approaches

Another avenue for sensitivity analysis opened by Charnes et al. (1992) bypasses the need for these kinds of algorithmic forays by turning to metric concepts. The basic idea is to use concepts such as "distance" or "length" (= norm of a vector) in order to determine "radii of stability" within which the occurrence of data variations will not alter a DMU's classification from efficient to inefficient status (or vice versa).

The resulting classifications obtained from these "radii of stability" can range from "unstable" to "stable" with the latter being identified by a radius of some finite value within which no reclassification will occur. For example, point like F in Figure 3-1 is identified as stable. A point like A, however, is unstable because an infinitesimal perturbation to the left of its present position would alter its status from inefficient to efficient.

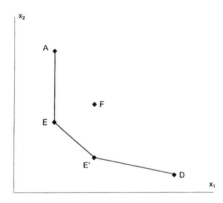

Figure 3-1. Stable and Unstable DMUs

A variety of metrics and models are examined by Charnes et al. Here, however, attention will be confined to the Chebychev (= l_∞) norm, as in the following model taken from Charnes et al. (1992, p.795),

$$\max \delta$$
$$\text{subject to}$$
$$y_{ro} = \sum_{j=1}^{n} y_{rj}\lambda_j - s_r^+ - \delta d_r^+, \quad r = 1,...,s$$
$$x_{io} = \sum_{j=1}^{n} x_{ij}\lambda_j + s_i^- + \delta d_i^-, \quad i = 1,...,m \qquad (3.1)$$
$$1 = \sum_{j=1}^{n} \lambda_j$$

with all variables (including δ) constrained to be nonnegative while the d_r^+ and d_i^- are fixed constants (or weights) which we now equate to unity. For instance, with all $d_i^- = d_r^+ = 1$ the solution to (3.1) may be written

$$\sum y_{rj}\lambda_j^* - s_r^{+*} = y_{ro} + \delta^*, \quad r = 1,...,s$$
$$\sum_{j=1}^{n} x_{ij}\lambda_j^* + s_i^{-*} = x_{io} - \delta^*, \quad i = 1,...,m \tag{3.2}$$

where "*" indicates an optimum value and the value of δ^* represents the maximum that this metric allows consistent with the solution on the left.

The formulation in (3.2) is for an inefficient DMU which continues to be inefficient for all data alterations from y_{ro} to $y_{ro} + \delta_r^*$ and from x_{io} to $x_{io} - \delta^*$. This is intended to mean that no reclassification to efficient status will occur within the open set defined by the value of $0 \le \delta^*$, which is referred to as a "radius of stability." See, for example, the point F in Figure 3-2 which is centered in the square (or box) which is referred to as a "unit ball" defined by this C (=Chebyshev) norm.

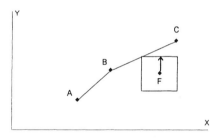

Figure 3-2. A Radius of Stability

The above model dealt with improvements in <u>both</u> inputs and outputs that could occur for an <u>inefficient</u> point before its status would change to efficient -- as in the upper left hand corner of the square surrounding F in Figure 3-2. The treatment of <u>efficient</u> points proceeds in the direction of "worsening" outputs and inputs as in the following model.

$$\min \delta$$
subject to
$$y_{ro} = \sum_{j=1, j\neq 0}^{n} y_{rj}\lambda_j - s_r^+ + \delta, \quad r = 1,...,s$$
$$x_{io} = \sum_{j=1, j\neq 0}^{n} x_{ij}\lambda_j + s_i^- - \delta, \quad i = 1,...,m \tag{3.3}$$
$$1 = \sum_{j=1, j\neq 0}^{n} \lambda_j$$

where, again, all variables are constrained to be nonnegative.

In this case $j \neq o$ refers to the efficient DMU_o that is being analyzed. Removal of this DMU_o is necessary because the solution will otherwise be $\delta^* = 0$ with $\lambda_o^* = 1$ and thereby indicate that the efficient DMU_o has a zero radius of stability.

We can generalize Charnes et al.'s (1992) instability property as follows

Definition 3.1: The coordinates of the point associated with an efficient DMU will always have both efficient and inefficient points within a radius of $\varepsilon > 0$ however small the value of ε.

Definition 3.2: Any point which has this property is unstable.

We can illustrate its applicability by noting that point A, which is not efficient in Figure 3-1, has this property, since a slight variation to the left will change its status from inefficient to efficient. In any case, a solution, δ^*, provides a radius in the Chebyshev norm that is to be attained before an efficient DMU is changed to inefficient status. We illustrate this with the following example,

Example: To portray what is happening, we synthesize an example from Figure 3-2 by assigning coordinates to these points in the order (x, y) as follows: A=(1,1), B=(3,4) and C=(4,5). To simplify matters we remove F and consider only the efficient points A and C to determine a radius of stability for B. This is accomplished by applying (3.3) as follows

$$
\begin{aligned}
&\min \delta \\
&\text{subject to} \\
&4 = 1\lambda_A + 5\lambda_C - s^+ + \delta \\
&3 = 1\lambda_A + 4\lambda_C + s^- - \delta \\
&1 = \lambda_A + \lambda_C.
\end{aligned}
$$

With all variables constrained to be non-negative the values $\lambda_A^* = 0.29$, $\lambda_C^* = 0.71$ and $\delta^* = 0.15$ with both slacks zero optimally satisfy all constraints.

With $\delta^* = 0.15$ as the radius of stability we have $4 - \delta^* = 3.85$ as the worsened output and $3 + \delta^* = 3.15$ as the worsened input values: The thus altered values of B remain efficient if no further worsenings occur. To confirm this latter point we may note that the line segment connecting A and C represents the efficient frontier for the reduced convex set that remains after B is removed. This segment is on the line $y = -\frac{1}{3} + \frac{4}{3}x$. Substituting $3 + \delta^* = x = 3.15$ in this expression yields $y = 3.85 = 4 - \delta^*$. Hence these worsened values of B place it on the efficient frontier of this reduced convex set where it retains its efficient status. Thus, as is evident in this example, the model (3.3) seeks to minimize the value of the norm that brings the thus removed point into contact with the frontier of the reduced convex set.

The above formulations are recommended by Charnes et al. (1992) only as the start for a sensitivity analysis. Because, inter alia, this (Chebychev) norm does not reflect nonzero slacks which may be present. However, Charnes et al. (1992) provide other models, such as additive model formulations which utilize the ℓ_1 metric, account for all such inefficiencies.

2.3   Multiplier Model Approaches

The two approaches described above -- i.e., the algorithmic and metric approaches -- use DEA "envelopment models" to treat one DMU at a time. Extensions are needed if we are to treat cases where the DMUs are numerous and where it is not clear which ones require attention. Ideally it should be possible to vary all data simultaneously for all DMUs until the status of at least one DMU is changed from inefficient to efficient or vice versa. An approach initiated in Thompson, Dharmapala and Thrall (1994) moves in this direction in a manner that we now describe.

For this purpose we record the following dual pair of problems from Thompson et al. (1996).

Envelopment Model	Multiplier Model	
minimize$_{\theta, \lambda}$ θ	maximize$_{u,v}$ $z = uy_o$	
subject to	subject to	
$Y\lambda \geq y_o$	$u \geq 0$	(3.4)
$\theta x_o - X\lambda \geq 0$	$v \geq 0$	
$\lambda \geq 0$	$uY - vX \leq 0$	
θ unrestricted	$vx_o = 1$	

where Y, X and y_o, x_o are data matrices and vectors of outputs and inputs, respectively, and λ, u, v are vectors of variables (λ: a column vector; u and v: row vectors). θ, a scalar, which can be positive, negative or zero in the envelopment model is the source of the condition $vx_o = 1$ which appears at the bottom of the multiplier model.

The above arrangement, we might note systematically relates the elements of these dual problems to each other in the following fashion. Non-negativity for the vectors u and v on the right in the multiplier model are associated with the inequality constraints on their left in the envelopment model. Similarly, non-negativity for the vector λ on the left is associated with the inequality conditions on the right. Finally, the unrestricted variable

θ in the envelopment model is associated with the condition $vx_o = 1$ on its right.

We now observe that no allowance for nonzero slacks is made in the objective of the above envelopment model. Hence the variables in the multiplier model are constrained only to be nonnegative. That is, the positivity requirement associated with the non-Archimedean element, ε described in chapter 1, is absent from both members of this dual pair. For present purposes, however, we only need to note that the sensitivity analyses we will now be considering are centered around the set of efficient extreme points and these points always have a unique optimum with nonzero slack solutions for the envelopment (but not the multiplier) model. (see Charnes, Cooper and Thrall (1991).)

The analysis used by Thompson, Dharmapala and Thrall (1994) is carried forward via the multiplier models. This makes it possible to exploit the fact that the values u^*, v^* which are optimal for the DMU being evaluated will remain valid over some (generally positive) range of variation in the data.

Following Thompson et al. we exploit this property by defining a new vector $w = (u, v)$ which we use to define a function $h_j(w)$ as follows,

$$h_j(w) = \frac{f_j(w)}{g_j(w)} = \frac{\sum_{r=1}^{s} u_r y_{rj}}{\sum_{i=1}^{m} v_i x_{ij}}. \tag{3.5}$$

Next, let

$$h_o(w) = \max_{j=1,\ldots,n} h_j(w) \tag{3.6}$$

so that

$$h_o(w) \geq h_j(w), \quad \forall j. \tag{3.7}$$

It is now to be noted that (3.5) returns matters to the nonlinear version of the CCR ratio form given by (1.1) and (IR) in chapter 1. Hence, we need not be concerned with continued satisfaction of the norm condition, $vx_o = 1$ in (3.4) when we study variations in the data in the manner now to be described.

When an optimal w^* does not satisfy (3.7), the DMU_o being evaluated is said to be "radial inefficient." The term is appropriate because this means that $\theta^* < 1$ will occur in the envelopment model. The full panoply of relations between the CCR ratio, multiplier and envelopment models is thus brought into play without any need for extensive computations or analyses.

Among the frontier points for which $\theta^* = 1$, attention is directed by Thompson et al. to "extreme efficient points." In particular, attention is centered on points in the set E for which, for some multiplier w^*,

$$h_o(w^*) > h_j(w^*) \quad \forall j \neq o. \tag{3.8}$$

This (strict) inequality will generally remain valid over some range of variation in the data. Hence, in more detail, we will have

$$h_o(w^*) = \frac{\sum\limits_{r=1}^{s} u_r^* y_{ro}}{\sum\limits_{i=1}^{m} v_i^* x_{io}} > \frac{\sum\limits_{r=1}^{s} u_r^* y_{rj}}{\sum\limits_{i=1}^{m} v_i^* x_{ij}} = h_j(w^*) \quad \forall j \neq o, \qquad (3.9)$$

which means that DMU_o is more efficient than any other DMU*j* and hence will be rated as fully efficient by DEA.

Thompson, Dharmapala and Thrall (1994) employ a ranking principle which they formulate as:

"If DMU_o is more efficient than all of the other DMUj relative to the vector w^*, then DMU_o is said to be top ranked."

Thus, holding w^* fixed, the data are varied and DMU_o is then said to be "top ranked" as long as (3.8) continues to hold for the data variations under consideration.

Thompson, Dharmapala and Thrall (1994) carry out experiments in which the data are allowed to vary in different ways -- including allowing the data variations to occur at random. Among these possibilities we examine only the following one: For DMU_o, which is extreme efficient, the outputs will all be decreased and the inputs will all be increased by a stipulated amount (or percentage). This same treatment is accorded to the other DMUs which are efficient. For the other DMUj, which are all inefficient, the reverse adjustment is made: All outputs are increased and all inputs are decreased in these same amounts (or percentages). In this way the value of the ratio will be decreased for DMU_o in (3.8) and the other extreme efficient DMUs while the ratios for the other DMUj will be increased. Continuing in this manner a reversal can be expected to occur at some point--in which case DMU_o will no longer be "top ranked" which means that it will then lose the status of being fully (DEA) efficient.

Table 3-1 taken from Thompson, Dharmapala and Thrall (1994) will be used to illustrate the procedure in a simple manner by varying only the data for the inputs x_1, x_2 in this table.

Table 3-1. Data for a Sensitivity Analysis

	E-Efficient*			Not Efficient		
DMU	1	2	3	4	5	6
Output: y	1	1	1	1	1	1
Input: x1	4	2	1	2	3	4
Input: x2	1	2	4	3	2	4

*E-Efficient = Extreme Point Efficient

To start the sensitivity analysis, Table 3-2 records the initial solutions obtained by applying the multiplier model in (3.4) to these data for each of DMU_1, DMU_2 and DMU_3 which are all E (=extreme point) efficient. As can be seen, these solutions show DMU_1, DMU_2 and DMU_3 to be top ranked in their respective columns.

Table 3-2. Initial Solutions

DMU	DMU1 $h_j(w^1)$	DMU2 $h_j(w^2)$	DMU3 $h_j(w^3)$
1	1.000	0.800	0.400
2	0.714	1.000	0.714
3	0.400	0.800	1.000
4	0.500	0.800	0.667
5	0.667	0.800	0.550
6	0.357	0.500	0.357

The gaps between the top and other ranks from these results show that some range of data variation can be undertaken without changing this top-ranked status in any of these three columns. To start we therefore follow Thompson et al. and introduce 5% increases in each of x_1 and x_2 for DMU_1, DMU_2 and DMU_3. Simultaneously, we decrease these inputs by 5% for the other DMUs to obtain Table 3-3.

Table 3-3. 5% Increments and Decrements

DMU	DMU1 $h_j(w^1)$	DMU2 $h_j(w^2)$	DMU3 $h_j(w^3)$
1	0.952	0.762	0.381
2	0.680	0.952	0.680
3	0.381	0.762	0.952
4	0.526	0.842	0.702
5	0.702	0.842	0.552
6	0.376	0.526	0.376

As can be seen in Table 3-3, each of DMU_1, DMU_2 and DMU_3 maintain their "top ranked status" and hence continue to be DEA fully efficient (relatively). Nor is this the end of the line. Continuing in this 5% increment-decrement fashion a 15% increment-decrement is needed, as Thompson, Dharmapala and Thrall (1994) report, for a first displacement in which DMU_2 is replaced by DMU_4 and DMU_5. Continuing further, a 20% increment-decrement is needed to replace DMU_4 with DMU_1 and, finally, still further incrementing and decrementing is needed to replace DMU₄ with DMU₁ as top ranked.

Note that the $h_j(w)$ values for all of the <u>efficient</u> DMUs decrease in every column when going from Table 3-2 to Table 3-3 and, simultaneously, the $h_j(w)$ values increase for the <u>inefficient</u> DMUs. The same behavior occurs for the other data variations.

As noted in Thompson, Dharmapala and Thrall (1994) this robust behavior is obtained for extreme efficient DMUs -- which are identified by their satisfaction of the Strong Complementary Slackness Condition for which a gap will appear like ones between the top and second rank shown in every column of Table 3-2. In fact, the choice of w^* can affect the degree of robustness as reported in Thompson et al. (1996) where use of an interior point algorithm produces a w^* closer to the "analytic center" and this considerably increases the degree of robustness for the above example.

As just noted, Thompson, Dharmapala and Thrall (1994) confine their analysis to points which are extreme efficient. They then utilize the "strong form of the complementary slackness principle' to ensure that all of the dual (=multiplier) values are positive -- as in Table 3-4, below -- and this avoids possible troubles associated with the appearance of zeros in the ratios represented in (3.9).

Table 3-4. Optimal Dual (=Multiplier) Values

	Outputs	Inputs	
	u^*	v_1^*	v_2^*
DMU1	1.0	0.10	0.60
DMU2	1.0	0.25	0.25
DMU3	1.0	0.60	0.10

Source: Thompson, Dharmapala and Thrall (1994).

To see what is happening with this approach we briefly review the principles involved. First we recall the "complementary slackness principle" of linear programming which we represent in the following form,

$$s_r^{+*}u_r^* = 0, \quad r = 1,...,s$$

$$s_i^{-*}v_i^* = 0, \quad i = 1,...,m, \tag{3.10}$$

where $s_r^{+*} \geq 0$ and $s_i^{-*} \geq 0$ are optimal output and input slacks, respectively, for the envelopment model, and u_r^* and v_i^* are the dual (=multiplier) values associated with the multiplier model in (3.4). Verbally, this means that at least one (and possibly both) variables in each pair must be zero in their corresponding optimal solutions.

The "strong principle of complementary slackness" may be represented

$$s_r^{+*} + u_r^* > 0, \quad r = 1,...,s$$

$$s_i^{-*} + v_i^* > 0, \quad i = 1,...,m. \tag{3.11}$$

In short, the possibility of both variables being zero is eliminated.

By restricting attention to the set of efficient points, Thompson et al. insure that only the "multiplier" variables will be positive. This occurs because the solutions associated with the extreme efficient points in the envelopment model are always unique with all slacks at zero. Conformance with the conditions in (3.11) therefore insures that all s+m multiplier values will be positive and the results portrayed in Table 3-4 are a consequence and this property. Having all of these variables positive is referred to as a solution with "full dimensionality" by Thompson, Dharmapala and Thrall (1994).

2.4   A Two-Stage Alternative

This use of strong complementary slackness involves recourse to special algorithms like the interior point methods developed in Thompson et al. (1996). To avoid this need we therefore suggest an alternative in the form of a two stage approach as follows: Stage one utilizes any of the currently available computer codes. This will yield satisfactory results in many (if not most) cases even though strong complementarity is not satisfied. Stage two is to be invoked only when these results are not satisfactory -- in which case recourse to special algorithms will be needed like those described in Thompson et al. (1996).

We illustrate this approach by using the data of Table 3-1 to apply the Thompson, Dharmapala and Thrall (1994) formulas in the manner that we described in the preceding section of this chapter. In this illustrative application we obtain solutions to the multiplier models for DMU_1, DMU_2, DMU_3 which we display in Table 3-5. As can be seen, some of these multiplier values differ from the ones exhibited in Table 3-4. In addition a zero value appears for v_2^* in the row for DMU_3. Nevertheless the results do not differ greatly from those reported by Thompson et al. and, in particular, the same kind of very robust results are secured.

Table 3-5. Multiplier Values

	Outputs	Inputs	
	u^*	v_1^*	v_2^*
DMU1	1	0.167	0.333
DMU2	1	0.333	0.167
DMU3	1	1.	0

To illustrate the Thompson et al. procedure we develop this example in more detail. For this purpose we focus on the multiplier values for DMU_1, which are $w^1 = (u^*, v_1^*, v_2^*) =$ (1, 0.167, 0.333) taken from Table 3-5. Applying these values to the data of Table 3-1 for each of these j =1, ..., 6 DMUs we obtain

$$h_1(w^1) = \frac{u^*}{v_1^* x_{11} + v_2^* x_{21}} = \frac{u^*}{4v_1^* + 1v_2^*} = \frac{1}{0.667 + 0.333} = 1$$

$$h_2(w^1) = \frac{u^*}{v_1^* x_{12} + v_2^* x_{22}} = \frac{u^*}{2v_1^* + 2v_2^*} = \frac{1}{0.333 + 0.667} = 1$$

$$h_3(w^1) = \frac{u^*}{v_1^* x_{13} + v_2^* x_{23}} = \frac{u^*}{1v_1^* + 4v_2^*} = \frac{1}{0.167 + 1.333} = 0.667$$

$$h_4(w^1) = \frac{u^*}{v_1^* x_{14} + v_2^* x_{24}} = \frac{u^*}{2v_1^* + 3v_2^*} = \frac{1}{0.333 + 1.0} = 0.75$$

$$h_5(w^1) = \frac{u^*}{v_1^* x_{15} + v_2^* x_{25}} = \frac{u^*}{3v_1^* + 2v_2^*} = \frac{1}{0.5 + 0.667} = 0.857$$

$$h_6(w^1) = \frac{u^*}{v_1^* x_{16} + v_2^* x_{26}} = \frac{u^*}{4v_1^* + 4v_2^*} = \frac{1}{0.667 + 1.333} = 0.5$$

In contrast to the results displayed in column 1 of Table 3-2, we find that DMU_2 has moved up to tie DMU_1 in top-rank position, since $h_1(w^1) = h_2(w^1) = 1$. Nevertheless, DMU_1 is not displaced. Since DMU_1 continues to maintain its efficient status, we consider the result to be satisfactory even though the previously clear separation of DMU_1 from all of the other DMUs is not maintained.

Undertaking the same operations with the values $w^2 = (1, 0.333, 0.167)$ and $w^3 = (1, 1, 0)$ as recorded for DMUs 2 and 3 in Table 3-5, we get the results displayed in Table 3-6.

Table 3-6. Initial Solutions

DMU	DMU_1 $h_j(w^1)$	DMU_2 $h_j(w^2)$	DMU_3 $h_j(w^3)$
1	1.000	0.667	0.250
2	1.000	1.000	0.500
3	0.667	1.000	1.000
4	0.750	0.857	0.500
5	0.857	0.750	0.333
6	0.500	0.500	0.250

As can be seen, a similar elevation to a top-rank tie occurs for DMU_2 but this does not displace DMU_2 for its top rank position. Finally the zero value for the v_2^* associated with DMU_3 does not cause any trouble and the very low value recorded for $h_6(w^3)$ is not very different from the value recorded in this same position in Table 3-2.

Following Thompson, Dharmapala and Thrall (1994), we now introduce a 5% increment to the input values for every one of the efficient DMUs and a 5% decrement to the input values for every one of the inefficient DMUs in Table 1. Also like Thompson, Dharmapala and Thrall (1994), we do not vary the outputs. The results are recorded in Table 3-7.

Table 3-7. Data Adjusted for 5% Increment and Decrement

	Efficient			Not Efficient		
DMU	1	2	3	4	5	6
Output	1.00	1.00	1.00	1.00	1.00	1.00
Input 1	4.20	2.10	1.05	1.90	2.85	3.80
Input 2	1.05	2.10	4.20	2.85	1.90	3.80

Applying the same procedure as before, we arrive at the new values displayed in Table 5-8. As was case for Table 3-3, none of the originally top-ranked DMUs are displaced after these 5% increment adjustments to the data are made. In fact, just as was reported in the discussion following Table 3-3, a 15% increment-decrement change is needed before a displacement occurs for any top-ranked DMU.

Table 3-8. Results from 5% Increment and Decrement

	DMU_1	DMU_2	DMU_3
DMU	$h_j(w^1)$	$h_j(w^2)$	$h_j(w^3)$
1	0.953	0.635	0.238
2	0.953	0.952	0.476
3	0.635	0.952	0.952
4	0.790	0.902	0.526
5	0.902	0.789	0.351
6	0.526	0.526	0.263

We do not pursue this topic in further detail. Instead we bring this discussion to a close by noting that no further changes occur in top rank until 20% increments for the inputs of the efficient DMUs and 20% decrements for the inefficient DMUs are made. The results are as displayed in Table 3-9 where, as can be seen, DMU_5 has displaced DMU_1 from its top rank and

DMU$_4$ has displaced DMU$_2$. The displacements differ from those obtained by Thompson, Dharmapala and Thrall (1994), (as described in the discussion following Table 3-3). However, approximately the same degree of robustness is maintained. Indeed, as was also true in the Thompson, Dharmapala and Thrall (1994) analysis even the 20% increments and decrements have not displaced DMU$_3$ from its top rank--as can be seen in column 3 of Table 3-9.

Table 3-9. Results from 20% Increments and Decrements

	DMU$_1$	DMU$_2$	DMU$_3$
DMU	$h_j(w^1)$	$h_j(w^2)$	$h_j(w^3)$
1	0.870	0.580	0.217
2	0.870	0.870	0.435
3	0.580	0.870	0.870
4	0.882	1.000	0.588
5	1.000	0.882	0.392
6	0.588	0.588	0.294

2.5   Envelopment Approach

The line of work we now follow extends the work of Charnes et al. (1992) to identify allowable variations in every input and output for every DMU before a change in status occurs for the DMU_o being analyzed. In contrast to the treatments in Thompson, Dharmapala and Thrall (1994), the focus shifts to the "envelopment" rather than the "multiplier model" portrayed in (3.4). This shift of focus helps to bypass concerns that might be noted which arise from the possibility of different degrees of sensitivity that may be associated with alternate optima.

An easy place to begin is with the following formulation as first given in Zhu (1996a) and Seiford and Zhu (1998a)

$$
\begin{array}{ll}
\underline{\text{Input Orientation}} & \underline{\text{Output Orientation}} \\
\beta_k^* = \min \beta_k \quad \text{for each } k = 1,...,m & \alpha_l^* = \max \alpha_l \quad \text{for each } l = 1,...,s \\
\sum_{j=1, j\neq 0}^{n} x_{kj}\lambda_j \leq \beta_k x_{ko} & \sum_{j=1, j\neq 0}^{n} y_{rj}\lambda_j \geq \alpha_\ell y_{lo} \\
\sum_{j=1, j\neq 0}^{n} x_{ij}\lambda_j \leq x_{io}, \quad i \neq k & \sum_{j=1, j\neq 0}^{n} y_{rj}\lambda_j \geq y_{ro}, \quad r \neq \ell \\
\sum_{j=1, j\neq 0}^{n} y_{rj}\lambda_j \geq y_{ro}, \quad r \neq 1...,s & \sum_{j=1, j\neq 0}^{n} x_{ij}\lambda_j \leq x_{io}, \quad i \neq 1...,m \\
\beta_k, \lambda_j \geq 0 & \alpha_\ell, \lambda_j \geq 0
\end{array}
\tag{3.12}
$$

In this and subsequent developments it is assumed that DMU_o is at least "weakly efficient" as defined in chapter 1 and, as in Charnes et al. (1992), this DMU is omitted from the sums to be formed on the left in these expressions. Then singling out an input k and an output ℓ we can determine the maximum proportional change that can be allowed before a change in status from efficient to inefficient will occur for DMU_o. In fact the values of β_k^* and α_l^* provide the indicated boundaries for this input and output. Moreover, continuing in this fashion one may determine the boundaries for all inputs and outputs for this DMU_o.

Seiford and Zhu (1998a) also extend (3.12) to the following modified version of the CCR model

$$\beta^* = \min \beta$$
$$\text{subject to}$$
$$\sum_{j=1,j\neq 0}^{n} x_{ij}\lambda_j \leq \beta x_{io}, \quad i \in I$$
$$\sum_{j=1,j\neq 0}^{n} x_{ij}\lambda_j \leq x_{io}, \quad i \notin I \qquad (3.13)$$
$$\sum_{j=1,j\neq 0}^{n} y_{rj}\lambda_j \geq y_{ro}, \quad r=1...,s$$
$$\beta, \lambda_j \geq 0.$$

Here the set $i \in I$ consists of inputs where sensitivity is to be examined and $i \notin I$ represents inputs where sensitivity is not of interest.

Seiford and Zhu next use this model to determine ranges of data variation when inputs are worsened for DMU_o in each of its x_{io} and improved for the x_{ij} of every DMU$_j$, $j = 1, ..., n$ in the set $i \in I$. We sketch the development by introducing the following formulation to determine the range of admissible variations,

$$\sum_{j=1,j\neq 0}^{n} x_{ij}\lambda_j \leq \delta x_{io}, \quad i \in I, \text{ where } 1 \leq \delta \leq \beta^* \qquad (3.14)$$

Now assume that we want to alter these data to new values $x_{io} \geq x_{io}$ and $x_{ij} \leq x_{ij}$ which will continue to satisfy these conditions. To examine this case we use

$$\sum_{j=1,j\neq o}^{n} \hat{x}_{ij}\lambda_j \leq \hat{\delta x}_{io} \text{ where } 1 \leq \delta \leq \beta^* \qquad (3.15.1)$$

and

$$\hat{\delta x}_{io} = x_{io} + \delta x_{io} - x_{io} = x_{io}+(\delta-1)x_{io}$$
$$\hat{x}_{ij} = \frac{x_{ij}}{\delta} = x_{ij} + \frac{x_{ij}}{\delta} - x_{ij} = x_{ij}-\left(\frac{\delta-1}{\delta}\right)x_{ij} \qquad (3.15.2)$$

for every $j = 1, ..., n$ in the set $i \in I$.

Thus $(\delta-1)$ represents the proportional <u>increase</u> to be allowed in each x_{io} and $(\delta-1)/\delta$ represents the proportional <u>decrease</u> in each x_{ij}. As proved by Seiford and Zhu, the range of variation that can be allowed for δ without altering the efficient status of DMU_o is given in the following,

Theorem 3.1 (Seiford and Zhu, 1998a)
If $1 \leq \delta \leq \sqrt{\beta^*}$ then DMU_o will remain efficient. That is, any value of δ within this range of proportional variation for both the x_{io} and x_{ij} will not affect the efficient status of DMU_o.

Here δ is a parameter with its values to be selected by the user. Theorem 3.1 asserts that no choice of δ within the indicated range to be reclassified as inefficient when the $\hat{x}_{io}$ and $\hat{x}_{ij}$ defined in (3.15.2) are substituted in (3.13), because the result will still give $\beta^* \geq 1$.

Seiford and Zhu supply a similar development for outputs and then join the two in the following model which permits simultaneous variations in inputs and outputs,

$$\gamma^* = \min \gamma$$
subject to
$$\sum_{j=1, j\neq 0}^{n} x_{ij}\lambda_j \leq (1+\gamma)x_{io}, \quad i \in I$$
$$\sum_{j=1, j\neq 0}^{n} x_{ij}\lambda_j \leq x_{io}, \quad i \notin I$$
$$\sum_{j=1, j\neq 0}^{n} y_{rj}\lambda_j \geq (1-\gamma)y_{ro}, \quad r \in s \qquad (3.16)$$
$$\sum_{j=1, j\neq 0}^{n} y_{rj}\lambda_j \geq y_{ro}, \quad r \notin s$$
$$\lambda_j \geq 0, \quad \gamma \text{ unrestricted.}$$

where $i \in I$ represents the input set for which data variations are to be considered and $r \in S$ represents the output set for which data variations are to be considered. Using δ to represent allowable input variations and τ to represent allowable output variations, Seiford and Zhu supply the following

Theorem 3.2 (Seiford and Zhu, 1998a)
If $1 \leq \delta \leq \sqrt{1+\gamma^*}$ and $\sqrt{1-\gamma^*} \leq \tau \leq 1$ then DMU_o will remain efficient.

As Seiford and Zhu note, for I = {1,...,m} and S = {1,..., s}, (3.16) is the same as the CCR correspond used in Charnes, Rousseau and Semple (1996). Seiford and Zhu have thus generalized the Charnes et al. (1992) results to allow simultaneous variations in all inputs and outputs for every DMU in the sets $i \in I$ and $r \in S$.

We now turn to Table 3-10 which Seiford and Zhu used to compare their approach with the Thompson, Dharmapala and Thrall (1994) approach. To interpret this Table we note that all results represent percentages in the allowed data variation by applying these two different approaches to the data of Table 3-1. The values in the rows labeled SCSC1 and SCSC2 are secured from two alternate optima which Thompson, Dharmapala and Thrall (1994) report as satisfying the strong complementary slackness condition. The parenthesized values of g_o and g at the top of Table 3-2 are reported by Seiford and Zhu as having been obtained by applying (3.13) to these same data.

Table 3-10. Comparison with Thompson et al.

	DMU1	DMU2	DMU3
(g_o, g)	(41,29)	(12,11)	(41,29)
SCSC1	20	14	20
SCSC2	32	9	32

Source: Seiford and Zhu (1998a). Here $g_o = \delta - 1$ and $g = \dfrac{\delta - 1}{\delta}$. See (3.15.1) and (3.15.2).

The results from Seiford and Zhu seem to be more robust than is the case for Thompson et al., at least for DMU_1 and DMU_3. This is not true for DMU_2, however, where a 14% worsening of its inputs and a 14 % improvement in the inputs of the non-efficient DMUs is required under the Thompson, Dharmapala and Thrall (1994) approach before DMU_2 will change from efficient to inefficient in its status. However, the Seiford and Zhu approach shows that DMU_2 will retain its efficient status until at least a 12% worsening of its 2 inputs occurs along with an 11% improvement in these same inputs for the inefficient DMUs. A range of 12% + 11% = 23% does not seem to be far out of line with the 2×14% = 28% or the 2×9%=18% reported by Thompson, Dharmapala and Thrall (1994). Moreover, as Seiford and Zhu note, their test is more severe. They match their worsening of DMU_2's inputs with improvement of the inputs of all of the other DMUs -- including the efficient DMU_1 and DMU_2.-whereas Thompson, Dharmapala and Thrall (1994) worsen the inputs of all of the efficient DMUs and improve the inputs of only the inefficient DMUs. (Note the fact that Seiford and Zhu deal only with "weak efficiency" is not pertinent here because DMU_1, DMU_2 and DMU_3 are all strongly efficient.)

Seiford and Zhu (1998b) also discuss situations where absolute (rather than proportional) changes in the data are of interest. This allows to study the sensitivity under the additive model discussed in chapter 2. We here briefly discuss this approach. The absolute data variation can be expressed as

For DMU_o

$$\begin{cases} \hat{x}_{io} = x_{io} + \alpha_i & \alpha_i \geq 0, i \in \mathbf{I} \\ \hat{x}_{io} = x_{io} & i \notin \mathbf{I} \end{cases} \text{ and } \begin{cases} \hat{y}_{ro} = y_{ro} - \beta_r & \beta_r \geq 0, r \in \mathbf{O} \\ \hat{y}_{ro} = y_{ro} & r \notin \mathbf{O} \end{cases}$$

For DMU_j $(j \neq o)$

$$\begin{cases} \hat{x}_{ij} = x_{ij} - \tilde{\alpha}_i , & \tilde{\alpha}_i \geq 0, i \in \mathbf{I} \\ \hat{x}_{ij} = x_{ij} & i \notin \mathbf{I} \end{cases} \text{ and } \begin{cases} \hat{y}_{rj} = y_{rj} + \tilde{\beta}_r , & \tilde{\beta}_r \geq 0, r \in \mathbf{O} \\ \hat{y}_{rj} = y_{rj} & r \notin \mathbf{O} \end{cases}$$

where ($^\wedge$) represents adjusted data. Note that the data changes defined above are not only applied to all DMUs, but also different in various inputs and outputs.

Based upon the above data variations, Seiford and Zhu (1998b) provide the following model

$$u^* = \min u$$
subject to

$$\sum_{j=1, j\neq 0}^{n} x_{ij}\lambda_j \leq x_{io} + u, \quad i \in I$$

$$\sum_{j=1, j\neq 0}^{n} x_{ij}\lambda_j \leq x_{io}, \quad i \notin I$$

$$\sum_{j=1, j\neq 0}^{n} y_{rj}\lambda_j \geq y_{ro} - u, \quad r \in s \qquad (3.17)$$

$$\sum_{j=1, j\neq 0}^{n} y_{rj}\lambda_j \geq y_{ro}, \quad r \notin s$$

$$\sum_{j=1, j\neq 0}^{n} \lambda_j = 1$$

$$u, \lambda_j \geq 0, \forall j.$$

If $\mathbf{I} = \{1, 2, ..., m\}$ and $\mathbf{O} = \{1, 2, ..., s\}$, then model (3.17) is used by Charnes et al. (1992) to study the sensitivity of efficiency classifications in the additive model via L_∞ norm when variations in the data are only applied to DMU_o.

Theorem 3.3 (Seiford and Zhu)
Suppose DMU_o is a frontier point. If $0 \leq \alpha_i + \tilde{\alpha}_i \leq u^*$ ($i \in \mathbf{I}$), $0 \leq \beta_r + \tilde{\beta}_r \leq \gamma^*$ ($r \in \mathbf{O}$), then DMU_o remains as a frontier point, where u^* is the optimal value to (3.17).

If we change the objective function of (3.17) to "minimize $\sum_{i \in \mathbf{I}} r_i^- + \sum_{r \in \mathbf{O}} \gamma_r^+$", we obtain the following model which studies the sensitivity of additive DEA models discussed in chapter 2.

$$\min \sum_{i \in \mathbf{I}} \gamma_i^- + \sum_{r \in \mathbf{O}} \gamma_r^+$$

subject to

$$\sum_{\substack{j=1 \\ j \neq o}}^{n} \lambda_j x_{ij} \leq x_{io} + \gamma_i^- \qquad i \in \mathbf{I}$$

$$\sum_{\substack{j=1 \\ j \neq o}}^{n} \lambda_j x_{ij} \leq x_{io} \qquad i \notin \mathbf{I}$$

$$\sum_{\substack{j=1 \\ j \neq o}}^{n} \lambda_j y_{rj} \geq y_{ro} - \gamma_r^+ \qquad r \in \mathbf{O} \qquad\qquad (3.18)$$

$$\sum_{\substack{j=1 \\ j \neq o}}^{n} \lambda_j y_{rj} \geq y_{ro} \qquad r \notin \mathbf{O}$$

$$\sum_{\substack{j=1 \\ j \neq o}}^{n} \lambda_j = 1$$

$$\gamma_i^-, \gamma_r^+, \lambda_j (j \neq o) \geq 0$$

Based upon model (3.18), we have

Theorem 3.4 (Seiford and Zhu, 1998b)
Suppose DMU_o is a frontier point. If $0 \leq \alpha_i + \tilde{\alpha}_i \leq \gamma_i^{-*}$ $(i \in \mathbf{I})$, $0 \leq \beta_r + \tilde{\beta}_r \leq \gamma_r^{+*}$ $(r \in \mathbf{O})$, then DMU_o remains as a frontier point, where γ_i^{-*} $(i \in \mathbf{I})$ and γ_r^{+*} $(r \in \mathbf{O})$ are optimal values in (3.18).

There are many more developments in these two approaches which we also do not cover here. We do need to note, however, that Seiford and Zhu (1998b,c) extend their results to deal with the infeasibility that can occur in such models as (3.12). Seiford and Zhu note that the possibility of infeasibility is not confined to the case when convexity is imposed. It can also occur when certain patterns of zeros are present in the data. They show that infeasibility means that the DMU_o being tested will preserve its efficient status in the presence of infinite increases in its inputs and infinite decreases in its outputs.

Although the above discussion is focused on changes in a specific (weakly) efficient DMU, Seiford and Zhu (1998b) and Zhu (2001) have extended the approach to situations where (unequal) changes in all DMUs (both efficient and inefficient DMUs) are considered. Further, Zhu (2002) provides a software that performs this type sensitivity analysis.

3. SUMMARY AND CONCLUSION

Using data from a study evaluating performances of Chinese cities (Charnes, Cooper and Li, 1989) and a Chinese textile company (Zhu, 1996b), Seiford and Zhu conclude that their results show DEA results to be robust. This is the same conclusion that Thompson, Dharmapala and Thrall (1994) arrive at from their sensitivity analysis of independent oil companies, Kansas farms and Illinois coal mines.

This brings us back to the opening discussion in this chapter which distinguished "substantive findings" and "methodological approaches" is the analysis of stability in DEA. These findings by Seiford and Zhu and by Thompson, Dharmapala and Thrall (1994) differ from substantive characterizations like those in Sexton, Silkman and Hogan (1986) which we cited earlier. The latter believe that results in DEA would generally fail to be robust because of its reliance on extreme value observations, but do not seem to have tested this claim with actual data. Seiford and Zhu and Thompson, Dharmapala and Thrall (1994) do at least supply the evidence (as well as rationales) for their conclusions. The evidence they adduce come from only limited bodies of data, and, in addition, these analyses are all limited by their focus on only changes in the efficient status of different DMUs. However, in any case, we now have a collection of methods which can be used to determine the robustness of results in any use of DEA.

As we have observed, the progress in the sensitivity analysis studies we discussed has effected improvements in two important directions. First, this work has moved from evaluating one input or one output at a time in one DMU and has proceeded into more general situations where all inputs and outputs for all DMUs can be simultaneously varied. Second, the need for special algorithms and procedures has been reduced or eliminated.

Finally, Zhu (2002) provide a detailed discussion on "envelopment approach" with Microsoft® Excel-based software. This greatly enhances the applicability of the DEA sensitivity analysis approaches. Further, Cooper, Seiford and Tone (2000) point out that sensitivity analysis are not restricted to the approaches discussed in this chapter. Window analysis can also be treated as a method for studying the stability of DEA results because such window analyses involve the removal of entire sets of observations and their replacement by other (previously not considered) observations.

REFERENCES

1. Ahn, T. And L.M. Seiford, 1993, Sensitivity of DEA to models and variable sets in an hypothesis test setting: The efficiency of university operations" in Y. Ijiri, ed. Creative and innovative Approaches to the Science of Management (new York: Quorum Books).
2. Banker, R.D., 2003, Maximum Likelihood, Consistency and Data Envelopment Analysis: A statistical Foundation, this volume.
3. Banker, R.D., H. Chang and W.W. Cooper, 1996, Simulation studies of efficiency, returns to scale and misspecification with nonlinear functions in DEA, Annals of Operations Research 66, 233-253.
4. Charnes, A. and W.W. Cooper, 1968, Structural sensitivity analysis in linear programming and an exact product form left inverse, Naval Research Logistics Quarterly 15, 517-522.
5. Charnes, A., W.W. Cooper and S. Li, 1989, Using DEA to evaluate relative efficiencies in the economic performance of Chinese cities, Socio-Economic Planning Sciences, 23 325-344.
6. Charnes, A. and L. Neralic, 1992, Sensitivity analysis of the proportionate change of inputs (or outputs) in data envelopment analysis" Glasnik Matematicki 27, 393-405.
7. Charnes, A., W.W. Cooper, A.Y. Lewin, R.C. Morey and J.J. Rousseau, 1985, Sensitivity and stability analysis in DEA, Annals of Operations Research 2, 139-150.
8. Charnes, A., S. Haag, P. Jaska and J. Semple, 1992, Sensitivity of efficiency calculations in the additive model of data envelopment Analysis, Journal of Systems Sciences 23, 789-798.
9. Charnes, A., J.J. Rousseau and J.H. Semple, 1996, Sensitivity and stability of efficiency classifications in DEA, Journal of Productivity Analysis 7, 5-18.
10. Charnes, A., W.W. Cooper and R.M. Thrall, 1991, A Structure for characterizing and classifying efficiencies in DEA, Journal of Productivity Analysis 3, 197-237.
11. Cooper, W.W., Seiford, L.M. and Tone, K., 2000, Data Envelopment Analysis: A Comprehensive Text with Models, Applications, References and DEA-Solver Software, Kluwer Academic Publishers, Boston.
12. Neralic, L., 1997, "Sensitivity in data envelopment analysis for arbitrary perturbations of data, Glasnik Matematicki 32, 315-335.
13. Seiford, L.M. and J. Zhu, 1998a, Stability regions for maintaining efficiency in data envelopment analysis, European Journal of Operational Research 108, 127-139.

14. Seiford, L.M. and J. Zhu, 1998b, Sensitivity analysis of DEA models for simultaneous changes in all of the data, Journal of the Operational Research Society 49, 1060-1071.
15. Seiford, L.M. and J. Zhu, 1998c, Infeasibility of super-efficiency data envelopment analysis models, INFOR 37, No. 2, 174-187.
16. Sexton, T.R., R.H. Silkman and R.H. Hogan, 1986, Measuring efficiency: An assessment of data envelopment analysis, new directions for program evaluations, in R.H. Silkman, ed., Measuring Efficiency: An Assessment of Data Envelopment Analysis. Publication No. 32 in the series New Directions for Program Evaluations. A publication of the American Evaluation Association (San Francisco: Jossy Bass).
17. Simar, L. and P. Wilson, 2003, DEA Bootstrapping, this volume.
18. Thompson, R.G., P.S. Dharmapala, J. Diaz, M.D. Gonzalez-Lima and R.M. Thrall, 1996, DEA multiplier analytic center sensitivity analysis with an illustrative application to independent oil Cos., Annals of Operations Research 66, 163-180.
19. Thompson, R.G., P.S. Dharmapala and R.M. Thrall, 1994, Sensitivity analysis of efficiency measures with applications to Kansas farming and Illinois coal mining, in A. Charnes, W.W. Cooper, A.Y. Lewin and L.M. Seiford, eds., Data Envelopment Analysis: Theory, Methodology and Applications (Norwell, Mass., Kluwer Academic Publishers) pp. 393-422.
20. Wilson, P.W., 1995, Detecting influential observations in data envelopment analysis, Journal of Productivity Analysis 6, 27-46.
21. Zhu, J., 1996a, Robustness of the efficient DMUs in data envelopment analysis, European Journal of Operational Research 90, 451-460.
22. Zhu, J., 1996b, DEA/AR analysis of the 1988-1989 performance of Nanjing textile corporation, Annals of Operations Research 66, 311-335.
23. Zhu, J., 2001, Super-efficiency and DEA sensitivity analysis, European Journal of Operational Research 129, No. 2, 443-455.
24. Zhu, J., 2002, Quantitative Models for Performance Evaluation and Benchmarking: Data Envelopment Analysis with Spreadsheets and DEA Excel Solver, Kluwer Academic Publishers, Boston.

Part of the material in this chapter is adapted from the article by Journal of Productivity Analysis, Cooper, W.W., Li, S., Seiford, L.M., Tone, K., Thrall, R.M. and Zhu, J, Sensitivity and Stability Analysis in DEA: Some Recent Developments", 2001, 15, 217-246, and is published with permission from Kluwer Academic Publishers.

Chapter 4

INCORPORATING VALUE JUDGMENTS IN DEA

Emmanuel Thanassoulis[1], Maria Conceição Portela[2], and Rachel Allen[3]
[1] Aston Business School, Aston Traingle, B4 7ET Birmingham, UK
email: e.thanassoulis@aston.ac.uk

[2] Portuguse Catholic University, Porto, Portugal   email: portemca@aston.ac.uk

[3] ABB, Zürich, Switzerland

Abstract: This chapter reviews various methodologies that have been used for incorporating value judgments in data envelopment analysis. We first raise some issues concerning the meaning of the input and output weights in DEA, and then point out the main reasons why value judgments may be required in an efficiency analysis. The methods we review include those that apply direct restrictions on the weights (absolute weights restrictions, assurance regions of type I and type II, and restrictions on virtual inputs and outputs) and those that change the data set in a way that aims at incorporating value judgments (in this category we include the cone ratio approach and the unobserved DMUs approach). We further highlight some issues concerning the interpretation of DEA results when value judgments are included in the assessment for each of the above mentioned methods, and put forward the implications on the returns to scale characteristics of the efficient frontier resulting from the existence of value judgments.

Keywords: Data envelopment analysis (DEA); Value judgments; Weights restrictions; Unobserved DMUs

1. INTRODUCTION

DEA efficiency can be computed either in a production (envelopment) or in a value (multiplier) framework. In the latter case the DEA efficiency measure of a DMU is the ratio of the weighted sum of outputs to the weighted sum of inputs. In original DEA formulations the assessed Decision Making Units (DMUs) can freely choose the weights or values to be assigned to each input and output in a way that maximises its efficiency, subject to this system of weights being feasible for all other DMUs. This freedom of choice shows the DMU in the best possible light, and is equivalent to assuming that no input or output is more important than any other.

The free imputation of input-output values can be seen as an advantage, especially as far as the identification of inefficiency is concerned. If a DMU is free to choose its own value system and some other DMU uses this same value system to show that the first DMU is not efficient, then a stronger statement is being made. The advantages of full flexibility in identifying inefficiency can be seen as disadvantages in the identification of efficiency. An efficient DMU may become so by assigning a zero weight to the inputs and/or outputs on which its performance is worst. This might not be acceptable by Decision Makers (DMs) as well as by the analyst, who after spending time in a careful selection of inputs and outputs sees some of them being completely neglected by DMUs.

DMs have in some contexts value judgments that can be formalised *a priori*, and therefore should be taken into account in the efficiency assessment. These value judgments can reflect known information about how the factors used by the DMUs behave, and/or 'accepted' beliefs or preferences on the relative worth of inputs, outputs or even DMUs.

This chapter examines ways in which the foregoing problem can be overcome and the difficulties, advantages and drawbacks of the relevant approaches available, with outlines on how to interpret the obtained results also put forward.

2. THE ROLE OF WEIGHTS IN DEA

The DEA weights model (multiplier) and its dual (envelopment model) are shown in (4.1) and (4.2). These models are output oriented and assume constant returns to scale (CRS). (CRS is the focus of the discussion in this chapter). Notation here is as introduced in Chapter 1.

Multiplier Model

$Min \; g = \sum_{i=1}^{m} v_i x_{io}$

s.t

$\sum_{i=1}^{m} v_i x_{ij} - \sum_{r=1}^{s} u_r y_{rj} \geq 0 \quad j = 1,\dots,n$

$\sum_{r=1}^{s} u_r y_{ro} = 1 \qquad\qquad (4.1)$

$u_r, v_i \geq \varepsilon$

Envelopment Model

$Max \; h = \theta + \varepsilon \left(\sum_{r=1}^{s} s_r^+ + \sum_{i=1}^{m} s_i^- \right)$

s.t.

$\sum_{j=1}^{n} \lambda_j x_{ij} + s_i^- = x_{io} \qquad i = 1,\dots,m$

$\sum_{j=1}^{n} \lambda_j y_{rj} - s_r^+ = \theta y_{ro} \quad r = 1,\dots,s \qquad (4.2)$

$\lambda_j, s_r^+, s_i^- \geq 0$

Model (4.1) is represented on a value-space where the variables u_r and v_i can be seen as imputed marginal values of output r and input i, respectively. The efficiency measure g represents the ratio of the total imputed value of inputs to the total imputed value of outputs, where the latter has been normalised to the value 1. Model (4.2) is defined on a production-space where the Production Possibility Set (PPS) is represented by linear combinations of DMUs' inputs and outputs. The efficiency measure h represents the maximum proportional expansion of outputs required to achieve the frontier of the PPS.

The role of the weights in model (4.1) will be illustrated through the example shown in Table 4-1 This table contains both the illustrative input and output data, and also the optimal solution to (4.1) for each DMU assessed. Note that factor's prices are also shown in Table 4-1, where p_r represents output prices and w_i represents input prices.

Table 4-1 Illustrative Example and Results from model (4.1)

DMU	y_1	y_2	x	p_1	p_2	w	$1/g^*$	u_1^*	u_2^*	v^*
A	1	8	2	7	4	5	100.0%	0.040	0.120	0.500
B	4	7	2	2	6	4	100.0%	0.169	0.046	0.500
C	15	10	6	1	5	4	100.0%	0.056	0.015	0.167
D	14	8	8	4	3	6	70.0%	0.071	ε	0.179
E	12	16	10	5	3	4	55.4%	0.061	0.017	0.181
F	1	10	5	7	5	6	50.0%	ε	0.100	0.400
G	12	24	12	7	3	5	56.0%	0.012	0.036	0.149
H	2	8	3	6	3	2	69.3%	0.038	0.115	0.481
I	3	6	2	5	3	4	84.0%	0.048	0.143	0.595

The efficiency measures in Table 4-1 are given by the inverse of g^* so that efficiency varies between zero and one. Some optimal weights in Table 4-1 are equal to ε. This value represents a non-Archimedean infinitesimal that prevents input and output weights from being strictly equal to zero [see for example Charnes et al. (1978) and Charnes et al. (1994)]. As ε is very small, it means that the corresponding factor is neglected in the efficiency assessment.

2.1   Marginal rates of substitution/transformation

Figure 4-1 shows the efficient frontier defined by DMUs A, B, and C of our illustrative example (outputs are normalised by the input).

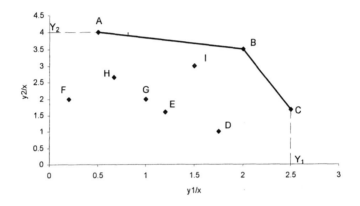

Figure 4-1. Graphical representation

Note that, in the optimal solution of (4.1) at least one constraint j is binding. Such a constraint assumes the form $\sum_{i=1}^{m} v_i^* x_{ij} - \sum_{r=1}^{s} u_r^* y_{rj} = 0$, defining a hyperplane (or facet) of the production frontier. Take the example of DMU A, whose hyperplane is given by $0.5x_A - 0.04\ y_{1A} - 0.12\ y_{2A} = 0$. This equation obviously reveals marginal rates of transformation between outputs, marginal rates of substitution between inputs, and marginal product between inputs and outputs (see Coelli et al. (1998) for details on these concepts). The marginal rate of transformation between outputs k and w is given by the ratio of total derivatives dy_k/dy_w, meaning the amount of output k that shall be sacrificed if output w is to be increased by one unit. For the hyperplane defined above for unit A, the marginal rate of transformation between output 1 and 2 is therefore given by $dy_1/dy_2 = -(\partial/\partial y_2)/(\partial/\partial y_1) = -u_2^*/u_1^* = -(0.12/0.04) = -3$ (Charnes et al., 1978), meaning that increasing one unit of output 2 implies a reduction of three units of output 1 (this value is directly seen when the hyperplane equation above is rearranged to become: $y_{1A} = 12.5x_A - 3y_{2A}$.). A ratio of two optimal output weights corresponds, therefore, to the marginal rate of transformation between these outputs. In the same way the marginal product of input and output 1 is 12.5.

DMUs A, B, and C are extreme points of the convex set represented in Figure 4-1 This means that there is not a unique hyperplane passing through these DMUs. This is equivalent to the existence of multiple optimal solutions to model (4.1) for extreme efficient DMUs. Although there are a wide diversity of weights shown in Table 4-1, they, in fact, represent only 4 hyperplanes. If we normalize the optimal output weights by the optimal input weight, as shown in Table 4-2, we conclude that model (4.1) can only identify four hyperplanes (as many as we can see in Figure 4-1), corresponding to four different rates of transformation/substitution.

Table 4-2. Hyperplane Equations (values are rounded)

DMU	A	B	C	D	E	F	G	H	I
$u_1{}^*$	0.08	0.34	0.34	0.40	0.34	0.00	0.08	0.08	0.08
$u_2{}^*$	0.24	0.09	0.09	0.00	0.09	0.25	0.24	0.24	0.24
v^*	1	1	1	1	1	1	1	1	1
Face	AB	BC	BC	CY_1	BC	Y_2A	AB	AB	AB

The above represent therefore alternative weighting schemes that would render efficiency scores as shown in Table 4-1. Note that for two facets (CY_1 and Y_2A) marginal rates of transformation between outputs cannot be defined. These facets are usually called weakly efficient. DMUs that are projected on these facets completely neglect one of the outputs in determining their efficiency rating (such is the case of DMUs D and F). In the envelopment model this fact is characterized by the existence of a non-zero slack associated to output 1 for DMU F, and to output 2 for DMU D. These slacks represent additional sources of inefficiency that are not captured by the radial efficiency scores (h or g).

The existence of zero weights can be problematic, not only because we may not accept that DMUs ignore some factors of production, but also because zero weights imply undefined marginal rates of transformation and substitution.

2.2 Interpreting and comparing weights

An important point concerning models (4.1) and (4.2) is that they are independent of units of measurement of inputs and outputs. This means that if some input i is scaled by a factor, say α_i, and/or some output r is scaled by a factor, say β_r, the resulting efficiency scores $g^* = h^*$ do not change. Furthermore, in the envelopment model (4.2) none of the variables change their value when units of measurement change (except the slacks that will be scaled accordingly). In the multiplier model (4.1), however, this is not so. The value of the weights change when units of measurement change, meaning that though the model is not sensitive to units of measurement, weights are (note that this can be easily checked by multiplying all inputs and outputs by α and β in models (4.1) and (4.2) above). If weights depend on units of measurement, then ratios of weights, and consequently marginal rates of substitution (or transformation), also depend on units of measurement.

Virtual inputs and outputs (the product of observed values and weights), however, do not depend on units of measurement. Consider virtual outputs of DMU o defined as $O_r = u_r y_{ro}$ and virtual inputs of DMU o defined as $I_i = v_i x_{io}$. As the final efficiency measure in (4.1) is in fact given by $\sum_{i=1}^{m} I_i / \sum_{r=1}^{s} O_r$ none of these components can change if the final efficiency measure does not change when measurement units are altered.

Care is, therefore, needed in interpreting and comparing weights as a larger or smaller weight does not necessarily mean large or small importance attached to a given input or output. A fair comparison between weights can be undertaken if data has previously been normalised (see Roll and Golany (1993)), or if virtual weights are compared rather than absolute weights (see Thanassoulis et al. (1987)).

3. INCORPORATING VALUE JUDGMENTS

Value judgments may appear in various forms, therefore requiring different treatments. A value judgment that is implicitly included in any DEA analysis is the choice of the number and type of inputs and outputs used and also the number and type of DMUs used (Allen et al., 1997).

The number of variables and DMUs used in a DEA assessment is directly linked with the discriminating power of DEA models and also with the potential number of zero weights. Clearly if the number of variables is very high the probability of a DMU finding at least one factor on which it performs well increases, and therefore the DMU has the chance of neglecting all other factors on which its performance is low, and eventually be rated efficient. At the same time if the number of DMUs under analysis is very small it is likely that each one specialises on a specific input/output mix not directly compared with the mix of other DMUs. In the absence of referents to which to be compared a DMU may be rated efficient simply by this fact (see for example Boussofiane et al. (1991) or Golany and Roll (1989) for rules of thumb that can be helpful in choosing the number of DMUs and variables).

A value judgment that analysts or DMs introduce in the analysis relates, therefore, with the choice of DMUs and variables. This chapter deals, however, with other types of value judgments that relate to the DMs preferences or prior views on the process.

3.1 Reasons for including Value Judgments

As pointed out by Allen et al. (1997) most of the developments in the incorporation of value judgments in DEA was led by real-life applications. Empirical applications have justified the incorporation of value judgments for a number of reasons, such as:

a) To capture prior views on the marginal rates of substitution and/or transformation of the factors of production
As seen previously ratios of optimal input-output weights yielded by a DEA model such as (4.1) are the *imputed* marginal rates of substitution or transformation between the inputs and/or outputs. Some of the imputed marginal rates of substitution or transformation may turn out to be ill-defined because some input and/or output weight takes the infinitesimal value ε, which in practice, works as if it was zero. Further, even when not ill-defined, the marginal rates of substitution or transformation may not be in line with the DM's prior views of the production process modelled. For example, when assessing the efficiency of police forces Thanassoulis (1995) arrived at marginal rates of transformation that in some cases were contrary to the intuitive feeling, that, clearing up a violent crime was more valuable than clearing up a burglary. This kind of situation may require the inclusion of additional information in DEA models, so that, resulting marginal rates of substitution/transformation are consistent with the DMs prior views.

b) To capture special interdependencies between the inputs and outputs of the production process being modelled
Standard DEA models assume that there is a positive relationship between inputs and outputs (under efficient production more inputs lead to more outputs), and that all chosen inputs/outputs are perfect substitutes. There may, however, exist other type of relationships between inputs/outputs that need to be accounted for in some way. For example, in the assessment of university efficiency Beasley (1990) establishes several relationships between the weights of inputs and outputs so that certain value judgments could be incorporated in the analysis. One such value judgment relates to the belief that the value of a postgraduate is higher for the university than the value of an undergraduate student, and therefore, the model should prevent universities from weighting undergraduates more than postgraduates. Other examples, can be found in Golany (1988) and Ali et al. (1991) where ordinal relationships between inputs or outputs are assured through ordinal relationships between their weights.

c) To arrive at some notion of 'overall' efficiency.

When price information is available, this can be incorporated in a DEA model, so that, the relationship between the weights emulate the relationship between the prices. In fact, the weights u_r and v_i represent values attached to inputs and outputs, and such a value, can be their price. When the weights reflect information on prices, a movement is made from technical efficiency to economic efficiency as we move from an unrestricted model such as (4.1) to a restricted model where price information is included in the analysis [see Charnes et al. (1990)]. Precise information on prices is not always available, especially in not-for-profit contexts and therefore, value judgments can be used as a proxy for such information. For example, we can define a range of prices for inputs or worth of outputs within the context of a value-based DEA model to assess something akin to overall efficiency. This will be discussed later.

d) To improve discrimination between efficient DMUs.

DEA-based efficiency results do not always reflect the desired degree of discrimination between DMUs. In some cases, this is due to the use of a small number of DMUs in relation to the number of inputs and outputs. A typical situation where discrimination between DMUs is important is when DEA is used in location problems where the objective is to choose the best location. Incorporation of value judgments by means of restrictions on the relative worth of inputs or outputs can aid in this choice. For example, Thompson et al. (1986) in an attempt to site nuclear physics facilities in Texas found a lack of discrimination as five out of six alternatives were found relatively efficient by the standard DEA model. The discrimination of DEA was improved by defining ranges of acceptable weights, namely 'assurance regions', which were then used to select the preferred efficient site.

Cook et al. (1992) also address the issue of discriminating between efficient DMUs and propose a set of methods to do so through the use of additional restrictions on the DEA weights.

Discrimination between efficient DMUs can also be done through other methods namely the super-efficiency model [Andersen and Petersen (1993)], the cross-evaluation procedure [see for example Green et al. (1996b), and Anderson et al. (2002)], and multiple objective linear programs [Li and Reeves (1999)]. Meza and Lins (2002) consider these methods in the class of those that can be used to improve the discrimination of DEA models without requiring *a priori* information.

e) To ensure that widely differing weights are not assigned to the same factor

This problem is related with extremely large or extremely small weights assigned to certain input or output factors. In such a case, it may be desirable to reduce the dispersion in the optimal weights assigned to each factor by each DMU. In the extreme case where no flexibility is allowed there is the Common Set of Weights (CSW) procedure of Roll et al. (1991). According to the authors the "difference between the efficiency measured with an 'individual' set of weights and that obtained with a CSW may indicate the effects of special circumstances under which a DMU operates", as the use of a CSW assumes that all DMUs face the same circumstances (same goals, policies, etc) (see also Roll and Golany (1993) or Cook et al. (1991), where the CSW methodology is further developed). Other approaches, that are less strict than the CSW, imposing some constraints on weights can also be used for the purpose of reducing the variance of unrealistic weights' distributions.

f) To establish preferences of the DM over the potential adjustments of inputs and outputs

In traditional radial models like (4.1) and (4.2) targets are established assuming equiproportinal reduction of inputs or equiproportional expansion of outputs is desired. As this might result in targets that are not the most preferred by the decision maker it is possible to integrate the DMs preferences and value judgments in DEA models that estimate targets according to these preferences. Examples of such models can be found in Thanassoulis and Dyson (1992) and in Zhu (1996a). When the preference information accords with factor prices the resulting models relate with the objective put forward in *c)* since the preference structure is equivalent to the establishment of restrictions on the weights as shown by Zhu (1996a) and also Seiford and Zhu (2002).

3.2 Methods for incorporating value judgments

There are various methods that can be used to incorporate value judgments in DEA, and to reduce the flexibility of DMUs in choosing their value system. The types dealt with in this chapter fall into two broad classes of methods, namely those that

- Apply restrictions on the DEA weights - weights restrictions;
- Change the comparative set of DMUs - changing the data set.

A number of other methods can, however, be used to reduce the flexibility of DMUs in choosing their weights that do not necessarily involve the inclusion of value judgments in a DEA model. An interesting example is the method of Olesen and Petersen (1996) that restricts facets of the efficient frontier where projections are allowed. As seen in Figure 4-1, zero weights in the multiplier model (or slacks in the envelopment model) correspond to projections on weak efficient parts of the efficient frontier. If such projections are not allowed then zero weights will not happen. The method of Olesen and Petersen (1996) [see also Green et al. (1996a)] forces DMUs to be projected on full dimension efficient facets (FDEF)[1]. One of the problems of this approach is the complexity of procedures that try to find FDEFs, and the fact that not always a reasonable number of FDEFs that envelop all the data can be determined, especially if there is little variation in the data [for details see Olesen and Petersen, (1996) and Olesen and Petersen (2002)].

Another method worth mentioning is that introduced by Bessent et al. (1988), which is referred as the constrained facet analysis (CFA). It aims at extending the existent efficient facets so that each inefficient DMU is 'fully enveloped' by efficient DMUs. This means that "composite referents lying outside of the initial feasible range of production possibilities are being invoked" (Lang et al., 1995, p. 478). This approach has been analysed and criticised by Lang et al. (1995) who advocate that the projection of DMUs on the artificially created extended facet is somewhat arbitrary, and it may not reflect meaningful trade-offs. Lang et al. (1995) proposed an alternative approach to the CFA called controlled envelopment analysis (CEA).

4. WEIGHTS RESTRICTIONS

Weights restrictions (WRs) may be applied directly to the DEA weights (in model (4.1) v_i and u_r) or to the product of these weights with the respective input or output level, referred to as *virtual input* or *virtual output*. We shall analyse each of these two cases separately.

4.1 Restrictions Applied to DEA Weights

In (4.3) we show the various types of WR that can be applied to model (4.1).

---

[1] These are facets with $s + m -1$ extreme points in the CRS case and $s + m$ extreme points in the VRS (variable returns to scale) case.

Absolute WRs

$$\delta_i \leq v_i \leq \tau_i \qquad (a_i) \qquad \rho_r \leq u_r \leq \eta_r \qquad (a_o)$$

Assurance regions of type I (Relative WRs)

$$\kappa_i v_i + \kappa_{i+1} v_{i+1} \leq v_{i+2} \quad (b_i) \qquad w_r u_r + w_{r+1} u_{r+1} \leq u_{r+2} \qquad (b_o)$$

$$\alpha_i \leq \frac{v_i}{v_{i+1}} \leq \beta_i \qquad (c_i) \qquad \theta_r \leq \frac{u_r}{u_{r+1}} \leq \zeta_r \qquad (c_o)$$

(4.3)

Assurance regions of type II (Input-Output WRs)

$$\gamma_i v_i \geq u_r \qquad (d)$$

The Greek letters $(\delta_i, \tau_i, \rho_r, \eta_r, \kappa_i, \omega_r, \alpha_i, \beta_i, \theta_r, \zeta_r, \gamma_i)$ are user-specified constants to reflect value judgments the DM wishes to incorporate in the assessment. They may relate to the perceived importance or worth of input and output factors. The restrictions (a) to (c) in (4.3) relate on the left hand side to input weights and on the right hand side to output weights. Constraint (d) links directly input and output weights. These constraints are classified as follows:

4.1.1   Absolute weights restrictions, e.g. (a$_i$) and (a$_o$) in (4.3)

Absolute WRs are the most immediate form of placing restrictions on the weights as they simply restrict them to vary within a specific range. This type of WR was first introduced by Dyson and Thanassoulis (1988) on an application to rates departments. Cook et al. (1991, 1994) also used these types of constraint to evaluate highway maintenance patrols.

One of the difficulties associated with absolute WRs is the meaning of the bounds $\delta_i, \tau_i, \rho_r, \eta_r$ since, in general, weights are significant on a relative basis. Depending on the context, however, some meaning can be attributed to these bounds as will be seen in section 4.3 Another difficulty relates with the potential infeasibility of DEA models with absolute WRs. Podinovski (2001, p. 575) refers and illustrates this fact, also claiming that restricted models "may not identify the maximum relative efficiency of the assessed DMU correctly". In a sense, this means that, the DEA weights under absolute WRs may not enable a DMU to appear in the best possible light relative to other DMUs [see also Podinovski (1999) and Podinovski and Athanassopoulos (1998)]. Podinovski (2001) proposes the replacement of traditional DEA objective functions (which measure absolute efficiency) by a measure of relative efficiency, thereby removing the possibility of absolute WRs leading to mis-representations of unit relative efficiencies. In Podinovski (2003) this analysis is taken further, and the author identifies certain types of absolute weights restrictions, which do not lead to mis-representations of unit relative efficiency.

In addition, there is a strong interdependence between the bounds on different weights. For example, setting an upper bound on an input weight imposes implicitly a lower bound on the total virtual input and this in turn has implications for the values the remaining input weights can take. When absolute WRs are used in a DEA model, switching from an input to an output orientation can produce different relative efficiency scores. Hence, the bounds need to be set in light of the model orientation used, which will flow out of the context of the DEA application and the degree of exogeneity of the input and the output variables.

4.1.2   Assurance regions of type I (ARI) e.g. (b) and (c) in (4.3)

These types of restriction link either only input weights (b_i and c_i) or only output weights (b_o and c_o), and were first used by Thompson et al. (1986).

Use of form (c) is more prevalent in practice, with various applications such as Thompson et al. (1992) to oil/gas producers, Schaffnit et al. (1997) to bank branches, Ray et al. (1998) to the Chinese iron and steel industry, Thanassoulis et al. (1995) to English perinatal care DMUs, and Olesen and Petersen (2002) to hospitals[2].

ARs are especially appropriate to be used when there is some price information and one wants to proceed from technical towards overall efficiency measures. When there is *a priori* information concerning marginal rates of technical substitution (transformation) between inputs (outputs) these are also the adequate WRs to use because they are based on ratios of weights that, as we saw previously, reflect these rates. ARIs have the advantage of providing the same results irrespective of the model orientation as long as CRS is assumed.

Lets assume a situation where prices are known with certainty and we add to model (4.1) constraints of the type: $u_1/u_2 = p_1/p_2$ (in our two output example). The addition of such restriction in (4.1) makes the following changes in the envelopment model:

[2] These authors used probabilistic bounds associated with ARs, that have been introduced in Olesen and Petersen (1999).

$$Max \quad h = \theta + \varepsilon \left(s_1^+ + s_2^+ + s^- \right)$$

s.t.

$$\sum_{j=1}^{n} \lambda_j x_j + s^- = x_o \ , \quad \sum_{j=1}^{n} \lambda_j y_{1j} - s_1^+ = \theta y_{1o} \ , + p_2 z \quad (4.4)$$

$$\sum_{j=1}^{n} \lambda_j y_{2j} - s_2^+ = \theta y_{2o} - p_1 z, \qquad \lambda_j, s_r^+, s^- \geq 0, z \text{ is free}$$

The ARI in the multiplier model implies the introduction of a new variable in the dual envelopment model. This new variable (z) reflects trade-offs between the outputs, since it is associated to both output constraints.

Solving model (4.4) or model (4.1) with the ARI restrictions (which are different for each DMU assessed as we assume different prices) results in the efficiency scores shown in Table 4-3, where we also show the efficiency scores resulting from model (4.1) or (4.2).

Table 4-3. Results from model (4.4)

	A	B	C	D	E	F	G	H	I
Eff (4.1)	100%	100%	100%	70%	55.4%	50%	56%	69.3%	84%
Eff (4.4)	69.6%	100%	52.9%	54.1%	52.7%	36.2%	53.1%	53.3%	80.5%

As efficiency scores cannot improve when WRs are added to a DEA model, all efficiency values resulting from (4.4) are equal or smaller than those resulting from (4.1) or (4.2). The efficiency scores resulting from model (4.4) are exactly equal to a revenue efficiency score as given by the ratio between observed revenue and maximum revenue, where the latter is the optimal solution of (4.5).

$$Max \ \left\{ R = \sum_{r=1}^{s} p_r y_r \ | \ \sum_{j=1}^{n} \lambda_j y_{rj} \geq y_r \ , \ \sum_{j=1}^{n} \lambda_j x_{ij} \leq x_{io} , \ \lambda_j \geq 0 \right\} \quad (4.5)$$

If we replace in (4.4) $\theta y_{1o} + p_2 z = y_1$ and $\theta y_{2o} - p_1 z = y_2$ we can take the value of z, which results in: $(y_1 - \theta y_{1o})/p_2 = (\theta y_{2o} - y_2)/p_1$. From this equality we can have $\theta = (p_1 y_1 + p_2 y_2)/(p_1 y_{1o} + p_2 y_{2o})$, that is, maximising θ in (4.4) is equivalent to maximising revenue, as in (4.5). The value of θ in (4.4) is therefore equal to the inverse of revenue efficiency, i.e. the ratio of maximum revenue to observed revenue. The same equivalence was proven by Schaffnit et al. (1997) for the case of cost efficiency and an input oriented ARI restricted DEA model.

Note that the equality between the weights restricted model and the economic model does not depend on assumptions concerning returns to scale, it is also valid for VRS.

Similar to the above analysis is the multiple criteria framework called Value Efficiency Analysis as described in Halme et al. (1999), where the DMs preferences are included in a value function estimated through the knowledge of the DMs' most preferred solution. This value function works in a similar way to a revenue or cost function except that it may reveal other preferences than those related with prices [see also Korhonen et al. (2002)].

4.1.3 Assurance regions type II (ARII) e.g. (d) within (4.3)

Thompson et al. (1990) termed relationships between input and output weights 'Type II Assurance Regions' (ARII) or linked cone Assurance regions (LC-ARs). Though in Thompson et al. (1990) both ARIs and ARIIs are put forward, only the former were applied to analysing the efficiency of Kansas farming, as they recognised the existence of some problems in ARIIs. These problems relate with "the questions of a single numeraire and absolute profitability" which were left unanswered in Thompson et al. (1990, p. 103). In addition, the authors also refer to infeasibility problems that may be associated with ARIIs resulting from these being not a cone ratio as ARIs (see also Thompson et al. (1995, p. 106)).

As ARIIs, link input and output weights, they should be related with profit efficiency (where profit is defined in absolute terms as revenues minus costs). Using a similar reasoning as before it can be proved that introducing the constraints $u_1/v = p_1/w$ and $u_2/v = p_2/w$ in the multiplier model results in a value of θ in the envelopment model that is shown in (4.6).

$$\theta = \frac{p_1 y_1 + p_2 y_2 - wx}{p_1 y_{1o} + p_2 y_{2o}} + \frac{wx_o}{p_1 y_{1o} + p_2 y_{2o}} . \tag{4.6}$$

Maximising θ is, therefore, equivalent to maximising profit (note that except for the numerator of the first term in (4.6) all remaining entries are constant), but the value of θ in (4.6) cannot be interpreted as a profit efficiency measure in an analogous way as seen previously for revenue efficiency.

In fact, in a profit context it is not clear that the ratio of actual profit to maximum profit represents a profit efficiency measure since maximum profit is zero under CRS, and observed profit may be negative (see Varian, 1992 and Portela and Thanassoulis, 2002). Thompson and Thrall (1994) draw extensively on this issue, referring that in DEA each DMU maximises its 'virtual profits' at zero. Efficiency in DEA, therefore, coincides with zero 'virtual profits' and inefficiency with negative 'virtual profits'[3]. This is a

---

[3] This is easily seen in model (4.1) if we consider that weights represent prices. Under this circumstance it is as if model (4.1) sought to minimise cost, subject to all units presenting negative profit (revenues minus costs).

common assumption in the economic theory for perfectly competitive markets, but when other than theoretical economic profit is being measured it might not be an acceptable assumption. For this reason, Thompson and Thrall (1994) and also Thompson et al. (1995) advocate the separate treatment between efficiency and profit analysis. This separate treatment is what Thompson et al. (1995) called "LC-AR profit ratio model" (see also Thompson et al. 1996, 1997). This methodology uses the multiplier model with ARII restrictions, but removes from the model all the constraints of type (4.7) (as they prevent virtual profits from being positive).

$$\frac{\sum_{r=1}^{s} u_r y_{rj}}{\sum_{i=1}^{m} v_i x_{ij}} \leq 1, \qquad j = 1, ..., n. \tag{4.7}$$

Replacing (4.7) by ARII restrictions Thompson et al. (1995, 1996) "provide a way to measure absolute profits, both maxima and minima, in contrast to the inability to say anything substantive about profits by use of the DEA CCR ratio (and BCC convex) models" (Thompson et al., 1996, p. 361). The LC-AR profit ratio model relates with profit potential or profit loss depending on the criteria applied to the objective function (max or min).

We apply the LC-AR profit ratio model to our illustrative example, where the restrictions imposed on weights are those shown in (4.8). Note that p_r^+ (or w_i^+) is the maximum price in Table 4-1 and p_r^- (or w_i^-) is the minimum price.

$$\frac{p_r^-}{p_k^+} \leq \frac{u_r}{u_k} \leq \frac{p_r^+}{p_k^-} \qquad and \qquad \frac{p_r^-}{w_i^+} \leq \frac{u_r}{v} \leq \frac{p_r^+}{w_i^-}, \tag{4.8}$$

Using these restrictions the profit ratio model applied to our example is:

$$Max \quad or \quad Min \quad u_1 y_{1o} + u_2 y_{2o}$$

s.t.

$$v x_o = 1 \tag{4.9}$$

$$\frac{1}{6} \leq \frac{u_1}{u_2} \leq \frac{7}{3}, \quad \frac{1}{6} \leq \frac{u_1}{v} \leq \frac{7}{2}, \quad \frac{3}{6} \leq \frac{u_2}{v} \leq \frac{6}{2}$$

When (4.9) is maximised the resulting objective function is MPR (maximum profit ratio), and when it is minimised the resulting objective function is mPR (minimum profit ratio). If MPR is greater than 1 then there is operational profit potential, whereas an MPR inferior to one assures operational losses. If mPR is greater than one then operational profits are assured, and there is loss potential if it is inferior to one (see Thompson et al. 1995, 1996). For our illustrative example the optimal solution of (4.9) is shown in Table 4-4, where operational profit is assured for all DMUs except D that exhibits potential for losses.

Table 4-4. LC-AR profit ratio model results

DMU	A	B	C	D	E	F	G	H	I
MPR	13.75	17.50	13.75	9.125	9	6.700	9.500	10.33	14.25
mPR	2.08	2.08	1.25	0.792	1	1.033	1.167	1.44	1.75

Note, that the DMU that has highest profit potential is DMU I, which is not technical efficient. As noted by Thompson et al. (1996), DEA-efficiency and LC-profit potential should be considered different characteristics of DMUs..

Applications of ARIIs are mostly based on the above profit ratio methodology (see apart from the above referred papers also Ray et al. (1998), and Taylor et al. (1997)). The use of ARIIs with traditional DEA models (that is a model such as (4.1) plus these constraints) is not common in the literature. An example is the application in Thanassoulis et al. (1995) to perinatal care DMUs in England. The authors incorporated a constraint in the DEA model forcing the weight on an input (number babies at risk) to be equal to the weight on an output (number of surviving babies at risk). This is obviously and AR type II, where, however, only one input and one output are being linked.

4.2 Restrictions on Virtual Inputs and Outputs

WRs may not only be applied to weights but also to virtual inputs and outputs. The virtual inputs and outputs can be seen as normalised weights reflecting the extent to which the efficiency rating of a DMU is underscored by a given input or output variable. The first study applying restrictions on virtual inputs/outputs was that of Wong and Beasley (1990). Such restrictions assume the form in (4.10), where the proportion of the total virtual output of DMU$_j$ accounted for by output r is restricted to lie in the range $[\phi_r, \psi_r]$. A similar restriction can be set on the virtual inputs.

$$\phi_r \le \frac{u_r y_{rj}}{\sum_{r=1}^{s} u_r y_{rj}} \le \psi_r, \qquad r = 1,...,s \qquad (4.10)$$

The range is normally determined to reflect prior views on the relative 'importance' of the individual outputs. Constraints such as (4.10) are DMU specific meaning that the DEA model with such constraints may become computationally expensive. Wong and Beasley (1990) suggest some modifications for implementing restrictions on virtual values:

– Add the restrictions only in respect of DMU$_o$ being assessed leaving free the relative virtual values of the comparative DMUs;

– Add the restrictions in respect of all the DMUs being compared. This is computationally expensive as the constraints added will be of the order of $2n(s + m)$;

- Add the restrictions (4.10) only in relation to the assessed DMU, and add constraints (4.11) with respect to the 'average' DMU, which has an average level of the rth output equal to $\sum_{j=1}^{n} y_{rj} / n$.

$$\phi_r \leq \frac{u_r \sum_{j=1}^{n} y_{rj} / n}{\sum_{r=1}^{s} u_r (\sum_{j=1}^{n} y_{rj} / n)} \leq \psi_r, \qquad r = 1,...,s \qquad (4.11)$$

Restrictions on the virtual input-output weights represent indirect absolute bounds on the DEA weights of the type shown in (a) in (4.3). For example a restriction such as (4.10) in our example would reduce to imposing $\phi_r \leq u_r y_{rj} \leq \psi_r$, as the denominator of (4.10) is 1 for each DMU j being assessed. The imposition of restrictions on virtual inputs or outputs is sensitive to the model orientation, and it may lead to infeasible solutions.

Pedraja-Chaparro et al. (1997) proposed the introduction of AR type restrictions to virtual inputs/outputs rather than to weights. In (4.12) such type of restrictions ('contingent weight restrictions') are shown.

$$\phi_r \leq \frac{u_r y_{rj}}{u_1 y_{1j}} \leq \psi_r, \qquad r = 1,...,s \qquad (4.12)$$

Constraints in (4.12) impose the proportion of virtual output r in relation to virtual output 1 to lie within a certain range. As y_{rj} and y_{1j} are constants, (4.12) in fact works as an ARI, except that, the ratio of output virtual weights are DMU specific. Therefore, (4.12) has the same computational problems as constraints (4.10). In fact, each DEA model requires as many constraints per output r (or input i) as the number of DMUs, or the double if both upper and lower bounds are specified. This problem can obviously be sorted out by imposing (4.12) only to DMU o, being assessed, while letting the ratio of virtual outputs to vary freely for the other DMUs.

As we have seen in section 2 virtual inputs and outputs, contrary to DEA weights, are independent on units of measurement. This apparently gives an advantage to restrictions on virtual inputs and outputs in relation to direct restrictions on the weights. Restrictions on the virtual input/output weights have, however, received relatively little attention in the DEA literature so far. For an exploration of their impact in DEA assessments see Sarrico (1999).

4.3   Estimating the parameters of WR

When using WRs it is important to estimate the appropriate values for the parameters in the restrictions, [e.g. values for δ_i, τ_i, ρ_r, η_r, κ_i, ω_r, α_i, β_i, θ_r, ζ_r, γ_i, in (4.3)]. A number of methods have been put forward to aid the estimation of such parameters. Although most of these were developed for establishing bounds in specific WR methodologies, in most cases they can be generalised. No method is all-purpose and different approaches may be appropriate in different contexts. We outline here some approaches.

a) Using information on prices and/or costs

When price/cost information is available it can be included in DEA models through weights restrictions. Usually such price information is not accurate and therefore a range of prices is used instead. Bounds based on prices are especially seen in AR applications. For example, Taylor et al. (1997) established AR bounds based on nominal interest rates of loans and deposits. Thompson et al. (1996) used different information sources to arrive at lower and upper AR bounds on the various inputs and outputs used. For example for the output 'crude oil produced' the authors used the smallest and largest monthly crude oil prices (see also Thompson et al. (1995) where a detailed exposition on using prices to define AR bounds is put forward).

b) Using unbounded DEA weights as reference levels

This approach is due to Roll et al. (1991) and Roll and Golany (1993). Initially an unbounded DEA model is run, a weights matrix compiled, and, if necessary, either the outlier weights or a certain percentage of the extreme weights are eliminated. Alternative optimal solutions generally exist, especially for the Pareto-efficient DMUs. Hence, there may exist alternative weights' matrices and a choice of which matrix to use is for the user to make. The mean weight for each factor is then calculated based on the selected weights matrix. A certain amount of allowable variation about each mean is subjectively determined, giving an upper and a lower bound for each factor weight. This approach was originally developed for establishing bounds in absolute WRs, but there is no reason why it cannot be adapted to be used in other types of WRs.

c) Using optimal weights of model DMUs

This approach is due to Charnes et al. (1990) and Brockett et al. (1997) who used cone ratios to analyse banks efficiency. The authors used a set of banks that were considered excellent to set the WRs (in this case in the form of cone ratios (CRs), which will be analysed in the next section). Excellent banks were tested for efficiency and the weights of those that were considered DEA efficient were used in constructing the CRs. In the study of Brockett et al. (1997) the excellent banks used were not part of the data set (Texas banks) being analysed, which means that these excellent banks were regarded as benchmarks to be followed by the banks under analysis.

d) Using Expert Opinion

This method consists in gathering different views from some people (experts) involved in the production process being analysed. For example, Cooper et al. (2000, Chap. 6.) refer to a decision process where a new city for locating Japanese governmental agencies and the Supreme Court was being chosen. For this purpose, some criteria were put forward for analysis (access, water supply, land acquisition, etc). Each expert evaluated each criterion on subjective grounds. These were then quantified using AHP (Analytic Hierarchy Process). As experts had conflicting evaluations it was important to derive some consensus among the evaluations. This could not be done through an average, for example, because the variety of opinions would not be considered. The way the authors solved the problem was through ARs where ratios of the weights assigned to each criterion were calculated for each expert and then the higher and lower values of these ratios were used as upper and lower bounds on AR restrictions (see also Zhu (1996b) who refers to the use of AHP for identifying bounds of AR restrictions). Expert opinion can be used alone (like in Beasley, 1990) or in combination with other type of information like price information (Thompson et al., 1990, 1992) or model DMUs' weights (Brockett et al., 1997).

e) Using estimated average marginal rates of transformation as reference levels.

This approach is due to Dyson and Thanassoulis (1988) and it is applicable only to the case where DMUs use a single input to secure multiple outputs or alternatively secure a single output using multiple inputs. Considering the case where the DMUs use a single resource, then the jth DMU uses x_j units to secure output levels y_{rj}, $(r = 1...s)$. Consider model (4.13) and its equivalent model (4.14) for the single-input multiple-output case.

$$Max \quad z = \sum_{r=1}^{s} u_r y_{ro}$$

$s.t$

$$\sum_{r=1}^{s} u_r y_{rj} - v x_j \leq 0 \quad j = 1,...,n \qquad (4.13)$$

$$v x_o = 1$$

$$u_r \geq \varepsilon$$

$$Max \quad z = \sum_{r=1}^{s} U_r y_{ro}$$

$s.t$

$$\sum_{r=1}^{s} U_r y_{rj} \leq x_j \quad j = 1,...,n \qquad (4.14)$$

$$u_r \geq \varepsilon$$

Model (4.14) is obtained from (4.13) by replacing $v = 1/x_o$, and by considering $u_r x_o = U_r$, $\forall r$. At the optimal solution to model (4.14) at least one constraint is binding. Let the binding constraint relate to DMU k so that we have the equality in (4.15), where the asterisk means optimal value from (4.14).

$$\sum_{r=1}^{s} U_r^* y_{rk} = x_k \tag{4.15}$$

We can interpret U_r^* within expression (4.15) as the level of input the DEA model allocates per unit of output r. This interpretation of the DEA weights makes possible the identification of reference levels for setting restrictions on the DEA weights. For example, regression analysis can be used to estimate the average level of resource per unit of output r. Let us assume that we estimate in the foregoing manner the regression equation

$$\overline{x} = \sum_{r=1}^{s} \phi_r y_o + \varsigma \tag{4.16}$$

where ϕ_r is the partial regression coefficient of output r, and ς is the regression constant. If $\varsigma \neq 0$ and is significant, then CRS is not a valid assumption and VRS should be used instead. If $\varsigma = 0$, or it is not statistically different from zero, then the ϕ_r can be interpreted as the estimated resource DMUs use on average per unit of output r. We can use the values of ϕ_r as reference levels for setting lower or upper bounds on the DEA weights. For example the user may decide to set lower bounds so that $U_r \geq 0.1\phi_r$ ($r = 1...s$). The argument would be that an efficient DMU cannot be so efficient as to use less than 10% of the resource level DMUs use on average per unit of the rth output. Alternatively, one may set upper bounds so that say $U_r \leq 10\phi_r$ ($r = 1...s$), where the argument would be analogous but with reference to inefficient DMUs. Relative rather than absolute WRs can also be imposed based on results from (4.16) as we can set for example $U_r/U_k \geq \phi_r/\phi_k$.

5.   CHANGING THE DATA SET

The above WR approaches incorporate value judgments in the analysis by acting directly on the weights (or virtual weights). There are, however, approaches that incorporate the DMs judgments through introducing some changes on the data set used in the DEA assessment. These changes can be of two types:
– Transforming the data so that the new data reflects the value judgments;
– Adding new DMUs that reflect the DM's value judgments to the data set.
    Each of these approaches will be summarised in the next two sections.

5.1   Transforming the data

The best known approach that acts on the data and transforms it, is by Charnes et al. (1989) and is called the Cone-Ratio (CR) approach. But

Golany (1988) and Ali et al. (1991), also used data transformations to replace ordinal relationships between weights in DEA.

Cone-Ratios are closely related with ARs, in the sense that an AR of type I is a CR. CRs, however, are more general than ARs. We shall draw on CRs by means of our illustrative example. Consider Figure 4-2 where the multiplier space is drawn considering $v = 1$.

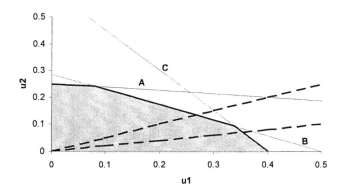

Figure 4-2. Multipliers space

Without any weight restriction the multiplier space is defined by the intersection of the hyperplanes defined by DMUs A, B, and C (the shaded region in Figure 4-2). With a WR $2 \leq u_1/u_2 \leq 5$, the feasible region is now the shaded region in Figure 4-2 between the dashed rays intersected with the hyperplanes of DMUs C and B. This feasible region is depicted in Figure 4-2 in intersection form (i.e. it is the result of the intersection on a number of inequalities). In other words, the output weights u_r belong to a cone U such that: $U = \{u \mid Cu \geq 0, u \geq 0\}$, where

$$C = \begin{bmatrix} -1 & 5 \\ 1 & -2 \end{bmatrix} \quad and \quad u = \begin{bmatrix} u_1 \\ u_2 \end{bmatrix} \Leftrightarrow U = \left\{ \begin{array}{c} -u_1 + 5u_2 \geq 0 \\ u_1 - 2u_2 \geq 0 \end{array} \right\}$$

U can be equivalently defined in sum form: $U = \{u \mid u = \sum_{k=1}^{K} \beta_k b_k \Leftrightarrow B^T\beta, \beta \geq 0\}$, where $B^T = (C^T C)^{-1} C^T$. In our specific case

$$B^T = \begin{bmatrix} 2/3 & 5/3 \\ 1/3 & 1/3 \end{bmatrix}$$

The matrix B can be used to transform our output data such that we have a new output vector $\mathbf{Y'} = \mathbf{BY}$. For example, for DMUs A and B the new output data are:

$$A: \begin{bmatrix} 2/3 & 1/3 \\ 5/3 & 1/3 \end{bmatrix} \times \begin{bmatrix} 1 \\ 8 \end{bmatrix} = \begin{bmatrix} 10/3 \\ 13/3 \end{bmatrix} \qquad B: \begin{bmatrix} 2/3 & 1/3 \\ 5/3 & 1/3 \end{bmatrix} \times \begin{bmatrix} 4 \\ 7 \end{bmatrix} = \begin{bmatrix} 5 \\ 9 \end{bmatrix}$$

The transformed data set resulting from applying the above transformation to our previous illustrative example is shown in Table 4-5.

Table 4-5. Transformed data of illustrative example

DMU	A	B	C	D	E	F	G	H	I
y_1	3.33	5	13.33	12	13.33	4	16	4	4
y_2	4.33	9	28.33	26	25.33	5	28	6	7
x	2	2	6	8	10	5	12	3	2

Solving an unrestricted DEA model like (4.1) on this new data set results in the same efficiency measures as applying model (4.1) plus the WR $2 \leq u_1/u_2 \leq 5$ to our original data set. This means that a data set can be transformed such that weights restrictions are not required since they become implicit in the data transformations. In general, given a matrix B used to transform output data and a matrix A used to transform input data, the CR procedure states the equivalence between model (4.17) and (4.18), which are represented in matrix form. (Note that matrix A may be obtained in an equivalent way as shown before for matrix B).

$$
\begin{aligned}
& Min\ z = v^T X_o \\
& s.t \\
& v^T X - u^T Y \geq 0 \\
& u^T Y_o = 1 \\
& v \in V, \quad u \in U
\end{aligned}
\qquad (4.17)
$$

$$
\begin{aligned}
& Min\ z = \alpha^T (AX_o) \\
& s.t \\
& \alpha^T (AX) - \beta^T (BY) \geq 0 \\
& \beta^T (BY_o) = 1 \\
& \alpha, \beta \geq 0
\end{aligned}
\qquad (4.18)
$$

Model (4.18) uses transformed data to capture value judgments and hence, ordinary DEA software can be used to solve CR models. The data transformation from the intersection to the sum form requires the computation of the inverse of a given matrix. It may happen, however, that this inverse is not defined, meaning that such data transformations are not always possible. Nevertheless, it is always possible to specify a matrix A and B such that model (4.18) can be applied to a data set, though this does not necessarily translate into a WR of the type we have seen previously. This is the reason why CRs are more general than ARs as they can be used to link any number of multipliers in a DEA model. Note that, data transformations in CRs may imply the elimination of constraints in the envelopment model as shown in Cooper et al. (2000, Chapter 6). The reader can verify this by using a data transformation where B = [2 1] to our illustrative example.

5.2 Adding new DMUs

As seen previously, the addition of weights restrictions in the multiplier model introduces new variables in the envelopment model. These new variables may be seen, under specific circumstances, as new DMUs added to the reference set [see Roll et al. (1991) and Thanassoulis and Allen (1998)].

The first study to use new DMUs in the reference set was that of Golany and Roll (1994), where 'standard' DMUs were introduced in a DEA assessment. Standard DMUs relate to benchmark practices that enlarge the size of the referent set, so that targets for originally efficient DMUs (deemed efficient before standards have been incorporated) can also be imposed. The main difficulty with this approach relates with the establishment of standards. The authors refer to this problem, but no guidelines on how these standards are actually to be generated were provided.

Allen and Thanassoulis (forthcoming) and Thanassoulis and Allen (1998) developed an approach that introduces Unobserved DMUs (UDMUs) into the reference set. Such approach was proved to be equivalent to the addition of weight restrictions to the multiplier model.

Consider Figure 4-1 of our illustrative example. The mathematical expressions of lines AB and BC are $x = 0.08y_1 + 0.24y_2$ and $x = 0.34y_1 + 0.09y_2$, respectively (see Table 4-2). Recall that these lines reveal information concerning the relationship between weights, which result in a ratio of weights $u_1/u_2 = 0.333$ for line AB, and a ratio of weights $u_1/u_2 = 3.778$ for line BC. This means that when an ARI restriction $u_1/u_2 \geq 1$ is included in the analysis all the DMUs in the segment AB or projected on this segment will change their efficiency score as they do not satisfy the marginal rate of transformation implicit in the AR.

Thanassoulis and Allen (1998) have shown that the introduction of an ARI restriction $u_1/u_2 \geq 1$ is equivalent to the addition of a set of DMUs whose input/output levels are equal to the radial targets obtained from the weights restricted model. That is, for our example the new Pareto-efficient frontier would be constituted by DMUs F'A'H'G'&I'BC shown in Figure 4-3.

The input and output values of these new DMUs are shown in Table 4-6, where the optimum radial expansion factor (θ^*) associated with the constrained model is also shown.

Table 4-6. Data for the new DMUs

DMU	A'	F'	G'	H'	I'
y_1	1.222	2.5	22	3.3	3.667
y_2	9.778	25	44	13.2	7.333
x	2	5	12	3	2
θ^*	1.222	2.5	1.833	1.65	1.222

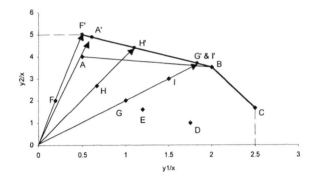

Figure 4-3. UDMUs and ARs

The weights restricted model provides exactly the same results as those obtained from a standard DEA model where DMUs F', A', H', G' and I' have been added to the reference set. The equivalence between the introduction of UDMUs and WRs is also valid for the case of VRS (as shown in Thanassoulis and Allen, 1998).

In Figure 4-3 the WR $u_1/u_2 \geq 1$ prevents DMU F from placing a zero weight (or an ε weight) on output 1 but does not prevent DMU D from neglecting output 2 in its assessment. The UDMUs methodology can be used to avoid this happening, by improving envelopment in DEA If the DM can trade-off between the output levels of say DMU A reducing the level of output 1 while raising the level of output 2 to compensate, a UDMU can be created at A" in Figure 4-4. Similarly C" could also be created expressing the DM's trade-off between increasing output 1 and decreasing output 2. This would extend the Pareto-efficient boundary to A"ABCC", and ε-weights would disappear for DMUs F and D.

Note that DMUs A and C in Figure 4-4 play a special role in the establishment of such trade-offs. Such DMUs are called in Thanassoulis and Allen (1998) *Anchor* DMUs (ADMUs). ADMUs are a sub-set of the Pareto-efficient DMUs, which normally, but not always, delineate the Pareto-efficient from the inefficient part of the PPS boundary. ADMUs are special because relatively minor (or local) adjustments to their input-output levels can reduce the DEA-inefficient part of the PPS and thereby increase the number of fully enveloped DMUs. A properly enveloped DMU is such that no ε-weights are assigned to each of its inputs and outputs.

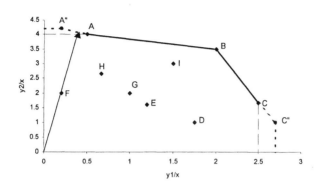

Figure 4-4. Improving Envelopment through UDMUs

5.2.1   The UDMUs approach

The ideas illustrated in Figure 4-4 are developed in Allen and Thanassoulis (forthcoming) into a general purpose procedure for '*improving envelopment*' in DEA. The procedure addresses the case where the DMUs operate in a single input multi-output CRS context. We sketch the procedure here [for mathematical proofs and details see Allen and Thanassoulis (forthcoming)].

(i)   Run an unrestricted assessment by DEA to identify the Pareto-efficient and non-enveloped DMUs. Stop if all DMUs are fully enveloped.

(ii) If any non-enveloped DMUs exist identify *Anchor* DMUs (ADMUs) from which to construct *Unobserved* DMUs (UDMUs).

(iii) In respect of each ADMU identify which· output(s) to adjust in order to construct suitable UDMUs.

(iv) Using the outputs in (iii) and DM value judgments construct UDMUs.

(v) Re-assess the observed DMUs by DEA after adding the UDMUs constructed. The number of fully enveloped observed DMUs will generally increase.

Steps (i) and (v) are performed using standard DEA models. Steps (ii), (iii), and (iv) require some further comments.

Step (ii): Determining Anchor DMUs (ADMUs)

In the general case it is necessary to solve a linear programming model to ascertain whether a Pareto-efficient DMU is an ADMU. The model solved is a variant of that introduced by Andersen and Petersen (1993). Let the set E consist of the Pareto-efficient DMUs identified in Step (i) and let E_o be the set E excluding DMU $o \in E$. In respect of each $o \in E$ solve model (4.19).

$$Min \ h_o' = \phi_o - \varepsilon \left(\sum_{r=1}^{s} g_r \right)$$

$$s.t.$$

$$\sum_{j \in E_o} \lambda_j y_{rj} - g_r = y_{ro} \qquad r = 1, \ldots, s \tag{4.19}$$

$$\sum_{j \in E_o} \lambda_j = \phi_o, \qquad g_r, \lambda_j \geq 0$$

Model (4.19) uses only the Pareto-efficient DMUs identified in Step (i), and excludes DMU o from the reference set. (The input is normalised to one unit.) DMU o is an ADMU if the optimal value of ϕ_o in (4.19) is $\phi_o > 1$ and there is at least one positive slack variable g_r. For example, in our illustrative case results from applying model (4.19), in Table 4-7, show that DMU A and C are Anchor DMUs, while B is not. More details on the rationale underlying ADMUs can be found in Allen and Thanassoulis (forthcoming).

Table 4-7. Anchor DMUs in illustrative example

DMU	ϕ_o	g_1	g_2
A	1.142857	1.785714	0
B	1.272727	0	0
C	1.25	0	2.708333

Step (iii): Determining where to impose values

After determining the set of ADMUs it is necessary to determine which output levels are to be adjusted. Let K_o be the set of outputs for which the instance of model (4.19) corresponding to ADMU o yields zero slack. For each $k \in K_o$ solve model (4.20), where ϕ_o^* is the optimal value of ϕ_o in (4.19). If there exists an optimal solution to the instance of (4.19) corresponding to ADMU o in which output k has a positive slack value, then g_k would be positive at the optimal solution to (4.20).

$$Max \ g_k$$

$$s.t. \hspace{6cm} (4.20)$$

$$\sum_{j \in E_o} \lambda_j y_{rj} - g_r = y_{ro} \qquad r = 1,...,s$$

$$\sum_{j \in E_o} \lambda_j = \phi_o^* , \qquad g_r , \lambda_j \geq 0$$

Step (iv): Constructing UDMUs

UDMUs are constructed by setting to a 'lowest permissible' level each output of ADMU o in turn for which either model (4.19) or (4.20) yields a positive slack value at its optimal solution. In the general case, the minimum permissible level my_r of output r per unit of input will reflect what the DM deems is feasible under both technological and policy constraints. In many cases $my_r = 0$, but in some practical contexts zero output levels are impossible or simply not acceptable. In our illustrative example the solution of model (4.20) did not show g_2 for DMU A and g_1 for DMU C different from zero, and as such the solution that will be used to construct UDMUs is that shown in Table 4-7.

For each ADMU the output level should be reduced as described above. A UDMU can now be constructed by suitably compensating for each output's reduction. This is done using implicit local trade offs between output levels given by the DM. The DM may have or not have preferences over the relative changes on the output levels. In each of these cases the type of interaction with the DM is obviously different (for details see Allen and Thanassoulis, (forthcoming). (It should be noted that, ABC (Activity Based Costing) information can readily be used to calculate UDMUs).

In our case UDMUs are constructed by setting output 1 of ADMU A, and output 2 of ADMU C to zero. Clearly, in this example, it is easy to see for the graph, that to extend the efficient frontier from DMU A, a new DMU with lower levels of output 1 and higher levels of output 2 needs to be introduced. (Also from the values from Table 4-7, i.e. DMU A has a positive g_1).

Assume that in our illustrative case the DM decides that a reduction of 0.5 in output 1 should be compensated by an equal increase of output 2 (UDMU created from the ADMU A), and that a reduction of 1.66667 of output 2 should be compensated by increasing 1 unit of output 1 (UDMU created from ADMU C). These trade-offs give rise to two UDMUs whose normalised output levels are (0 4.5) and (3.5, 0). So, our new frontier, constructed with these two UDMUs, is shown in Figure 4-5, where all DMUs are now fully enveloped.

The number of observed DMUs that are fully enveloped would normally be larger than in the absence of the UDMUs. For the proof of this, see Appendix 2 in Allen and Thanassoulis (forthcoming). The increase in the number of fully enveloped DMUs obtained will depend largely on how successful the DM has been in giving trade-offs between the output levels of ADMUs, which lead to Pareto-efficient UDMUs.

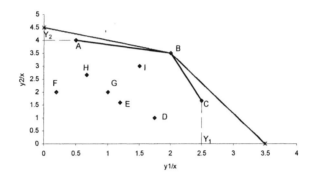

Figure 4-5. Efficient frontier with UDMUs

6.   INTERPRETING RESULTS OF DEA MODELS WITH VALUE JUDGMENTS

The interpretations of the results of DEA models do not in general carry over between '*unrestricted*' DEA models and 'weights-restricted' models or models that change the data set in some way.

When all DEA weights are unrestricted [as in model (4.1)], the technical output efficiency rating of a DMU represents the maximum radial expansion of its outputs that is feasible given its input levels. Similarly, in an input oriented model the efficiency score of a DMU represents the minimum radial contraction to its input levels that is feasible given its output levels.

In unrestricted DEA models input and output targets are given by (4.21), where * represents the optimal solution of the envelopment model.

$$\text{Target input } i: \ \sum_{j=1}^{n} \lambda_{j}^{*} x_{ij} = x_{io} - s_{i}^{-*}$$

$$\text{Target output } r: \ \sum_{j=1}^{n} \lambda_{j}^{*} y_{rj} = \theta_{o}^{*} y_{ro} + s_{r}^{+*}$$

(4.21)

Radial targets result from ignoring slacks in the right hand side of (4.21), and therefore radial input targets are equal to observed inputs, and radial output targets are equal to the product of θ^* and observed outputs[4]. In restricted DEA models targets as defined in (4.21) can no longer be used. The efficiency score also changes both in terms of value and in terms of interpretation from unrestricted to restricted DEA models. In restricted DEA models (whatever the method used to incorporate value judgments) the efficiency score can never be higher than that obtained using the unrestricted version of those models (but see Podinovski (2003) for the possibility this may not be so for *relative* efficiencies as defined there). Concerning the interpretation of targets and efficiency scores we consider the impacts of the different types of methods for including value judgments.

(i) ARI

Consider DMU H of our illustrative example. This DMU has an unrestricted efficiency score of 69.33% meaning that outputs 1 and 2 should increase by 1/0.6933 so that the efficient frontier is reached. This happens at point H' = (y_1, y_2, x) = (2.885, 11.54, 3), which is Pareto-efficient. Normalising the outputs by the input value we have a projection point H' that is shown in Figure 4-6.

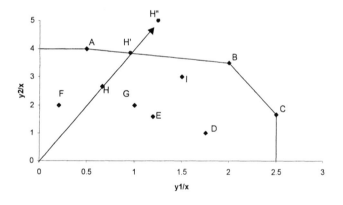

Figure 4-6. Projection of DMU H on the frontier

If we now add to the multiplier model of DMU H an ARI of $u_1/u_2 \geq 2$, the optimal solution of the envelopment model results in $1/\theta^* = 1/1.875 = 53.33\%$. If we multiply the observed outputs of DMU H by the value of θ^* the resulting point is (y_1, y_2, x) = (3.75, 15, 3). In Figure 4-6 we represented this point as H", after the two outputs have been normalised by the input. Point H" lies outside the PPS, which means that θ^* can no longer be interpreted radially. Thus, the measure loses the type of operational meaning

---

[4] For input oriented models radial input targets are given by observed inputs multiplied by θ^*, and radial output targets are equal to observed outputs.

it had in the absence of ARs. If the restricted efficiency measure leads to a point outside the PPS in the manner suggested above, it implies that the DMU concerned needs to alter its input or its output mix to attain 100% efficiency within the PPS, under the ARs. (Mix refers here to the ratios the input and output levels are to each other.) Targets in the presence of ARs are, therefore, influenced by additional variables that are added to the envelopment model. In this particular case targets are given by (4.22), where α is the coefficient of the corresponding weight in the AR, and z is the envelopment variable associated with the AR in the multiplier model.

$$\text{Target input } i: \quad \sum_{j=1}^{n} \lambda_j^* x_{ij} = x_{io} - s_i^{-*}$$

$$\text{Target output } r: \quad \sum_{j=1}^{n} \lambda_j^* y_{rj} = \theta_o^* y_{ro} + s_r^{+*} - \alpha z \tag{4.22}$$

Recall that the optimal solution of the restricted model for DMU H is θ^* = 1.875, $s_1^{+*} = s_2^{+*} = s^{-*} = 0$, $\lambda_B^* = 1.5$, and $z^* = 2.25$. This solution gives the targets shown in (4.23).

$$\text{Target input} \quad : 1.5 \times 2 = 3$$

$$\text{Target output 1}: 1.5 \times 4 = 1.875 \times 2 - (-1) \times 2.25 = 6 \tag{4.23}$$

$$\text{Target output 2}: 1.5 \times 7 = 1.875 \times 8 - 2 \times 2.25 = 10.5$$

The normalisation of these output targets by the input gives exactly point B in Figure 4-6, meaning that in order to be efficient DMU H should change its output mix to that of DMU B, increasing both outputs accordingly to (4.23).

The basic conclusion that the efficiency measure, θ, does not reflect radial expansions of outputs (or radial contractions of inputs) that are feasible under efficient operations is valid whatever the type of weights restrictions used [for details see Allen et al. (1997) and Chapter 8 of Thanassoulis (2001)].

(ii) Transforming the data set

Although the above conclusions are also valid for the case of Cone Ratios, as data are transformed prior to the use of DEA it is important to re-transform the data so that targets are meaningful. Such data transformations are put forward by Brockett et al. (1997) and Cooper et al. (2000). Note that the resulting slacks from the transformed CR envelopment model are, in matrix form, given by (4.24).

$$s' = (BY)\lambda - \theta(BY_o) \quad \text{and} \quad e' = (AX_o) - (AX)\lambda \tag{4.24}$$

These can be directly transformed in their original values by multiplying each term by the inverse of B and A, respectively:

$$B^{-1}s' = Y\lambda - \theta Y_o \quad \text{and} \quad A^{-1}e' = X_o - X\lambda \tag{4.25}$$

Once (4.25) has been used to determine the slacks based on the original data, targets can be calculated in the usual way by applying (4.21). It should be noted that, although slacks on the transformed data will always be non-negative, the slacks on the original data maybe negative, thus reflecting the trade-offs implicit in the CRs.

(iii) Absolute weights restrictions and virtual restrictions

The introduction of absolute WRs on the multiplier model implies the addition of new variables in the envelopment model not only on its constraints but also on its objective function. Under this circumstance, though the objective function of the primal (multiplier) and dual (envelopment) models is the same, the value of θ in the dual no longer equals the ratio of virtual inputs and virtual outputs in the primal, as customary in unrestricted DEA models (See Allen et al. (1997) for details).

(iv) Adding unobserved DMUs

If the DM is happy to have targets that are based on unobserved points, then radial targets can be set. However, if the DM only wants to use observed DMUs as targets, then the same holds here as in the ARI case.

(v) General guidelines

In general in the presence of WRs, targets:

- May involve substantial changes to the current mix of input and outputs of a given DMU;
- May involve deterioration in some observed input or output level.

These two features do not happen in unrestricted DEA models where the mix of inputs is preserved under pure radial models (note that there are non-radial models that do not consider this mix preservation as being important), and where target inputs are never higher than observed inputs and target outputs are never lower than observed outputs (inputs and outputs can only improve and never deteriorate).

It should be noted that under WRs these features are perfectly in line with intuition. As we now have prior views about the relative worth of inputs and outputs it is quite acceptable that for a DMU to attain maximum efficiency it may for example have to change the relative volume of its activities (i.e. change its output mix). Further, the value judgments incorporated within the model may mean that by worsening the level of one output some other output can rise so as to more than compensate the loss of value due to the worse level on the former output.

Restricted DEA models may also provide inefficient DMUs with peers that may offer a different input-output mix to that of the inefficient DMU being assessed. This does not happen in general under unrestricted models, as the peers are those DMUs rated efficient under the weights system of the DMU being assessed. In this sense peer DMUs and assessed DMUs have in general identical strengths and weakness concerning certain inputs and outputs, meaning that their mix will not be very dissimilar [for details see Allen et al. (1997) and Thanassoulis, (1997)].

Note that the efficiency rating yielded by a DEA model under binding WRs can still be interpreted as a ratio of total output to total input 'value' (or vice versa depending on the model orientation) or, alternatively, as a measure of the shortfall between observed input/output levels and 'radial targets'. The total input or output value derived now uses marginal values of inputs and outputs, which obey the weights restrictions. In the envelopment framework under binding WRs the radial targets corresponding to the derived efficiency score may or may not lie within the PPS. The DEA model solved will always yield targets within the PPS but those targets may or may not represent a radial expansion or contraction of the observed input/output levels of the DMU concerned. One can always use the efficiency score yielded by the DEA model under WR to estimate radial targets for a DMU (see Thanassoulis and Allen, 1998) if one is prepared to accept that sometimes such targets will lie outside the PPS.

7.   WEIGHTS RESTRICTIONS VERSUS CHANGING THE DATA SET

The choice between using weights restrictions or approaches that change the data set depends mainly on the existing information concerning the DMs value judgments. Nevertheless there are advantages and disadvantages of each approach that one should be aware of.

a) Local vs global trade offs

In the UDMUs approach the DM is asked on the trade-offs between inputs and outputs on the basis of observed DMUs. This preference information is therefore local to some DMU rather than global. Global preferences in linear form as used in WR are too restrictive and may not capture marginal rates of substitution or transformation, which vary within the PPS. This is likely to be especially true in VRS technologies where we expect a inter-dependence between marginal rates of substitution or transformation and scale. Under VRS the possibility of specifying local rather than global trade-offs might be, therefore, an important aspect favouring the use of the UDMUs approach.

b) Change in radial nature of efficiency measures.

As seen in the previous section, under WRs and CRs the targets estimated for an inefficient DMU may reflect changes in the mix of its inputs or outputs and may mean certain input levels need to rise or output levels need to fall. Contrary, the introduction of UDMUs retains the radial nature of targets albeit at the expense that such targets may lie on a part of the efficient boundary that makes reference to UDMUs. Such a boundary contains points that are *judgmentally* feasible but not feasible by the normal definition of the PPS (in the UDMUs approach the feasibility of the new PPS is in fact assessed by the DM, while in the WR approach the new PPS is implicit in the weights restrictions and its feasibility is not ascertained by the DM). Radial targets of this nature offer an alternative route to efficiency for an inefficient DMU. That is, the DMU can attain efficiency by improving the absolute levels of its input-output bundle but keeping the mix constant, rather than by needing to alter the mix and the absolute levels of its input-output bundle.

c) Computational time

The UDMUs approach is very demanding of DMs time. This depends on the size of the problem, but in principle it may be necessary to construct many UDMUs. The construction of each UDMU requires the involvement of the DM, who may thus find the process elaborate and time consuming. In addition the DM may find it difficult to provide the information required, even though it is local and it is supported by data on anchor DMUs. WR and CR approaches, on the other hand, are less time consuming. In computational terms CRs are in fact solved readily through available software since models with transformed data are standard DEA models. Nevertheless, data transformation at the beginning and then the re-transformation of results, so that these can be interpreted in light of meaningful values, may be time consuming activities.

8.   RETURNS TO SCALE (RTS)

Most of the approaches on WRs and the approaches that change the PPS are defined in respect to CRS technologies. The exception is the ARs that, since the beginning, were defined by Thompson et al. (1990) both under CRS and VRS technologies. These authors also calculated the type of Returns To Scale (RTS) applying at each part of the frontier. In Thompson et al. (1992) the issue of calculating DEA-AR efficiency ratios both in relation to CRS and VRS technologies is addressed. In the AR context the authors call the characteristic associated to RTS of '*AR-viability*' therefore avoiding classifying projections on regions of increasing returns to scale (IRS), constant returns to scale (CRS), or decreasing returns to scale (DRS) (see also Ray et al. (1998) who used the approach of Thompson et al. (1992) on AR-viability). According to this approach, DMUs are classified in strongly

or weakly AR-viable, or alternatively in AR-inviable. Strong AR viable
DMUs are those whose RTS interval lower bound is higher than 1, and weak
AR-viable DMUs are those whose RTS interval includes 1. All remaining
DMUs are AR-inviable (for details see Thompson et al. (1992)).

Recently Tone (2001) has dealt with the issue of defining RTS under
WRs, but used only ARs to do so.

Allen et al. (1997) restricted their literature review to CRS models and
Thanassoulis and Allen (1998, p. 587) point out that under VRS "where
DMUs come in a range of scales of operation, WRs and the global marginal
rates of substitution they represent may be inappropriate". In fact, the trade-
offs between inputs and outputs that are made explicit by WRs may not be
equally adequate for every scale of operation.

Note that the facets of a CRS or VRS frontier reflect trade-offs between
or within inputs and outputs under efficient operation. Such trade-offs are
closely linked with the RTS characteristics applying at each facet. This
means that if we impose a constraint on the trade-off between inputs and
outputs (ARII) we are implicitly forcing a certain type of RTS in relation to
which all DMUs are to be assessed. Consider, the single input/output
example shown in Figure 4-7.

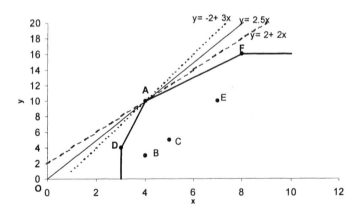

Figure 4-7. Single Input/Output Example

Without WRs a VRS output oriented multiplier model identifies three
possible facets for projection: (DA) $y = -14 + 6x$, (AF) $y = 4 + 1.5x$, and
(OA) $y = 2.5x$. (These facets are identified as shown in section 2.1 of this
chapter, i.e. collecting all optimal weights resulting from the multiplier
model and reducing them to a number of facets as shown in Table 4-2). The
first facets lie on the VRS frontier and the latter facet corresponds to the
CRS frontier that passes through the origin and point A.

The intercepts and inclination of these lines clearly have a RTS interpretation. A negative intercept implies IRS, a zero intercept CRS and a positive intercept implies DRS (see also Thanassoulis (2001), section 6.6.2.). The slope of the lines reflects how much of the output can be gained by increasing one unit of input. When this value is higher than 2.5 (as defined by the CRS line) it means that we have IRS and when it is below 2.5 it means that we have DRS. Clearly the line DA presents IRS, while the line AF presents DRS.

If we now impose a WR to this model stating that $v/u \geq 2$, the line AF can no longer define the frontier because part of the efficient frontier (to the right of point A) is now defined by the line $y = 2 + 2x$ (this line is derived from the optimal weights as explained in section 2.1). Therefore, DMU F loses its 100% efficiency status. The frontier has still a portion to the left of A with IRS, and a portion to the right of A with DRS, but the latter has changed its magnitude from a rate of input/output transformation of 1.5 to 2. Hence, WRs can change the RTS characteristics of the frontier. In fact, if our WR is $v/u \geq 3$, then the new frontier is now defined by DA to the left of A and by the line segment $y = -2 + 3x$ to the right of A. Clearly IRS prevails on the whole new frontier.

This example shows that RTS and weights are intimately linked because WRs may affect the RTS characteristics of the frontier. This is true for the ARII constraints linking input and output weights, and also for other types of WRs, CRs, or the UDMUs approach. For example, relative WRs (ARI) by linking input or output weights, reflect trade-offs between inputs or outputs. When restrictions on these trade-offs are imposed, the whole or part of the frontier changes, and the RTS characteristics on the frontier may also change. Note that, as we highlighted before, the inclusion of price information through ARIs implies a movement from technical to economic efficiency. This means that the measurement of RTS under ARIs might imply, in the extreme case, where prices are perfectly known, the calculation of RTS characteristics across a cost or revenue plane. The meaning of such a calculation is obviously dubious, as RTS is a technological characteristic and not a cost or revenue related characteristic.

Hence, it is necessary to question the inclusion of value judgments in light of the changes these may introduce on the efficient frontier, and on its RTS properties. Further, WRs can be used to model different RTS properties of the frontier. Imagine for example that for a given production process it is known that the technology exhibits always DRS. The imposition of suitable WRs can be used as a means to arrive at an efficient frontier with global DRS. Clearly the link between WRs and RTS is one where further research is in need.

9.   CONCLUSION AND DEVELOPMENTS

The incorporation of value judgments in DEA models may be motivated by several goals and may be applied through very different methods. None of the methods outlined in this chapter is all-purpose and none is free of disadvantages. In the review of Allen et al. (1997) it was pointed out that the mathematical and managerial implications of introducing value judgments have yet to be explored in full. This statement has not lost currency, and more is still to be done concerning this issue. In particular the interpretation of the efficiency rating resulting from weights restricted models, or related models that pass through changing the data set, is still a contemporary issue, as the loss of a radial meaning makes these scores difficult to communicate to managers. Some WRs approaches have been fully explored and applied in the literature, such as ARs, CRs and absolute WRs. Others, such as restrictions on virtual inputs and outputs, have not been fully applied or explored. The UDMUs approach has the advantage over other approaches of involving the DM in the efficiency assessment process. This encourages the DM to regard UDMUs as genuinely attainable DMUs that can be used to set more demanding targets for DMUs.

The choice of the method to be used for including value judgments should be done in light of the objectives of the analysis. If the objective is simply to prevent DMUs from assigning a zero or ε-value to some variables, then the analyst should bear in mind that weights restrictions do not necessarily prevent this from happening. In this context, it is important to note that other methods are at hand to handle this problem, which do not require the explicit participation of the DM. However, involving the DM in the process of analysing unrestricted results and reaching some conclusions about ways to restrict the models, for example, to reach more discrimination between DMUs, is a valuable process where both the analyst and the DM have much to gain.

REFERENCES

1. Allen, R., A. Athanassopoulos, R. G. Dyson and E. Thanassoulis, 1997, Weights restrictions and value judgments in data envelopment analysis: Evolution, development and future directions, *Annals of Operations Research* 73, 13-34.
2. Allen, R. and E. Thanassoulis, Improving envelopment in data envelopment analysis, *Forthcoming in EJOR*.
3. Ali, A. I., W. D. Cook, L. M. Seiford, 1991, Strict vs. weak ordinal relations for multipliers in data envelopment analysis, *Management Science* 37, 733-738.
4. Andersen, P. and N. Petersen, 1993, A procedure for ranking efficient units in data envelopment analysis, *Management Science* 39/10, 1261-1264.

5. Anderson, T., K. Hollinsgsworth, L. Inman, 2002, The fixed weighting nature of a cross-evaluation model, *Journal of Productivity Analysis* 17/3, 249-255.
6. Beasley, J. E., 1990, Comparing University departments, *Omega, The International Journal of Management Science* 18/2, 171-183.
7. Bessent A., W. Bessent, J. Elam and T. Clark, 1988, Efficiency frontier determination by constrained facet analysis, *Journal of Operational Research Society* 36/5, 785-796.
8. Boussofiane, A., R. G. Dyson and E. Thanassoulis, 1991, Applied data envelopment analysis, *European Journal of Operational Research* 52/1, 1-15.
9. Brockett, P. L., A. Charnes, W. W. Cooper, Z. M. Huang, D. B. Sun, 1997, Data transformations in DEA cone ratio envelopment approaches for monitoring bank performances, *European Journal of Operational Research* 98, 250-268.
10. Charnes, A., W. W. Cooper, and E. Rhodes, 1978, Measuring efficiency of decision making units, *European Journal of Operational Research* 2, 429-444.
11. Charnes, A., W. W. Cooper, Q. L. Wei and Z. M. Huang, 1989, Cone ratio data envelopment analysis and multi-objective programming. *International Journal of Systems Science* 20/7, 1099-1118.
12. Charnes, A., W. W. Cooper, Z. M. Huang, and D. B. Sun, 1990, Polyhedral Cone-Ratio DEA models with an illustrative application to large industrial Banks, *Journal of Econometrics* 46, 73-91.
13. Charnes, A., W. W. Cooper, A. Y. Lewin, and L. W. Seiford, 1994, Data Envelopment Analysis: Theory, Methodology and Applications, Kluwer Academic Publishers, Dordrecht.
14. Cooper, W. W., L. M. Seiford and K. Tone, 2000, Data Envelopment Analysis: A comprehensive text with models, applications, references and DEA-Solver software, Kluwer Academic Publishers.
15. Coelli, T., D. S. P. Rao and G. E. Battese, 1998, An Introduction to Efficiency and Productivity Analyisis, Kluwer Academic Publishers, Boston.
16. Cook, W. D., A. Kazakov, Y. Roll and L. M. Seiford, 1991, A data envelopment analysis approach to measuring efficiency: Case analysis of highway maintenance patrols, *The Journal of Socio-Economics* 20/1, 83-103.
17. Cook, W. D., M. Kress and L. M. Seiford, 1992, Prioritisation models for frontier decision making units in DEA, *European Journal of Operational Research* 59, 319-323.
18. Cook, W. D., A. Kazakov, Y. Roll, 1994, On the measurement and monitoring of relative efficiency of Highway maintenance patrols. In *Data Envelopment Analysis, Theory, Methodology and Applications*, Charnes A., Cooper W. W. Lewin A.Y. and Seiford L.M. eds, Kluwer Academic Publishers, 195-210.
19. Dyson R. G. and E. Thanassoulis, 1988, Reducing weight flexibility in data envelopment analysis, *Journal of the Operational Research Society* 39/6, 563-576.
20. Golany, B., 1988, A note on including ordinal relations among multipliers in data envelopment analysis, *Management Science* 34, 1029-1033.
21. Golany, B. and Y. Roll, 1989, An application procedure for DEA, *Omega: The International Journal of Management Science* 17/3, 237-250.
22. Golany, B. and Y. Roll, 1994, Incorporating Standards in DEA. In *Data Envelopment Analysis: Theory, Methodology and Applications*, Charnes A., Cooper W. W., Lewin A. Y. and Seiford L. W. eds, Kluwer Academic Publishers, 393-422.

23. Green, R.H., J. R. Doyle and W. D. Cook, 1996a, Efficiency bounds in data envelopment analysis, *European Journal of Operational Research* 89, 482-490.
24. Green, R. H, J. R. Doyle and W. D. Cook, 1996b, Preference voting and project ranking using DEA and cross-evaluation, *European Journal of Operational Research* 90, 461-472.
25. Halme, M., T. Joro, P. Korhonen, S. Salo and J. Wallenius, 1999, A value efficiency approach to incorporating preference information in data envelopment analysis, *Management Science* 45/1, 103-115
26. Korhonen, P., M. Soismaa and A. Siljamäki, 2002, On the use of value efficiency analysis and some further developments, *Journal of Productivity Analysis* 17, 49-65.
27. Lang, P., O. R. Yolalan and O. Kettani, 1995, Controlled envelopment by face extension in DEA, *Journal of Operational Research Society* 46/4, 473-491.
28. Li, X.B., G. R. Reeves, 1999, A multiple criteria approach to data envelopment analysis, *European Journal of Operational Research* 115/3, 507-517
29. Meza, L. A. and M. P. E. Lins, 2002, Review of methods for increasing discrimination in data envelopment analysis, *Annals of Operational Research* 116, 1-4, 225-242.
30. Olesen, O.B. and N. C. Petersen, 1996, Indicators of ill-conditioned data sets and model misspecification in data envelopment analysis: An extended facet approach, *Management Science* 42/2, 205-219.
31. Olesen, O. B. and N. C. Petersen, 1999, Probabilistic bounds on the virtual multipliers in data envelopment analysis: Polyhedral cone constraints, *Journal of Productivity Analysis* 12, 103-133.
32. Olesen, O.B. and N. C. Petersen, 2002, Identification and use of efficient faces and facets in DEA, *Forthcoming in Journal of Productivity Analysis.*
33. Olesen, O. B. and N. C. Petersen, 2002, The use of data envelopment analysis with probabilistic assurance regions for measuring hospital efficiency, *Journal of Productivity Analysis* 17, 83-109.
34. Pedraja-Chaparro, F., J. Salinas-Jimenez and P. Smith, 1997, On the role of weight restrictions in data envelopment analysis, *Journal of Productivity Analysis* 8, 215-230.
35. Podinovski, V. V. and A. D. Athanassopoulos, 1998, Assessing the relative efficiency of decision making units using DEA models with weights restrictions, *Journal of the Operational Research Society* 49/5, 500-508.
36. Podinovski, V.V., 1999, Side effects of absolute weight bounds in DEA models, *European Journal of Operational Research* 115/3, 583-595.
37. Podinovski, V.V., 2001, DEA models for the explicit maximisation of relative efficiency, *European Journal of Operational Research* 131,572-586.
38. Podinovski, V.V., 2003, Suitability and redundancy of non-homogeneous weight restrictions for measurinf the relative efficiency in DEA, *European Journal of Operational Research,* forthcoming.
39. Portela, M.C.A.S. and E. Thanassoulis, 2002, Profit efficiency in DEA. Aston Business School Research Paper RP 0206, ISBN 1 85449 502 X, University of Aston, Aston Triangle, Birmingham B4 7ET, UK.
40. Ray, S. C., L. M. Seiford and J. Zhu, 1998, Market Entity behavior of Chinese state-owned enterprises, *Omega, The International Journal of Management Science* 26/2, 263-278.
41. Roll, Y., W. D. Cook, and B. Golany, 1991, Controlling factor weights in data envelopment analysis, *IIE Transactions* 23/1, 2-9.

42. Roll, Y. and B. Golany, 1993, Alternate methods of treating factor weights in DEA, *Omega, The International Journal of Management Science* 21/1, 99-109.
43. Sarrico, C. S., 1999, Performance measurement in UK universities: Bringing in the stakeholders' perspectives using Data Envelopment Analysis, PhD thesis, Warwick Business school, University of Warwick, Coventry CV4 7AL, UK.
44. Seiford, L. M. and J. Zhu, 2002, Value judgment versus allocative efficiency: a case of Tenesse county jails, *The Journal of Management Sciences &Regional Development* 4, 89-98.
45. Schaffnit, C., D. Rosen and J. C. Paradi, 1997, Best practice analysis of bank branches: An application of DEA in a large Canadian bank, *European Journal of Operational Research* 98, 269-289.
46. Taylor, W.M., R. G. Thompson, R. M. Thrall and P. S. Dharmapala, 1997, DEA/AR efficiency and profitability of Mexican banks. A total income model, *European Journal of Operational Research* 98, 346-363.
47. Thanassoulis, E., R. G. Dyson, and M. J. Foster, 1987, Relative efficiency assessments using data envelopment analysis: An application to data on rates departments, *Journal of Operational Research Society* 38/5, 397-411.
48. Thanassoulis, E., and R. G. Dyson, 1992, Estimating preferred input-output levels using data envelopment analysis, *European Journal of Operational Research* 56, 80-97.
49. Thanassoulis, E., 1995, Assessing Police forces in England and Wales using data envelopment analysis, *European Journal of Operational Research* 87, 641-657.
50. Thanassoulis, E., A. Boussofiane, R. G. Dyson, 1995, Exploring output quality targets in the provision of perinatal care in England using data envelopment analysis, *European Journal of Operational Research* 80, 588-607.
51. Thanassoulis, E., 1997, Duality in Data Envelopment Analysis under constant returns to scale. *IMA Journal of Mathematics Applied in Business and Industry* 8/3, 253-266.
52. Thanassoulis, E. and R. Allen, 1998, Simulating weights restrictions in data envelopment analysis by means of unobserved DMUs, *Management Science* 44/4, 586-594.
53. Thanassoulis, E., 2001, Introduction to the theory and application of Data Envelopment analysis: A foundation text with integrated software. Kluwer Academic Publishers.
54. Thompson, R. G., F. D. Singleton, Jr, R. M. Thrall, and B.A. Smith, 1986, Comparative site evaluations for locating a high-energy physics lab in Texas. *Interfaces* 16, 35-49.
55. Thompson, R. G., L. N. Langemeier, C. Lee, E. Lee, and R. M. Thrall, 1990, The role of multiplier bounds in efficiency analysis with application to Kansas farming, *Journal of Econometrics* 46, 93-108.
56. Thompson, R. G., E. Lee and R. M. Thrall, 1992, DEA/AR-efficiency of U.S. independent oil/gas producers over time, *Computers and Operations Research* 19/5, 377-391.
57. Thompson, R. G. and R. M. Thrall, 1994, Polyhedral assurance regions with linked constraints. In *New Directions in Computational Economics*, Cooper, W.W. and Whinston, A.B. eds. Kluwer Academic Publishers, 121-133.
58. Thompson, R. G., P. S. Dharmapala and R. M. Thrall, 1995, Linked-cone DEA profit ratios and technical efficiency with application to Illinois Coal mines, *International Journal of Production Economics* 39, 99-115.

59. Thompson, R. G., P. S. Dharmapala, L. J. Rothenberg, and R. M. Thrall, 1996, DEA/AR efficiency and profitability of 14 major oil companies in US exploration and production, *Computers & Operations Research* 23/4, 357-373.

60. Thompson, R. G., E. J. Brinkmann, P. S. Dharmapala, M. D. Gonzalez-Lima, R. M. Thrall, 1997, DEA/AR profit ratios and sensitivity of 100 large U.S. banks, *European Journal of Operational Research* 98, 213-229.

61. Tone, K., 2001, On returns to scale under weight restrictions in data envelopment analysis, *Journal of Productivity Analysis* 16/1, 31-47.

62. Varian, H.R., 1992, Microeconomic analysis, W.W. Norton and Company, 3rd edition.

63. Wong, Y-H. B., and J. E. Beasley, 1990, Restricting weight flexibility in data envelopment analysis, *Journal of Operational Research Society* 41/9, 829-835.

64. Zhu, J., 1996a, Data Envelopment Analysis with preference structure, *Journal of the Operational Research Society* 47, 136-150.

65. Zhu, J., 1996b, DEA/AR analysis of the 1988-1989 performance of the Nanjing Textiles Corporation, *Annals of Operations Research* 66, 311-335.

Chapter 5

DISTANCE FUNCTIONS
With Applications to DEA

Rolf Färe[1], Shawna Grosskopf[2] and Gerald Whittaker[3]

[1] *Department of Economics and Department of Agricultural Economics, Oregon State University, Corvallis, OR 97331 USA*

[2] *Department of Economics, Oregon State University, Corvallis, OR 97331 USA email: shawna.grosskopf@orst.edu*

[3] *National Forage Seed Production Research Center, Agricultural Research Sevice, USDA, Corvallis, OR 97331 USA*

Abstract: Duality between distance functions and support functions is shown to be the basis for performance measures and their decompositions. DEA may be used to evaluate the measures.

Key words: Data envelopment analysis (DEA); Distance functions; Support functions

1.   INTRODUCTION

Distance functions can be viewed as theoretical tools in duality theory and as applied tools in performance measurement. As an example we have the input distance function which, as Shephard[1] showed, is dual to the cost function. This function or its reciprocal is also known as that Debreu /Farrell[2] measure of technical efficiency. In DEA terminology it is the input oriented measure of technical efficiency.

---

[1] Shephard (1953).
[2] Debreu (1951), Farrell (1957).

In DEA one frequently distinguishes between input and output orientations, i.e., between input and output distance functions. These functions are reciprocal to each other if and only if the parent technology exhibits constant returns to scale[3]. In general though, they are associated, through duality theory, with two different optimization problems. The input distance function is associated with cost minimization and output distance function with revenue maximization. These associations make them component measures in the cost and revenue measures of overall efficiency. In particular the cost measure of overall efficiency can be multiplicatively decomposed into an input technical (input distance function) and an input allocative component[4]. A similar decomposition holds for the revenue measure.

Until recently, however, there was no known simple dual relationship between distance functions and the profit function. Profit as the difference between revenue and cost did not have the simple multiplicative structure of cost, revenue and traditional input and output distance functions. The missing link is what we call the directional distance function. This function has the hybrid structure that provides the dual to the profit function and the basis for a decomposition into allocative and technical efficiency components. It also yields the traditional distance functions as special cases.

The direction in which the traditional input and output distance functions evaluate efficiency is determined by the input-output data from an observation or DMU (decision making unit) itself. In contrast, for the directional distance function, the direction in which DMU's are to be evaluated is a choice variable, i.e., these distance functions allow the researcher to choose in which direction to estimate technical efficiency. This may be an important consideration in some applications. For example, when desirable and undesirable (polluting) outputs are jointly produced, a natural choice would be in the direction of more desirable output and less polluting output. This option is easily modeled with a directional distance function.

2.   DISTANCE FUNCTIONS AND DUALITY

Shephard (1953) was the first economist to recognize the duality between a distance function and support function[5]. In particular, he proved that input distance function and the cost function are duals. In practice this means that all information about technology can be derived from either of the two

---

[3] Färe and Lovell (1978).
[4] These functions are contained in what Luenberger calls the shortage function (Luenberger, 1992, 1995).
[5] See Mahler (1939).

functions, given some regularity conditions. Shephard, however, did not attempt to use the distance functions as an estimate of technical inefficiency. This was left to Debreu (1951), Farrell (1957) and Charnes, Cooper and Rhodes (1978).

In this section we follow Shephard and discuss duality. In the next section we show how these theories apply to the estimation of efficiency. Let $x \in \Re_+^N$ denote inputs and $y \in \Re_+^M$ outputs. The technology is given by

$$T = \{(x, y) : x \text{ can produce } y\} \tag{5.1}$$

We assume that T meets the standard set of axioms sufficient for duality theory. This means that T is a closed convex set with inputs and outputs freely disposable[6].

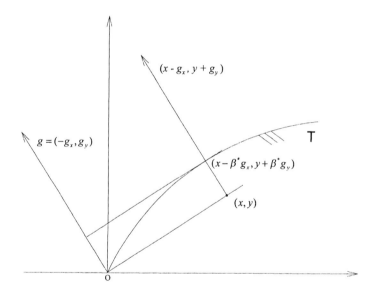

Figure 5-1. The directional distance function with a single input and output.

Let $g = (-g_x, g_y)$ be a directional vector with $-g_x \in \Re_-^N$ and $g_y \in \Re_+^M$. Then the directional technology distance function is defined as

$$\overrightarrow{D_T}\left(x, y; -g_x, g_y\right) = \max\{\beta : \left(x - \beta g_x, y + \beta g_y\right) \in T\} \tag{5.2}$$

[6] See Färe and Primont (1995) for details.

if there exists a $\beta \in \Re$ such that $(x - \beta g_x, y + \beta g_y) \in T$ and $+\infty$ otherwise[7]. This function is illustrated in Figure 5-1.

The technology is represented by T, and the directional vector $g = (-g_x, g_y)$ is in the fourth quadrant. The distance function projects the input-output vector (x, y) onto the frontier of T at $(x - \beta^* g_x, y + \beta^* g_y)$, where $\overrightarrow{D_T}(x, y; -g_x, g_y) = \beta^*$. This distance function contracts input and expands output, which differs from the input and output distance functions. They separately contract inputs or expand outputs.

The directional distance function inherits its properties from the technology T, and under the above conditions represents T in the sense of

$$\overrightarrow{D_T}\left(x, y; -g_x, g_y\right) \geq 0 \text{ if and only if } (x, y) \in T \tag{5.3}$$

That is, the distance function is nonnegative provided the input-output vector (x, y) is feasible.

From its definition, the distance function possesses the translation property, i.e.,

$$\overrightarrow{D}\left(x - \alpha g_x, y + \alpha g_y; -g_x, g_y\right) = \overrightarrow{D_T}\left(x, y; -g_x, g_y\right) - \alpha, \quad \alpha \in \Re. \tag{5.4}$$

This condition states that if the input output vector (x, y) is translated to $(x - \alpha g_x, y + \alpha g_y)$, then the value of the function decreases by α. This property is the analog to homogeneity of the input and output distance functions[8].

To explore the duality between the directional technology distance function and the profit function, let $p \in \Re_+^M$ denote output prices and $w \in \Re_+^N$ input prices. The profit function is defined as[9]

$$\Pi(p, w) = \max\{py - wx : (x, y) \in T\}. \tag{5.5}$$

When it exists, it models maximal feasible profit for the price vector (p, w). This means that

$$\Pi(p, w) \geq py - wx \text{ for all } (x, y) \in T. \tag{5.6}$$

---

[7] This distance function was introduced by Luenberger (1992) where he called it the *shortage function*. Here we follow the terminology of Chambers, Chung and Färe (1998).

[8] See Färe and Grosskopf (2000) for details.

[9] For details see Färe and Primont (1995).

Recall from the definition of the distance function that

$$\left(x - \overrightarrow{D_T}\left(x, y; -g_x, g_y\right)g_x, y + \overrightarrow{D_T}\left(x, y; -g_x, g_y\right)g_y\right) \in T. \quad (5.7)$$

Thus by combining the last two expressions we obtain

$$\frac{\Pi(p, w) - (py - wx)}{pg_y + wg_x} \geq \overrightarrow{D_T}\left(x, y; -g_x, g_y\right). \quad (5.8)$$

Following Chambers, Chung and Färe (1998), we term the LHS of (5.8) the Nerlovian indicator of profit efficiency, and the RHS is the Nerlovian technical efficiency component. We may close the inequality by adding an allocative efficiency component $\overrightarrow{AE_T}$ to obtain

$$\frac{\Pi(p, w) - (py - wx)}{pg_y + wg_x} = \overrightarrow{D_T}\left(x, y; -g_x, g_y\right) + \overrightarrow{AE_T}. \quad (5.9)$$

Thus profit efficiency can be decomposed into a technical and allocative component, where the technical component is the directional distance function and the allocative component is the residual that takes (5.8) into the equality expression (5.9). As an efficiency measure we note that

$$\frac{\Pi(p, w) - (py - wx)}{pg_y + wg_x}$$

is non-negative with efficiency signalled when the measure equals zero. The same interpretations can be given to the technical component $\overrightarrow{D_T}(x, y; -g_x, g_y)$ and the allocative component $\overrightarrow{AE_T}$.

In a strict sense duality theory tells us that the primal and dual expressions of the technology can be derived from each other. In our case the profit function should be derived from the distance function, which in turn should be recovered from the profit function. To show that this is the case, apply (5.8) twice;

$$\Pi(p, w) = \max_{(x, y)} p\left(y + \overrightarrow{D_T}\left(x, y; -g_x, g_y\right)g_y\right) \\ - w\left(x - \overrightarrow{D_T}\left(x, y; -g_x, g_y\right)g_x\right) \quad (5.10a)$$

$$\overrightarrow{D_T}\left(x, y; -g_x, g_y\right) = \min_{(p,w)} \frac{\Pi(p,w) - (py + wx)}{pg_y + wg_x}. \qquad (5.10b)$$

The first of these expressions shows how the profit function is derived from the distance function, and the second shows how the distance function is recovered from the profit function. Thus we see that duality theory is the foundation for the Nerlovian indicator, and its decomposition.

Let us next restrict the directional vector by first choosing it to be $g = (-g_x, 0)$ and then $g = (-x, 0)$. In the first case with $g = (-g_x, 0)$, the directional input distance function is generated. This function is defined by

$$\overrightarrow{D_i}\left(x, y; -g_x\right) = \max\left\{\beta; (x - \beta g_x, y) \in T\right\}. \qquad (5.11)$$

It is a special case of the technology distance function, namely

$$\overrightarrow{D_i}(x, y; -g_x) = \overrightarrow{D_T}(x, y; -g_x, 0). \qquad (5.12)$$

If we set $g = (-x, 0)$, then we get the input distance function

$$\overrightarrow{D_i}\left(x, y; -x\right) = 1 - 1/D_i(y, x), \qquad (5.13)$$

where the input distance function is

$$D_i(y, x) = \max\left\{\lambda : (x/\lambda, y) \in T\right\}. \qquad (5.14)$$

To prove (5.13), we notice first that the input distance function (5.14) is homogeneous of degree +1 in inputs and that it represents technology, i.e.,

$$D_i\left(y, x\right) \geq 1 \text{ if and only if } (x, y) \in T. \qquad (5.15)$$

Using homogeneity, by definition (5.11) and the representation condition (5.15) we find that

$$\vec{D}_i(x,y:-x) = \max\{\beta;(x-\beta x,y)\in T\}$$

$$= \max\{\beta:D_i(y,x(1-\beta))\geq 1\}$$

$$= \max\{1-1+\beta:D_i(y,x)(1-\beta)\geq 1\} \qquad (5.16)$$

$$= 1-\min\{1-\beta:1-\beta\geq 1/D_i(y,x)\}$$

$$= 1-1/D_i(y,x),$$

i.e., (5.13) holds.

The last two distance functions in conjunction with the inequality (5.8) yield two additional measures of efficiency, one additive and one multiplicative. The additive measure we call the *cost indicator* of efficiency, and multiplicative measure we call the *cost index* of efficiency[10]. The cost indicator is given by

$$\frac{wx-C(y,w)}{wg_x} \geq \vec{D}_i(x,y;-g_x) \qquad (5.17)$$

where the cost function $C(y,w)$ is defined as

$$C(y,w) = \min\{wx:(x,y)\in T\}. \qquad (5.18)$$

To see that the cost indicator is a special case of the profit indicator, denote

$$\Pi(p,w) = py^* - wx^*,$$

where "*" denotes optimality. If the observed output y is also optimal, and $g_y = 0$, then it follows from (5.8) and (5.12) that

$$\frac{py^* - wx^* - py^* + wx}{wg_x} \geq \vec{D}_i(x,y;-g_x), \qquad (5.19)$$

and with $C(y,w) = wx^*$, (5.17) follows. We may close the inequality (5.17) by adding an allocative efficiency component, i.e.,

---

[10] The terminology, *indicator* and *index*, is due to Diewert (1998).

$$\frac{wx - C(y, w)}{w g_x} = \overrightarrow{D_i}(x, y; -g_x) + \overrightarrow{AE_i}. \qquad (5.20)$$

Thus we have derived a cost efficiency indicator with an additive decomposition. Recall that the Farrell decomposition is multiplicative. We next show how it is related to (5.17).

Take $g_x = x$, then by (5.16) and (5.17) we find

$$\frac{wx - C(y, w)}{wx} \geq 1 - 1/D_i(y, x) \qquad (5.21)$$

or

$$\frac{C(y, w)}{wx} \leq 1/D_i(y, x). \qquad (5.22)$$

The last inequality may be closed by multiplying (5.22) with an allocative efficiency component, i.e.,

$$\frac{C(y, w)}{wx} = \frac{1}{D_i(y, x)} \cdot AE_i \qquad (5.23)$$

This expression is the original Farrell measure of cost efficiency decomposed multiplicatively into a technical component $(1/D_i(y, x))$ and an allocative residual AE_i. We refer to them as efficiency indexes.

By now it is clear that the allocative efficiency components in (5.9), (5.20) and (5.23) are just residuals that close the associated inequalities. Thus, let us focus on the inequalities, in particular let us compare (5.17) and (5.22). The latter is referred to as a Mahler inequality. The apparent observation is that the cost function has two duals, the directional and the Shephard input distance functions. The first has an additive structure while the second is multiplicative. This opens the door to two alternative approaches to efficiency estimation - one additive and one multiplicative. The researcher needs to make the choice, which of course depends on the research question at hand.

On the output side we also have a choice between an additive and multiplicative approach. To verify this let the directional vector be $g = (0, g_y)$ and $g = (0, y)$ respectively. Then from the directional technology distance function we obtain

$$\overrightarrow{D}_o\left(x,y;g_y\right)=\overrightarrow{D}_T\left(x,y;0,g_y\right) \tag{5.24}$$

and

$$\overrightarrow{D}_o\left(x,y;y\right)=1/D_o\left(x,y\right)-1. \tag{5.25}$$

These expressions, together with the assumption[11] that observed cost wx is minimum cost, can be substituted into our Nerlovian profit inequality to yield

$$\frac{R(x,p)-py}{pg_y}\geq\overrightarrow{D}_o\left(x,y;g_y\right) \tag{5.26}$$

and if $g_y = y$,

$$\frac{R(x,p)}{py}\geq 1/D_o\left(x,y\right) \tag{5.27}$$

where the revenue function $R(x,p)$ is defined as

$$R(x,p)=\max\left\{py:\left(x,y\right)\in T\right\} \tag{5.28}$$

and the output distance function $D_o(x,y)$ is

$$D_o\left(x,y\right)=\min\left\{\theta:\left(x,y/\theta\right)\in T\right\}, \tag{5.29}$$

If we add an allocative component to (5.26) and a multiplicative allocative component to (5.22), then we get a pair of output oriented measures of revenue efficiency. The first is the additive indicator and the second the multiplicative index;

$$\frac{R(x,p)-py}{pg_y}=\overrightarrow{D}_o\left(x,y;g_y\right)+\overrightarrow{AE}_o \tag{5.30}$$

and

---

[11] Of course, one may directly derive the two expression without this assumption.

$$\frac{R(x,p)}{py} = \frac{1}{D_o(x,y)} \cdot AE_o . \qquad (5.31)$$

The multiplicative index parallels the traditional Farrell decomposition developed on the cost side. The additive indicator provides an alternative which gives the analyst more flexibility through the choice of g_y.

3.   DEA ESTIMATIONS OF DISTANCE FUNCTIONS

We next turn to the DEA estimations of the various distance functions discussed in Section 2. For completeness, we show how their support functions - profit, revenue and cost - also may be estimated via DEA. Thus in this section we provide the framework for estimating the efficiency indicators and indexes introduced in Section 2.

To formulate our DEA model, suppose there are $k = 1,...,K$ observations, DMU's (Decision Making Units) of inputs $x^k = (x_{k1},...,x_{kN}) \in \Re_+^N$ and outputs $y^k = (y_{k1},...,y_{kM}) \in \Re_+^N$. The technology T may be built from the data as

$$T = \{(x,y): \sum_{k=1}^{K} z_k y_{km} \geq y_m, \qquad m = 1,...,M,$$

$$\sum_{k=1}^{K} z_k x_{kn} \leq x_n, \qquad n = 1,...,N, \qquad (5.32)$$

$$z_n \geq 0, \quad k = 1,...,K \},$$

where the data satisfies the Kemeny, Thompson and Morgenstern (1953) conditions

$$i. \quad \sum_{k=1}^{K} x_{kn} > 0, \qquad n = 1,...,N,$$

$$ii. \quad \sum_{n=1}^{N} x_{kn} > 0, \qquad k = 1,...,K,$$

$$\qquad (5.33)$$

$$iii. \quad \sum_{k=1}^{K} y_{km} > 0, \qquad m = 1,...,M,$$

$$iv. \quad \sum_{m=1}^{M} y_{km} > 0, \qquad k = 1,...,K,$$

In words, (i) says that each input must be used by at least one DMU, and each DMU or activity must use at least one input (ii). Conditions (iii) and (iv) mimic (i) and (ii) for outputs.

The technology (5.32) satisfies strong disposability of inputs, i.e.,

$$\text{if } (x, y) \in T \text{ and } x' \geq x \text{ then } (x', y) \in T$$

and strong disposability of outputs, i.e.,

$$\text{if } (x, y) \in T \text{ and } y' \leq y \text{ then } (x, y') \in T.$$

Moreover T is convex, i.e.

$$(x, y) \text{ and } (x', y') \in T \text{ imply } (\lambda x + (1 - \lambda)x', \lambda y + (1 - \lambda)y') \in T.$$

It exhibits constant returns to scale, i.e., $\lambda T = T$, $\lambda > 0$, and it is a closed set.

Since under constant returns to scale, profit is zero, we may wish to relax the returns to scale by restricting the intensity variables z_k, $k = 1, ..., K$. In the case that

$$\sum_{k=1}^{K} z_k = 1, \tag{5.34}$$

we say that the technology satisfies variable returns to scale and if

$$\sum_{k=1}^{K} z_k \leq 1 \tag{5.35}$$

it satisfies non-increasing returns to scale. In the first case profit may be negative, zero or positive, in the non-increasing case profit is non-negative.

We start with the estimation of distance functions. The directional technology distance function is estimated for DMU k' by solving the linear programming problem[12]

$$\overrightarrow{D_T}(x^{k'}, y^{k'}; -g_x, g_y) = \max \beta$$

subject to

$$\sum_{k=1}^{K} z_k y_{km} \geq y_{k'm} + \beta g_{y_m}, \quad m = 1, ..., M, \tag{5.36}$$

$$\sum_{k=1}^{K} z_k x_{km} \leq x_{k'n} - \beta g_{x_n}, \quad n = 1, ..., N,$$

$$z_k \geq 0, \quad k = 1, ..., K.$$

---

[12] The software *OnFront 3* does these estimations, see www.Appliedeconomics.se.

In addition, if so required, one may add the restrictions (5.34) or (5.35). The directional vector $g = (-g_x, g_y)$ should be specified, e.g., one may take $g = (-1,1)$.

If we next choose $g = (-g_x, 0)$ then (5.36) turns into the estimation of (5.12) - the directional input distance function. If we set $g = (0, g_y)$ then (5.36) is the estimation of the directional output distance function (5.24). Again, g_x and g_y must be specified by the analyst.

The standard Shephard input function is estimated as

$$\left(D_i(y^{k'}, x^{k'})\right)^{-1} = \min \lambda$$

subject to

$$\sum_{k=1}^{K} z_k y_{km} \geq y_{k'm}, \qquad m = 1,...,M \tag{5.37}$$

$$\sum_{k=1}^{K} z_k x_{kn} \leq \lambda x_{k'n}, \qquad n = 1,...,N$$

$$z_n \geq 0, \quad k = 1,...,K$$

along with the appropriate scale restrictions on the intensity variables.

The output distance function is estimated as

$$\left(D_o(x^{k'}, y^{k'})\right)^{-1} = \max \theta$$

subject to

$$\sum_{k=1}^{K} z_k y_{km} \geq \theta y_{k'm}, \qquad m = 1,...,M \tag{5.38}$$

$$\sum_{k=1}^{K} z_k x_{kn} \leq x_{k'n}, \qquad n = 1,...,N$$

$$z_k \geq 0, \quad k = 1,...,K,$$

with suitable restrictions on z_k.

Suppose we know the input and output prices (w^k, p^k) respectively. Then we may estimate maximal profit, maximal revenue and mininal cost, again using DEA. Profit maximization is given by

$$\Pi\left(p^{k'}, w^{k'}\right) = \max_{(y_m, x_n)} \sum_{m=1}^{M} p_m^{k'} y_m - \sum_{n=1}^{N} w_n^{k'} x_n$$

subject to

$$\sum_{k=1}^{K} z_k y_{km} \geq y_m, \qquad m = 1, ..., M \tag{5.39}$$

$$\sum_{k=1}^{K} z_k x_{kn} \leq x_n, \qquad n = 1, ..., N$$

$$z_k \geq 0, \quad k = 1, ..., K,$$

and appropriate restrictions on z_k. Notice that the RHS of the output and input inequalities now are decision variables rather than "data". Thus (y^*, x^*) will be part of the solution to this problem.

Revenue maximization is estimated viz,

$$R(x^{k'}, p^{k'}) = \max_{(y_m)} \sum_{m=1}^{M} p_m^{k'} y_m$$

subject to

$$\sum_{k=1}^{K} z_k y_{km} \geq y_m, \qquad m = 1, ..., M, \tag{5.40}$$

$$\sum_{k=1}^{K} z_k x_{kn} \geq x_{k'n}, \qquad n = 1, ..., N,$$

$$z_k \geq 0, \quad k = 1, ..., K$$

with appropriate restrictions on z_k. Hence y_m (but not x) is a choice variable.

Cost minimization is estimated by

$$C(y^{k'}, w^{k'}) = \min_{(y_m)} \sum_{n=1}^{N} w_m^{k'} x_m$$

subject to

$$\sum_{k=1}^{K} z_k y_{km} \geq y_{k'm}, \qquad m = 1, ..., M, \tag{5.41}$$

$$\sum_{k=1}^{K} z_k x_{kn} \geq x_n, \qquad n = 1, ..., N,$$

$$z_k \geq 0, \quad k = 1, ..., K,$$

and, as usual, appropriate restrictions on z_k. In this problem, x_m (but not y) is a choice variable. The DEA models (5.36)-(5.41) allow us to estimate all indicators and indexes introduced in Section 2. The allocative efficiency components may be estimated as residuals.

REFERENCES

1. Chambers, R.G., Y. Chung and R. Färe, 1998, Profit, directional distance functions, and Nerlovian efficiency, Journal of Optimization Theory and Applications 98, 351-364.
2. Charnes, A., W.W. Cooper and E. Rhodes, 1978, Measuring the efficiency of decision making units, European Journal of Operational Research 2, 429-444.
3. Debreu, G., 1951, The coefficient of resource utilization, Econometrica 19, 237-292.
4. Diewert, W.E., 1998, Index number theory using differences rather than ratios, Discussion Paper 98-10 (Department of Economics, University of British Columbia, Vancouver).
5. Färe, R. and S. Grosskopf, 2000, Theory and application of directional distance functions, Journal of Productivity Analysis 13, 93-103.
6. Färe, R. and D. Primont, 1995, Multi-output Production and Duality: Theory and Applications, Kluwer Academic Publishers, Boston.
7. Färe, R. and C.A.K. Lovell, 1978, Measuring the technical efficiency of production, Journal of Economic Theory 19, 150-162.
8. Farrell, M.J., 1957, The measurement of productive efficiency, Journal of the Royal Statistical Society, Series A, General, 120, Part3, 253-281.
9. Kemeny, J.G., O. Morgenstern and G.L. Thompson, 1956, A generalization of the von Neumann model of an expanding economy, Econometrica 24, 115-135.
10. Luenberger, D.G., 1992, New optimality principles for economic efficiency and equilibrium, Journal of Optimization Theory and Applications 75, 221-264.
11. Luenberger, D.G., 1995, Microeconomic Theory, McGraw Hill, New York.
12. Mahler, K., 1939, Ein übertragungs prinzip für konvexe körper, Casopis pro Pěstovănl Matematiky a Fusiky 54, 93-102.
13. Shephard, R.W., 1953, Cost and Production Functions, Princeton University Press, Princeton.
14. Shephard, R.W., 1970, Theory of Cost and Production Functions, Princeton University Press, Princeton.

Chapter 6

QUALITATIVE DATA IN DEA

Wade D. Cook
Schulich School of Business, York University, Toronto, Canada, M3J 1P3
email: wcook@schulich.yorku.ca

Abstract: In many real world applications involving performance measurement, it is
             necessary to deal with qualitative data factors. This chapter discusses the
             modeling of such factors within the DEA structure.

Key words: Data envelopment analysis (DEA); Efficiency; Rank position; Ordinal data;
             Qualitative data

1.   INTRODUCTION

In a wide range of problem settings to which DEA can be applied, particularly in not-for-profit cases, *qualitative* factors are often present. In some situations such factors may be legitimately "quantifiable," but very often such quantification is superficially forced, as a modeling convenience. Typically, a qualitative factor such as management competence, for example, is captured either on a Likert scale, or is represented by some quantitative *surrogate* such as plant downtime or percentage sick days by employees.

It can be the case as well, that purely *quantitative* variables may be such that accurate data is not available, hence figures provided are often rough estimates of the actual data values. In a number of studies of bank and bank branch efficiency, for example, discretionary inputs such as "percentage of high value customers" in the customer base, can be an important influence variable vis-à-vis performance. It reflects investment potential on the part of the customer. See Cook, Hababou and Tuenter (2000) and Cook and Hababou (2001). This variable is, however, generated from disposable income of the customer, for which accurate data is seldom available. For existing branches, a surrogate for such a variable is the level of investment of the customer. For new (planned) branches, the level of investment that would be created can be predicted from income demographics for the customer base for that branch. Such income data is, however, often unreliable.

In situations such as those described, the "data" for certain influence factors (inputs and outputs) might better be represented as rank positions in an ordinal, rather than numerical sense. Refer again to the management competence example. In certain circumstances, the information available may permit one *only* to put each DMU into one of L categories or groups (e.g. 'high', 'medium' and 'low' competence). In other cases, one may be able to provide a complete rank ordering of the DMUs on such a factor.

This chapter examines the modeling of qualitative data in the DEA structure. The following Section (2) discusses two practical problem settings in which qualitative data occurs naturally. In the first, we examine a problem of R&D project ranking and selection, where various non-quantifiable factors need to be considered. In the context of DEA, the projects represent the decision making units. This example is adopted from Cook et al. (1996). In the second example, due to Kim et al. (1999), and Zhu (2003), a mix of ordinal and numerical factors are evaluated. Section 3 examines the radial projection DEA model in the context of ordinal data. Section 4 discusses the application of this ordinal DEA model to the two presented problems. In Section 5, various settings involving ordinal data are discussed. Conclusions and further directions are presented in Section 6.

2. PROBLEM SETTINGS INVOLVING ORDINAL DATA

2.1 Ordinal Data in R&D Project Selection

Consider the problem of selecting R&D projects in a major public utility corporation with a large research and development branch. Research activities are housed within several different divisions, for example, thermal, nuclear, electrical, and so on. In a budget constrained environment in which such an organization finds itself, it becomes necessary to make choices among a set of potential research initiatives or projects that are in competition for the limited resources. To evaluate the impact of funding (or not funding) any given research initiative, two major considerations generally must be made. First, the initiative must be viewed in terms of more than one factor or criterion. Second, some or all of the criteria that enter the evaluation may be qualitative in nature. Even when clearly quantitative factors are involved, such as long term savings to the organization, it may be extremely difficult to obtain even a crude estimate of the value of that factor. The most that one can do in many such situations is to classify the project (according to this factor) on some scale (high/medium/low or say a 5-point scale).

Let us assume that for each qualitative criterion, each initiative is rated on a 5-point scale, where the particular point on the scale is chosen through a consensus on the part of executives within the organization. Table 6-1 presents an illustration of how the data might appear for 10 projects, 3 qualitative output criteria (benefits), identified as 1, 2, and 3, and 3 qualitative input criteria (cost of resources), identified as 4,5, and 6. In the actual setting examined, a number of potential benefit and cost criteria were considered as displayed in Tables 6-2 and 6-3.

We use the convention that for both outputs and inputs, a rating of 1 is "best", and 5 "worst". For outputs, this means that a DMU ranked at position 1 generates *more* output than is true of a DMU in position 2, and so on. For inputs, a DMU in position 1 consumes *less* input than one in position 2.

Table 6-1. Ratings by Criteria

Project No.	Outputs			Inputs		
	1	2	3	4	5	6
1	2	4	1	5	2	1
2	1	1	4	3	5	2
3	1	1	1	1	2	1
4	3	3	3	4	3	2
5	4	3	5	5	1	4
6	2	5	1	1	2	2
7	1	4	1	5	4	3
8	1	5	3	3	3	3
9	5	2	4	4	2	5
10	5	4	4	5	5	5

Regardless of the manner in which such a scale rating is arrived at, the conventional DEA model is capable only of treating the information as if it has cardinal meaning (e.g. something which receives a score of 4 is evaluated as being twice as important as something that scores 2). There are a number of problems with this approach. First and foremost, the projects' original data in the case of some criteria may take the form of an ordinal ranking of the projects. Specifically, the most that can be said about two projects i and j is that i is preferred to j. In other cases it may only be possible to classify projects as say 'high', 'medium' or 'low' in importance on certain criteria. When projects are rated on, say, a 5-point scale, it is generally understood that this scale merely provides a relative positioning of the projects. In a number of agencies investigated (for example, hydro electric and telecommunications companies), 5-point scales are common for evaluating alternatives in terms of qualitative data, and are often accompanied by interpretations such as:

1 = Extremely important
2 = Very important
3 = Important
4 = Low in importance
5 = Not important,

which are easily understood by management. While it is true that market researchers often treat such scales in a numerical (i.e. cardinal) sense, no one seriously believes that an 'extremely important' classification for a project should be interpreted literally as meaning that this project rates three times better than one which is only classified as 'important.' The key message here is that many, if not all criteria, used to evaluate R&D projects are qualitative in nature, and should be treated as such. The model presented in the following sections extends the DEA idea to an ordinal setting, hence accommodating this very practical consideration.

Table 6-2. Potential Benefits

Criteria	Sub-criteria or Interpretation
1. Enhancement of energy efficiency	-development of high yield technologies
	-initiatives which will reduce energy demand
	-development of technologies for utilizing residues
2. Enhancement of diversification/alternative energy sources	-initiatives which provide or strive for new energy sources
	-provide for flexibility in or adaptability of existing and new facilities
3. $Saved internal to organization	-cost reduction devices
	-new technology to replace obsolete equipment
4. Impact on environment	-reduction of emissions into water and atmosphere
	-reduction of risk of nuclear accidents
5. Enhancement to internal technical capability and research profile	- provides training and develops expertise
	-provides technical resources (software, equipment, etc.)
	-builds linkages to external research community.
6. Enhancement to research profile as viewed by the external community	-impact on research status among other utility companies
	-impact on profile abroad
7. Economic impact on external community	-job creation outside organization
	-$ savings to public and industry created by energy efficiency devices
8. Impact on nuclear performance	-influence on nuclear station maintenance, etc.

Table 6-3. Potential Costs

Criteria	Sub-criteria or Interpretation
1. Technical expertise available internally	
2. Technical expertise available externally	consultants
	other research centres
3. Technology available	equipment
	software

2.2   Efficiency Performance of Korean Telephone Offices

Kim et al. (1999) examine 33 telephone offices in Korea and use the following factors to develop performance measures.

Inputs
(1) manpower
(2) operating costs
(3) number of telephone lines

Outputs
(1) local revenues
(2) long distance revenues
(3) international revenues
(4) operation/maintenance level
(5) customer satisfaction.

All inputs and outputs (1),(2),(3) are quantitative, and can be used in the DEA framework in the usual way. Output #4 is, however, ordinal and provides a complete ranking of the 33 DMUs. Output #5 is a categorization of the DMUs on a 5-point Likert scale. Table 6-4 displays the data.

In the section to follow the conventional DEA structure is adapted to accommodate variables measured on an ordinal scale.

3.   MODELING ORDINAL DATA

The above problems typify situations in which pure ordinal data or a mix of ordinal and numerical data are involved in the performance measurement exercise. There appear to be two general approaches in the literature to the handling of ordinal/qualitative data within the DEA framework. The first effort was presented in Cook et al. (1993), (1996). The general approach given below leads ultimately to their model. The second and related effort is that due to Cooper et al. (1999), under the title *imprecise data*. Again, using the general structure given below, one arrives at their model. Rather than adopting, outright, one or the other of these approaches, let us cast the ordinal data problem in a general DEA format. Specifically, consider the situation in which a set of N decision making units (DMUs), k=1,...N are to be evaluated in terms of R_1 numerical outputs, R_2 ordinal outputs, I_1 numerical inputs, and I_2 ordinal inputs. Let $Y_k^1 = (y_{rk}^1)$, $Y_k^2 = (y_{rk}^2)$ denote the R_1- dimensional and R_2-dimensional vectors of outputs, respectively.

Table 6-4. Data for Telephone Offices

DMU No	X1	X2	X3	Y1	Y2	Y3	Y4	Y5
1	239	7.03	158	47.1	16.67	34	28	2
2	261	3.94	163	37.5	14.11	20	26	3
3	170	2.1	90	20.7	6.8	12.6	19	3
4	290	4.54	201	41.8	11.07	6.27	23	4
5	200	3.99	140	33.4	9.81	6.49	30	2
6	283	4.65	214	42.4	11.34	5.16	21	4
7	286	6.54	197	47	14.62	13	9	2
8	375	6.22	314	55.5	16.39	7.31	14	1
9	301	4.82	257	49.2	16.15	6.33	8	3
10	333	6.87	235	47.1	13.86	6.51	6	2
11	346	6.46	244	49.4	15.88	8.87	18	2
12	175	2.06	112	20.4	4.95	1.67	32	5
13	217	4.11	131	29.4	11.39	4.38	33	2
14	441	7.71	214	61.2	25.59	33	16	3
15	204	3.64	163	32.3	9.57	3.65	15	4
16	216	2.24	154	32.8	11.46	9.02	25	2
17	347	5.65	301	59	17.82	8.19	29	1
18	288	4.66	212	42.3	14.52	7.33	24	4
19	185	3.37	178	33	9.46	2.91	7	2
20	242	5.12	270	65.1	24.57	20.7	17	1
21	234	2.52	126	31.6	8.55	7.27	27	2
22	204	4.24	174	32.5	11.15	2.95	22	3
23	356	7.95	299	66	22.25	14.9	13	2
24	292	4.52	236	50	14.77	6.35	12	3
25	141	5.21	63	21.5	9.76	16.3	11	2
26	220	6.09	179	47.9	17.25	22.1	31	2
27	298	3.44	225	42.4	11.14	4.25	4	2
28	261	4.3	213	41.7	11.13	4.68	20	5
29	216	3.86	156	31.6	11.89	10.5	3	3
30	171	2.45	150	24.1	9.08	2.6	10	5
31	123	1.72	61	12	4.78	2.95	5	1
32	89	0.88	42	6.4	3.18	1.48	2	5
33	109	1.35	57	10.6	3.43	2	1	4

Similarly, let $X_k^1 = (x_{ik}^1)$ and $X_k^2 = (x_{ik}^2)$ be the I_1-dimensional and I_2-dimensional vectors of inputs, respectively.

In the situation where all factors are quantitative, the conventional radial projection model for measuring DMU efficiency is expressed by the ratio of weighted outputs to weighted inputs. Adopting the general variable returns to scale (VRS) model of Banker, Charnes and Cooper (1984), and stating it in ratio form, the efficiency of DMU "o" follows from the solution of:

$$e_o = \max \left(\mu_o + \sum_{r \in R_1} \mu_r^1 y_{ro}^1 + \sum_{r \in R_2} \mu_r^2 y_{ro}^2 \right) / \left(\sum_{i \in I_1} \upsilon_i^1 x_{io}^1 + \sum_{i \in I_2} \upsilon_i^2 x_{io}^2 \right)$$

$$\text{s.t.} \left(\mu_o + \sum_{r \in R_1} \mu_r^1 y_{rk}^1 + \sum_{r \in R_2} \mu_r^2 y_{rk}^2 \right) / \left(\sum_{i \in I_1} \upsilon_i^1 x_{ik}^1 + \sum_{i \in I_2} \upsilon_i^2 x_{ik}^2 \right) \le 1, \text{ all } k$$

$$\mu_r^1, \mu_r^2, \upsilon_i^1, \upsilon_i^2 \ge \varepsilon, \text{ all } r, i \qquad (6.1)$$

where $\varepsilon > 0$ is the "non-archimedian infinitesimal" described after (1.2) in Chapter 1.

Problem (6.1) is convertible to the linear programming format:

$$e_o = \max \mu_o + \sum_{r \in R_1} \mu_r^1 y_{ro}^1 + \sum_{r \in R_2} \mu_r^2 y_{ro}^2$$

$$\text{s.t.} \sum_{i \in I_1} \upsilon_i^1 x_{io}^1 + \sum_{i \in I_2} \upsilon_i^2 x_{io}^2 = 1 \qquad (6.2)$$

$$\mu_o + \sum_{r \in R_1} \mu_r^1 y_{rk}^1 + \sum_{r \in R_2} \mu_r^2 y_{rk}^2 - \sum_{i \in I_1} \upsilon_i^1 x_{ik}^1 - \sum_{i \in I_2} \upsilon_i^2 x_{ik}^2 \le 0, \text{ all } k$$

$$\mu_r^1, \mu_r^2, \upsilon_i^1, \upsilon_i^2 \ge \varepsilon, \text{ all } r, i,$$

whose dual is given by

$$\min \theta - \varepsilon \sum_{r \in R_1 U R_2} s_r^+ - \varepsilon \sum_{i \in I_1 U I_2} s_i^-$$

$$\text{s.t.} \quad \sum_{k=1}^{N} \lambda_k y_{rk}^1 - s_r^+ = y_{ro}^1, \ r \in R_1$$

$$\sum_{n=1}^{N} \lambda_k y_{rk}^2 - s_r^+ = y_{ro}^2, \ r \in R_2$$

$$\theta x_{io}^1 - \sum_{k=1}^{N} \lambda_k x_{ik}^1 - s_i^- = 0, i \in I_1 \qquad (6.2')$$

$$\theta x_{io}^2 - \sum_{k=1}^{N} \lambda_k x_{ik}^2 - s_i^- = 0, \ i \in I_2$$

$$\sum_{k=1}^{N} \lambda_k = 1$$

$$\lambda_k, s_r^+, s_i^- \ge 0, \text{ all } k, r, i, \theta \text{ unrestricted}$$

For the problem settings described in the previous section, precise values for outputs in R_2 and inputs in I_2 are not available. Cooper et al. (1999), (2001), and Zhu (2003) refer to this as an example of *imprecise* DEA or IDEA. To place the problem in a general framework, assume that for each ordinal factor ($r \in R_2$, $i \in I_2$), a DMU k can be assigned to one of L rank positions, where $L \le N$. As discussed earlier, L=5, is an example of an appropriate number of rank positions in many practical situations. We point out that in certain application settings, different ordinal factors may have different L-values associated with them. For example, in the problem described in subsection 2.2, 'customer satisfaction,' y_5 is measured on a 5-point scale, while 'operation/maintenance level,' y_4 provides for a full ranking of all 33 DMUs (L=33). For exposition purposes, we assume a common L-value throughout. We demonstrate later that this provides no loss of generality.

In the development below it is assumed that a "full ranking" of all DMUs is available for each ordinal factor. That is, each DMU is assumed to occupy a rank position on each ordinal factor, as opposed to there being only a *partial ranking* of the DMUs on some factor. In Section 5 we discuss a situation where such partial ranking does occur.

One can view the allocation of a DMU to a rank position ℓ on an output r, for example, as having assigned that DMU an output *value* or *worth* $y_r^2(\ell)$. The implementation of the DEA model (6.1) (and (6.2)) thus involves determining two things:

(1) multiplier values μ_r^2, v_i^2 for outputs $r \in R_2$ and inputs $i \in I_2$;
(2) rank position values $y_r^2(\ell)$, $r \in R_2$, and $x_i^2(\ell)$, $i \in I_2$, all ℓ.

Cooper et al. (1999) use a similar format to the one presented here, and approach this problem in a two-stage manner. Their approach for handling *imprecise data* first derives appropriate values (in our notation) for the $y_r^2(\ell)$ and $x_i^2(\ell)$ (i.e., they resolve item (2) above). These values having now been quantified, the conventional DEA model (6.2) can be solved. In this section we show that the problem can be reduced to the standard VRS model by considering items (1) and (2) simultaneously. Further mention of IDEA appears later.

To facilitate development herein, define the L-dimensional unit vectors $\gamma_{rk} = (\gamma_{rk}(\ell))$, and $\delta_{ik} = (\delta_{ik}(\ell))$ where

$$\gamma_{rk}(\ell) = \begin{cases} 1 & \text{if DMU } k \text{ is ranked in } \ell \text{ th position on output } r \\ 0, & \text{otherwise} \end{cases}$$

$$\delta_{ik}(\ell) = \begin{cases} 1 & \text{if DMU } k \text{ is ranked in } \ell \text{ th position on input } i \\ 0, & \text{otherwise} \end{cases}$$

For example, if a 5-point scale is used, and if DMU #1 is ranked in $\ell = 3^{rd}$ place on ordinal output r=5, then $\gamma_{51}(3) = 1$, $\gamma_{51}(\ell) = 0$, for all other rank ?. Thus, y_{51}^2 is assigned the value $y_5^2(3)$, the *worth* to be credited rank position on output factor 5. It is noted that y_{rk}^2 can be in the form

$$y_{rk}^2 = y_r^2(\ell_{rk}) = \sum_{\ell=1}^{L} y_r^2(\ell)\, \gamma_{rk}(\ell),$$

where ℓ_{rk} is the rank position occupied by DMU k on output r. Hence, model (6.2) can be rewritten in the more representative format.

$$e_o = \max \mu_o + \sum_{r \in R_1} \mu_r^1 y_{ro}^1 + \sum_{r \in R_2} \sum_{\ell=1}^{L} \mu_r^2 y_r^2(\ell)\, \gamma_{ro}(\ell)$$

$$\text{s.t.} \quad \sum_{i \in I_1} \upsilon_i^1 x_{io}^1 + \sum_{i \in I_2} \sum_{\ell=1}^{L} \upsilon_i^2 x_i^2(\ell)\, \delta_{io}(\ell) = 1$$

$$\mu_o + \sum_{r \in R_1} \mu_r^1 y_{rk}^1 + \sum_{r \in R_2} \sum_{\ell=1}^{L} \mu_r^2 y_r^2(\ell)\, \gamma_{rk}(\ell) - \sum_{i \in I_1} \upsilon_i^1 x_{ik}^1 -$$

$$\sum_{i \in I_2} \sum_{\ell=1}^{L} \upsilon_i^2 x_i^2(\ell)\, \delta_{ik}(\ell) \le 0, \text{ all k} \qquad (6.3)$$

$$\{Y_r^2 = (y_r^2(\ell)),\ X_i^2 = (x_i^2(\ell))\} \in \Psi$$

$$\mu_r^1,\ \upsilon_i^1 \ge \varepsilon$$

In (6.3) we use the notation Ψ to denote the set of *permissible worth vectors*. We discuss this set below.

It must be noted that the same infinitesimal ε is applied here for the various input and output multipliers, which may, in fact, be measured on scales that are very different from another. If two inputs are, for example, x_{i1k}^1 representing 'labor hours', and x_{i2k}^1 representing 'available computer technology', the scales would clearly be incompatible. Hence, the likely sizes of the corresponding multipliers υ_{i1}^1, υ_{i2}^1 may be similarly different. Thrall (1996) has suggested a mechanism for correcting for such scale incompatibility, by applying a *penalty vector* G to augment ε, thereby creating differential lower bounds on the various υ_i, μ_r. Proper choice of G can effectively bring all factors to some form of common scale or unit. For simplicity of presentation we will assume the cardinal scales for all $r \in R_1$, $i \in I_1$ are similar in dimension, and that G is the unit vector. The more general case would proceed in an analogous fashion.

Permissible Worth Vectors

The values or worths $\{y_r^2(\ell)\}$, $\{x_i^2(\ell)\}$, attached to the ordinal rank positions for outputs r and inputs i, respectively, must satisfy the minimal requirement that it is *more* important to be ranked in P^{th} position than in the $(\ell+1)^{st}$ position on any such ordinal factor. Specifically, $y_r^2(\ell) > y_r^2(\ell+1)$ and $x_i^2(\ell) < x_i^2(\ell+1)$. That is, for outputs, one places a higher weight on being ranked in ℓ^{th} place than in $(\ell+1)^{st}$ place. For inputs, the opposite is true. A set of linear conditions that produce this realization is defined by the set Ψ, where

$$\Psi = \{(Y_r^2, X_i^2) \mid y_r^2(\ell) - y_r^2(\ell+1) \geq \varepsilon, \ \ell=1, \ldots L\text{-}1, \ y_r^2(L) \geq \varepsilon,$$
$$x_i^2(\ell+1) - x_i^2(\ell) \geq \varepsilon, \ \ell=1, \ldots L\text{-}1, x_i^2(1) \geq \varepsilon \}.$$

Arguably, ε could be made dependent upon ℓ (i.e. replace ε by ε_ℓ). It can be shown, however, that all results discussed below would still follow. For convenience, we, therefore, assume a common value for ε.

We now demonstrate that the nonlinear problem (6.3) can be written as a linear programming problem.

Theorem 6.1
Problem (6.3), in the presence of the permissible worth space Ψ, can be expressed as a linear programming problem.

Proof: In (6.3), make the change of variables
$$w_{r\ell}^1 = \mu_r^2 y_r^2(\ell), \ w_{i\ell}^2 = \upsilon_i^2 x_i^2(\ell)$$
It is noted that in Ψ, the expressions
$$y_r^2(\ell) - y_r^2(\ell+1) \geq \varepsilon, \ y_r^2(L) \geq \varepsilon$$
can be replaced by
$$\mu_r^2 y_r^2(\ell) - \mu_r^2 y_r^2(\ell+1) \geq \mu_r^2 \varepsilon, \ \mu_r^2 y_r^2(L) \geq \mu_r^2 \varepsilon,$$
which becomes
$$w_{r\ell}^1 - w_{r\ell+1}^1 \geq \mu_r^2 \varepsilon, \ w_{rL}^2 \geq \mu_r^2 \varepsilon.$$
A similar conversion holds for the $x_i^2(\ell)$.

Problem (6.3) now becomes

$$e_0 = \max \ \mu_0 + \sum_{r \in R1} \mu_r^1 y_{ro}^1 + \sum_{r \in R2} \sum_{\ell=1}^{L} w_{r\ell}^1 \gamma_{ro}(\ell)$$

$$\text{s.t.} \ \sum_{i \in I1} \upsilon_i^1 x_{io}^1 + \sum_{i \in I2} \sum_{\ell=1}^{L} w_{i\ell}^2 \delta_{io}(\ell) = 1$$

$$\mu_0 + \sum_{r \in R1} \mu_r^1 y_{rk}^1 + \sum_{r \in R2} \sum_{\ell=1}^{L} w_{r\ell}^1 \gamma_{rk}(\ell) - \sum_{i \in I1} \upsilon_i^1 x_{ik}^1 -$$

$$\sum_{i \in I2} \sum_{\ell=1}^{L} w_{i\ell}^2 \delta_{ik}(\ell) \leq 0, \text{ all k} \qquad (6.4)$$

$$w_{r\ell}^1 - w_{r\ell+1}^1 \geq \mu_r^2 \varepsilon, \ \ell=1,\ldots L\text{-}1, \text{ all } r \in R_2$$

$$w_{rL}^1 \geq \mu_r^2 \varepsilon, \text{ all } r \in R_2$$

$$w_{i\ell+1}^2 - w_{i\ell}^2 \geq \upsilon_i^2 \varepsilon, \ \ell=1, \ldots L\text{-}1, \text{ all } i \in I_2$$

$$w_{i1}^2 \geq \upsilon_i^2 \varepsilon, \text{ all } i \in I_2$$

$$\mu_r^1, \upsilon_i^1 \geq \varepsilon, \text{ all } r \in R_1, i \in I_1$$

$$\mu_r^2, \upsilon_i^2 \geq \varepsilon, \text{ all } r \in R_2, i \in I_2$$

Problem (6.4) is clearly in linear programming problem format.

We state without proof the following theorem.

Theorem 6.2
At the optimal solution to (6.4), $\mu_r^2 = \upsilon_i^2 = \varepsilon$ for all $r \in R_2$, $i \in I_2$.

Problem (6.4) can then be expressed in the form:

$$e_o = \max \mu_o + \sum_{r \in R_1} \mu_r^1 y_{ro}^1 + \sum_{r \in R2} \sum_{\ell=1}^{L} w_{r\ell}^1 \gamma_{ro}(\ell)$$

$$\text{s.t.} \sum_{i \in I1} \upsilon_i^1 x_{io}^1 + \sum_{i \in I2} \sum_{\ell=1}^{L} w_{i\ell}^2 \delta_{io}(\ell) = 1$$

$$\mu_o + \sum_{r \in R1} \mu_r^1 y_{rk}^1 + \sum_{r \in R2} \sum_{\ell=1}^{L} w_{r\ell}^1 \gamma_{rk}(\ell) -$$

$$\sum_{i \in I1} \upsilon_i^1 x_{ik}^1 - \sum_{i \in I2} \sum_{\ell=1}^{L} w_{i\ell}^2 \delta_{ik}(\ell) \le 0, \text{ all k} \qquad (6.5)$$

$$- w_{r\ell}^1 + w_{r\ell+1}^1 \le - \varepsilon^2, \quad \ell=1,\dots L-1, \text{ all } r \in R_2$$

$$- w_{rL}^1 \le - \varepsilon^2, \text{ all } r \in R_2$$

$$- w_{i\ell+1}^2 + w_{i\ell}^2 \le - \varepsilon^2, \quad \ell=1, \dots, L-1, \text{ all } i \in I_2$$

$$- w_{i1}^2 \le - \varepsilon^2, \text{ all } i \in I_2$$

$$\mu_r^1, \upsilon_i^1 \ge \varepsilon, r \in R_1, i \in I_1$$

It can be shown that (6.5) is equivalent to the *standard* VRS model. First we form the dual of (6.5).

$$\min \theta - \varepsilon \sum_{r \in R1} s_r^+ - \varepsilon \sum_{i \in I1} s_i^- - \varepsilon^2 \sum_{r \in R2} \sum_{\ell=1}^{L} \alpha_{r\ell}^1 - \varepsilon^2 \sum_{i \in I2} \sum_{\ell=1}^{L} \alpha_{i\ell}^2$$

$$\text{s.t.} \sum_{k=1}^{N} \lambda_k y_{rk}^1 - s_r^+ = y_{ro}^1, r \in R_1$$

$$\theta x_{io}^1 - \sum_{k=1}^{N} \lambda_k x_{ik}^1 - s_i^- = 0, i \in I_1$$

$$\qquad (6.5')$$

$$\left.\begin{array}{l} \sum_{k=1}^{N} \lambda_k \gamma_{rk}(1) - \alpha_{r1}^1 = \gamma_{ro}(1) \\ \sum_{k=1}^{N} \lambda_k \gamma_{rk}(2) + \alpha_{r1}^1 - \alpha_{r2}^1 = \gamma_{ro}(2) \\ \vdots \\ \sum_{k=1}^{N} \lambda_k \gamma_{rk}(L) + \alpha_{rL-1}^1 - \alpha'_{rL} = \gamma_{ro}(L) \end{array}\right\} r \in R_2$$

$$\left.\begin{array}{c} \delta_{io}\,(L)\,\theta - \sum_{k=1}^{N}\lambda_k\,\delta_{ik}\,(L) - \alpha_{iL}^2 = 0 \\[2mm] \delta_{io}\,(L-1)\theta - \sum_{k=1}^{N}\lambda_k\,\delta_{ik}\,(L-1) + \alpha_{iL}^2 - \alpha_{iL-1}^2 = 0 \\ \vdots \\ \delta_{io}\,(1)\,\theta - \sum_{k=1}^{N}\lambda_k\,\delta_{ik}\,(1) + \alpha_{i2}^2 - \alpha_{i1}^2 = 0 \end{array}\right\}\ i \in I_2$$

$$\sum_{k=1}^{N}\lambda_k = 1$$

$$\lambda_k,\, s_r^+,\, s_i^-,\, \alpha_{r\ell}^1,\, \alpha_{i\ell}^2 \ge 0$$

$$\theta \text{ unrestricted.}$$

Here, we use $\{\lambda_k\}$ as the standard dual variables associated with the N ratio constraints, and the variables $\{\alpha_{i\ell}^2,\ \alpha_{r\ell}^1\}$ are the dual variables associated with the rank order constraints defined by Ψ. The slack variables $s_r^+,\, s_i^-$ correspond to the lower bound restrictions on $\mu_r^1,\, \upsilon_i^1$.

Now, perform simple row operations on (6.5') by replacing the ℓth constraint by the sum of the first ℓ constraints. That is, the second constraint (for those $r \in R_2$ and $i \in I_2$) is replaced by the sum of the first two constraints, constraint 3 by the sum of the first three, and so on. Letting

$$\overline{\gamma}_{rk}\,(\ell) = \sum_{n=1}^{\ell}\ \gamma_{rk}\,(n) = \gamma_{rk}(1) + \gamma_{rk}(2) + \ldots + \gamma_{rk}(\ell),$$

and

$$\overline{\delta}_{ik}\,(\ell) = \sum_{n=\ell}^{L}\ \delta_{ik}\,(n) = \delta_{ik}\,(L) + \delta_{ik}\,(L-1) + \ldots + \delta_{ik}\,(\ell),$$

problem (6.5') can be rewritten as:

$$\min\ \theta - \varepsilon\ \sum_{r \in R1}s_r^+ - \varepsilon\ \sum_{i \in I1}\ s_i^- - \varepsilon^2\ \sum_{r \in R2}\sum_{\ell=1}^{L}\alpha_{r\ell}^1 - \varepsilon^2\ \sum_{i \in I2}\sum_{\ell=1}^{L}\alpha_{i\ell}^2$$

$$\text{s.t.}\ \ \sum_{k=1}^{N}\lambda_k\,y_{rk}^1 - s_r^+ = y_{ro}^1,\ r \in R_1$$

$$\theta\,x_{io}^1 - \sum_{k=1}^{N}\lambda_k\,x_{ik}^1 - s_i^- = 0,\ i \in I_1 \qquad\qquad (6.6')$$

$$\sum_{k=1}^{N}\lambda_k\,\overline{\gamma}_{rk}\,(\ell) - \alpha_{r\ell}^1 = \overline{\gamma}_{ro}\,(\ell),\ r \in R_2,\ \ell = 1,\ldots,L$$

$$\theta\,\overline{\delta}_{io}(\ell) - \sum_{k=1}^{N}\lambda_k\,\overline{\delta}_{ik}(\ell) - \alpha_{il}^2 = 0,\ i \in I_2,\ \ell = 1,\ldots L$$

$$\sum_{k=1}^{N}\lambda_k = 1$$

$$\lambda_k,\, s_r^+,\, s_i^-,\, \alpha_{r\ell}^1,\, \alpha_{i\ell}^2 \ge 0,\ \text{all } i,\, r,\ \ell,\, k,\ \theta \text{ unrestricted in sign.}$$

The dual of (6.6') has the VRS format:

$$e_o = \max \; \mu_o + \sum_{r \in R1} \mu_r^1 \, y_{ro}^1 + \sum_{r \in R2} \sum_{\ell=1}^{L} w_{r\ell}^1 \, \overline{\gamma}_{ro} (\ell)$$

$$\text{s.t.} \quad \sum_{i \in I1} \upsilon_i^1 x_{io}^1 + \sum_{i \in I2} \sum_{\ell=1}^{L} \overline{w}_{i\ell}^2 \, \overline{\delta}_{io} (\ell) = 1 \qquad (6.6)$$

$$\mu_o + \sum_{r \in R1} \mu_r^1 y_{rk} + \sum_{r \in R2} \sum_{\ell=1}^{L} w_{r\ell}^1 \, \overline{\gamma}_{rk} (\ell) - \sum_{i \in I1} \upsilon_i^1 x_{ik}^1 - \sum_{i \in I2} \sum_{\ell=1}^{L} w_{i\ell}^2 \, \overline{\delta}_{ik} (\ell) \le 0,$$

$$\text{all k}$$

$$\mu_r^1, \upsilon_i^1 \ge \varepsilon, \; w_{r\ell}^1, \; w_{i\ell}^2 \ge \varepsilon^2,$$

which is a form of the VRS model. The slight difference between (6.6) and the conventional VRS model of Banker et al. (1984), is the presence of a different ε (i.e., ε^2) relating to the multipliers $w_{r\ell}^1$, $w_{i\ell}^2$, than is true for the multipliers μ_r^1, υ_i^1. It is observed that in (6.6') the common L-value can easily be replaced by criteria specific values (e.g. L_r for output criterion r). The model structure remains the same, as does that of model (6.6). Of course, since the intention is to have an infinitesimal lower bound on multipliers (i.e., $\varepsilon > 0$), one can, from the start, restrict

$$\mu_r^1, \upsilon_i^1 \ge \varepsilon^2$$

and

$$\mu_r^2, \upsilon_i^2 \ge \varepsilon.$$

This leads to a form of (6.6) where all multipliers have the same infinitesimal lower bounds, making (6.6) precisely a VRS model in the spirit of Banker et al. (1984).

It is interesting to note that the IDEA approach of Cooper et al (1999) essentially involves tackling problem (6.2) by first attributing values to the imprecise data (rank positions), and second, optimizing (in the DEA structure) to arrive at optimal multipliers. The Cook et al (1993), (1996) approach to (6.2) is somewhat the reverse of this. It amounts ultimately to attributing values to the multipliers, and then letting the DEA optimization derive the values for the rank positions. Thus, these seemingly quite different approaches would appear to arrive at approximately the same final point.

Criteria Importance

The presence of ordinal data factors results in the need to *impute* values $y_r^2(\ell)$, $x_i^2(\ell)$ to outputs and inputs, respectively, for DMUs that are ranked at positions ℓ on an L-point Likert or ordinal scale. Specifically, all DMUs ranked at that position will be credited with the same "amount" $y_r^2(\ell)$ of output r (r $\in$ R$_2$) and $x_i^2(\ell)$ of input i (i $\in$ I$_2$).

A consequence of the change of variables undertaken above, to bring about linearization of the otherwise nonlinear terms, e.g., $w_{r\ell}^1 = \mu_r^2 y_r^2(\ell)$, is that at the optimum, all $\mu_r^2 = \varepsilon^2$, $\upsilon_i^2 = \varepsilon^2$. Thus, all of the ordinal criteria are relegated to the status of being of *equal importance*. Arguably, in many situations, one may wish to view the relative importance of these ordinal criteria (as captured by the μ_r^2, υ_i^2) in the same spirit as we have viewed the data values {y_{rk}^2}. That is, there may be sufficient information to be able to *rank* these criteria. Specifically, suppose that the R$_2$ output criteria can be grouped into L$_1$ categories and the I$_2$ input criteria into L$_2$ categories.

Now, replace the variables μ_r^2 by $\mu^2(m)$, and υ_i^2 by $\upsilon^2(n)$, and restrict:

$$\mu^2(m) - \mu^2(m+1) \geq \varepsilon, \; m=1,\ldots L_1-1$$
$$\mu^2(L_1) \geq \varepsilon$$

and

$$\upsilon^2(n) - \upsilon^2(n+1) \geq \varepsilon, \; n=1,\ldots,L_2-1$$
$$\upsilon^2(L_2) \geq \varepsilon.$$

Letting m$_r$ denote the rank position occupied by output r $\in$ R$_2$, and n$_i$ the rank position occupied by input i $\in$ I$_2$, we define the change of variables

$$w_{r\ell}^1 = \mu^2(m_r) \, y_r^2(\ell)$$
$$w_{i\ell}^2 = \upsilon^2(n_i) \, x_i^2(\ell)$$

The corresponding version of model (6.4) would see the lower bound restrictions μ_r^2, $\upsilon_i^2 \geq \varepsilon$ replaced by the above constraints on $\mu^2(m)$ and $\upsilon^2(n)$. Again, arguing that at the optimum in (6.4), these variables will be forced to their lowest levels, the resulting values of the $\mu^2(m)$, $\upsilon^2(n)$ will be

$$\mu^2(m) = (L_1+1-m)\,\varepsilon, \; \upsilon^2(n) = (L_2+1-n)\,\varepsilon.$$

This implies that the lower bound restrictions on $w_{r\ell}^1$, $w_{i\ell}^2$ become

$$w_{r\ell}^1 \geq (L_1+1-m_r)\,\varepsilon^2, \; w_{i\ell}^2 \geq (L_2+1-n_i)\,\varepsilon^2.$$

We now apply the above concepts to the data for the two problem settings discussed earlier.

4.   SOLUTIONS TO APPLICATIONS

4.1   R&D Project Efficiency Evaluation

When model (6.6') is applied to the data of Table 6-1, the efficiency scores obtained are as shown in Table 6-5.

Table 6-5. Efficiency Scores (Non-ranked Criteria)

Project	1	2	3	4	5	6	7	8	9	10
Score	0.76	0.73	1.00	0.67	1.00	0.82	0.67	0.67	0.55	0.37

Here, projects 3 and 5 turn out to be 'efficient', while all other projects are rated well below 100%. In this particular analysis, ε was chosen as 0.03. In another run (not shown here) where ε = 0.01 was used, projects 3, 5 and 6 received ratings of 1.00, while all others obtained somewhat higher scores than those shown in Table 6-5. When a very small value of ε (ε =0.001) was used, all except one of the projects was rated as efficient.

Clearly this example demonstrates the same degree of dependence on the choice of ε as is true in the standard DEA model. See Ali and Seiford (1993).

From the data in Table 6-1 it might appear that only project 3 should be efficient since 3 dominates project 5 in all factors except for input 5 where project 3 rates fourth while project 5 rates fifth. As is characteristic of the standard ratio DEA model, a single factor can produce such an outcome. In the present case this situation occurs because w_{25}^2 = 0.03 while w_{24}^2 = 0.51. Consequently, project 5 is accorded an 'efficient' status by permitting the gap between w_{24}^2 and w_{25}^2 to be (perhaps unfairly) very large. Actually, the set of multipliers which render project 5 efficient also constitute an optimal solution for project 3.

If we further constrain the model by implementing criteria importance conditions as defined in the previous section, the relative positioning of some projects change as shown in Table 6-6.

Table 6-6. Efficiency Scores (Ranked Criteria)

Project	1	2	3	4	5	6	7	8	9	10
Score	0.71	0.72	1.00	0.60	1.00	0.80	0.62	0.63	0.50	0.35

Hence, criteria importance restrictions can have an impact on the efficiency status of the projects.

4.2 Evaluation of Telephone Office Efficiency

The data of Table 6-4 has been evaluated using Model (6.6'). Both CRS and VRS models were applied, the results of which are presented in Table 6-7.

Table 6-7. Efficiency Scores

DMU#	CRS Score	VRS Score	VRS Score-constrained
1	1	1	1
2	1	1	1
3	1	1	1
4	.927	1	.973
5	1	1	.921
6	.907	.994	.906
7	.848	.849	.823
8	.668	.670	.644
9	.848	.970	.885
10	.617	.747	.731
11	.763	.815	.716
12	1	1	.915
13	1	1	1
14	1	1	1
15	1	1	1
16	1	1	.886
17	.898	1	1
18	.928	1	.935
19	.993	.993	.961
20	1	1	1
21	1	1	1
22	1	1	1
23	.846	1	1
24	.918	1	.904
25	1	1	1
26	1	1	.955
27	.824	.937	.926
28	.954	1	.919
29	.949	1	1
30	1	1	1
31	1	1	.907
32	1	1	1
33	.962	1	1

Initially, in applying DEA in this application, no attempt was made to impose constraints on multipliers. Under the CRS structure, approximately half of the offices (17 of the 33) are declared efficient. With the VRS model, the number of efficient units climbs to 25 out of 33. When criteria importance is introduced, the efficiency status (efficient versus inefficient) changes for some units. As well, the relative sizes of efficiency scores change. Note, for example, that the relative positions of offices 10 and 11 are reversed under the constrained VRS model versus those assumed in the unconstrained model. As well, only 15 of the offices (rather than 25) are rated as being efficient.

5. PROBLEM SETTINGS AND ISSUES INVOLVING QUALITATIVE DATA

Qualitative data arises in a multitude of problem settings. As well, some of these practical settings can involve complex issues that are not immediately compatible with the standard model as presented above. Thus, handling the realities of ordinal data in a practical setting often does not automatically simply entail applying the models of Section 3. In this section we examine some case examples of situations involving qualitative data, and explore some of the aforementioned complexities.

5.1 Implementation of Robotics: Identifying Efficient Implementers

Cook, Johnston and McCutcheon (1992) examine robotics implementations in 30 companies, seeking to determine which among these are the most efficient implementers. The study is based upon three types of variables; (1) inputs or initial conditions prior to installation; (2) outputs or outcomes of the end of the installation; and (3) environmental factors that were treated here as control variables *after* the DEA analysis was completed.

The initial conditions at the start of the project (*inputs*), were:

- System Complexity: A count of four components of robotics systems;
- Previous Experience with Technology: A summed score of four 5-point Likert scale measures
- Novelty of the Application: A Likert scale measure of the installation's innovativeness.

The *outputs* in this study are three outcomes at the end of the project's installation, namely:

- Startup Time: Number of weeks required to take the technology from physical installation to routine use.
- Uptime: Percent of total production time, facility is available

- Management Satisfaction: A perceptual measure on a 5-point Likert scale.

A major difficulty in applying DEA in this setting was developing a rationalization of multiple factors, with some being quantitative and some having Likert scale measures. In the case of the input "Previous Experience," which contained *multiple* Likert scale parameters, the sum of the scale values was finally treated as a quantitative variable.

Part of this case study, as well, was the analysis of efficiency in terms of various *control variables*, namely, supplier management, plant size, and urgency. By examining the average ratings for a subgroup of installations when the DEA model was run only for that subset (e.g. those in small plants), versus the average for that same group as part of the total set of 30 plants, one is able to draw conclusions about the most effective settings in which to undertake robotics installations.

5.2   A Fair Model for Aggregating Preferential Votes

Cook and Kress (1990) examine the use of DEA to prioritize candidates, (the DMUs), in a preferential election. Here, each voter is asked to select a subset R of K candidates, and to rank order these candidates from most to least preferred. By its very nature, the data in this setting is of the ordinal or rank order type. Such a voting format is common in municipal elections where a number of candidates are required to fill various positions.

In this setting, each candidate receives some number v_{1k} of first place votes, v_{2k} of second place votes, ..., v_{Rk} of Rth place votes. The objective is to derive a score e_k for each candidate k which rationally accounts for the numbers of 1st, 2nd,..., Rth place votes received by that candidate. The complexity here was that each voter provided a *partial ranking* only of the candidates. Unlike the previous application on robotics, where each DMU is evaluated in terms of several factors (inputs and outputs), in the current application, the voters may be thought of as the evaluation factors. Unfortunately, in most voting situations the number of such factors (voters) is very large. Thus, from a practical standpoint, the model of Section 3 cannot be applied directly. The approach taken here was to *total* the number of "hits" received by a candidate at each ordinal rank position (1,2,...R). In the context of the usual DEA structure, this is equivalent to saying that all voters get the same weight.

The model solved in Cook and Kress (1990) for deriving e_k was

$$e_o = \max \sum_{r=1}^{R} w_r \, v_{ro}$$

$$\text{s.t. } \sum_{r=1}^{R} w_r \upsilon_{rk} \le 1, \ \forall k$$

$$w_r - w_{r+1} \ge d\,(r, \ \varepsilon), r = 1, \ldots, \text{R-1}$$

$$w_r \ge d\,(r, \ \varepsilon),$$

where d(r, ε), the *discrimination intensity function*, is intended to reflect the minimum allowable gap between the worths associated with the rth and r+1st rank positions. The above constraints involving d (r, ε) constitute (in this application) the set Ψ as discussed in Section 3.

5.3 Multiple Criteria Decision Modelling: Ordinal Data, Criteria Importance and Criteria Clearness

Certain types of DEA problems may involve *only outputs* (no inputs). The previous application in ranking candidates in a preferential voting situation is one such example. A special class of such problems is the set of multiple criteria decision problems. In multiple criteria decision settings, the decision maker is generally required to evaluate each member of a set K of alternatives in terms of various criteria. The objective is normally to create a final overall ranking of the alternatives, or at least to pick a winner. In cases where ordinal data are present, at least two issues arise. One of these issues, as addressed in the previous example on preferential voting, involves incorporating known information on the relative importance of the criteria. The AR (assurance region) restrictions in Ψ , as discussed in Section 4.3 of Chapter 1, are designed to capture such information. In many settings, an additional feature involving the criteria may be present, namely, a ranking of those criteria in terms of the *degree of clearness* whereby the alternatives can be evaluated in terms of such criteria. That is, for a criterion that is very *fuzzy*, the resulting ranking of the alternatives (in terms of that criterion) would be less reliable, hence less important than would be true of a ranking in terms of a more clear (less fuzzy) criterion.

Cook and Kress (1991) (and later (1994)) present a multiple criteria model for ranking alternatives when ordinal data are present, and where criteria can be ordinally ranked, both in terms of importance and clearness. Their paper models the ranking problem utilizing the DEA philosophy of treating each alternative as a decision making unit, and maximizes the overall score of each alternative. AR constraints are specifically imposed that take criteria importance and criteria clearness into account, hence enlarging the worth vector restriction set as reflected in Ψ .

The following are some case examples where ordinal data arises naturally, and where both criteria importance and clearness are an issue.

5.3.1   Evaluating Vendors for Complex Systems

Cook and Johnston (1992) examine the prioritization of a set of six venders who have submitted bids on the development of a piece of automatic testing equipment (ATE) that was to be designed for Nortel Incorporated (then called Northern Telecom). Forty different criteria were being used to evaluate the proposals; these were divided into three classes – vender issues, ATE specifications, and delivery factors.

Historically, Nortel had, in such situations, scored vendors on a Likert scale, and rated criteria on a 10-point scale. A weighted score would then be computed for each alternative (e.g. vendor) under consideration. In the particular vendor application under discussion, the company wanted to approach the vendor choice issue in a somewhat more scientific way, given that certain criteria (e.g. vendor reliability) were less "clear," hence less reliable, than was true of other factors such as experience in designing ATE products. As a result, the aforementioned structure with the enlarged AR set Ψ, and treating the vendors as the DMUs, was applied.

5.3.2   Country Risk Evaluation

Cook and Hebner (1993) examine the use of this same model structure to evaluate and prioritize 100 countries (the DMUs) in terms of their risk level relating to investment. The data were made available by the Japan Bond Research Institute. Fourteen criteria were used that reflect a country's risk, including (1) social stability, (2) political stability, (3) fiscal policy, ...etc.

The actual data available on countries' risk positions for any criterion were given in the form of percentages (e.g., one of the criteria was the growth potential expressed as the expected rate of real per capita GNP growth in the coming year). To accomplish an overall ranking of countries in this setting, the percentage data for each criterion was translated first to an ordinal ranking of the countries. Then, the actual numerical data were used to generate criteria importance discrimination factors to allow for appropriate gaps between rank positions. Again, the decision maker was asked to rank criteria both in terms of importance and clearness, hence creating the appropriate AR restrictions defining Ψ.

5.3.3   Mutual Fund Selection

Cook and Hebner (1993) developed a multiple criteria selection model for prioritizing a set of mutual funds, treating the funds as DMUs and the selection criteria as the outputs. The criteria include both quantitative indicators such as front end loading fees, and the standard deviation of a fund's rate of return, as well as qualitative factors such as the service quality of the fund manager, in terms of his/her understanding of financial markets. The approach here was similar to that used in the country risk setting. Specifically, output data were represented by the rank position occupied by each fund on each criterion. Available cardinal data for those criteria such as front end fees, were used to generate criteria discrimination parameters, leading to assurance region restrictions.

5.3.4 Ordinal Data in Multicriteria Modelling: Evaluation in Terms of Subsets of Criteria

Returning again to the special class of DEA models involving multiple criteria evaluation of a set of alternatives, it is often the case that for any given DMU (alternative), only a proper subset of criteria may be relevant to that evaluation. Consider the example of the ranking of a set of R&D projects in a *pure output* setting (no inputs involved). Some projects may, for example, involve the development of nuclear technology while others may not. If a given criterion relates to environmental impacts, such as risk of a nuclear accident, such a criterion would be relevant only for projects concerning nuclear aspects. Specifically, a project that had no such involvement must be rated on only a *subset* of the criteria.

Cook, Doyle, Green and Kress (1997) examine the modeling of DEA problems in the presence of ordinal factors when DMUs may be rated only on a proper subset of the outputs (criteria). The standard model of Section 3 is not immediately applicable, and the suggested approach uses the concept of the *ideal* rank position (w_{r1}) on each criterion r applicable to the DMU in question. That is, rank position "1" is the ideal or best possible ranking achievable.

Specifically, let $R_k \subseteq R$, denote the set of outputs or criteria against which DMU k will be evaluated. Then, solve the ratio problem:

$$e_o = \max \; e_o = \max \sum_{r \in R_o} \sum_{\ell=1}^{L} w_{r\ell} \gamma_{ro}(\ell) / \sum_{r \in R_o} w_{r1}$$

$$\text{s.t.} \; \sum_{r \in R_k} \sum_{\ell=1}^{L} w_{r\ell} \gamma_{rk}(\ell) / \sum_{r \in R_k} w_{r1} \leq 1, \; \forall \; k$$

$$w_{r\ell} \in \Psi$$

This problem is translatable into a linear programming framework in the usual manner.

For ordinal data problems, particularly those of the *pure output* type, this concept of measuring efficiency of a DMU relative to the ideal position of that DMU, provides an innovative way of tackling problems where the DMU has "missing" data.

6. DISCUSSION

We have examined in this chapter the issue of performance measurement in the presence of qualitative data. The methodology presented herein demonstrates that when the idea of rank position data is introduced within the DEA structure, the resulting model can be transformed to a version of the conventional VRS model. This implies that all of the output results from standard DEA models apply. The CRS and VRS scores achieved using the model (6.6') are close to those obtained using the alternative IDEA structure of Cooper et al. (1999). This hints at the potential equivalence of the two approaches.

An important observation regarding radial projection, both here and in the IDEA approach of Cooper et al. (2001), is that one assumes that a $(1-\theta)$ x 100% reduction in a rank order position for an inefficient DMU, results in a legitimate (projected) rank order position. Of course, since radial projection treats all scales as continuous, not discrete, it would rarely be the case that projected points on the frontier would in fact correspond to discrete (Likert scale) positions. Hence, efficiency scores obtained by model (6.6') really represent lower bounds (on θ), and would in practice need to be adjusted upward to bring the projected positions to points that are allowable in Likert scale sense. We do not pursue herein how such adjustments would be made, but point to this as an interesting direction for future research.

REFERENCES

1. Ali, A.I. and L. Seiford , 1993, Computational accuracy and infinitesimals in data envelopment analysis. Working paper, University of Massachusetts at Amherst, MA.
2. Banker, R.D., A. Charnes and W.W. Cooper, 1984, Some models for technical and scale efficiencies in Data Envelopment Analysis, *Management Science*, 30, 1078-1092.
3. Cook, W.D., J. Doyle, R. Green and M. Kress, (1997), *European Journal of Operational Research*, 93(3), 602-609.
4. Cook, W.D., M. Hababou, 2001, Sales performance measurement in bank branches, *OMEGA – International Journal in Management Science*, 29, 299-307.
5. Cook, W.D., M. Hababou and H. Tuenter, 2000, Multicomponent efficiency measurement and shared inputs in Data Envelopment Analysis: An application to sales and service performance in bank branches, *Journal of Productivity Analysis*, 14(3), 2000.

6. Cook, W.D., and K. Hebner, 1993, A multicriteria approach to country risk evaluation: With an example employing Japanese data, *International Review of Economics and Finance*, 2(4), 327-348.

7. Cook, W.D., and K. Hebner, 1993, A multicriteria approach to mutual fund selection, *Financial Services Review*, 2(1), 1-20.

8. Cook, W.D., D. A. Johnston and D. McCutcheon, 1992, Implementation of robotics: Identifying efficient implementers, *OMEGA-International Journal in Management Science*.

9. Cook, W.D., and D. A. Johnston, 1992, Evaluating suppliers of complex systems: A multiple criteria approach, *Journal of the Operational Research Society*, 43(11), 1055-1061.

10. Cook, W.D., and M. Kress, 1990, A Data Envelopment model for aggregating preference rankings, *Management Science*36(11), 1302-1310.

11. Cook, W.D., and M. Kress, 1991, A multiple criteria decision model with ordinal preference data, *European Journal of Operational Research*, 54(2), 191-198.

12. Cook, W.D., and M. Kress, 1994, A multiple criteria composite index model for quantitative and qualitative data, *European Journal of Operational Research*, 78(3), 367-379.

13. Cook, W.D., M. Kress and L. Seiford, 1993, On the use of ordinal data in Data Envelopment Analysis, *Journal of the Operational Research Society*, 44(2), 133-140.

14. Cook, W.D., M. Kress and L. Seiford, 1996, Data Envelopment Analysis in the presence of both quantitative and qualitative factors, *Journal of the Operational Research Society*, 47, 945-953.

15. Cooper, W.W., K.S. Park and G. Yu, 1999, IDEA and AR-IDEA: Models for dealing with imprecise data in DEA, *Management Science*, 45(4), 597-607.

16. Cooper, W.W., K.S. Park and G. Yu, 2001, IDEA (Imprecise Data Envelopment Analysis) with CMDs (column maximum decision making units), *Journal of the Operational Research Society* 52, 176-181.

17. Kim, S.H., C.G. Park and K.S. Park, 1999, An application of Data Envelopment Analysis in telephone offices evaluation with partial data, *Computers & Operations Research*, 26, 59-72.

18. Thrall, R.M, 1996, Duality, classification and slacks in DEA, *Annals of Operations Research*, 66, 109-138.

19. Zhu, J. 2003, Imprecise Data Envelopment Analysis (IDEA): A review and improvement with an application, *European Journal of Operational Research*, 144, 513-529.

Chapter 7

CONGESTION
Its Identification and Management with DEA

William W. Cooper[1], Honghui Deng[1], Lawrence M. Seiford[2] and Joe Zhu[3]
[1] Red McCombs School of Business, University of Texas at Austin, Austin, TX 78712 USA email: cooperw@mail.utexas.edu

[2] Department of Industrial and Operations Engineering, University of Michigan at Ann Arbor, Ann Arbor, MI 48102 USA email: seiford@umich.edu

[3] Department of Management, Worcester Polytechnic Institute, Worcester, MA 01609 USA email: jzhu@wpi.edu

Abstract: Congestion is a term that is applicable in a variety of disciplines which range from medical science to traffic engineering. It also has many uses in practical everyday life. This brings with it a certain looseness in usage. We therefore expand (and refine) our discussion of congestion with reference to its use in economics where we have access to a precise meaning which we can develop in this chapter. This chapter covers the standard approaches used for treating congestion in DEA

Key words: Data envelopment analysis (DEA); Efficiency; Performance, Congestion

1.   CONGESTION

Congestion is a term that is applicable in a variety of disciplines which range from medical science to traffic engineering. It also has many uses in practical everyday life. This brings with it a certain looseness in usage. We therefore expand (and refine) our discussion of congestion with reference to its use in economics where we have access to a precise meaning which we can develop as follows.

Figure 7-1, below, will help us to see what is involved. Drawing on the classical theory of production in economics, this figure is intended to portray what is involved in a relatively simple way by restricting attention to the case of one output and one input. The horizontal axis represents the input amounts, x, and the vertical axis represents the output amounts, y. The curve portrayed in this figure relates these input amounts to the corresponding output amounts which they can produce. It is therefore referred to as a "production function."

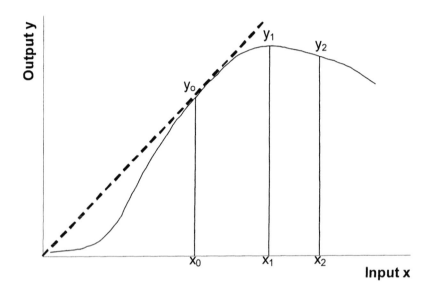

Figure 7-1. Example of a Production Function

The concept of a production function in economics as graphed in Figure 7-1 has a special meaning. It is used to relate input to output in a manner that <u>maximizes</u> the output for each input amount that may be used[1]. This implies that no activity will occur below the "frontier" defined by this function. It is also assumed that the production surface is in the form of a smooth curve with a single peak like the one in Figure 7-1[2]. Hence, frontier points which are technically inefficient in the sense of Chapter 1 are

---

[1] See the text by H. Varian (1984). The classic testament can be found in Chapter IV of P. Samuelson (1947).

[2] Empirical applications generally use models such as Cobb-Douglas functions--a log linear function--which are ever increasing in all inputs and hence have no peaks.

assumed not to occur. (See the discussion of the horizontal segment in Figure 1-4). Congestion, however, can occur as represented by points on the curve to the right of the peak in Figure 7-1.

Notice that production continues to be maximal even on the declining portion of this curve. That is, at each input value beyond x_1, any point on the solid curve continues to depict the maximum output amount that can be attained. This leads to the following definition

> **Definition 7-1**: <u>Congestion</u> is said to occur when the output that is <u>maximally</u> possible can be increased by reducing one or more inputs without improving any other input or output. Conversely, congestion is said to occur when some of the outputs that are maximally possible are reduced by increasing one or more inputs without improving any other input or output. In Figure 7-1, for example, this would mean that the loss in output is to be identified with the difference between y_1 and y_2 while the congesting amount of input is x_2-x_1.

For a concrete example we may think of coal miners operating in an underground mine which we can relate to Figure 7-1 as follows. Starting with input (in the form of number of miners) at $x=0$ the output, y, measured in tons of coal, can be increased at an increasing rate until x_o is reached at output y_o. This can occur, for instance, because an increase in the number of miners makes it possible to form "teams" to perform tasks in a manner that would not be possible with a smaller number of miners. From x_0 to x_1, however, total output continues to increase, but at a decreasing rate, until the maximum possible output is reached at y_1. Using more input results in a decrease from this maximum so that at x_2 we have $y_2<y_1$ and y_1-y_2 is the amount of output lost due to congestion.

It is to be noted that these differences are all measured relative to the frontier. However, as we shall later see, it is possible to extend these concepts to identify such phenomena even when additional inefficiencies result in observations that lie below the frontiers associated with production functions.

In contrast to a smooth (everywhere differentiable) curve like the one in Figure 7-1, the production surfaces used in DEA are piecewise linear (and continuous). Figure 7-2 provides an example which we can compare with Figure 7-1 as follows. The single smooth curve exhibited in Figure 7-1 is replaced by a series of piecewise linear segments to form a continuous boundary derived from the data in the manners that were earlier described in Chapter 1.

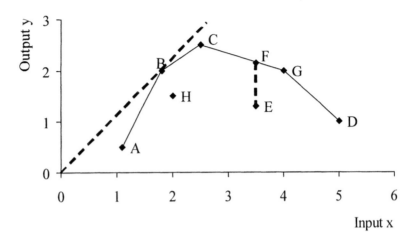

Figure 7-2. A DEA Production Surface

Being piecewise linear the surface in Figure 7-2 fails to be everywhere differentiable since, in particular, derivatives fail to exist at the points where the segments are joined. Hence we need to replace the concept of a "derivative" and the associated concept of a tangent with the more general concept of a "supporting hyperplane." An example of such a hyperplane (here a line) is exhibited by the broken line from the origin which touches the production surface at B in a manner analogous to the tangent plane (here a line) that represented the point of maximal returns to scale at y_o in Figure 7-1. Hence, we can, if we wish, preserve the concept of "returns to scale"-- which, like "congestion," is also a frontier concept in that it is associated with the frontiers of the set of production possibilities.

In Figure 7-2 we refer to the "production surfaces" as "production boundaries" or "production frontiers" in place of the "production function" concept which we employed for Figure 7-1. This is because in place of a single production function for a particular firm we can regard the production boundaries in Figure 2 as a surface generated from a series of supporting hyperplanes that touch the surface of at least one production function in a collection of production functions--which functions may differ for each of the different firms (=DMUs) represented in the data.

In DEA we do not require all points to lie on the surface, as is the case in economics. Instead we allow points like H and E to occur, which lie below the surface, as in Figure 7-2. This means that H and E fall short of the output they could have produced with the inputs they used. Nevertheless, we shall continue to use the economic concept of congestion as in Definition 7-1. This will help us to make distinctions between "congestion" and other types of inefficiency such as "technical inefficiency," etc., in our development.

The points A, B and C and the segments connecting them in Figure 7-2 represent the efficiency frontier. In the approaches we use, it is this portion of the frontier which is used to evaluate the efficiency of all of the other points in the Production Possibility Set (PPS). The segments connecting CFGD represent the congested part of the frontier. Notice that on the segments connecting A, B and C, which together represent the efficiency frontier, it is not possible to increase (i.e., improve) the output at any point without increasing (i.e., worsening) the input. The congestion frontier has the opposite property: It is not possible to increase (i.e., improve) the output on this frontier without also decreasing (i.e., improving) the input.

The points H and E will help us to understand some of the properties that will be of interest. Both points lie below the frontier. Projecting H vertically until it reaches the efficiency frontier on the segment connecting B and C shows that the point H is technically inefficient because, as evidence from the data shows, it is possible to increase the output of H without increasing its input. Alternatively, a horizontal projection from H to the segment of the efficiency frontier between A and B also shows that H is "technically inefficient" because it is possible to decrease the input without decreasing the output. These vertical and horizontal projections reflect results from "output oriented" and "input oriented" objectives respectively in the models we shall be using.

Now consider E. A horizontal projection to the segment of the frontier between A and B shows that it is possible to decrease the input at E without worsening the output. This projection, as previously noted, reflects an "input oriented" objective. It does not permit us to make distinctions like those we want to make between "technical inefficiency" and "congestion." We therefore turn to an "output oriented" objective which effects the projection from E to F. This shows that it is possible to increase the output for E without changing its input. However, this does not end the matter since we can further increase (i.e., improve) the output by decreasing (i.e., improving) the input amount for F. Indeed this increase in output along the frontier can be continued until the maximal output possible is reached, as occurs when the input is reduced to a point directly below C.

The difference between the input at E and the input at C represents the amount of congestion in the input amount used by E, while the difference between the output coordinate at C and the output coordinate at E represents the output lost due not only to congestion but also to the way it was managed. We will later use the difference between the ordinates at C and F as our measure of the output loss due to congestion and the difference between the ordinates on E and F as the measure of output loss due to the way this input was managed. This will not only keep us in touch with economic theory, it will also allow us to deal with problems that are of managerial interest. We may note, for instance, that identifying C in this (frontier) manner also produces a byproduct in the form of an estimate of the capacity that is available with fully efficient performance by the entity associated with E.

We do not pursue this topic further at this point other than to present the following as an example of managerial inefficiency in managing a congesting input,

Example: Excess inventory cluttering a factory floor in a way that interferes with production. By simply reconfiguring this excess inventory it may be possible to increase output without reducing inventory. This improvement represents the elimination of inefficiency that is caused by the way excess inventory is managed. It is the kind of inefficiency which can be represented by the movement from E to F in Figure 2, and which we will identify as "managerial inefficiency." The movement from F to C will then give us a measure of the output lost due to congestion. In either case the amount of the congesting input will be measured by the difference in the value of x for E and the value of x for C.

Having now clarified what we mean by "congestion" we should probably also illustrate some of the respects in which a treatment of this topic might be of interest. A case in point is provided in the doctoral thesis of H. Deng (2003)[3] which is directed to the treatment of congestion and its management in Chinese production. As Deng (2003) shows, congestion is used by the Chinese government in order to deal with the problem of providing employment for a huge labor force--with some 16,000,000 to 18,000,000 new entrants every year. However, congestion is not confined to such examples. It also appears in other contexts such as when the inflow from one department leads to large excesses of "in process" inventories in another department of the same company. See Balakrishnan and Soderstron (2000). In either case it may be possible to improve its management and thereby improve output performances along lines like those indicated by the movement from E to F in Figure 7-2.

We now note that we are dealing only with congesting inputs, partly because that is where the main managerial interest lies, and partly because it is not clear what is to be meant by congesting outputs. The topic of output congestion is dealt with mathematically by Färe, Grosskopf and Lovell

---

[3] This chapter represents an adaptation of material taken from the thesis of Honghui Deng (2003).

(1985,1994). However, their treatment is restricted to a formal development without reference to any example of possible occurrences.

Models for distinguishing between different characterizations of inefficiency such as "congestion" and "technical inefficiency" will shortly be introduced. These concepts will also be developed further as the different applications we consider are brought into view. However, before undertaking to do this, we will first briefly review the DEA literature that has treated congestion in order to obtain additional perspective on the topics we will be covering.

2. COMPARISON OF TWO LITERATURES ON CONGESTION

Congestion has been an under-researched topic in western economics partly because G.J. Stigler (1976), a Nobel Laureate economist, questioned whether "congestion" as a topic of research should have any place in economics in his review of the "X-Efficiency" concept of H. Leibenstein (1966, 1976). However, after a long period of neglect in the economics literature, Färe and Svensson (1980) initiated new research in this area by reformulating some of the concepts associated with congestion. Färe and Grosskopf (1983) then gave these concepts operational form. Färe, Grosskopf and Lovell (1985) subsequently finalized the models (and methods of analysis) that they used to analyze congestion and accorded them a form that would now be identified with DEA (Data Envelopment Analysis). This approach was the only one available in the DEA literature and was therefore employed in all of the research into congestion in the numerous applications that were then undertaken. Cooper, Thompson and Thrall (CTT, 1996), however, formulated an alternative approach which has also begun to see various extensions and applications.

Interest in the development of alternative approaches has now begun to result in additions and extensions to CTT (1996) in a variety of ways. This has been advantageous because new alternatives provide perspective on shortcomings as well as advantages in the use of existing models. This was exhibited, for instance, in the exchanges between Färe and Grosskopf (1998, 2000 a and b) and Brockett, Cooper, Shin and Wang (1998) and Cooper, Seiford and Zhu (2000, 2001c). Shortcomings in the Färe, Grosskopf and Lovell approach were then identified by Cooper, Gu and Li (2001b) in a manner that led to the exchanges between Cherchye, Kuosmanen and Post (2001) and Cooper, Gu and Li (2001a).

Many new applications have been reported in various fields. Numerous references which involve such applications may be found in the bibliography for DEA compiled by Seiford (1994). This bibliography has now been extended and incorporated in a CD-ROM that accompanies the textbook by Cooper, Seiford and Tone (2000). Additional models have also come into being which we survey in the sections that follow after we start with FGL (Färe, Grosskopf and Lovell 1985, 1994) because it has been the longest standing and most used approach to congestion in the DEA literature

3.   FÄRE, GROSSKOPF AND LOVELL (FGL) APPROACH

The FGL approach proceeds in two stages. The first stage uses an "input-oriented" model as follows (Färe et al. 1985):[4]

$$\theta^* = \min \ \theta$$
$$\text{subject to}$$

$$\theta x_{io} \geq \sum_{j=1}^{n} x_{ij} \lambda_j , \quad i=1,2,\ldots,m \tag{7.1}$$

$$y_{ro} \leq \sum_{j=1}^{n} y_{rj} \lambda_j , \quad r=1,2,\ldots,s$$

$$\lambda_j \geq 0, \qquad j=1,2,\ldots,n$$

where $j=1,\ldots,n$ indexes the set of DMUs (Decision Making Units) which are of interest. Here x_{ij} is the observed amount of input $i=1,\ldots,m$ used by DMU_j and y_{ro} is the observed amount of output $r=1,\ldots,s$ produced by DMU_j. The x_{io} and y_{ro} represent the amounts of inputs $i=1,\ldots,m$ and outputs $r=1,\ldots,s$ associated with DMUo, where DMUo is the $DMU_j=DMU_o$ to be evaluated relative to all DMU_j (including itself) via (7.1). The objective is to minimize all of the inputs of DMUo in the proportion θ^* where, because the $x_{io} = x_{ij}$ and $y_{ro} = y_{rj}$ for $DMU_j = DMU_o$ appear on both sides of the constraints in (1), the optimal $\theta = \theta^*$ does not exceed unity and the non-negativity of the λ_o, x_{ij} and y_{ij} implies that the value of θ^* will not be negative under the optimization in (7.1). Hence,

$$0 \leq \text{Min } \theta = \theta^* \leq 1. \tag{7.2}$$

We now have the following definition of technical efficiency and inefficiency,

Definition 7-2 FGL Technical Efficiency:
(i) Technical efficiency is achieved by DMUo if and only if $\theta^* = 1$.
(ii) Technical inefficiency is present in the performance of DMUo if and only if $0 \leq \theta^* < 1$.

---

[4] An output oriented version is also supplied. See p.110 in Färe, et al. (1994).

This definition ignores the possible presence of non-zero slacks even when the solution of (7.1) shows them to be present. We therefore say that this definition refers to "weak" technical efficiency. This is the term used in the operation research literature. In the economics literature it is referred to as the assumption of "strong disposal." In any case, FGL then go on to the following second stage model,

$$\beta^* = \min \ \beta$$

subject to

$$\beta x_{io} = \sum_{j=1}^{n} x_{ij}\lambda_j , \quad i=1,2,\ldots,m \tag{7.3}$$

$$y_{ro} \le \sum_{j=1}^{n} y_{rj}\lambda_j , \quad r=1,2,\ldots,s$$

$$\lambda_j \ge 0 , \quad j=1,2,\ldots,n.$$

Note that the first i=1,...,m inequalities in (7.1) are replaced by equations in (7.3). Thus slack is not possible in the inputs. The fact that only the output can yield non-zero slack is then referred to as "weak disposal" by Färe et al. (1985).

Now we note that (7.3) is more restricted than (7.1) by virtue of replacing inequalities with equations. Hence, we have $0 \le \theta^* \le \beta^*$. FGL use this property to develop a "measure" of congestion,

$$0 \le C(\theta^*, \beta^*) = \frac{\theta^*}{\beta^*} \le 1. \tag{7.4}$$

Combining models (7.1) and (7.3) in a two-stage manner, FGL utilize this measure to identify congestion in terms of the following conditions,

(i) Congestion is identified as present in the performance of DMUo if and only if $C(\theta^*, \beta^*) < 1$ (7.5)

(ii) Congestion is identified as not present in the performance of DMUo if and only if $C(\theta^*, \beta^*) = 1$.

Table 7-1, taken from Färe, Grosskopf and Lovell (1985 p.160), provides a numerical example which involves 7 DMUs in a two input and one output case.

To help explain this approach, we can also utilize Figure 7-3, taken from FGL (1985) which uses the x_1, x_2 values in Table 7-1 to represent these DMUs geometrically in a Cartesian coordinate system. The line segments in Figure 7-3 form what is called an "isoquant" (level line) in economics. These segments are obtained by running a plane through the 3 dimensional production surface at the level y=2 recorded in the output (=y) column of Table 7-1. After projecting these results down into a 2-dimensional representation we obtain Figure 7-3 in which all of the points shown, including those not on the isoquant represent x_1, x_2 coordinate values associated with the output values of y=2 recorded in Table 7-1.

Table 7-1. Congestion Example

DMU	Output y	Input x_2	Input x_1
1	2	1	2
2	2	2	2
3	2	2	1
4	2	1	3
5	2	1	4
6	2	3	1.25
7	2	4	1.25

Source: Färe, Grosskopf and Lovell (1985, p.160)

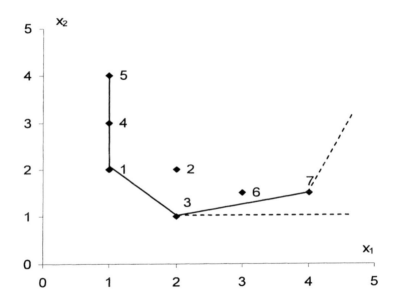

Figure 7-3. Representation of Points in Table 7-3

The line segment connecting points 3 and 7 is referred to as "backward bending" by FGL. To see what this means and how it is associated with congestion, consider point 7. Moving to the left along the level line from 7 to 3 is associated with reductions in both inputs while maintaining output at y=2. This suggests the presence of congestion which is further confirmed by noting that horizontal movement to the left from 7 represents a reduction in x_1 that provides access to an output level on the surface of the production possibility set which is greater than y=2. Hence, as required for congestion to be present, a reduction in input is thereby associated with an increase in the output that is maximally possible.

To see how FGL use models (7.1) and (7.3) and the measure supplied by (7.4) to determine whether congestion is present, we return to the data in Table 7-1 which provides the coordinates of the points in Figure 1. Applying models (7.1) and (7.3) in the two stage manner we have already described yields the values shown in Table 2.

In accordance with (7.5), only DMUs 6 and 7, where the value for $C(\theta^*, \beta^*) = \theta^*/\beta^*$ is less than 1, display congestion in their performance. All of the other DMUs are identified as at least "weakly efficient" except DMU2 which is technically inefficient, because $\theta^* < 1$, but nevertheless has a value of $\theta^*/\beta^* = 1$.

Table 7-2. Input Efficiency and Congestion in the Example of Table 7-1

DMU	θ^*	β^*	$\dfrac{\theta^*}{\beta^*}$
1	1.0	1.0	1.0
2	0.75	0.75	1.0
3	1.0	1.0	1.0
4	1.0	1.0	1.0
5	1.0	1.0	1.0
6	0.8	0.86	0.93
7	0.8	1.0	0.8

Source: Färe, Grosskopf and Lovell (1985, p.76)

To relate the solutions in Table 7-2 to the diagram in Figure 7-3 we focus on the points for DMUs 6 and 7, which are the only ones that satisfy the conditon for congestion specified in (7.5). Using the coordinate for $x_2 = 1.25$, as given for DMUs 6 and 7 in the last column of Table 7-1 and coupling this value with $\theta^* = 0.8$ in Table 7-2 we obtain 0.8(1.25)=1.00. This positions both of these points on the horizontal line emanating from DMU$_3$ with coordinates (x_1, x_2)=(2,1) in Figure 7-1. Evidently DMU$_3$ is the only point on this line which is efficient so, at this point, we must have non-zero slack. Therefore there must be slack removal in input 1 for DMUs 6 and 7. To confirm this we note that the intersection point with DMU$_3$ of this coordinate for DMU$_6$ is 0.8(3)=2.4 and for DMU$_7$ it is 0.8(4)=3.2. Thus, 0.4 unit of slack in input 1 must be removed for DMU$_6$ and 1.2 units must be removed for DMU$_7$ in order to obtain coincidence with DMU$_3$ at (x_1, x_2)=(2,1). Hence the assumption of "weak efficiency" must be dropped in order to fully portray this solution.

Having examined the solution for (7.1) in terms of θ^* we next turn to the solution for β^* in (7.3). The equality condition for β^* in (7.3) requires solutions to lie on the line segment connecting the points for DMU$_3$ and DMU$_7$ and the algebraic expression which corresponds to this line is $x_2 = 0.75 + 0.125x_1$. As is evident from Figure 7-3, DMU$_7$ is already on this line, which is confirmed by substituting the value $x_1 = 4$ in this algebraic expression to obtain $x_2 = 1.25$. For DMU$_6$ we use $\beta^* = 0.86$ to obtain the pertinent coordinates $\hat{x}_1 = 0.86(3) = 2.58$ and $\hat{x}_2 = 0.86(1.25) = 1.075$ for substitution in this algebraic expression to obtain $1.075 \approx 0.75 + 0.125(2.58)$ where we have used the approximation symbol "$\approx$" to allow for roundoff error in the value of $\beta^* = 0.86$ reported by FGL in Table 7-2.

The rationale underlying the use of the measure in (7.4) in the manner specified by (7.5) to determine whether congestion is present can now be made clear from the following considerations. The horizontal broken line represents a boundary (=frontier) of the production possibility set as determined from (7.1). (There will always be such a boundary.) Hence a value of $\theta^*/\beta^* < 1$ can occur only if the isoquant generated from some observation such as DMU$_7$ is "backward bending."

We now return to the issue of non-zero slacks that was noted in connection with our discussion of the θ^* value obtained from (7.1). The presence of such non-zero slack plays a critical role in the FGL development not only in this case but also in the case of movement to the right from this backward bending segment (such movements are associated with output reductions) or movements above this segment (such movements are associated with output increases). FGL therefore associate congestion with mix inefficiencies.[5] This condition is limitational rather than general, however, since one can readily supply examples where it does not hold. For instance, returning to the example of an underground mine one can have situations in which congestion is due to too many miners and too much equipment. Withdrawing these two inputs without changing their proportion can therefore serve to eliminate this congestion without altering the mix proportion.

Troubles are also encountered in cases where one wants to identify the sources and amounts of congesting inputs. The methods supplied for treating this problem, as described by FGL (1994, pp. 76-77), involve solving for each of the pertinent partition of the m inputs. This can be onerous, so when the identification of such inputs and their congesting amounts is of interest it is better to turn to one of the other models we shall shortly describe.

Other deficiencies are described in Cooper, Gu and Li (2001a, b) who show the following: (1) The FGL approach can show congestion to be

---

[5] "Mix" refers to the proportions in which inputs are used or outputs are produced. See the discussion in Cooper, Park and Pastor (1999).

present when this is not the case and (2) it can fail to show congestion to be present even when this is the case.

Undoubtedly these defects in the FGL approach can be eliminated or alleviated but the research needed to do this has not yet been done. Fortunately, however, other models are now available that we will next describe which can be used while this research is being undertaken.

4. COOPER, THOMPSON AND THRALL (CTT) APPROACH

Cooper, Thompson and Thrall (1996) introduced another model which was extended by Brockett et al. (1998) in their study of congestion in Chinese production. See the further developments on the use of these results for policy guidance in Cooper, Deng, Gu, Li and Thrall (2000a). This alternate approach also proceeds in a two-stage manner with the following "output oriented" model used in the first stage.

$$\max \ \phi_o + \varepsilon(\sum_{r=1}^{s} s_r^+ + \sum_{i=1}^{m} s_i^-)$$

subject to

$$x_{io} = \sum_{j=1}^{n} x_{ij}\lambda_j + s_i^- \qquad i=1,2,\ldots,m \qquad (7.6)$$

$$\phi_o y_{ro} = \sum_{j=1}^{n} y_{rj}\lambda_j - s_r^+ \qquad r=1,2,\ldots,s$$

$$1 = \sum_{j=1}^{n} \lambda_j$$

$$0 \leq \lambda_j, s_r^+, s_i^- \qquad j=1,\ldots,n; \ i=1,\ldots,m; \ r=1,\ldots,s.$$

Comparison with the first stage of the FGL approach in (7.3) shows the following three differences: (i) the convexity condition $\sum \lambda_j = 1$ in (7.6) is added to (7.3); (ii) the objective in (7.6) is for an "output-oriented" model whereas an "input orientation" is used in the objective of (7.3); (iii) The model (7.6) introduces slacks into the objective multiplied by a value $\varepsilon > 0$ which insures that some nonzero slack is not overlooked, possibly in an alternative optimum, as described for (1.2) in Chapter 1.

The second stage of this CTT approach, as taken from Brockett et al. (1998), can be represented as follows:

$$\text{Max } \sum_{i=1}^{m} \delta_i^-$$

Subject to

$$\hat{x}_{io} = \sum_{j=1}^{n} x_{ij} \hat{\lambda}_j - \delta_i^- \qquad i=1,2,\ldots,m \qquad (7.7)$$

$$\hat{y}_{ro} = \sum_{j=1}^{n} y_{rj} \hat{\lambda}_j \qquad r=1,2,\ldots,s$$

$$1 = \sum_{j=1}^{n} \hat{\lambda}_j$$

$$s_i^{-*} \geq \delta_i^- \qquad i=1,2,\ldots m$$

$$0 \leq \hat{\lambda}_j, \delta_i^- \quad \forall i,j,$$

where, $\hat{y}_{ro}$, $\hat{x}_{io}$ represent coordinates of a point on the efficiency frontier. These $\hat{y}_{ro}$, $\hat{x}_{io}$ are the coordinates of the point used to evaluate DMUo. They are obtained from the solution of (7.6) and have the following values.

$$\hat{x}_{io} = x_{io} - s_i^{-*} \qquad i=1,2,\ldots,m \qquad (7.8)$$

$$\hat{y}_{ro} = \phi_o^* y_{ro} + s_r^{+*} \qquad r=1,2,\ldots,s.$$

Remark: Comparison with (2.3) in Chapter 2 shows that (7.8) is also a projection formula which can be used in place of the input-oriented model projection associated with (2.3).

Notice that the inequalities implied for the inputs by the first $i=1,\ldots,m$ constraints in (7.7) are reversed from the usual form as exhibited in (7.6) (cf. the change in sign for the slacks). The objective in (7.7) is to maximize the sum of the input slacks[6] with the additional constraint $s_i^{-*} \geq \delta_i^-$ limiting each slack to the maximum value obtained in the preceding solution to (7.6). The difference may be represented as

$$s_i^{-c} = s_i^{-*} - \delta_i^{-*}, \quad i=1,\ldots, m \qquad (7.9)$$

where, for each $i =1,\ldots,m$, δ_i^{-*} is obtained by solving (7.7) after s_i^{-*} has been subtracted from x_{io} as in (7.8).

These s_i^{-C} values, when positive, represent congesting amounts in each of the $i=1,\ldots,m$ inputs while the $\delta_i^{-*} \geq 0$ represent corresponding technical inefficiency components. Thus referring to s_i^{-*} as "total slack" (as obtained

---

[6] If desired, these slacks may be stated in a ratio form relative to the observed x_{io}, as in Cooper, Seiford and Zhu (2000), in order to obtain a measure of congestion which is invariant to the units of measure used. See (7.10), below.

from (7.6)), we have, $s_i^{-*} = s_i^{-C} + \delta_i^{-*}$ i =1,...,m. That is, each "total slack", as obtained from (7.6) in stage one, is decomposed into two components via (7.7) in stage two. These two components consist of (1) ordinary technical inefficiency in amount δ_i^{-*} and (2) congesting amount s_i^{-c} as defined in (7.9).

4.1   A Numerical Example

Figure 7-4, below, can help to explain what is intended. Note, for instance, that point C with coordinates $(x,y) = (3,2)$ is inefficient relative to B because B produced the same 2 units of output as C but did so with only 2 units of input. Hence, compared to B, C used an excess of 1 unit of input. However, there is no output reduction associated with this input excess, so the resulting inefficiency is "purely technical,"--i.e., no congestion is present at C.

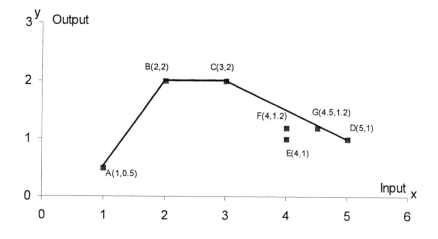

Figure 7-4. A Numerical Example

Source: Brockett et al. (1998)

While no evidence of congestion is present in the performance of C, all points to the right of C, do, in fact, exhibit congestion. For illustration, we apply models (7.6) and (7.7) to E in the two-stage manner that we just described. Application of (7.6) to the evaluation of E, with coordinate (x, y) = (4, 1), yields the following model:

$$\max \phi + \varepsilon(s^+ + s^-)$$
subject to
$$1\phi = 0.5\lambda_A + 2\lambda_B + 2\lambda_C + 1\lambda_D + 1\lambda_E + 1.2\lambda_F + 1.2\lambda_G - s^+$$
$$4 = 1\lambda_A + 2\lambda_B + 3\lambda_C + 5\lambda_D + 4\lambda_E + 4\lambda_F + 4.5\lambda_G + s^-$$
$$1 = \lambda_A + \lambda_B + \lambda_C + \lambda_D + \lambda_E + \lambda_F + \lambda_G$$
$$0 \le \lambda_A, ..., \lambda_G, s^+, s^-.$$

This has as its solution $\phi^* = 2$, $\lambda_B^* = 1$, $s^{-*} = 2$. Thus E is inefficient with both (i) $\phi^* > 1$ and (ii) some slacks are positive.

To ascertain whether congestion is present, we turn to (7.7) and use the results we have just secured to replace the preceding model with

$$\max \delta^-$$
subject to
$$2 = 1\lambda_A + 2\lambda_B + 3\lambda_C + 5\lambda_D + 4\lambda_E + 4\lambda_F + 4.5\lambda_G - \delta^-$$
$$2 = 0.5\lambda_A + 2\lambda_B + 2\lambda_C + 1\lambda_D + 1\lambda_E + 1.2\lambda_F + 1.2\lambda_G$$
$$1 = \lambda_A + \lambda_B + \lambda_C + \lambda_D + \lambda_E + \lambda_F + \lambda_G$$
$$2 \ge \delta^-$$
$$0 \le \lambda_A, ..., \lambda_G, \delta^-.$$

Here, $\hat{x} = 2$ and $\hat{y} = 2$, as represented on the left in this model, are obtained by using the preceding solution to obtain the values $\hat{x} = 2$, $\hat{y} = 2$ by means of (7.8). Similarly, $s^{-*} = 2$ (from the preceding solution) bounds the admissible values of δ^- via the last constraint.

The solution to the just stated problem is $\lambda_C^* = 1$, $\delta^{-*} = 1$ and all other variables at zero. Substitution of δ^{-*} in (7.9) therefore gives $s^{-c} = 2 - 1 = 1$ as the congesting amount of this input that is identified in the performance of E. Hence, the total slack, at $s^{-*} = 2$, as obtained from the first stage, is decomposed in this solution to a value of $\delta^{-*} = 1$ for "purely technical inefficiency" and $s^{-c} = 1$ for the "congesting" part of this "total slack," so that, as required, $s_i^{-*} = s_i^{-C} + \delta_i^{-*}$.

All of this information, which is automatically available, can be related to Figure 7-4 by noting that $s^{-*} = 2$ is the reduction in the x=4 units of input used by E which is necessary to obtain coincidence with the input coordinate of the efficient point represented by B. At the same time, $s^{-c} = 1$ is the reduction in the input of E needed for coincidence with the input coordinate of C. Further, $\delta^{-*} = 1$ is the amount (here = one unit) of technical inefficiency that needs to be removed from the input of C to attain coincidence with the input of B. Finally, comparing the two units of output at C with the one unit of observed output for E shows the amount of reduction (=one unit) in output.[7]

---

[7] As will be seen in section 6 there is also a managerial inefficiency component to be accounted for in this output reduction, which is represented by the fact that output fails to

5. A UNIFIED ADDITIVE MODEL

The FGL approach, as described earlier, uses a radial measure of inefficiency in both stages one and two. The CCT approach, however, uses a radial measure only in stage one. In stage two, on the other hand, CCT use a modified version of an "additive model" which measures the amount of inefficiency in terms of "ℓ_1 measure."[8]

The recent publication, by Cooper, Seiford and Zhu (CSZ 2000) replaces the mixture of measures (radial and ℓ_1) used in the CTT approach. The version of the additive model used by CSZ unifies matters by applying the same "ℓ_1 measure" in both stages. Hence we now turn to this CSZ approach to describe how this is accomplished.

The first stage model in this approach is formulated by CSZ in the following manner,

$$Max \frac{\sum\limits_{r=1}^{s} \frac{s_r^+}{y_{ro}}}{s} + \varepsilon \frac{\sum\limits_{i=1}^{m} \frac{s_i^-}{x_{io}}}{m}$$

subject to

$$\sum_{j=1}^{n} \lambda_j x_{ij} + s_i^- = x_{io} \quad i = 1,\ldots,m \qquad (7.10)$$

$$\sum_{j=1}^{n} \lambda_j y_{ij} - s_r^+ = y_{ro} \quad r = 1,\ldots s$$

$$\sum_{j=1}^{n} \lambda_j = 1$$

$$\lambda_j, s_i^-, s_r^+ \geq 0$$

where $\varepsilon > 0$ is the non-Archimedean element, smaller than any positive real number, that was described earlier in the discussion of (7.6). Also as discussed in association with (1.2) in Chapter 1, this use of $\varepsilon > 0$ accords preemptive priority to maximizing $\sum\limits_{r=1}^{s} \frac{s_r^+}{y_{ro}}$.

The results from (7.10) are used to form the following model,

---

achieve the boundary (=maximal output value) that is attainable with the 4 units of input used at E.

[8] The "ℓ_1 measure" is also called the "city block measure" of distance in the operations research literature. A comprehensive treatment ℓ_1 and other measures used in mathematics may be found in Appendix A of Charnes and Cooper (1961).

$$Max \frac{\sum\limits_{i=1}^{m} \frac{s_i^-}{x_{io}}}{m}$$

subject to

$$\sum_{j=1}^{n} \lambda_j x_{ij} + s_i^- = x_{io} \quad i=1,\ldots,m \tag{7.11}$$

$$\sum_{j=1}^{n} \lambda_j y_{ij} = \hat{y}_{ro} \qquad r=1,\ldots s$$

$$\sum_{j=1}^{n} \lambda_j = 1$$

$$\lambda_j, \ s_i^- \geq 0$$

where $\hat{y}_{ro} = y_{ro} + s_r^{+*}$ so that $\hat{y}_{ro}$ with the slacks, s_r^{+*}, representing output slack, r=1,...,s, is obtained from (7.10).

The solution of (7.11) yields a new set of maximal input slacks consistent with the thus adjusted outputs. We next 'back out' the maximal inputs. This backing out is accomplished by means of the following modification of (7.7):

$$Max \frac{\sum\limits_{i=1}^{m} \frac{\delta_i^-}{x_{io}}}{m}$$

subject to

$$\sum_{j=1}^{n} \lambda_j x_{ij} - \delta_i^- = \hat{x}_{io} \quad i=1,\ldots,m \tag{7.12}$$

$$\sum_{j=1}^{n} \lambda_j y_{ij} = \hat{y}_{ro} \qquad r=1,\ldots s$$

$$\sum_{j=1}^{n} \lambda_j = 1$$

$$\delta_i^- \leq s_i^{-*}$$

$$\lambda_j, \ \delta_i^- \geq 0$$

where $\hat{x}_{io} = x_{io} - s_i^{-*}$ with s_i^{-*} representing the optimal slacks obtained from (7.11) in the second stage optimization associated with $\varepsilon > 0$ in (7.10).

This returns us to (7.9) to obtain the values of $s_i^{-c} = s_i^{-*} - \delta_i^{-*} \geq 0$ where s_i^{-c} represents the amount of congestion in each input i =1,...,m. In many cases the output reductions resulting from congestion will be apparent from inefficiency. For a formal development that will handle all cases, however, we now replace (7.11) with

$$Max \ \frac{\sum_{r=1}^{s} \frac{\delta_r^+}{y_{ro}}}{m}$$

subject to

$$\sum_{j=1}^{n} \lambda_j x_{ij} + s_i^- = x_{io} \quad i=1,\dots,m \qquad (7.13)$$

$$\sum_{j=1}^{n} \lambda_j y_{ij} + \delta_r^+ = \hat{y}_{ro} \quad r=1,\dots,s$$

$$\sum_{j=1}^{n} \lambda_j = 1$$

$$\lambda_j, \ \delta_r^+ \geq 0$$

where $\hat{y}_{ro}$ is defined as in (7.11) and the x_{io} are original data as in the inputs for (7.10) and (7.11). When the optimal solution for (7.10) is unique, the solution to (7.13) will simply reproduce the original data via $\hat{y}_{ro} - \delta_r^{+*} = y_{ro}$. When not unique, however, other possibilities may be present.

6. ESTIMATING THE OUTPUT EFFECTS OF CONGESTION

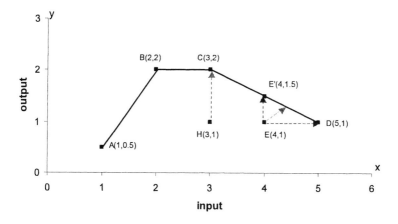

Figure 7-5. Congestion and Its Effects

Figure 7-5, above, is a modified version of Figure 7-4 that can help us to determine the effects of congestion. As shown in the case of Figure 7-4 we can use (7.7) and (7.9) to determine the congesting amount of input as

$s^{-C} = 1$, which represents the difference between $x_E = 4$ at E and $x_H = 3$ at H, the point where technical inefficiency gives way to congestion. Thus the amount of congesting input is $x_E - x_H = 1$. However, the decrease in output is given by the coordinate value of $y_{E'} = 1.5$ at E', according to definition 1, and not by the coordinate value of $y_E = 1$ since the latter reflects a managerial inefficiency as well as a congestion component.

To estimate the amount of managerial inefficiency we use the following version of our output oriented "additive" model as taken from Cooper, Deng, Gu, Li and Thrall (2001a),

$$
\max \sum_{r=1}^{s} s_r^+
$$
subject to
$$
\begin{aligned}
y_{ro} &= \sum_{j=1}^{n} y_{rj}\lambda_j - s_r^+ & r = 1,2,...,s; \\
x_{io} &= \sum_{j=1}^{n} x_{ij}\lambda_j - s_i^- & i = 1,2,...,m; \\
1 &= \sum_{j=1}^{n} \lambda_j & \\
0 &\le \lambda_j, s_i^-, s_r^+, \quad \forall i,j,r.
\end{aligned} \tag{7.14}
$$

To see what is involved, we note that the input (like the output) constraints take the form $x_{io} < \sum x_{ij}\lambda_j$, with the slacks $s_i^- \ge 0$ used to replace the inequalities with equivalent equations. Hence, the solution values of the inputs, just like the outputs, all need to equal or exceed these observed values, x_{io}, for this DMUo, so that no input reductions are allowed in this model. In this adaptation of the additive models, the outputs are to be maximized without reducing any inputs. Formally, in (7.14) the inputs may be increased provided this does not reduce the maximum output values that might otherwise be available.

To illustrate what is involved we use the coordinate values in Figure 7-5 and employ the model in (7.14) to evaluate E as DMUo. This produces,

$$
\max s^+
$$
subject to
$$
\begin{aligned}
1 &= 0.5\lambda_A + 2\lambda_B + 2\lambda_C + 1\lambda_D + 1\lambda_E + 1\lambda_H - s^+ \\
4 &= 1\lambda_A + 2\lambda_B + 3\lambda_C + 5\lambda_D + 4\lambda_E + 3\lambda_H - s^- \\
1 &= \lambda_A + \lambda_B + \lambda_C + \lambda_D + \lambda_E + \lambda_H \\
0 &\le \lambda_A,..., \lambda_H, s^+, s^-.
\end{aligned}
$$

This problem has as its solution
$$
\lambda_C^* = \lambda_D^* = 1/2, \quad s^{+*} = 1/2,
$$

and all other variables zero. Hence we have $y_o + s^{+*} = 1 + 1/2 = 1.5$, so the coordinates of E=(4,1) give way to E'=(4, 1.5). Thus, without any reduction in this congesting amount of input we get an increase in output. Finally, to show that this point is on the line segment connecting C and D in Figure 7.5, we need merely substitute $x_{E'} = 4$ in $y = 7/2 - 1/2 * x$, the equation that corresponds to this line segment, to obtain y = 7/2−4/2 =3/2 and thereby confirm the previously obtained 1/2 unit increase that replaces y = 1 with y = 3/2 via the just displayed solution.

We can now use the following formulas to estimate the output reductions due to congestion

$$\hat{y}_{ro} - \breve{y}_{ro} \geq 0, \; r = 1,...,s, \tag{7.15}$$

where the $\hat{y}_{ro}$ are obtained from (7.8) or (7.14)--depending on which model was used--and the $\breve{y}_{ro}$ are obtained from the solution to (7.14) via the following formulas

$$\breve{y}_{ro} = y_{ro} + s_r^{+*}, \; r = 1,...,s, \tag{7.16}$$

with the s_r^{+*} obtained from (7.14) after modifying its constraints by replacing $x_{io} \leq \sum x_{ij}\lambda_j$ with $x_{io} = \sum x_{ij}\lambda_j$, i=1,2,...,m.

For illustration we note that $s^{+*} = 1/2$ is the only non-zero slack in the solution to the preceding numerical example and that y=1. Hence, using (7.16), $\breve{y}_{ro} = y + s^{+*} = 1\frac{1}{2}$ is found to be the ordinate for E' in Figure 7-5. We next subtract this result from the value of $\hat{y} = 2$ which was obtained by applying (7.7) and (7.8) to the evaluation of E. This gives $\hat{y} - \breve{y} = 2 - 1\frac{1}{2} = \frac{1}{2}$ as the output lost because of the congesting input.

Finally, to see that the inequality represented in (7.15) is justified we note that the optimization obtained from the modification of (7.14)[9] is restricted to a very small subset of the region available to the models used to obtain the $\hat{y}_{ro}$. Reference to the definition of congestion given in (7.1) then shows that no increases in any output can accompany the decreases in some outputs[10] as a result of the congestion associated with some of the inputs.

7.   EXTENTIONS

We have now covered the standard approaches used for treating congestion in DEA. Progress continues to be made with other models that have recently appeared in the literature. We do not cover these models in detail because experience with their use has not accumulated to a point where this appears to be warranted. We simply refer to them with a few brief remarks and provide references for readers who wish to study them further.

---

[9] That is, the modification that replaces the input inequalities with equations in (7.14).

[10] This condition is necessary to distinguish the effects of congestion from the case where output substitutions with accompanying changes in inputs are made in which some output decreases are needed to secure increases in other outputs. Thus, if $\hat{y} - \breve{y} < 0$ for any r=1,..., s, then "substitution" rather than "congestion" is present.

One such model due to Cooper, Deng, Huang and Li, (2002a) is referred to as the "one-model approach" because it replaces the use of two models, as in the other approaches we have covered. This model can be written

$$\max \phi + \varepsilon \left(\sum_{r=1}^{s} s_r^+ - \sum_{i=1}^{m} s_i^{-c} \right)$$

subject to

$$\phi \, y_{r0} - \sum_{j=1}^{n} y_{rj} \lambda_j + s_r^+ = 0, \quad r = 1, \ldots, s, \qquad (7.17)$$

$$\sum_{j=1}^{n} x_{ij} \lambda_j + s_i^{-c} = x_{i0}, \quad i = 1, \ldots, m,$$

$$\sum_{j=1}^{n} \lambda_j = 1,$$

$$0 \le \lambda_j; \; s_i^{-c}, j = 1, \ldots, n; \; i = 1, \ldots, m.$$

As can be seen, the constraints in this model are the same as for (7.6), but (i) the objective of (7.6) is modified by replacing $+s_i^-$ with $-s_i^{-c}$ for each i=1,...,m and (ii) the input constraints replace the usual slacks s_i^- with s_i^{-c}.

A series of theorems proved by Cooper, Deng, Huang and Li (2002a) are followed by examples that guide the use of (7.17). We do not reproduce these theorems here, however, other than to note that an optimum with $s^{-C^*} > 0$ is required to show that congestion is present in the performance of the DMUo being evaluated. However, access to the decomposition represented in (7.9) is lost so that identification of technical inefficiency is not available. Other information on efficient vs. inefficient behavior is also absent from this "one-model" approach. Hence, if this information is wanted, recourse may be made to one of the models described earlier in this chapter.

This is, of course, not the end of the line for new openings to research in congestion and its management. Brockett et al. (2003) extend the study of congestion to include other ways to improve its management. This includes providing ways to estimate tradeoffs between inputs and outputs that extend the use of concepts such as "marginal rates of transformation" (from economics) in ways that can further improve the output values. This includes, for instance, the use of such estimates to assign congesting inputs to different plants or departments to further reduce its effects on outputs. This could prove useful to the Chinese government, for instance, by allowing it to deal with its huge and rapidly growing labor force in ways that can reduce the diminutions in output that might otherwise occur when its labor assignments result in congestion.

Finally, we note that work on introducing stochastic elements into the treatment of congestion, and other parts of DEA, has also begun to appear. See Cooper, Deng, Huang and Li (2002b, 2003) who treat both technical inefficiencies and congestion with chance constrained programming models.

See Chapter 9 in this handbook for other uses of Chance Constrained Programming in DEA.

REFERENCES

1. Balakrishnan, R. and N.S. Soderstrom, 2000, "The cost of system congestion: Evidence from the health care sector," *Journal of Management Accounting Research* 12, 97-114.
2. Brockett, P.L., W.W. Cooper, H. Deng, L.L. Golden and T.W. Ruefli, 2003, "Using DEA to identify and manage congestion," *Journal of Productivity Analysis* (submitted).
3. Brockett, P.L., W.W. Cooper, H.C. Shin and Y. Wang, 1998, "Inefficiency and congestion in Chinese production before and after the 1978 economic reforms," *Socio-Economic Planning Sciences*, 32, 1-20.
4. Charnes, A. and W.W. Cooper, 1961, *Management Models and Industrial Applications of Linear Programming.* John Wiley and Sons, Inc. New York.
5. Cherchye, L., T. Kuosmanen and T. Post, 2001, "Alternative treatments of congestion in DEA: A rejoinder to Cooper, Gu and Li," *European Journal of Operational Research* 132, 75-80.
6. Cooper, W.W., H. Deng, B. Gu, S. Li and R.M. Thrall, 2001a, "Using DEA to improve the management of congestion in Chinese industry (1981-1997)" *Socio-Economic Planning Sciences* 35, 1-16.
7. Cooper, W.W., H. Deng, Z. Huang and S.X. Li, 2003, "Chance Constrained Programming approaches to congestion in stochastic Data Envelopment Analysis." *European Journal of Operational Research*, (to appear).
8. Cooper, W.W., H. Deng, Z. Huang and S.X. Li, 2002a, "A one-model approach to the analysis of congestion in Data Envelopment Analysis," *Socio-Economic Planning Sciences* 36, 231-238.
9. Cooper, W.W., H. Deng, Z. Huang and S.X. Li, 2002b, "Chance Constrained Programming approaches to technical efficiencies and inefficiencies in stochastic Data Envelopment Analysis." *Journal of the Operational Research Society*, 53, 1-10.
10. Cooper, W.W., B. Gu, and S. Li, 2001a, "Alternative treatments of congestion in DEA: A response to the Cherchye, Kuosmanen and Post critique," *European Journal of Operational Research* 132 81-87.
11. Cooper, W.W., B. Gu and S. Li, 2001b, "Comparisons and evaluations of alternative approaches to evaluating congestion in DEA," *European Journal of Operational Research* 32/1, 1-13.

12. Cooper, W.W., K.S. Park and J.T. Pastor, 1999, "RAM: A range adjusted measure of inefficiency for use with additive models and relations to other models and measures in DEA", *Journal of Productivity Analysis*, 11, 5-42.

13. Cooper, W.W., L.M. Seiford and K. Tone, 2000, *Data Envelopment Analysis: A Comprehensive Text with Uses, Example Applications, References and DEA-Solver Software.* Kluwer Academic Publishers, Norwell, Mass.

14. Cooper, W.W., L.M. Seiford and J. Zhu, 2000, "A unified additive model approach for evaluating inefficiency and congestion with associated measures in DEA," *Socio-Economic Planning Sciences* 34, 1-25.

15. Cooper, W.W., L.M. Seiford and J. Zhu, 2001c, "Slacks and congestion: A response to comments by Färe and Grosskopf," *Socio-Economic Planning Sciences,* 35, 1-11.

16. Cooper, W.W., R.G. Thompson and R.M. Thrall, 1996, "Introduction: Extensions and new developments in DEA," *Annals of Operations Research* 66, 3-45.

17. Deng, H.H. (2003) Congestion and its management in Chinese production. Ph.D. Thesis, Austin, Texas: The Red McCombs School of Business, University of Texas at Austin. Also available from University Microfilms, Inc., Ann Arbor, Mich.

18. Färe, R. and S. Grosskopf, 1998, "Congestion: A note" *Socio-Economic Planning Sciences* 33, 21-23.

19. Färe, R. and S. Grosskopf, 2000a, "Congestion: A response," *Socio-Economic Planning Sciences* 34, 35-50.

20. Färe, R. and S. Grosskopf, 1983, "Measuring congestion in production," *Zeitschrift Für Nationalökonomie* 43, 251-271.

21. Färe, R. and S. Grosskopf, 2000b, "Slacks and congestion: A comment," *Socio-Economic Planning Sciences* 34, 27-33.

22. Färe, R., S. Grosskopf and C.A.K. Lovell, 1994, *Production Frontiers,* Cambridge University Press, Cambridge, England.

23. Färe, R., S. Grosskopf and C.A.K. Lovell, 1985, *The Measurement of Efficiencies of Production,* Kluwer-Nihoff Publishing, Boston, Mass.

24. Färe, R., S. and L. Svensson, 1980, "Congestion of factors of production," *Econometrica* 48 1743-1753.

25. Leibenstein, H., 1966, "Allocative efficiency vs. X-Efficiency," *American Economic Review* 56, 392-415.

26. Leibenstein, H., 1976, *Beyond Economic Man*, Harvard University Press, Cambridge, Mass.

27. Samuelson, 1947, *Foundations of Economics*, Harvard University Press, Cambridge, Mass.

28. Seiford, L.M., 1994, "References" in A. Charnes, W.W. Cooper, and A.Y. Lewin eds, *Data Envelopment Analysis: Theory, Methodology and Applications*, Kluwer Academic Publisher, Norwell, Mass.
29. Stigler, G.J., 1976, "The X-istence of X-efficiency," *American Economic Review* 66, 213-216.
30. Varian, H., 1984, *Microeconomic Analysis*, 2nd ed. W.W. Norton, Inc., New York, N.Y.

Chapter 8

MALMQUIST PRODUCTIVITY INDEX
Efficiency Change Over Time

Kaoru Tone
National Graduate Institute for Policy Studies, 2-2 Wakamatsu, Shinjuku, Tokyo 162-8677, Japan   email: tone@grips.ac.jp

Abstract: The Malmquist index (MI) evaluates the efficiency change over time. In the non-parametric framework, it is measured as the product of catch-up (or recovery) and frontier-shift (or innovation) terms, both coming from the DEA technologies. We introduce three different approaches for measuring the Malmquist index along with scale efficiency related subjects.

Key words: Malmquist index, catch-up, recovery, frontier-shift, innovation, radial, non-radial, slacks-based measure, oriented, non-oriented

1.   INTRODUCTION

The concept of Malmquist productivity index was first introduced by Malmquist (1953), and has further been studied and developed in the non-parametric framework by several authors. See for example, among others, Caves, Christensen and Diewert (1982), Färe and Grosskopf (1992), Färe, Grosskopf, Lindgren and Roos (1989, 1994), Färe, Grosskopf and Russell (1998b) and Thrall (2000). It is an index representing Total Factor Productivity (TFP) growth of a Decision Making Unit (DMU), in that it reflects progress or regress in efficiency along with progress or regress of the frontier technology over time under the multiple inputs and multiple outputs framework. In this chapter we first introduce the concept of Malmquist productivity index, and then present three different approaches for its measurement: radial, non-radial, and non-radial and non-oriented approaches. Scale efficiency change is discussed in Section 4. We demonstrate with numerical examples for comparing different approaches in Section 5. Concluding remarks follow in Section 6.

2. DEALING WITH PANEL DATA

The Malmquist index evaluates the productivity change of a DMU between two time periods. It is defined as the product of "Catch-up" and "Frontier-shift" terms. The catch-up (or recovery) term relates to the degree that a DMU attains for improving its efficiency, while the frontier-shift (or innovation) term reflects the change in the efficient frontiers surrounding the DMU between the two time periods. We deal with a set of n DMUs $(x_j, y_j)(j = 1, \cdots, n)$ each having m inputs denoted by a vector $x_j \in R^m$ and q outputs denoted by a vector $y_j \in R^q$ over the periods 1 and 2. We assume $x_j > 0$ and $y_j > 0 (\forall j)$. The notations $(x_o, y_o)^1$ and $(x_o, y_o)^2$ are employed for designating DMU$_o$ ($o = 1, \cdots, n$) in the periods 1 and 2 respectively.

The production possibility set $(X,Y)^t$ (t = 1 and 2) spanned by $\{(x_j, y_j)^t\}$ ($j = 1, \cdots, n$) is defined by:

$$(X,Y)^t = \left\{ (x, y) \middle| x \geq \sum_{j=1}^{n} \lambda_j x_j^t, 0 \leq y \leq \sum_{j=1}^{n} \lambda_j y_j^t, L \leq e\lambda \leq U, \lambda \geq 0 \right\},$$

where e is the row vector with all elements being equal to one and $\lambda \in R^n$ is the intensity vector, and L and U are the lower and upper bounds of the sum of the intensity respectively. $(L,U)=(0, \infty)$, $(1,1)$, $(1, \infty)$ and $(0,1)$ correspond to the CCR (Charnes-Cooper-Rhodes or Constant Returns to Scale = CRS), BCC (Banker-Charnes-Cooper or Variable Returns to Scale = VRS), IRS (Increasing Returns to Scale) and DRS (Decreasing Returns to Scale) models respectively. The production possibility set $(X,Y)^t$ is characterized by frontiers that are composed of $(x, y) \in (X,Y)^t$ such that it is not possible to improve any element of the input x or any element of the output y without worsening some other input or output. See Cooper et al. (1999) for further discussion on this and related subjects. We call this frontier set the *frontier technology* at the period t. In the Malmquist index analysis, the efficiencies of DMUs $(x_o, y_o)^1$ and $(x_o, y_o)^2$ are evaluated by the frontier technologies 1 and 2 in several ways.

2.1 Catch-up effect

The catch-up effect is measured by the following formula.

$$\text{Catch} - \text{up} = \frac{\text{Efficiency of } (x_o, y_o)^2 \text{ w.r.t. the period 2 frontier}}{\text{Efficiency of } (x_o, y_o)^1 \text{ w.r.t. the period 1 frontier}}. \quad (8.1)$$

We evaluate catch-up of each DMU defined in the above formula by appropriate DEA models. A simple example in the case of a single input and output technology is illustrated in Figure 8-1.

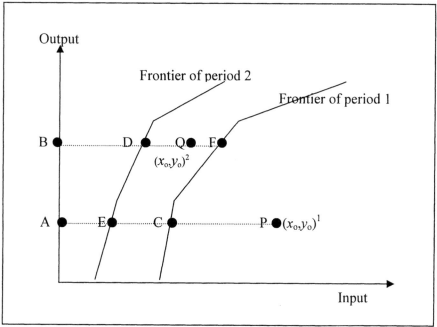

Figure8-1.: Catch-up

The catch-up effect (in input-orientation) can be computed as:

$$\text{Catch} - \text{up} = \frac{\text{BD}}{\text{BQ}} \bigg/ \frac{\text{AC}}{\text{AP}}.$$

(8.2)

(Catch-up) > 1 indicates progress in the relative efficiency from period 1 to 2, while (Catch-up) = 1 and (Catch-up) < 1 indicate respectively no change and regress in efficiency.

2.2 Frontier-shift effect

In addition to the catch-up term, we must take account of the frontier-shift (innovation) effect in order to fully evaluate the productivity change since the catch-up effect is determined by the efficiencies being measured by the distances from the respective frontiers. In Figure 8-1 case, this can be explained as follows. The reference point C of $(x_o, y_o)^1$ moved to E on the frontier of period 2. Thus, the frontier-shift effect at $(x_o, y_o)^1$ is evaluated by

$$\varphi_1 = \frac{\text{AC}}{\text{AE}}.$$

(8.3)

This is equivalent to

$$\varphi_1 = \frac{\dfrac{AC}{AP}}{\dfrac{AE}{AP}} = \frac{\text{Efficiency of } (x_o, y_o)^1 \text{ w.r.t. the period 1 frontier}}{\text{Efficiency of } (x_o, y_o)^1 \text{ w.r.t. the period 2 frontier}}. \qquad (8.4)$$

The numerator of the right hand side of (8.4) is already captured in (8.1). The denominator is measured as the efficiency score of $(x_o, y_o)^1$ relative to the period 2 frontier. Similarly, the frontier-shift effect at $(x_o, y_o)^2$ is expressed by

$$\varphi_2 = \frac{\dfrac{BF}{BQ}}{\dfrac{BD}{BQ}} = \frac{\text{Efficiency of } (x_o, y_o)^2 \text{ w.r.t. the period 1 frontier}}{\text{Efficiency of } (x_o, y_o)^2 \text{ w.r.t. the period 2 frontier}}. \qquad (8.5)$$

Using φ_1 and φ_2, we define "Frontier-shift" effect by their geometric mean, i.e.,

$$\text{Frontier - shift} = \varphi = \sqrt{\varphi_1 \varphi_2}. \qquad (8.6)$$

(Frontier-shift) > 1 indicates progress in the frontier technology around the DMU_0 from period 1 to 2, while (Frontier-shift) = 1 and (Frontier-shift) < 1 indicate respectively the status quo and regress in the frontier technology.

2.3   Malmquist index

The Malmquist index (MI) is computed as the product of (Catch-up) and (Frontier-shift), i.e.,

$$MI = (\text{Catch - up}) \times (\text{Frontier - shift}). \qquad (8.7)$$

We now employ the following notation for the efficiency score of DMU $(x_o, y_o)^{t_1}$ measured by the frontier technology t_2.

$$\delta^{t_2}((x_o, y_o)^{t_1}) \qquad (t_1 = 1, 2 \text{ and } t_2 = 1, 2). \qquad (8.8)$$

Using this notation, the catch-up effect (C) in (8.1) can be expressed as

$$C = \frac{\delta^2((x_o, y_o)^2)}{\delta^1((x_o, y_o)^1)}. \qquad (8.9)$$

The frontier-shift effect is described as

$$
F = \left[\frac{\delta^1((x_o, y_o)^1)}{\delta^2((x_o, y_o)^1)} \times \frac{\delta^1((x_o, y_o)^2)}{\delta^2((x_o, y_o)^2)} \right]^{1/2} . \qquad (8.10)
$$

As the product of C and F, we obtain the following formula for the computation of MI, the Malmquist Index,

$$
MI = \left[\frac{\delta^1((x_o, y_o)^2)}{\delta^1((x_o, y_o)^1)} \times \frac{\delta^2((x_o, y_o)^2)}{\delta^2((x_o, y_o)^1)} \right]^{1/2} . \qquad (8.11)
$$

This last expression gives an another interpretation of MI, i.e., the geometric means of the two efficiency ratios: the one being the efficiency change measured by the period 1 technology and the other the efficiency change measured by the period 2 technology.

As can be seen from these formulas, the MI consists of four terms: $\delta^1((x_o, y_o)^1)$, $\delta^2((x_o, y_o)^2)$, $\delta^1((x_o, y_o)^2)$ and $\delta^2((x_o, y_o)^1)$. The first two are related to the measurements *within* the same time period, while the last two are for *intertemporal* comparison.

MI >1 indicates progress in the total factor productivity of the DMU$_o$ from period 1 to 2, while MI $= 1$ and MI < 1 indicate respectively the status quo and decay in the total factor productivity.

3. MEASUREMENT OF MALMQUIST INDEX

In the non-parametric framework the Malmquist index (MI) is constructed by means of DEA technologies. There are a number of ways to compute MI. First, Färe, Grosskopf, Lindgren and Roos (1989, 1994) utilized the input and output oriented radial DEA model to compute MI. However, the radial model suffers from one shortcoming, i.e., neglect of slacks. To overcome this shortcoming, MI can be computed using the "slacks-based" non-radial and oriented DEA model. Alternatively MI can be computed from the non-radial and non-oriented DEA model.

3.1 The radial MI

The input-oriented radial MI measures the *within* and *intertemporal* scores by the linear programs given below.

[*Within* score in input-orientation]

$$\delta^s((x_o, y_o)^s) = \min_{\theta, \lambda} \theta \qquad (8.12)$$
$$\text{subject to } \theta\, x_o^s \geq X^s \lambda$$
$$y_o^s \leq Y^s \lambda$$
$$L \leq e\lambda \leq U$$
$$\lambda \geq 0,$$

where $X^s = (x_1^s, \cdots, x_n^s)$ and $Y^s = (y_1^s, \cdots, y_n^s)$ are respectively input and output matrices (observed data) at the period s. We solve this program for s = 1 and 2. It holds that $\delta^s((x_o, y_o)^s) \leq 1$, and $\delta^s((x_o, y_o)^s) = 1$ indicates $(x_o, y_o)^s$ being on the technically efficient frontiers of $(X, Y)^s$.

[*Intertemporal* score in input-orientation]

$$\delta^s((x_o, y_o)^t) = \min_{\theta, \lambda} \theta \qquad (8.13)$$
$$\text{subject to } \theta\, x_o^t \geq X^s \lambda$$
$$y_o^t \leq Y^s \lambda$$
$$L \leq e\lambda \leq U$$
$$\lambda \geq 0.$$

We solve this program for the pairs $(s, t) = (1, 2)$ and $(2, 1)$. If $(x_o, y_o)^t$ is not enveloped by the technology at the period s, the score $\delta^s((x_o, y_o)^t)$, if it exists, results in the value greater than 1. This corresponds to the concept of super-efficiency proposed by Andersen and Petersen (1993).

Although the above schemes are input-oriented, we can develop the output-oriented MI as well by means of the output-oriented radial DEA models. This is explained below.

[*Within* score in output-orientation]

$$\delta^s((x_o, y_o)^s) = \min_{\theta, \lambda} \theta \qquad (8.14)$$
$$\text{subject to } x_o^s \geq X^s \lambda$$
$$(1/\theta)\, y_o^s \leq Y^s \lambda$$
$$L \leq e\lambda \leq U$$
$$\lambda \geq 0.$$

[*Intertemporal* score in output-orientation]

$$\delta^s((x_o, y_o)^t) = \min_{\theta, \lambda} \theta \qquad (8.15)$$
$$\text{subject to } x_o^t \geq X^s \lambda$$
$$(1/\theta)\, y_o^t \leq Y^s \lambda$$
$$L \leq e\lambda \leq U$$
$$\lambda \geq 0.$$

[*Remark* 8.1: Inclusive or Exclusive scheme]

In evaluating the *within* score $\delta^s((x_o, y_o)^s)$, there are two schemes: one 'inclusive' and the other 'exclusive'. 'Inclusive' scheme means that, when we evaluate $(x_o, y_o)^s$ with respect to the technology $(X, Y)^s$, the DMU $(x_o, y_o)^s$ is always included in the evaluator $(X, Y)^s$, thus resulting in a score not greater than 1. 'Exclusive' scheme employs a method in which the DMU $(x_o, y_o)^s$ is removed from the evaluator group $(X, Y)^s$. This method of evaluation is equivalent to that of super-efficiency evaluation, and the score, if it exists, may be greater than 1. The *intertemporal* comparisons naturally utilize this 'exclusive' scheme. So adoption of this scheme even in the *within* evaluations is not unnatural and promotes the discrimination power. Refer to Section 5.2 for further discussions on this subject.

[*Remark* 8.2: Infeasible LP issues]

In the BCC (VRS) model [$(L, U) = (1,1)$: variable returns to scale], it may occur that the intertemporal LP (8.13) has no solution in its input and output orientations. In the case of input-oriented model, (8.13) has no feasible solution if there exists i such that $y'_{io} > \max_j \{y^s_{ij}\}$ whereas in the output-oriented case, (8.15) has no feasible solution if there exists i such that $x'_{io} < \min_j \{x^s_{ij}\}$. In the IRS model [$(L, U) = (1, \infty)$: increasing returns to scale], it may occur that the output-oriented intertemporal LP has no solution, while the input-oriented case is always feasible. In case of DRS model [$(L, U) = (0,1)$: decreasing returns to scale], it might be possible that the input-oriented (8.13) has no solution, while the output-oriented model is always feasible. However, the CRS (CCR) model does not suffer from such trouble in its intertemporal measurement. Refer to Seiford and Zhu (1999) for infeasibility issues of radial super-efficiency models. One solution for avoiding this difficulty is to assign 1 to the score, since we have no means to evaluate the DMU within the evaluator group.

3.2   The non-radial and slacks-based MI

As noted earlier, the radial approaches suffer from the neglect of slacks. In an effort to overcome this problem, several authors, e.g., Zhu (1996, 2001), Tone (2001, 2002) and Chen (2003) have developed non-radial measures of efficiency and super-efficiency. We develop here the non-radial and slacks-based MI.

First, we introduce the input-oriented SBM (slacks-based measure) and -Super-SBM (Tone (2001, 2002)). The SBM evaluates the efficiency of the examinee $(x_o, y_o)^s$ ($s = 1, 2$) with respect to the evaluator set $(X, Y)^t$ ($t = 1$, 2) with help of the following LP:

[SBM-I]

$$\delta'((x_o, y_o)^s) = \min_{\lambda, s^-} \quad 1 - \frac{1}{m}\sum_{i=1}^{m} s_i^- / x_{io}^s \tag{8.16}$$
$$\text{subject to } x_o^s = X'\lambda + s^-$$
$$y_o^s \leq Y'\lambda$$
$$L \leq e\lambda \leq U$$
$$\lambda \geq 0, s^- \geq 0.$$

Or equivalently,
[SBM-I]

$$\delta'((x_o, y_o)^s) = \min_{\theta, \lambda} \frac{1}{m}\sum_{i=1}^{m} \theta_i \tag{8.17}$$
$$\text{subject to } \theta_i x_{io}^s \geq \sum_j \lambda_j x_{ij}' \quad (i = 1, \cdots, m)$$
$$y_o^s \leq Y'\lambda$$
$$\theta_i \leq 1 (\forall i)$$
$$L \leq e\lambda \leq U$$
$$\lambda \geq 0,$$

where the vector $s^- \in R^m$ denotes input-slacks. The equivalence between (8.16) and (8.17) can be shown as follows: Let $\theta_i = (1 - s_i^- / x_{io}^s)$. Then it holds that $\theta_i \leq 1 (\forall i)$ and the equivalence follows straightforward. This model takes input slacks (surpluses) into account but no output slacks (shortfalls). Notice that, under the 'inclusive' scheme (see *Remark* 8.2 above), [SBM-I] is always feasible for the case where $s = t$. However, under the 'exclusive' scheme, we remove $(x_o, y_o)^s$ from the evaluator group $(X, Y)^s$ and hence [SBM-I] may have no feasible solution even for the case $s = t$. In this case, we solve the [Super-SBM-I] below:

[Super-SBM-I]

$$\delta'((x_o, y_o)^s) = \min_{\lambda, s^-} \quad 1 + \frac{1}{m}\sum_{i=1}^{m} s_i^- / x_{io}^s \tag{8.18}$$
$$\text{subject to } x_o^s \geq X'\lambda - s^-$$
$$y_o^s \leq Y'\lambda$$
$$L \leq e\lambda \leq U$$
$$\lambda \geq 0, s^- \geq 0.$$

Or equivalently,
[Super-SBM-I]

$$\delta'((x_o, y_o)^s) = \min_{\theta, \lambda} \frac{1}{m}\sum_{i=1}^{m} \theta_i \tag{8.19}$$
$$\text{subject to } \theta_i x_{io}^s \geq \sum_j \lambda_j x_{ij}' \quad (i = 1, \cdots, m)$$
$$y_o^s \leq Y'\lambda$$

$$\theta_i \geq 1 \ (\forall i)$$
$$L \leq e\lambda \leq U$$
$$\lambda \geq 0.$$

In this model, the score, if exists, satisfies $\delta'((x_o, y_o)^s) \geq 1$.
  In the output-oriented case, we solve the following LPs.

[SBM-O]

$$\delta'((x_o, y_o)^s) = \min_{\lambda, s^+} \ 1 \Big/ (1 + \tfrac{1}{q} \sum_{i=1}^{q} s_i^+ / y_{io}^s) \tag{8.20}$$

$$\text{subject to } x_o^s \geq X'\lambda$$

$$y_o^s = Y'\lambda - s^+$$

$$L \leq e\lambda \leq U$$

$$\lambda \geq 0, s^+ \geq 0,$$

where the vector $s^+ \in R^q$ denotes the output-slacks.

Or equivalently,
[SBM-O]

$$\delta'((x_o, y_o)^s) = \min_{\lambda, \phi} \ 1 \Big/ (\tfrac{1}{q} \sum_{i=1}^{q} \phi_i) \tag{8.21}$$

$$\text{subject to } x_o^s \geq X'\lambda$$

$$\phi_i y_o^s \leq \sum_{j=1}^{n} y_{ij}^t \lambda_j \quad (i = 1, \ldots, q)$$

$$\phi_i \geq 1 \quad (\forall i)$$

$$L \leq e\lambda \leq U$$

$$\lambda \geq 0.$$

[Super-SBM-O]

$$\delta'((x_o, y_o)^s) = \min_{\lambda, s^+} \ 1 \Big/ (1 - \tfrac{1}{q} \sum_{i=1}^{q} s_i^+ / y_{io}^s) \tag{8.22}$$

$$\text{subject to } x_o^s \geq X'\lambda$$

$$y_o^s \leq Y'\lambda + s^+$$

$$L \leq e\lambda \leq U$$

$$\lambda \geq 0, s^+ \geq 0.$$

Or equivalently,

[Super-SBM-O]

$$\delta'((x_o, y_o)^s) = \min_{\lambda, \phi} 1 \Big/ \big(\tfrac{1}{q}\sum_{i=1}^{q} \phi_i\big) \qquad (8.23)$$

$$\text{subject to } x_o^s \geq X'\lambda$$

$$\phi_i y_o^s \leq \sum_{j=1}^{n} y'_{ij}\lambda_j \quad (i = 1, \cdots, q)$$

$$0 \leq \phi_i \leq 1 \quad (\forall i)$$

$$L \leq e\lambda \leq U$$

$$\lambda \geq 0.$$

The output-oriented models take all output slacks (shortfalls) into account but no input slacks (surpluses).

The non-radial and slacks-based MI evaluates the four elements of MI: $\delta^1((x_o, y_o)^1)$, $\delta^2((x_o, y_o)^2)$, $\delta^1((x_o, y_o)^2)$ and $\delta^2((x_o, y_o)^1)$ by means of the LPs [SBM-I] and [Super-SBM-I]. This is now explained below.

3.2.1   The inclusive scheme case

1. The within scores $\delta^1((x_o, y_o)^1)$ and $\delta^2((x_o, y_o)^2)$ are measured by solving [SBM-I] or [SBM-O]. They are not greater than 1.
2. For the intertemporal scores $\delta^1((x_o, y_o)^2)$ and $\delta^2((x_o, y_o)^1)$, we solve the corresponding [SBM-I] or [SBM-O]. If they are feasible, the optimal value is the score that does not exceed 1. Otherwise, if they are infeasible, we apply [Super-SBM-I] or [Super-SBM-O] for measuring the score. In this case, the score, if exists, is always greater than 1.

3.2.2   The exclusive scheme case

1. To compute each component of MI, we apply [SBM-I] or [SBM-O] under the exclusive scheme. If they are feasible, the optimal value is the score that will never exceed 1.
2. Otherwise, if they are infeasible, we solve [Super-SBM-I] or [Super-SBM-O] to compute the score. In this case, the score, if exists, is always greater than 1.

[*Remark* 8.3: Another Non-radial model]

Chen (2003) proposed a non-radial Malmquist productivity index in which free slacks s^- in (8.14) are allowed. This model is also able to incorporate the decision maker's preference over performance improvement as introduced by Zhu (1996). Chen's model is described, in the non-preference case, as follows:

$$\delta'((x_o, y_o)^s) = \min_{\theta, \lambda} \frac{1}{m} \sum_{i=1}^{m} \theta_i \qquad (8.24)$$

$$\text{subject to } \theta_i x_{io}^s \geq \sum_{j=1}^{n} \lambda_j x_{ij}^t \qquad (i = 1, \cdots, m)$$

$$y_o^s \leq Y' \lambda$$

$$\theta_i : \text{free } \forall i$$

$$\lambda \geq 0.$$

[SBM-I] in (8.17) and [Super-SBM-I] in (8.19) correspond to the cases with the additional conditions $\theta_i \leq 1 (\forall i)$ and $\theta_i \geq 1 (\forall i)$ to (8.24), respectively.

[*Remark* 8.4: Infeasible LP issues]
These models may suffer from the same infeasible troubles as the radial ones may encounter.

3.3 The non-radial and non-oriented MI

The models in this category deal with input and output slacks. The [SBM] and [Super-SBM] models used for computing $\delta'((x_o, y_o)^s)$ are represented by the following fractional programs:

[SBM]

$$\delta'((x_o, y_o)^s) = \min_{\lambda, s^-, s^+} \left(1 - \frac{1}{m} \sum_{i=1}^{m} s_i^- / x_{io}^s \right) \Big/ \left(1 + \frac{1}{q} \sum_{i=1}^{q} s_i^+ / y_{io}^s \right) \qquad (8.25)$$

$$\text{subject to } x_o^s = X' \lambda + s^-$$

$$y_o^s = Y' \lambda - s^+$$

$$L \leq e\lambda \leq U$$

$$\lambda \geq 0, s^- \geq 0, s^+ \geq 0.$$

Or equivalently,
[SBM]

$$\delta'((x_o, y_o)^s) = \min_{\theta, \phi, \lambda} \left(\frac{1}{m} \sum_{j=1}^{m} \theta_i \right) \Big/ \left(\frac{1}{q} \sum_{i=1}^{q} \phi_i \right) \qquad (8.26)$$

$$\text{subject to } \theta_i x_{io}^s \geq \sum_{j=1}^{n} x_{ij}^t \lambda_j \quad (i=1,\cdots,m)$$

$$\phi_i y_{io}^s \leq \sum_{j=1}^{n} y_{ij}^t \lambda_j \quad (i=1,\cdots,q)$$

$$\theta_i \leq 1 (\forall i), \quad \phi_i \geq 1 (\forall i)$$

$$L \leq e\lambda \leq U$$

$$\lambda \geq 0.$$

[Super-SBM]

$$\delta'((x_o,y_o)^s) = \min_{\lambda,s^-,s^+} \left(\frac{1}{m} \sum \overline{x}_i / x_{io}^s \right) \Big/ \left(\frac{1}{q} \sum_{i=1}^{q} \overline{y}_i / y_{io}^s \right) \quad (8.27)$$

$$\text{subject to } \overline{x} \geq X'\lambda$$

$$\overline{y} \leq Y'\lambda$$

$$\overline{x} \geq x_o^s, \overline{y} \leq y_o^s$$

$$L \leq e\lambda \leq U$$

$$\overline{y} \geq 0, \lambda \geq 0.$$

Or equivalently,
[Super-SBM]

$$\delta'((x_o,y_o)^s) = \min_{\theta,\phi,\lambda} \left(\frac{1}{m} \sum_{i=1}^{m} \theta_i \right) \Big/ \left(\frac{1}{q} \sum_{i=1}^{q} \phi_i \right) \quad (8.28)$$

$$\text{subject to } \theta_i x_{io}^s \geq \sum_{j=1}^{n} x_{ij}^t \lambda_j \quad (i=1,\cdots,m)$$

$$\phi_i y_{io}^s \leq \sum_{j=1}^{n} y_{ij}^t \lambda_j \quad (i=1,\cdots,q)$$

$$\theta_i \geq 1 (\forall i), \quad 0 \leq \phi_i \leq 1 (\forall i)$$

$$L \leq e\lambda \leq U$$

$$\lambda \geq 0.$$

These fractional programs can be transformed into LPs. See Tone (2002) for detailed discussions. This model under the exclusive scheme (see *Remark 8.1*) evaluates the four components of MI: $\delta^1((x_0,y_0)^1)$, $\delta^2((x_o,y_o)^2)$, $\delta^1((x_o,y_o)^2)$ and $\delta^2((x_o,y_o)^1)$ using [SBM], and, if the corresponding LP is found infeasible, we then apply [Super-SBM].

[*Remark* 8.5: Infeasible LP issues]

For this non-oriented model, [Super-SBM] is always feasible and has a finite minimum in any RTS environment, but under some mild conditions, i.e., for each output i (= $1, \cdots, q$) at least two DMUs have positive values. This can be seen from the constraints in (8.28). See Tone (2002) for the details.

4. SCALE EFFICIENCY CHANGE

Changes in efficiency may also be associated with returns to scale efficiencies. We therefore treat this topics as follows:

Let $\delta_C^{t_1}((x_o, y_o)^{t_2})$ and $\delta_V^{t_1}((x_o, y_o)^{t_2})$ denote the scores $\delta^{t_1}((x_o, y_o)^{t_2})$ (8.8) that are obtained under CRS and VRS environments respectively. It holds, for any combination of t_1 and t_2, that

$$\delta_C^{t_1}((x_o, y_o)^{t_2}) \le \delta_V^{t_1}((x_o, y_o)^{t_2}) . \qquad (8.29)$$

We define the *scale efficiency* of $(x_o, y_o)^{t_2}$ relative to the technology $(X, Y)^{t_1}$ by

$$\sigma^{t_1}(x_o, y_o)^{t_2} = \delta_C^{t_1}((x_o, y_o)^{t_2})/\delta_V^{t_1}((x_o, y_o)^{t_2}) . \qquad (8.30)$$

From (8.29) it holds that $\sigma^{t_1}(x_o, y_o)^{t_2} \le 1$, and, if $\sigma^{t_1}(x_o, y_o)^{t_2} = 1$, it means that $(x_o, y_o)^{t_2}$ is positioned in the *most productive scale size* (MPSS) region of the technology $(X, Y)^{t_1}$, as named by Banker (1984) and Banker et al. (1984). Hence, $\sigma^{t_1}(x_o, y_o)^{t_2}$ is an efficiency indicator showing how far $(x_o, y_o)^{t_2}$ deviates from the point of technically optimal scale of operation. Several authors (Färe et al. (1994), Lovell and Grifell-Tatje (1994), Ray and Desli (1997), and Balk (2001)) have made attempts to enhance the Malmquist index by taking into account scale efficiency change effects. We introduce here two of them.

4.1 Proposal by Ray and Delsi (1997)

The Malmquist index in the CRS environment, MI_C, can be rewritten via the Malmquist index in VRS, MI_V, as follows:

$$MI_C = MI_V \times \left[\frac{\sigma^1(x_o, y_o)^2}{\sigma^1(x_o, y_o)^1} \times \frac{\sigma^2(x_o, y_o)^2}{\sigma^2(x_o, y_o)^1} \right]^{1/2} \qquad (8.31)$$

Ray and Delsi interpret the term in the square parenthesis as the geometric mean of scale efficiency changes of (x_o, y_o) evaluated by the frotiers at the time periods 1 and 2. This can be illustrated in Figure 8-2 using the simple single-input and single-output case. The CRS frontier at the time period 1 is the dotted line (ray) through the origin as denoted by CRS^1, while the CRS frontier at 2 is the dotted line CRS^2. The VRS frontiers at 1(2) are the broken lines VRS^1 (VRS^2). The points $P(x_o, y_o)^1$ and $Q(x_o, y_o)^2$ designate the concerned DMU at 1 and 2 respectively. The numerator of the first fractional term in the parenthesis (8.31) above is (FI/FQ)/(FJ/FQ)=FI/FJ, while the denominator is (AE/AP)/(AC/AP)=AE/AC. Hence the first fractional term turns out to be (FI/FQ)/(AE/AC), i.e., the scale efficiency change from $P(x_o, y_o)^1$ to $Q(x_o, y_o)^2$ as evaluated by the $(X,Y)^1$ technology. Likewise, the second fractional term corresponds to the scale efficiency change evaluated by the $(X,Y)^2$ technology. Thus, their geometric mean indicates the average scale efficiency change of (x_o, y_o) from 1 to 2.

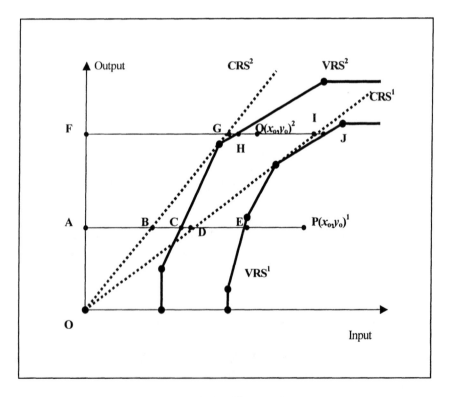

Figure 8-2. Scale efficiency change

Ray and Delsi decompose MI_C in the following manner.

$$MI_C = \text{Catch - up}(V) \times \text{Frontier - shift}(V) \times \text{Scale efficiency change} \qquad (8.32)$$
where

$$\text{Catch - up}(V) = \frac{\delta_V^2((x_o, y_o)^2)}{\delta_V^1((x_o, y_o)^1)} \qquad (8.33)$$

$$\text{Frontier - shift}(V) = \left[\frac{\delta_V^1((x_o, y_o)^1)}{\delta_V^2((x_o, y_o)^1)} \times \frac{\delta_V^1((x_o, y_o)^2)}{\delta_V^2((x_o, y_o)^2)}\right]^{1/2} \qquad (8.34)$$

and

$$\text{Scale - efficiency change} = \left[\frac{\sigma^1(x_o, y_o)^2}{\sigma^1(x_o, y_o)^1} \times \frac{\sigma^2(x_o, y_o)^2}{\sigma^2(x_o, y_o)^1}\right]^{1/2}. \qquad (8.35)$$

4.2   Proposal by Balk (2001)

Balk (2001) introduced two fictitious DMUs R $= (x_o^2, y_o^1)$ and S $= (x_o^1, y_o^2)$ in addition to P $= (x_o^1, y_o^1)$ and Q $= (x_o^2, y_o^2)$. R (S) differs from P (Q) in the input-mix. The (input oriented) scale efficiency relative to the $(X, Y)^1$ technology is defined by

$$SEC^1(x_o^1, y_o^1, y_o^2) = \frac{\sigma^1(x_o^1, y_o^2)}{\sigma^1(x_o^1, y_o^1)}. \qquad (8.36)$$

The numerator differs from Ray and Delsi's $\sigma^1(x_o^2, y_o^2)$. The rationale here is that, if this ratio is larger (smaller) than 1, the output combination y_o^2 lies closer to (farther away from) the point of technically optimal scale than y_o^1 did, distances being measured in the $x_o^1 / \|x_o^1\|$-direction. Likewise, the (input-oriented) scale efficiency change relative to $(X, Y)^2$ technology is defined by

$$SEC^2(x_o^2, y_o^1, y_o^2) = \frac{\sigma^2(x_o^2, y_o^2)}{\sigma^2(x_o^2, y_o^1)}.$$ (8.37)

The scale efficiency change can be measured as the geometric mean of (8.36) and (8.37). Balk further introduced the *input-mix effect* by

$$ME^1(x_o^1, x_o^2, y_o^2) = \frac{\sigma^1(x_o^2, y_o^2)}{\sigma^1(x_o^1, y_o^2)}.$$ (8.38)

This measures how the relative distance of y_o^2 to the frontier of the CRS technology changes at period 1 when the input-mix changes. Similarly, the input-mix effect relative to the period 2 technology is measured by

$$ME^2(x_o^1, x_o^2, y_o^1) = \frac{\sigma^2(x_o^2, y_o^1)}{\sigma^2(x_o^1, y_o^1)}.$$ (8.39)

The product of these four factors, (8.36), (8.37), (8.38) and (8.39), results in the Ray and Delsi's scale-efficiency change (8.35). Hence, the latter is a combination of the scale efficiency change and the input-mix effect proposed by Balk. Eventually, Balk's decomposition turns out to be:

$$MI_C = MI_V \times \left[SEC^1(\cdot) \times SEC^2(\cdot)\right]^{1/2} \times \left[ME^1(\cdot) \times ME^2(\cdot)\right]^{1/2}.$$ (8.40)

5.   ILLUSTRATIVE EXAMPLES FOR COMPARISONS OF MODELS

We here demonstrate and compare several models introduced in the preceding section using simple examples. The numerical results are obtained by using DEA-Solver Pro Version 4.0 (2003).

5.1   Radial vs. Non-radial

Table 8-1 exhibits a small example with two DMUs, A and B, each having two inputs x_1 and x_2, and a single output y, over two periods 1 and 2. At Period 1 DMU B has 1 unit of slacks in x_2 against A that is removed at Period 2.

For brevity, we use the notation $\delta^s(A^t)$ for designating the efficiency score of DMU A at Period t relative to the production frontier at Period s. The scores measured by the input-oriented radial CRS and the non-radial CRS models under the inclusive scheme are displayed in Table 8-2. The radial model records all scores as 1, whereas the non-radial (slacks-based) model returns $\delta^1(B^1)$ =0.75 and $\delta^2(B^1)$ =0.75 reflecting the existence of slacks in x_2 at Period 1.

Table 8-1. Example 1

	Period 1			Period 2		
DMU	x_1	x_2	y	x_1	x_2	y
A	1	1	1	1	1	1
B	1	2	1	1	1	1

Table 8-2. Input-oriented scores

	$\delta^1(A^1)$	$\delta^1(B^1)$	$\delta^1(A^2)$	$\delta^1(B^2)$	$\delta^2(A^1)$	$\delta^2(B^1)$	$\delta^2(A^2)$	$\delta^2(B^2)$
Radial	1	1	1	1	1	1	1	1
Non-radial	1	0.75	1	1	1	0.75	1	1

From Table 8-2, we obtained the catch-up (C), the frontier-shift (F) and the Malmquist index (MI) of each DMU from Period 1 to Period 2. These are recorded in Table 8-3.

Table 8-3. Catch-up, Frontier-shift and Malmquist index

	$C(A)$	$C(B)$	$F(A)$	$F(B)$	$MI(A)$	$MI(B)$
Radial	1	1	1	1	1	1
Non-radial	1	1.33	1	1	1	1.33

As can be seen from Table 8-1, DMU B has 1 unit of slacks in input x_2 against A at Period 1, while the slacks are removed at Period 2. This effort is well captured in the non-radial model $(C(B)=1.33)$, but the radial model neglects this removal, as well as the existence, of the slacks. In consequence, DMU B is seen as improved in its productivity by 1.33 from Period 1 to Period 2 in the non-radial model, while the radial model reports no change in productivity, i.e., $MI(B)=1$.

5.2 Inclusive vs. Exclusive

In *Remark* 8.1, we noted that, 'inclusive' means that, when we evaluate $(x_o, y_o)^s$ with respect to the technology $(X,Y)^s$, the DMU $(x_o, y_o)^s$ is always included in the evaluator $(X,Y)^s$, thus resulting in the distance score being not greater than 1, while 'exclusive' employs the scheme that the DMU $(x_o, y_o)^s$ is removed from the evaluator group $(X,Y)^s$. Here we compare the two schemes using an example. Table 8-4 exhibits data set comprising three DMUs, A, B and C, with each one using a single input x to produce a single output y. Figure 8-3 shows the diagrammatical representation of this data set, where the legend A^1 indicates DMU A at the period 1.

Table 8-4. Example 2

	Period 1		Period 2	
	x	y	x	y
A	5	5	3	5
B	5	2	3	4
C	8	5	4	6

From the figure we can see that, under the CRS assumption, DMU A has advantage against B and C in both periods. However, the degree of superiority is less at the period 2 compared with that at period 1, implying that DMU A was *caught up* by its competitors.

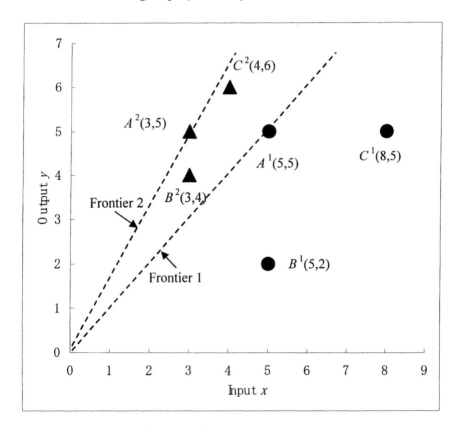

Figure 8-3. Three DMUs in two periods

First, we observe the catch-up effect of DMU A. The 'inclusive' CRS models assign 1 to both $\delta^1(A^1)$ and $\delta^2(A^2)$, since A is on the efficient frontier of the CRS models at both periods. Hence, there is no catch-up effect, i.e., $C(A)=1$. On the other hand, the input-oriented 'exclusive' non-radial CRS model evaluates A^1's superiority over B^1 and C^1 as $\delta^1(A^1) = 1+3/5 = 1.67$ and A^2's superiority over B^2 and C^2 as $\delta^2(A^2) = 1+0.33/3 = 1.11$. Apparently, A's superiority is diminished in the period 2. The 'exclusive' model thus renders $C(A) = 1.11/1.67=0.69$ which is less than 1. DMU A lost its superiority by about 40% at the period 2. Second, both the 'inclusive' and 'exclusive' CRS models yield the intertemporal distance scores: $\delta^1(A^2) = 1.67$ and $\delta^2(A^1) = 0.6$. Using the frontier-shift formula (8.10), we have $F(A) = 1.67$ for the 'inclusive' model and $F(A) = 2$ for the 'exclusive' model. The difference is due to those in $\delta^1(A^1)$ and $\delta^2(A^2)$ by the two schemes. So, we go further into the difference. We use the suffixes *incl* and *excl* for distinguishing the two schemes. Then, we have

$$\frac{F_{incl}(A)}{F_{excl}(A)} = \left[\left(\frac{\delta^1_{incl}(A^1)}{\delta^1_{excl}(A^1)}\right) \bigg/ \left(\frac{\delta^2_{incl}(A^2)}{\delta^2_{excl}(A^2)}\right)\right]^{1/2} \tag{8.41}$$

Both the terms in the parenthesis on the right of (8.41) are not greater than 1, since it holds $\delta^s_{incl}(A^s) \le \delta^s_{excl}(A^s)$ for $s = 1, 2$. If the first term is greater (less) than the second, then we have $F_{incl}(A) \ge (\le) F_{excl}(A)$. In case of our example, $1.67 = F_{incl}(A) < F_{excl}(A) = 2$.

Apart from this example, let us observe each component on the right of (8.41). If A^1 is positioned in the relative interior of $(X,Y)^1$, then we have $\delta^1_{incl}(A^1) = \delta^1_{excl}(A^1) < 1$. Hence no difference arises between the two. If A^1 is positioned on the efficient frontier of $(X,Y)^1$, then we have $\delta^1_{incl}(A^1) = 1$ and $\delta^1_{excl}(A^1) \ge 1$. In this case, $\delta^1_{excl}(A^1)$ is measured as the distance from A^1 to the set $(X,Y)^1$ excluding A^1 that may be called the distance from A^1 to the *second* best frontier of $(X,Y)^1$ around A^1. Similarly, when A^2 is positioned on the efficient frontier of $(X,Y)^2$, then $\delta^2_{excl}(A^2)$ indicates the distance from A^2 to the second best frontier of $(X,Y)^2$ around A^2. Let us define $\alpha(\ge 1)$ and $\beta(\ge 1)$ by

$$\alpha = \frac{\delta^1_{excl}(A^1)}{\delta^1_{incl}(A^1)} \quad \text{and} \quad \beta = \frac{\delta^2_{excl}(A^2)}{\delta^2_{incl}(A^2)}$$

From (8.41), we have

$$F_{excl}(A) = F_{incl}(A) \times (\alpha/\beta)^{1/2} \tag{8.42}$$

If $\alpha/\beta > 1 (<1)$, it means that A's superiority is decreased (increased) against the second best frontier, and hence, the frontier-shift effect around A increased (decreased) from Period 1 to Period 2 compared to those under the inclusive scheme evaluation.

The Malmquist indexes under these two schemes are respectively obtained as

$$MI_{incl}(A) = C_{incl}(A) \times F_{incl}(A) = 1 \times 1.67 = 1.67 \text{ and}$$
$$MI_{excl}(A) = C_{excl}(A) \times F_{excl}(A) = 0.69 \times 2 = 1.38.$$

As can be seen, in the exclusive scheme, A's superiority over B and C is lost while its frontier-shift effect is increased. So, the Malmquist index value in the exclusive scheme is lower for DMU A compared to that in inclusive scheme, indicating a good follow-up by competitors in Period 2.

5.3   Oriented vs. Non-oriented

Table 8-5 exhibits three DMUs, A, B and C, each having two inputs (x_1 and x_2) and two outputs (y_1 and y_2) for two time periods. As can be seen, DMU A outperforms B and C in Period 1 since DMU B has one unit of slacks in x_2 and DMU C one unit of slacks in y_1 against A. These slacks are removed in Period 2, thus enabling them to reach an even status.

Table 8-5. Example 3

	Period 1				Period 2			
	x_1	x_2	y_1	y_2	x_1	x_2	y_1	y_2
A	2	2	2	2	2	2	2	2
B	2	3	2	2	2	2	2	2
C	2	2	1	2	2	2	2	2

Using this data set, we compare four oriented models (input-oriented radial, input-oriented non-radial, output-oriented radial and output-oriented non-radial) and one non-oriented model (non-oriented non radial) under CRS environment. Table 8-6 displays the results. As expected, both radial models report no change as is seen from the second and fourth columns of this table, since all DMUs are on the efficient frontiers from the radial standpoint. The input-oriented non-radial model assigns 1.2 to $C(B)$ reflecting the removal of input slacks of DMU B in x_2 at Period 2. However, this model gives a score of 1 (no change) to $C(C)$, $F(C)$ and $MI(C)$ ignoring C's improvement in output y_1 at Period 2. The output-oriented non-radial model assigns a score of 1.5 to $C(C)$, thus evaluating the removal of output shortfalls of C in y_1 at Period 2 but gives a score of 1 (no change) to $C(B)$, $F(B)$ and $MI(B)$ neglecting B's effort for removing input slacks in x_2. Finally, the non-oriented non-radial model takes account of all existing slacks, i.e., input surpluses and output shortfalls, and results in $MI(A)=0.89$, $MI(B)=1.2$ and $MI(C)=1.5$. As this example shows, the value of Malmquist index is conditional upon the model selection, i.e., radial or non-radial, or, oriented or non-oriented. So, due care is needed for deciding upon the selection of an appropriate model.

Table 8-6. Comparisons

	Input-oriented	Input-oriented	Output-oriented	Output-oriented	Non-oriented
	Radial	Non-radial	Radial	Non-radial	Non-radial
$C(A)$	1	0.8	1	0.75	0.8
$C(B)$	1	1.2	1	1	1.2
$C(C)$	1	1	1	1.5	1.5
$F(A)$	1	1.12	1	1.15	1.12
$F(B)$	1	1	1	1	1
$F(C)$	1	1	1	1	1
$MI(A)$	1	0.89	1	0.87	0.89
$MI(B)$	1	1.2	1	1	1.2
$MI(C)$	1	1	1	1.5	1.5

The legends $C(A)$, $F(A)$ and $MI(A)$ denote the catch-up, frontier-shift and Malmquist index, respectively.

5.4   Infeasibility LP issues

As noted in *Remarks* 8.2 and 8.4, the oriented models in the VRS environment may suffer from the problem of infeasibility in measurement of the super-efficiency. However, the non-oriented VRS model is free from this trouble. We now demonstrate this fact with the help of a small data set reported in Table 8-7.

Table 8-7. Example 4

	Period 1		Period 2	
	x	y	x	y
A	3	1	1	3
B	4	2	2	4

In the oriented VRS model, we must solve the intertemporal scores $\delta^2(A^1)$, $\delta^2(B^1)$, $\delta^1(A^2)$ and $\delta^1(B^2)$ defined by (8.13), (8.15), (8.19) or (8.23). We encounter difficulties in solving LPs in this example. Specifically, in the input-oriented case, (8.13) and (8.19) have no feasible solution for $\delta^1(A^2)$ and $\delta^1(B^2)$, since $y_A^2 > \max\{y_A^1, y_B^1\}$ and $y_B^2 > \max\{y_A^1, y_B^1\}$. In the output-oriented case, (8.15) and (8.23) have no feasible solution for $\delta^1(A^2)$ and $\delta^1(B^2)$, since $x_A^2 < \min\{x_A^1, x_B^1\}$ and $x_B^2 < \min\{x_A^1, x_B^1\}$. Thus, in both cases, we cannot evaluate the frontier-shift effects and hence the Malmquist indexes are *not available* for all DMUs A and B. The non-oriented VRS model (8.28), however, can always measure these scores, as is seen from Table 8-8.

Table 8-8. Results by the non-oriented model

	$\delta^1(\cdot^1)$	$\delta^1(\cdot^2)$	$\delta^2(\cdot^1)$	$\delta^2(\cdot^2)$	Catch-up	Frontier	Malmquist
A	1.33	6	0.11	2	1.5	6	9
B	2	4	0.17	1.33	0.67	6	4

6.   CONCLUDING REMARKS

In this chapter we have discussed Malmquist productivity index, and its measurements along with scale efficiency change and related subjects. All of the approaches we have covered in our opening paragraph utilize the ratio forms required in (8.11) with the exception of Thrall (2000) who provides an additive model. Thrall's formulation might best be regarded as an alternative to the MI and has therefore not been covered in this chapter. We conclude this chapter by pointing to several crucial issues involved in applying the Malmquist index model.

6.1   Selection of types of returns to scale

We need to exercise due care in selecting an appropriate RTS type. This identification problem is common to all DEA applications, since DEA is a data-oriented and non-parametric method. If the data set includes numeric values with a large difference in magnitude, e.g., comparing big companies with small ones, the VRS model may be a choice. However, if the data set consists of normalized numbers, e.g., per capita, acre and hour, the CRS model might be an appropriate candidate.

6.2   Radial or non-radial

The radial Malmquist index is based upon the radial DEA scores and hence remaining non-zero slacks are not counted in the scores. If slacks are not freely disposable, the radial Malmquist index cannot fully characterize the productivity change. In contrast, the input (output) oriented non-radial Malmquist index takes into account the input (output) slacks. This results in a smaller objective function value than in the radial model both in the *Within* and *Intertemporal* scores. Chen (2003) applied the non-radial Malmquist model to measure the productivity change of three major Chinese industries: Textiles, Chemicals and Metallurgicals. The result indicates that it is important to consider the possible non-zero slacks in measuring the productivity change. The radial model does not explain and hence cannot fully capture what the data say because of the large amount of slacks ignored.

6.3 Infeasibility and its solution

As pointed out in *Remarks* 8.2 and 8.4, the oriented VRS, IRS and DRS models suffer from the problem of infeasible solution in the *Intertemporal* score (super-efficiency) estimation. This is not a rare case, since the Malmquist model deals with data sets over time that display large variations in magnitude and hence the super-efficiency model suffers from many infeasible solutions. However, the non-radial and non-oriented model is free from such problems. In the radial framework, Bjurek (1996) has proposed the Malmquist total factor productivity index (MTFP) which computes an output-oriented index and an input-oriented index independently, and combines the two indexes into one. The MTFP is free from the infeasible problem in the VRS case.

6.4 Empirical studies

The Malmquist indexes have been used in a variety of studies. Färe et al. (1998a) surveyed empirical applications comprising various sectors such as agriculture, airlines, banking, electric utilities, insurance companies, and public sectors. However, most of these studies utilized the oriented-radial models. Thus, they suffer from the neglect of slacks and infeasibility. Empirical studies are at a nascent stage as of now.

REFERENCES

1. Andersen, P. and Petersen, N.C., 1993, A procedure for ranking efficient units in data envelopment analysis, *Management Science* 39, 1261-1264.
2. Balk, B.M., 2001, Scale efficiency and productivity change, *Journal of Productivity Analysis* 15, 159-183.
3. Banker, R.D., 1984, Estimating the most productive scale size using data envelopment analysis, *European Journal of Operational Research* 17, 35-44.

4. Banker, R.D., Charnes, A. and Cooper, W.W., 1984, Some models for estimating technical and scale inefficiencies in data envelopment analysis, *Management Science* 30, 1078-1092.

5. Bjurek, H., 1996, The Malmquist total factor productivity index, *Scandinavian Journal of Economics* 98, 303-313.

6. Caves, D.W., Christensen, L.R. and Diewert, W.E., 1982, The economic theory of index numbers and the measurement of input, output and productivity, *Econometrica* 50, 1393-1414,

7. Chen, Y., 2003, Non-radial Malmquist productivity index with an illustrative application to Chinese major industries, *International Journal of Production Economics* 83, No. 1, 27-35.

8. Cooper, W.W., Seiford, M.L. and Tone, K., 1999, *Data Envelopment Analysis: A Comprehensive Text with Models, Applications, References and DEA-Solver Software*, Boston, Kluwer Academic Publishers.

9. DEA-Solver Pro Version 4.0, 2003, Saitech-Inc, www.saitech-inc.com.

10. Färe, R. and Grosskopf, S., 1992, Malmquist indexes and Fisher ideal indexes, *The Economic Journal* 102, 158-160.

11. Färe, R., Grosskopf, S, Lindgren, B. and Roos, P., 1989, 1994, Productivity change in Swedish hospitals: a Malmquist output index approach, in Charnes, A., Cooper, W.W., Lewin, A.Y. and Seiford, M.L. (eds.) *Data Envelopment Analysis: Theory, Methodology and Applications,* Boston, Kluwer Academic Publishers.

12. Färe, R., Grosskopf, S, Norris, M. and Zhang, Z., 1994, Productivity growth, technical progress, and efficiency change in industrialized countries, *The American Economic Review* 84, 66-83.

13. Färe, R., Grosskopf, S. and Roos, P., 1998a, Malmquist productivity indexes: a survey of theory and practice, in *Index Numbers: Essays in Honour of Sten Malmquist*, Kluwer Academic Publishers, 127-190.

14. Färe, R., Grosskopf, S. and Russell, R., 1998b, *Index Numbers: Essays in Honour of Sten Malmquist*, Kluwer Academic Publishers.

15. Lovell, C.A.K. and Grifell-Tatje, E., 1994, A generalized Malmquist productivity index, Paper presented at the Georgia Productivity Workshop at Athens, GA, October 1994.

16. Malmquist, S., 1953, Index numbers and indifference surfaces, *Trabajos de Estadistica* 4, 209-242.

17. Ray, S.C. and Delsi, E., 1997, Productivity growth, technical progress, and efficiency change in industrialized countries: comment, *The American Economic Review* 87, 1033-1039.

18. Seiford, M.L. and Zhu, J., 1999, Infeasibility of super-efficiency data envelopment analysis, *INFOR* 37, 174-187.

19. Thrall, R.M., 2000, Measures in DEA with an application to the Malmquist index, *Journal of Productivity Analysis* 13, 125-137.

20. Tone, K., 2001, A slacks-based measure of efficiency in data envelopment analysis, *European Journal of Operational Research* 130, 498-509.
21. Tone, K., 2002, A slacks-based measure of super-efficiency in data envelopment analysis, *European Journal of Operational Research* 143, 32-41.
22. Zhu, J., 1996, Data envelopment analysis with preference structure, *Journal of the Operational Research Society* 47, 136-150.

Acknowledgements

The author would like to extend his sincere gratitude to the Editors and Biresh K Sahoo for their careful reading of the manuscript.

Chapter 9

CHANCE CONSTRAINED DEA

William W. Cooper[1], Zhimin Huang[2] and Susan X. Li[2]
[1] Red McCombs School of Business, University of Texas at Austin, Austin, TX 78712 USA
email: cooperw@mail.utexas.edu

[2] School of Business, Adelphi University, Garden City, NY 11530 USA
email: huang@adelphi.edu   li@adelphi.edu

Abstract: Incorporation of random variations into DEA analysis has received significant attention in recent years. This chapter describes some of these developments and offers examples of possible uses in the area of chance constrained programming models in DEA.

Key words: Chance constrained DEA; Stochastic; Joint Probability; Congestion; Satisficing.

1.   INTRODUCTION

This chapter deals with chance constrained programming extensions of the usual deterministic DEA formulations. This kind of approach makes it possible to replace deterministic characterizations in DEA, such as "efficient" and "not efficient," with characterizations such as "probably efficient" and "probably not efficient." Indeed, it is possible to go still further into characterizations such as "sufficiently efficient," with associated probabilities of not being correct in making inferences about the performance of a DMU.

It is also possible to extend the deterministic objectives usually used in DEA with additional alternatives. For instance, one may use the "E-model" of chance constrained programming to obtain an "expected value" approach.

However, this expected value objective is not the only possibility. One may also use the "P-model" of chance constrained programming to obtain the "most probable" occurrences, perhaps in order to determine whether this probability is sufficiently high. Indeed, one can extend this by incorporating constraints (also probabilistic in character) to insure that the resulting solutions are satisfactory.

These chance constrained formulations provide new ways to incorporate new concepts into the DEA literature such as the "satisficing concepts" of H. A. Simon (1957). Originally formulated for use in social psychology these satisficing concepts have now spread to other disciplines such as economics and so it is natural to extend them for use in DEA as is done later in this chapter. See also Cooper, Huang and Li (1996).

These are the kinds of ideas and extensions that will be covered in this chapter. The purpose of this chapter, however, is to provide a systematic presentation of major developments of chance constrained DEA models that have appeared in the literature. The results to be covered are fairly recent so that there is not much to report in the way of significant applications. Indeed, the situation is analogous to the state of game theory and DEA combinations as far as actual applications are concerned. See, for instance, Charnes et al. (1989). As we shall see in the course of developing these chance constrained approaches to DEA, there is more research to be done (e.g., in the way of developing more efficient algorithms) so this chapter is oriented accordingly.

In the next section, which is section 2 of this chapter, we present basic concepts of efficiency and efficiency dominance as well as models to implement these concepts. In section 3, we provide a detailed introduction to "joint chance constrained" efficiency and mathematical formulas. Potential uses and deterministic equivalents of the models immediately follow. In section 4, we utilize "E-model" (expected value) formulations to discuss DEA efficiency and its relationship with "sensitivity analysis" in stochastic situations. In section 5, we briefly summarize another type of chance constrained DEA models, which are referred to as "P-model" formulations in chance constrained programming. We then use this class of models to incorporate the "satisficing" concepts of H. A. Simon (1959) for use with DEA. Concluding remarks are in section 6.

2. EFFICIENCY AND EFFICIENCY DOMINANCE

We start with some concepts of "efficiency dominance" (Bowlin et al. (1984)) for which we introduce the following notation. Let $x_j = (x_{1j}, ..., x_{mj})^T$ and $y_j = (y_{1j}, ..., y_{rj})^T$ represent input and output vectors, respectively, for j^{th} Decision Making Unit (DMU), j = 1, ..., n. The superscript T represents transpose. The DMU to be evaluated is designated as DMU_0 and its input-output vector is denoted (x_0, y_0).

Let us consider a discrete production set which consists of only actually observed input-output vectors, (x_j, y_j), j = 1, ..., n, as follows

$$T_0 = \{(x_j, y_j)\}_{1 \le j \le n} . \tag{9.1}$$

Definition 9.1 (Efficiency Dominance): DMU_j dominates DMU_0 with respect to T_0 if and only if $x_j \le x_0$ and $y_j \ge y_0$ with strict inequality holding for at least one of the components in the input or output vector.

Thus, DMU_0 is <u>not</u> dominated in its efficiency if and only if there is <u>no</u> DMU_j which exhibits a performance that satisfies the above definition.

One approach to implement definition 9.1 utilizes the additive model with integer constraints as in Bowlin *et al.* (1984) as follows:

<u>Dominance Model</u>

$$\text{Max} \sum_{r=1}^{s} s_r^+ + \sum_{i=1}^{m} s_i^-$$

s.t.

$$x_{i0} = \sum_{j=1}^{n} x_{ij}\lambda_j + s_i^- , \quad i = 1, ..., m \tag{9.2}$$

$$y_{r0} = \sum_{j=1}^{n} y_{rj}\lambda_j - s_r^+ , \quad r = 1, ..., s$$

$$1 = \sum_{j=1}^{n} \lambda_j ,$$

$$\lambda_j \in \{0,1\}, \ s_i^-, \ s_r^+ \ge 0, \ , j = 1, ..., n; i = 1, ..., m; r = 1, ..., s.$$

Solutions with any slacks at non-zero values show the sources and amounts of inefficiency for DMU_0 relative to the DMU_j for which $\lambda_j = 1$. The maximization of the slacks in the objective ensures that a "most dominant" DMU_j will be designated for effecting the evaluations. Thus, if the slacks are all zero in a solution to (9.2) then there is <u>no</u> DMU_j that dominates DMU_0 in the sense of definition 9.1, above, and this in turn implies that DMU_0 operated efficiently relative to all of the other DMU_j.

We can bring this all together by assuming that DMU_k is found to be "most dominant" and then writing the solution to (9.2) as follows,

$$x_{i0} - s_i^{+*} = \sum_{j=1}^{n} x_{ij}\lambda_j = x_{ik}, \quad i = 1, \ldots, m, \tag{9.3}$$

$$y_{r0} + s_r^{-*} = \sum_{j=1}^{n} y_{rj}\lambda_j = y_{rk}, \quad r = 1, \ldots, s. \tag{9.4}$$

Using "*" to represent an optimal solution we have let $\lambda_k^* = 1$ with all other $\lambda_j^* = 0$ reflect the fact that an optimal solution to (9.2) has designated DMU_k as "most dominant." DMU_k will dominate DMU_0 in the efficiency of its observed performance, however, if and only if <u>any</u> s_i^{+*}, s_r^{-*} is not zero. Conversely, DMU_0 is not dominated by <u>any</u> DMU_j if and only if all $s_i^{+*} = s_r^{-*} = 0$ in an optimal solution for (9.2).

Let us generalize this efficiency dominance to a continuous production possibility set T as follows

$$T = \left\{ (x, y) : x = \sum_{j=1}^{n} x_j\lambda_j, y = \sum_{j=1}^{n} y_j\lambda_j, \sum_{j=1}^{n} \lambda_j = 1, \lambda_j \geq 0, \forall j \right\} \tag{9.5}$$

Definition 9.2 (General Efficiency Dominance). Let $(x', y') \in T$ and $(x'', y'') \in T$. We say that (x', y') dominates (x'', y'') with respect to the production possibility set T if and only if $x' \leq x''$ and $y' \geq y''$ with strict inequality holding for at least one of the components in the input or the output vector.

Thus, a point in T is not dominated if and only if there is no other point in T which satisfies the definition. This leads to the following definition of efficiency:

Definition 9.3 (Efficiency). DMU_0 is efficient with respect to T if and only if there is no $(x, y) \in T$ such that (x_0, y_0) is dominated by (x, y).

A variety of mathematical models are available to implement Definition 9.3. Two models, which are typically employed in the DEA literature are the BCC (Banker, Charnes and Cooper (1984)) and the Additive model (Charnes et al. (1985)). Here we treat only BCC models but the results also hold for Additive models—and also other DEA models with their associated production possibility sets. Let us consider the following BCC model:

BCC Model

$$\text{Max } \phi + \varepsilon \left(\sum_{r=1}^{s} s_r^+ + \sum_{i=1}^{m} s_i^- \right)$$

s.t.

$$\phi \, y_{r0} - \sum_{j=1}^{n} y_{rj} \lambda_j + s_r^+ = 0, \quad r = 1, \ldots, s, \tag{9.6}$$

$$\sum_{j=1}^{n} x_{ij} \lambda_j + s_i^- = x_{i0}, \quad i = 1, \ldots, m,$$

$$\sum_{j=1}^{n} \lambda_j = 1,$$

$$\lambda_j \geq 0, \ s_r^+ \geq 0, \ s_i^- \geq 0, j = 1, \ldots, n; \ i = 1, \ldots, m; \ r = 1, \ldots, s.$$

Definition 9.4 (BCC Model). DMU_0 is DEA efficient with respect to T if and only if the following two conditions are both satisfied in model (9.6)

(i) $\quad \phi^* = 1,$

(ii) $\quad s_r^{+*} = s_i^{-*} = 0, \ \forall \ i, r,$

where "*" represents an optimum.

As in Charnes, Cooper and Thrall (1986, 1991), the performances of all *DMUs* can be partitioned into the following four classes

$$E, E', F, N,$$

where E is a set of efficient *DMUs* which are also extreme points and E' is a set of efficient *DMUs* which are not extreme points. F is a set of points which are on a part of the frontier that is not efficient. Finally, N consists of all points which are not on a frontier and hence are inefficient.

3.   STOCHASTIC DOMINANCE AND JOINT CHANCE CONSTRAINED EFFICIENCY

We follow the notation conventions in Cooper et al. (1996 & 1998) with $\tilde{x}_j = (\tilde{x}_{1j},...,\tilde{x}_{mj})^T$ and $\tilde{y}_j = (\tilde{y}_{1j},...,\tilde{y}_{rj})^T$ to represent random behavior for the output and input vectors for DMU_j, $j = 1, ...n$. For the jth DMU, we also let $x_j = (x_{1j},...,x_{mj})^T$ and $y_j = (y_{1j},...,y_{rj})^T$ stand for the expected input and output vector values, respectively. The probability distributions of $\tilde{x}_{ij}$ and $\tilde{y}_{rj}$ will usually be determined by historical data on the inputs and outputs but we may replace some or all of these historically determined probability distributions by theoretical probability distributions, as we shall do when this serves our purposes.

For ease of reference, let

$\tilde{Y} = (\tilde{y}_1,...,\tilde{y}_n)$ be the $(s \times n)$ "output" matrix,

$Y = (y_1,...,y_n)$ be the $(s \times n)$ expected "output" matrix,

$_k\tilde{y} = (\tilde{y}_{k1},...,\tilde{y}_{kn})$ be the kth row of the "output" matrix $\tilde{Y}$, $k = 1, ..., s$,

$_ky = (y_{k1},...,y_{kn})$ be the kth row of the expected "output" matrix Y,

$\qquad\qquad k = 1, ..., s,$

$\tilde{X} = (\tilde{x}_1,...,\tilde{x}_n)$ be the $(m \times n)$ "input" matrix,

$X = (x_1,...,x_n)$ be the $(m \times n)$ expected "input" matrix,

$_i\tilde{x} = (\tilde{x}_{i1},...,\tilde{x}_{in})$ be the ith row of the "input" matrix $\tilde{X}$, $i = 1, ..., m$,

$_ix = (x_{i1},...,x_{in})$ be the ith row of the expected "input" matrix X,

$\qquad\qquad i = 1, ..., m.$

Using this notation we can extend our characterizations to "stochastic efficiency dominance" which, for any DMU_o, can be obtained from the <u>joint</u> probabilistic comparisons of its outputs and inputs with every other observed DMU. Thus, informally, if $\tilde{y}_0$ and $\tilde{x}_0$ are the output and input vectors of the DMU_0 to be tested relative to all DMU_j, $j = 1, ..., n$, then we will say that DMU_0 is stochastically not dominated in its efficiency if it is stochastically impossible to augment any output without increasing any input and without decreasing any other output, or if it is stochastically impossible to decrease any input without augmenting any other input and without decreasing any output. This is intended as a stochastic generalization of efficiency dominance as defined in the preceding section. It is also an adaptation of Pareto-Koopmans efficiency to stochastic situations with the discrete production set $\tilde{T}_0 = \left\{ (\tilde{x}_j, \tilde{y}_j) \right\}_{1 \le j \le n}$. This direct generalization to stochastic situations could be very restricted because of random variations in inputs and outputs. Therefore, we could incorporate a tolerance or risk level to the definition. For a given scalar α $(0 \le \alpha < 1)$, DMU_0 is not stochastically dominated in its efficiency if and only if there is a joint probability less than or equal to α that some other observed DMU displays efficiency dominance relative to DMU_0. Formally (Cooper et al. (1998)),

Definition 9.5. DMU_0 is not stochastically dominated in its efficiency with respect to $\tilde{T}_0$ if and only if for all λ satisfying $e^T \lambda = 1$, $\lambda_j \in \{0,1\}$--i.e., the components of λ are bivalent—we have

$$P \left\{ \bigcap_{i=1}^{m} ({}_i \tilde{x} \lambda \le \tilde{x}_{i0}) \bigcap_{r=1}^{s} ({}_r \tilde{y} \lambda \ge \tilde{y}_{r0}) \right\}$$

$$= P \left\{ \sum_{j=1}^{n} \lambda_j \tilde{x}_{ij} \le \tilde{x}_{i0}, \sum_{j=1}^{n} \lambda_j \tilde{y}_{rj} \ge \tilde{y}_{r0}, i = 1,...,m, r = 1,...,s \right\} \le \alpha. \quad (9.7)$$

Note that our definition can be applied to any probability distribution of inputs and outputs for the $DMUs$ to be considered. Also note that if $\tilde{y}_j$ and $\tilde{x}_j$ follow a continuous joint probability distribution, the requirement of "at least one strict inequality" in the above definition is not necessary.

The model in (9.2) is deterministic, so we now provide a stochastic alternative via the following formulation,

$$\text{Max } P \left\{ \bigcap_{i=1}^{m} ({}_i \tilde{x} \lambda \le \tilde{x}_{i0}) \bigcap_{r=1}^{s} ({}_r \tilde{y} \lambda \ge \tilde{y}_{r0}) \right\} = \beta . \quad (9.8)$$

Now let $\lambda_k = 1$ be a maximal choice satisfying $e^T \lambda = 1$, $\lambda_j \in \{0,1\}$, $\forall$ j. DMU_0 is then to be regarded as dominated stochastically if and only if $\beta > \alpha$—in which event DMU_k is designated as the DMU which has the highest probability of dominating DMU_0 in the efficiency of its performance.

We illustrate from a simple example involving four $DMUs$ using one input and one output with known uniform distributions which are also assumed to be independent in order to show more concretely what our definition of stochastic efficiency dominance means. We represent a uniformly distributed random variable $\tilde{z}$ over an interval $a \le z \le b$, by $\tilde{z} \approx$ Uni[a, b]. Hence we write

$$\tilde{y}_1 \approx Uni[1.5, 2.5], \ \tilde{y}_2 \approx Uni[3,4], \ \tilde{y}_3 \approx Uni[4,5], \ \tilde{y}_4 \approx Uni[2.2, 3.2],$$

$$\tilde{x}_1 \approx Uni[0.5, 1.5], \ \tilde{x}_2 \approx Uni[2,3], \ \tilde{x}_3 \approx Uni[4.5, 5.5], \ \tilde{x}_4 \approx Uni[3.5, 4.5],$$

to mean that these variables are each uniformly distributed as indicated.

For a given α between 0 and 0.5, suppose it is desired to determine whether DMU_1 is dominated stochastically by any other DMU_j. Substitution in (9.7) produces

$$P\{\lambda_1 \tilde{x}_1 + \lambda_2 \tilde{x}_2 + \lambda_3 \tilde{x}_3 + \lambda_4 \tilde{x}_4 \le \tilde{x}_1, \lambda_1 \tilde{y}_1 + \lambda_2 \tilde{y}_2 + \lambda_3 \tilde{y}_3 + \lambda_4 \tilde{y}_4 \ge \tilde{y}_1\}$$
$$= P\{\tilde{x}_2 \le \tilde{x}_1, \tilde{y}_2 \ge \tilde{y}_1\}$$

when we set $\lambda_2 = 1$ and all other $\lambda_j = 0$. The domain of $(\tilde{x}_1, \tilde{x}_2)$ is [0.5, 1.5]$\times$ [2, 3]. This does not have any overlap with the area of $\{(x_1, x_2) : x_2 \le x_1\}$, so we have

$$P\{\tilde{x}_2 \le \tilde{x}_1\} = \iint\limits_{x_2 \le x_1} f_1(x_1) f_2(x_2) dx_1 dx_2 = 0.$$

Turning to the outputs, the comparison with DMU_2 in the domain of $(\tilde{y}_1, \tilde{y}_2)$ is [1.5, 2.5]$\times$ [3, 4]. All values of y_2 exceed every possible value of y_1. Hence we have the domain of $(\tilde{y}_1, \tilde{y}_2)$ contained in the area of $\{(y_1, y_2) : y_1 \le y_2\}$ and therefore

$$P\{\tilde{y}_2 \ge \tilde{y}_1\} = \iint\limits_{y_2 \ge y_1} g_1(y_1) g_2(y_2) dy_1 dy_2 = 1.$$

The distributions in our example are all independent. Because of the independence of $\tilde{x}_1, \tilde{x}_2, \tilde{y}_1, \tilde{y}_2$, the joint probability reduces to the product of the probabilities and we have

$$P\{\tilde{x}_2 \le \tilde{x}_1, \tilde{y}_2 \ge \tilde{y}_1\} = P\{\tilde{x}_2 \le \tilde{x}_1\} P\{\tilde{y}_2 \ge \tilde{y}_1\} = 0.$$

Therefore DMU_1 is not dominated stochastically in its efficiency by DMU_2.

Applying the same procedures to compare DMU_1 with DMU_3 and DMU_4, respectively, we also have

$$P\{\tilde{x}_3 \le \tilde{x}_1, \tilde{y}_3 \ge \tilde{y}_1\} = 0,$$
$$P\{\tilde{x}_4 \le \tilde{x}_1, \tilde{y}_4 \ge \tilde{y}_1\} = 0.$$

Therefore, by Definition 9.5, DMU_1 is <u>not</u> dominated stochastically in its efficiency.

It is easy to check that DMU_2 and DMU_3 are also stochastically not dominated in the efficiency of their performances. However, DMU_4 is found to be stochastically dominated in its efficiency because there is a very high probability (98%) that the output of DMU_2 will exceed the output of DMU_4 <u>and</u> the probability is unity that the input of DMU_2 will be less than that of DMU_4.

Potential Uses

As described in Sinha (1996), competition in high tech industries can be fast and fierce. Merchant semiconductor manufacturers, for example, face a constant array of new products and processes coming on-stream from competitors and changing demands from users. Merchant semi-conductor manufacturers must therefore constantly search for new products of their own to enable them to occupy a new niche, at least for a time, or at least they must develop new processes that will enable them to reduce costs and prices and, in general, they must do both—i.e., develop new products and introduce new processes to constantly reduce prices—as a condition of survival. It is extremely important for the management of such a firm to match itself against the <u>best</u> of its competitors, and this must generally be done in the presence of uncertainty as to who these "best" competitors will be, or what they will have to offer in the way of product capabilities, prices and costs. Thus, we may visualize one potential use of our formulations in situations where such a manufacturer is considering a path of future development in order to continue to survive against competitors for whom, at best, he will know only the probability distributions of pertinent inputs and outputs. Indeed, this manufacturer will know his own output and input prospects only probabilistically since he is considering whether to undertake <u>proposed</u> developments. The above formulations can help this manufacturer to locate the potential "best competitors"—i.e., those which are likely to be most efficient with input and output mixes similar to those required for the niche being considered.

As another example, we turn to the 5-year "field experiment" conducted by R.L. Clarke (1989) which is summarized in Charnes, Clarke and Cooper (1989). The main objective of the study was directed to examining the possible effects of repeated uses of DEA to evaluate performances when used over an extended period of time. This was done in a context that involved evaluating the performance of vehicle maintenance units located at each of several different bases in the US Air Force's Tactical Air Command. Under orders from Central Headquarters the commanders at each base were required to report results each month in a form suited to DEA evaluations and in ways that conformed to the desired study conditions with the understanding that their vehicle maintenance operations were likely to be subject to on-site inspections if their performance was out of line (i.e., less efficient) than the performances of the maintenance units at other bases.

Noting that our development, as given above, admits a use of degenerate distributions, we can visualize a situation in which a base commander knows his own performance and wants to determine the likelihood of an inspection resulting from performances reported to headquarters by other bases. The performances of <u>other</u> bases are not known exactly but all commanders have records of past performance from these other bases which can be used to synthesize probability distributions. Given exact knowledge of performance at his own base, it is then a simple matter for each commander to calculate the probability that one or more of the other bases will report a better performance record in one or more of its inputs or outputs. Conversely, we may think of Central Headquarters as using such probability distributions to help plan its inspection actions.

Other examples could include reanalyses of earlier DEA studies as exemplified by Land, Lovell and Thore (1993) in their chance-constrained programming re-evaluation of the earlier deterministic treatment by Charnes, Cooper and Rhodes (1981) of the (huge) Program-Follow-Through experiment in public school education conducted by the US Office (now Department) of Education. Such analyses can become quite complex, of course, but others can be treated in very simple and straightford ways.

The formulations for stochastic efficiency dominance used above are based on the discrete stochastic production set

$$\widetilde{T}_0 = \left\{ (\widetilde{x}, \widetilde{y}) : \widetilde{x} = \widetilde{X}\lambda, \widetilde{y} = \widetilde{Y}\lambda, e^T\lambda = 1, \lambda_j \in \{0,1\}, j = 1,...,n \right\}$$
$$= \left\{ (\widetilde{x}_j, \widetilde{y}_j) \right\}_{1 \le j \le n} \tag{9.9}$$

where e is a $(n \times 1)$ vector with all elements equal to unity. We extend this to the continuous stochastic production set $\widetilde{T}$, in which the bivalency conditions on the variables λ_j are relaxed. These variables are now allowed to be continuous so the required evaluation can be effected in terms of convex combinations of observed *DMUs*. Therefore $\widetilde{T}$ can be written as

$$\widetilde{T} = \left\{ (\widetilde{x}, \widetilde{y}) : \widetilde{x} = \widetilde{X}\lambda, \widetilde{y} = \widetilde{Y}\lambda, e^T\lambda = 1, \lambda \ge 0 \right\}. \tag{9.10}$$

One of the associated stochastic production possibility sets of $\widetilde{T}$ which is also considered here can be defined as

$$\widetilde{T}_1 = \left\{(\widetilde{x}, \widetilde{y}) : \widetilde{x} = \widetilde{X}\lambda + s^+, \widetilde{y} = \widetilde{Y}\lambda - s^-, e^T\lambda = 1, \lambda \geq 0, s^+ \geq 0, s^- \geq 0\right\}. (9.11)$$

We now note that this brings us into contact with other parts of the DEA literature. $\widetilde{T}$ and $\widetilde{T}_1$ are stochastic generalizations of the production possibility sets defined in Charnes *et al.* (1985) and Banker *et al.* (1984), respectively, where the "Additive" and "BCC" models of DEA were first introduced into the literature. To see this, let us represent the general stochastic production possibility set as:

$$\widetilde{T}_2 = \{(\widetilde{x}, \widetilde{y}) : \widetilde{y} \text{ can be produced from } \widetilde{x} \}$$

We next postulate the following properties for the production possibility set (Cooper et al. (1998)), $\widetilde{T}_2$:

Postulate 1. *Convexity.* If $(\widetilde{x}_j, \widetilde{y}_j) \in \widetilde{T}_2$, j = 1, ..., n, and $\lambda_j \geq 0$ are non-negative scalars such that $e^T\lambda = 1$, then $(\widetilde{X}\lambda, \widetilde{Y}\lambda) \in \widetilde{T}_2$.

Postulate 2. *Inefficiency Postulate.* (a) If $(\widetilde{x}, \widetilde{y}) \in \widetilde{T}_2$ and $\widetilde{x}^* = \widetilde{x} + s^+$ with $s^+ \geq 0$, then $(\widetilde{x}^*, \widetilde{y}) \in \widetilde{T}_2$. (b) If $(\widetilde{x}, \widetilde{y}) \in \widetilde{T}_2$ and $\widetilde{y}^* = \widetilde{y} - s^-$ with $s^- \geq 0$, then $(\widetilde{x}, \widetilde{y}^*) \in \widetilde{T}_2$.

Postulate 3. *Minimum Intersection.* $\widetilde{T}_2$ is the intersection set of all $\hat{T}$ satisfying Postulates 1 and 2, and subject to the condition that each of the vectors $(\widetilde{x}_j, \widetilde{y}_j) \in \hat{T}$, j = 1, ..., n.

$$\widetilde{T}_1 = \left\{(\widetilde{x}, \widetilde{y}) : \widetilde{x} = \widetilde{X}\lambda + s^+, \widetilde{y} = \widetilde{Y}\lambda - s^-, e^T\lambda = 1, \lambda \geq 0, s^+ \geq 0, s^- \geq 0\right\}$$

is thus a stochastic production possibility set satisfying the above three postulates. Furthermore, if we omit the convexity condition $e^T\lambda = 1$ in $\widetilde{T}_1$ the production possibility set becomes a stochastic generalization of the production possibility set for the CCR model introduced in Charnes *et al.* (1978). Cooper et al. (1998) have shown that both $\widetilde{T}$ and $\widetilde{T}_1$ have the same efficiency properties. Therefore, here we only discuss some major results on $\widetilde{T}$.

We can now replace "stochastic efficiency dominance" with a more general concept of "stochastic efficiency" on $\widetilde{T}$,

Definition 9.6. For $0 \le \alpha < 1$, $(\tilde{x}^*, \tilde{y}^*) \in \tilde{T}$ is "α-stochastically efficient" with respect to $\tilde{T}$ if for any λ satisfying $e^T \lambda = 1$ and $\lambda \ge 0$, we have

$$P\left\{\bigcap_{i=1}^{m}(_i\tilde{x}\lambda \le \tilde{x}_i^*) \bigcap_{r=1}^{s}(_r\tilde{y}\lambda \ge \tilde{y}_r^*)\right\} \le \alpha. \tag{9.12}$$

Definition 9.7. The α-stochastically efficient frontier of $\tilde{T}$ is defined as the set of α-stochastically efficient points for which there exists a λ satisfying $e^T \bar{\lambda} = 1$ with $\bar{\lambda} \ge 0$ such that equality holds in (9.12).

Definition 9.8. DMU_0 is α-stochastically efficient if for any λ satisfying $e^T\lambda = 1$ and $\lambda \ge 0$, we have

$$P\left\{\bigcap_{i=1}^{m}(_i\tilde{x}\lambda \le \tilde{x}_{i0}) \bigcap_{r=1}^{s}(_r\tilde{y}\lambda \ge \tilde{y}_{r0})\right\} \le \alpha. \tag{9.13}$$

The rest of this section is devoted to sharpening our characterizations of α-stochastic efficiencies. Since

$$\left(\bigcap_{i=1}^{m}(_i\tilde{x}\lambda \le \tilde{x}_{i0}) \bigcap_{r=1}^{s}(_r\tilde{y}\lambda \ge \tilde{y}_{r0})\right) \subset \left\{e^T \tilde{X}\lambda - e^T \tilde{Y}\lambda < e^T \tilde{x}_0 - e^T \tilde{y}_0\right\},$$

it follows that

$$P\left(\bigcap_{i=1}^{m}(_i\tilde{x}\lambda \le \tilde{x}_{i0}) \bigcap_{r=1}^{s}(_r\tilde{y}\lambda \ge \tilde{y}_{r0})\right) \le P\left\{e^T \tilde{X}\lambda - e^T \tilde{Y}\lambda < e^T \tilde{x}_0 - e^T \tilde{y}_0\right\}.$$

Therefore, $P\left\{e^T \tilde{X}\lambda - e^T \tilde{Y}\lambda < e^T \tilde{x}_0 - e^T \tilde{y}_0\right\} \le \alpha$ is sufficient for DMU_0 to be α-stochastically efficient. Next we let ε be the non-Archimedean positive infinitesimal. The following theorem develops a necessary condition for a DMU to be α-stochastically efficient.

Theorem 9.1 (Cooper, Huang, Lelas, Li, and Olesen, 1998). Let DMU_0 be α-stochastically efficient. Then for any λ which satisfies

$$P\{_i\tilde{x}\lambda < \tilde{x}_{i0}\} \ge 1 - \varepsilon, \quad i = 1, ..., m, \tag{9.14}$$

$$P\{_k\tilde{y}\lambda > \tilde{y}_{k0}\} \ge 1 - \varepsilon, \quad k = 1, ..., s, \tag{9.15}$$

$$e^T \lambda = 1, \lambda \ge 0, \tag{9.16}$$

we have

$$P\left\{e^T \tilde{X}\lambda - e^T \tilde{Y}\lambda < e^T \tilde{x}_0 - e^T \tilde{y}_0\right\} \leq \alpha. \tag{9.17}$$

Theorem 9.1 allows us to develop an extension of (9.2) for use in evaluating stochastic efficiency because it implies that if DMU_0 is α-stochastically efficient, then the maximum value of the chance functional $P\left\{e^T(\tilde{X}\lambda - \tilde{x}_0) + e^T(\tilde{y}_0 - \tilde{Y}\lambda) < 0\right\}$, subject to constraints (9.14)–(9.17), is less than or equal to the specified risk level α. It is obvious that if the maximum value of the chance functional $P\left\{e^T(\tilde{X}\lambda - \tilde{x}_0) + e^T(\tilde{y}_0 - \tilde{Y}\lambda) < 0\right\}$ exceeds α, then DMU_0 is not α-stochastically efficient. Since ε is a positive non-Archimedean infinitesimal, we call (9.14) and (9.15) "almost 100% confidence" chance constraints. This is reasonable because we are almost 100% sure that for any point in $\tilde{T}$, which satisfies constraints (9.14) and (9.15), its individual inputs and outputs are less than and greater than the corresponding inputs and outputs of DMU_0, respectively. Hence, at least in principle, the determination of α-stochastic efficiency of all $DMUs$ can be characterized by solving a series of almost 100% confidence chance constrained programming problems.

To represent this explicitly, we introduce the "almost 100% confidence" chance constrained problem represented in (9.18)–(9.22),

$$\text{Max } P\left\{e^T(\tilde{X}\lambda - \tilde{x}_0) + e^T(\tilde{y}_0 - \tilde{Y}\lambda) < 0\right\} \tag{9.18}$$

s.t.

$$P\{_i\tilde{x}\lambda < \tilde{x}_{i0}\} \geq 1 - \varepsilon, \quad i = 1, ..., m, \tag{9.19}$$

$$P\{_k\tilde{y}\lambda > \tilde{y}_{k0}\} \geq 1 - \varepsilon, \quad k = 1, ..., s, \tag{9.20}$$

$$e^T \lambda = 1, \lambda \geq 0, \tag{9.21}$$

$$\lambda \geq 0. \tag{9.22}$$

From the above discussions we then have the following,

Theorem 9.2 (Cooper, Huang, Lelas, Li, and Olesen, 1998). (i) Let DMU_0 be α-stochastically efficient. Then the optimal objective value of the "almost 100% confidence" chance constrained programming problem (9.18)–(9.22) is less than or equal to α. (ii) If the optimal objective value of (9.18) exceeds α, then DMU_0 is not stochastically efficient.

We now undertake further developments which depend on explicit assumptions for the types of probability distributions to be used. A simple, frequently used approach, is to suppose that $\tilde{x}_{ij}$, $\tilde{y}_{kj}$ follow a multivariate normal distribution with means and a covariance matrix as follows

$$E(\tilde{x}_{ij}) = x_{ij}, \tag{9.23}$$

$$E(\tilde{y}_{kj}) = y_{kj}, \tag{9.24}$$

$$\Delta = \begin{pmatrix} \left(\Delta_{ij}^{II}\right)_{m \times m} & \left(\Delta_{ik}^{IO}\right)_{m \times s} \\ \left(\Delta_{kj}^{OI}\right)_{s \times m} & \left(\Delta_{ij}^{OO}\right)_{s \times s} \end{pmatrix} \tag{9.25}$$

where

$$\Delta_{ij}^{II} = \left(Cov(\tilde{x}_{iq}, \tilde{x}_{jr})\right)_{n \times n}, \quad 1 \le i, j \le m, \tag{9.26}$$

$$\Delta_{ik}^{IO} = \left(Cov(\tilde{x}_{iq}, \tilde{y}_{kr})\right)_{n \times n}, \quad 1 \le i \le m, 1 \le k \le s, \tag{9.27}$$

$$\Delta_{ij}^{OO} = \left(Cov(\tilde{y}_{iq}, \tilde{y}_{jr})\right)_{n \times n}, \quad 1 \le i, j \le s, \tag{9.28}$$

$$\Delta_{ik}^{IO} = \Delta_{ki}^{OI}, \quad 1 \le i \le m, 1 \le k \le s. \tag{9.29}$$

In order to simplify our model development, we also introduce new notations as follows:

$$\left(\sigma_i^I(\lambda)\right)^2 = V(\tilde{x}_{i0} - {}_i\tilde{x}\lambda) = \lambda^T(\Delta_{ii}^{II})\lambda - 2\sum_{j=1}^n \lambda_j Cov(\tilde{x}_{ij}, \tilde{x}_{i0}) + V(\tilde{x}_{i0}) \tag{9.30}$$

$$\left(\sigma_k^O(\lambda)\right)^2 = V(\tilde{y}_{k0} - {}_k\tilde{y}\lambda) = \lambda^T(\Delta_{kk}^{OO})\lambda - 2\sum_{i=1}^n \lambda_j Cov(\tilde{y}_{ki}, \tilde{y}_{k0}) + V(\tilde{y}_{k0}) \tag{9.31}$$

$$(\sigma(\lambda))^2 = V(e^T(\tilde{X}\lambda - \tilde{x}_0) + e^T(\tilde{y}_0 - \tilde{Y}\lambda))$$

$$= \lambda^T \left[\sum_{i=1}^m \sum_{j=1}^m \Delta_{ij}^{II} + \sum_{k=1}^s \sum_{j=1}^s \Delta_{kj}^{OO} - 2\sum_{i=1}^m \sum_{k=1}^s \Delta_{ik}^{IO} \right] \lambda$$

$$+ 2\sum_{p=1}^n \lambda_p \left[\sum_{i=1}^m \sum_{k=1}^s Cov(\tilde{x}_{i0}, \tilde{y}_{kp}) + \sum_{i=1}^m \sum_{k=1}^s Cov(\tilde{x}_{ip}, \tilde{y}_{k0}) \right.$$

$$\left. - \sum_{i=1}^m \sum_{j=1}^m Cov(\tilde{x}_{i0}, \tilde{x}_{jp}) - \sum_{k=1}^s \sum_{i=1}^s Cov(\tilde{y}_{kp}, \tilde{y}_{i0}) \right]$$

$$+ \left[\sum_{i=1}^m \sum_{j=1}^m Cov(\tilde{x}_{i0}, \tilde{x}_{j0}) - 2\sum_{i=1}^m \sum_{k=1}^s Cov(\tilde{x}_{i0}, \tilde{y}_{k0}) \right]$$

$$+ \sum_{k=1}^{s} \sum_{j=1}^{s} Cov(\tilde{y}_{k0}, \tilde{y}_{j0}) \Bigg]. \tag{9.32}$$

The assumption of multivariate normality implies that the production possibility set and its efficient frontier will vary randomly in a symmetric manner across *DMUs*, and in this manner reflects the results of events such as bad weather and poor luck, etc. and it also permits data measurement and other errors to occur symmetrically.

This interpretation is similar to the two-sided error assumptions used by Aigner, Lovell and Schmidt (1977) for the estimation of single output parametric stochastic frontier production functions. There is another one-sided error in their work, which represents a component that reflects an assumption that each *DMU*'s output must lie on or below the stochastic frontier function if it is to represent "inefficiency." Although we do not consider this one-sided disturbance explicitly in our stochastic DEA model, we do need to note that the structure of our stochastic production possibility set implicitly allows for one-sided disturbances from the efficient frontier due to possible *DMU* inefficiencies and this is reflected in our chance constraints being oriented in the direction where inefficiencies might occur. We restrict our consideration to the class of "zero-order decision rules" in chance constrained programming to achieve a deterministic equivalent for the problem (9.18)–(9.22) as follows:

$$\text{Min } e^{T}(X\lambda - x_0) + e^{T}(y_0 - Y\lambda) + \sigma(\lambda)\Phi^{-1}(\alpha) \tag{9.33}$$

s.t.

$$y_{k0} \leq_k y\lambda + \sigma_k^O(\lambda)\Phi^{-1}(\varepsilon), \quad k = 1, ..., s, \tag{9.34}$$

$$_i x\lambda \leq x_{i0} + \sigma_i^I(\lambda)\Phi^{-1}(\varepsilon), \quad i = 1, ..., m, \tag{9.35}$$

$$e^{T}\lambda = 1, \tag{9.36}$$

$$\lambda \geq 0. \tag{9.37}$$

for which we have the following theorem,

Theorem 9.3 (Cooper, Huang, Lelas, Li, and Olesen, 1998). (i) Let DMU_0 be α-stochastically efficient. Then the optimal objective value of problem (9.33)-(9.37) is greater than or equal to zero. (ii) If the optimal objective value of (9.33) is less than zero, then DMU_0 is not α-stochastically efficient.

4.   STOCHASTIC EFFICIENCY IN MARGINAL CHANCE CONSTRAINED MODELS

Land, Lovell and Thore (1993) introduced a formal "E-model" form of marginal chance constrained DEA model in CCR (Charnes, Cooper, and Rhodes (1978)) form as follows

$$\text{Min } \theta \qquad\qquad\qquad (9.38)$$

s.t.

$$P\left\{\theta\tilde{x}_{i0} \geq \sum_{j=1}^{n}\tilde{x}_{ij}\lambda\right\} \geq 1 - \alpha, \quad i = 1, ..., m,$$

$$P\left\{\sum_{j=1}^{n}\tilde{y}_{rj}\lambda_j \geq \tilde{y}_{r0}\right\} \geq 1 - \alpha, \quad r = 1, ..., s,$$

$$\lambda_j \geq 0, j = 1, ..., n.$$

The meaning of the chance constraints is that they should not be violated with probability at most *a*.

Olesen and Petersen (1995) also utilized marginal chance constrained programming theory to develop an "E-model" for use in DEA by introducing confidence regions for all *DMUs*. For given probability level γ,

$$D_j(\gamma) = \left\{(x, y) : (x^T - E(\tilde{x}_j)^T, y^T - E(\tilde{y}_j)^T)\sum_j{}^{-1}\begin{pmatrix} x - E(\tilde{x}_j) \\ y - E(\tilde{y}_j) \end{pmatrix} \leq c^2 \right\}$$

is called the confidence region of *DMU_j*, where $\sum_j$ is the covariance matrix of $(\tilde{x}_j, \tilde{y}_j)$, c is determined by $P\left(\chi^2_{(n)} \leq c^2\right) = \gamma$, and $\chi^2_{(n)}$ is the Chi square random variable with n degrees of freedom.

Letting $\alpha = 1 - \Phi(c)$, the chance constrained constrained DEA model is:

$$\text{Max } u^T y_0 - v^T x_0 \qquad\qquad\qquad (9.39)$$

s.t.

$$P(u^T y_j \leq v^T x_j) \geq 1 - \alpha, j = 1, ..., n,$$

$$u \geq \varepsilon e, v \geq \varepsilon e,$$

where ε is the non-Archimedean positive infinitesimal defined scalar and e is a vector of ones.

Differences between models in (9.38) and (9.39) are: (a) the model in (9.38) generalized the CCR envelopment form to marginal chance constrained formulations, while the model in (9.39) extended CCR multiplier models to marginal chance constrained formulations; (b) the scalar α was predetermined directly by the user in model (9.38), but in model (9.39) the scalar α was determined by another scalar γ through confidence regions of *DMUs*.

We here consider another version of "E-model" form for marginal chance constrained DEA models (Cooper et al. (2002 & 2003)), which we will use to extend the concepts of "DEA efficiency" and congestion of BCC models to a chance constrained programming context,

$$\text{Max } \phi$$

s.t.

$$P\left\{ \sum_{j=1}^{n} \tilde{y}_{rj}\lambda_j \geq \phi \tilde{y}_{r0} \right\} \geq 1-\alpha, \text{r} = 1, \ldots, \text{s}, \tag{9.40}$$

$$P\left\{ \sum_{j=1}^{n} \tilde{x}_{ij}\lambda_j \leq \tilde{x}_{i0} \right\} \geq 1-\alpha, \quad \text{i} = 1, \ldots, \text{m},$$

$$\sum_{j=1}^{n} \lambda_j = 1,$$

$$\lambda_j \geq 0, \text{j} = 1, \ldots, \text{n}.$$

Here α is a predetermined number between 0 and 1.

Definition 9.9 (Chance Constrained Efficiency). *DMU₀* is stochastic efficient if and only if the following two conditions are both satisfied:

> (i) $\phi^* = 1$;
> (ii)   Slack values are all zero for **all** optimal solutions.

Here (ii) refers to all alternate optima because the second stage optimization associated with $\varepsilon > 0$ is not used in (9.40).

Since j = o is one of the n *DMUⱼ*, we can always get a solution with $\phi = 1$, $\lambda_0 = 1$ and $\lambda_j = 0$ (j ≠ o) and all slacks zero. However, this solution need not be maximal. It follows that a maximum with $\phi^* > 1$ in (9.40) for any sample of $j = 1, \ldots, n$ observations means that the DMU_0 being evaluated is not efficient because, to the specified level of probability defined by α, the evidence will then show that all outputs of DMU_0 can be increased to $\phi^* \tilde{y}_{r0} > \tilde{y}_{r0}$, $r = 1, \ldots, s$, by using a convex combination of other $DMUs$ which also satisfy

$$P\left\{\sum_{j=1}^{n} \tilde{x}_{ij} \lambda_j \leq \tilde{x}_{i0}\right\} \geq 1-\alpha, \quad i = 1, \ldots, m. \tag{9.41}$$

Hence, as required by definition 9.3, no output or input is worsened by this increase so, in effect, we have added a stochastic element to the deterministic formulations in Definitions 3 and 4.

Now suppose $\zeta_r > 0$ is the "external slack" for the rth output. By "external slack" we refer to slack outside the braces. We can choose the value of this external slack so it satisfies

$$P\left\{\sum \tilde{y}_{rj} \lambda_j - \phi\tilde{y}_{r0} \geq 0\right\} = (1-\alpha) + \zeta_r. \tag{9.42}$$

There must then exist a positive number $s_r^+ > 0$ such that

$$P\left\{\sum \tilde{y}_{rj} \lambda_j - \phi\tilde{y}_{r0} \geq s_r^+\right\} = 1 - \alpha. \tag{9.43}$$

This positive value of s_r^+ permits a still further increase in $\tilde{y}_{r0}$ for any set of sample observations without worsening any other input or output. It is easy to see that $\zeta_r = 0$ if and only if $s_r^+ = 0$.

In a similar manner, suppose $\xi_i > 0$ represents "external slack" for the ith input chance constraint. We choose its value to satisfy

$$P\left\{\sum_{j=1}^{n} \tilde{x}_{ij} \lambda_j - \tilde{x}_{i0} \leq 0\right\} = (1-\alpha) + \xi_i. \tag{9.44}$$

There must then exist a positive number $s_i^- > 0$ such that

$$P\left\{\sum_{j=1}^{n} \tilde{x}_{ij} \lambda_j + s_i^- \leq \tilde{x}_{i0}\right\} = 1 - \alpha. \tag{9.45}$$

Such a positive value of s_i^- permits a decrease in $\tilde{x}_{i0}$ for any sample without worsening any other input or output to the indicated probabilities. It is easy to show that $\xi_i = 0$ if and only if $s_i^- = 0$.

We again introduce the non-Archimedean infinitesimal, $\varepsilon > 0$, and extend (9.40) so that stochastic efficiencies and inefficiencies can be characterized by the following model:

$$\text{Max } \phi + \varepsilon \left(\sum_r s_r^+ + \sum_i s_i^- \right)$$

s.t.

$$P\left\{ \sum \tilde{y}_{rj} \lambda_j - \phi \tilde{y}_{r0} \geq s_r^+ \right\} = 1 - \alpha, \quad r = 1, \ldots, s, \qquad (9.46)$$

$$P\left\{ \sum_{j=1}^{n} \tilde{x}_{ij} \lambda_j + s_i^- \leq \tilde{x}_{i0} \right\} = 1 - \alpha, \quad i = 1, \ldots, m,$$

$$\sum_{j=1}^{n} \lambda_j = 1,$$

$$\lambda_j \geq 0, \ s_r^+ \geq 0, \ s_i^- \geq 0, \quad j = 1, \ldots, n; \ r = 1, \ldots, s; \ i = 1, \ldots, m.$$

This leads to the following modification of Definition 9.9:

Definition 9.10. DMU_0 is stochastic efficient if and only if the following two conditions are both satisfied

      (i) $\phi^* = 1$,

      (ii) $s_r^{+*} = 0, \ s_i^{-*} = 0, \ \forall \ i, r.$

This definition aligns more closely with Definition 9.4 since the $\varepsilon > 0$ in the objective of (9.46) makes it unnecessary to refer to "all optimal solutions," as in Definition 9.9. It differs from Definition 9.4, however, in that it refers to stochastic characterizations. Thus, even when the conditions of Definition 9.10 are satisfied there is a chance (determined by the choice of α) that the thus characterized DMU_0 is not efficient.

The stochastic model in (9.46) is evidently a generalization of the BCC model in (9.4). Assume that inputs and outputs are random variables with a multivariate normal distribution and known parameters. The deterministic equivalent for model (9.46) is as follows:

$$\text{Max } \phi + \varepsilon \left(\sum s_r^+ + \sum s_i^- \right)$$

s.t.

$$\phi y_m - \sum y_{rj} \lambda_j + s_r^+ - \Phi^{-1}(\alpha)\sigma_r^o(\phi, \lambda) = 0, \text{r} = 1, \ldots, \text{s}, \quad (9.47)$$

$$\sum x_{ij} \lambda_j + s_i^- - \Phi^{-1}(\alpha)\sigma_j^I(\lambda) = x_{i0}, \qquad \text{i} = 1, \ldots, \text{m},$$

$$\sum_j \lambda_j = 1,$$

$$\lambda_j \geq 0, \ s_r^+ \geq 0, \ s_i^- \geq 0, \text{j} = 1, \ldots, \text{n}; \text{r} = 1, \ldots, \text{s}; \text{i} = 1, \ldots, \text{m},$$

were Φ is the standard normal distribution function and Φ^{-1}, its inverse, is the so-called "fractile function," Finally,

$$\left(\sigma_r^o(\phi, \lambda)\right)^2 = \sum_{i \neq 0} \sum_{j \neq 0} \lambda_i \lambda_j Cov(\tilde{y}_{ri}, \tilde{y}_{rj}) + 2(\lambda_0 - \phi) \sum_{i \neq 0} \lambda_i Cov(\tilde{y}_{ri}, \tilde{y}_{r0})$$

$$+ (\lambda_0 - \phi)^2 Var(\tilde{y}_{r0})$$

and

$$\left(\sigma_i^I(\lambda)\right)^2 = \sum_{j \neq 0} \sum_{k \neq 0} \lambda_j \lambda_k Cov(\tilde{x}_{ij}, \tilde{x}_{ik}) + 2(\lambda_0 - 1) \sum_{j \neq 0} \lambda_j Cov(\tilde{x}_{ij}, \tilde{x}_{i0})$$

$$+ (\lambda_0 - 1)^2 Var(\tilde{x}_{i0}),$$

where we have separated out the terms for DMU_0 because they appear on both sides of the expressions in (9.46) ff. Thus, ϕ^*, s_r^{+*}, and s_i^{-*} can be determined from (9.47) where the data (means and variances) are all assumed to be known.

Let us simplify our assumptions in a manner that will enable us to relate what we are doing to other areas such as the sensitivity analysis research in DEA that is reported in Cooper et al. (2001). Therefore, assume that only DMU_0 has random variations in its inputs and outputs and they are statistically independent. In this case, model (9.47) can be written in the following simpler form:

$$\text{Max } \phi + \varepsilon \left(\sum s_r^+ + \sum s_i^- \right)$$

s.t.

$$\phi y_{r0}' - \sum y_{rj}' \lambda_j + s_r^+ = 0, \text{r} = 1, \ldots, \text{s}, \qquad (9.48)$$

$$\sum x_{ij}' \lambda_j + s_i^- = x_{i0}', \text{I} = 1, \ldots, \text{m},$$

$$\sum \lambda_j = 1,$$

$$\lambda_j \geq 0, \ s_r^+ \geq 0, \ s_i^- \geq 0, \ j = 1, \ldots, n; \ r = 1, \ldots, s; \ i = 1, \ldots, m,$$

where

$$y'_{r0} = y_{r0} - \sigma_{r0}^o \Phi^{-1}(\alpha), \ r = 1, \ldots, s,$$

$$y'_{rj} = y_{rj}, \ j \neq o, \ r = 1, \ldots, s,$$

$$x'_{i0} = x_{i0} + \sigma_{i0}^I \Phi^{-1}(\alpha), \ i = 1, \ldots, m,$$

$$x'_{ij} = x_{ij}, \ j \neq o, \ i = 1, \ldots, m,$$

$$\sigma_{r0}^o = \sqrt{Var(y_{r0})},$$

$$\sigma_{r0}^I = \sqrt{Var(x_{i0})}.$$

Reasons for us to consider random variations only in DMU_0 are as follows: First, treating more than one DMU in this manner leads to deterministic equivalents with the more complicated relations that have been discussed in detail in Cooper et al. (2002 & 2003). The simpler approach used here allows us to arrive at analytical results and characterizations in a straightforward manner. Second it opens possible new routes for effecting "sensitivity analyses." We are referring to the "sensitivity analyses" that are to be found in Charnes and Neralic (1990), Charnes, Haag, Jaska and Semple (1992), Charnes, Rousseau and Semple (1996) and Seiford and Zhu (1998). In the terminology of the survey article by Cooper et al. (2001), these sensitivity analyses are directed to analyzing allowable limits of data variations for only one DMU at a time and hence contrast with other approaches to sensitivity analysis in DEA that allow all data for all DMUs to be varied simultaneously until at least one *DMU* changes its status from efficient to inefficient, or vice versa. These sensitivity analyses are entirely deterministic. Our chance constrained approach can be implemented by representations that are similar in form to those used in sensitivity analysis but the conceptual meanings are different. A chance constrained programming problem can be solved by a deterministic equivalent, as we have just shown, but the issue originally addressed in the chance constrained formulation is different and this introduces elements, such as the risk associated with α, that are nowhere present in these sensitivity analyses.

This can be illustrated by Figure 9-1 where point C is technically inefficient but does not display the congestion that is associated with the negative sloped segment that connects point C to point D. For further details on this Figure, see the chapter on congestion in this handbook. The point C is evidently very sensitive to changes in its output (but not its input) data. In the terminology of Charnes et al. (1996), it has a "zero radius of stability." If its output is raised in any positive amount, C becomes efficient. Alternatively, if C is lowered it becomes an example of congestion. All these arguments are from a sensitivity analysis point of view. If we consider changes in point C to involve only random variations, these characterizations will change. They will depend not on the change in output level, but will depend rather on the specified probability level α. When $\alpha = 0.5$, random variations in the coordinates of point C do not have any impact on its efficiency, inefficiency or congestion characterizations. Hence it is satisfactory to employ the deterministic model (9.6) since, with this choice of α, the user is indifferent to the possible presence of inefficiency (or congestion) stochastically. This is different from the sensitivity analysis results. When α is taken between 0 and 0.5, point C will be efficient in the stochastic sense irrespective of the random variations (See Theorem 9.5(b), below). This is again different from the result of the sensitivity analysis discussed above. Finally, if α is assigned a value between 0.5 and 1, point C will be inefficient in the stochastic sense--no matter what the direction of random variations (See Theorem 9.6(b), below). Thus, in all cases the choice of α plays the critical role.

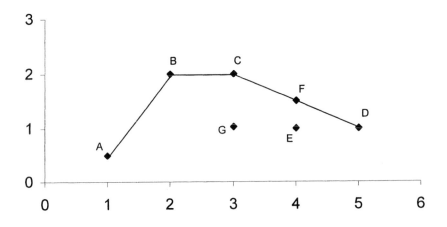

FIGURE 9-1.   TECHNICAL INEFFICIENCY

Source: Cooper et al. (2202)

Theorem 9.4 (Cooper, Deng, Huang, and Li, 2002). For $\alpha = 0.5$. The inefficiency vs. efficiency classification of *DMUo* in input-output mean model (9.6) is the same as in stochastic model (9.46).

Theorem 9.5 (Cooper, Deng, Huang, and Li, 2002). For $0 < \alpha < 0.5$.

    (a) Suppose DMUo is efficient with $DMU_0 \in E \cup E'$ in input-output mean model (9.6), then $DMU_0 \in E$ in stochastic model (9.46);

(b) Suppose $DMU_0 \in$ F in input-output mean model (9.6), then $DMU_0 \in$ E in stochastic model (9.46);

(c) Suppose $DMU_0 \in$ N in input-output mean model (9.6), then $DMU_0 \in$ N in stochastic model (9.46) if $\sigma_{i0}^I < \beta_i^{-*} / (-\Phi^{-1}(\alpha))$ and $\sigma_{r0}^o < \beta_r^{+*} / (-\Phi^{-1}(\alpha))$, where, for $\alpha < 0.5$ we have $\Phi^{-1}(\alpha) < 0$. Here $\sum_{r=1}^{s} \beta_r^{+*} + \sum_{i=1}^{m} \beta_i^{-*}$ is the optimal value of

$$\text{Max} \sum_{r=1}^{s} \beta_r^+ + \sum_{i=1}^{m} \beta_i^-$$

s.t.

$$\sum_{j=1}^{n} y_{rj}\lambda_j - \beta_r^+ \geq y_{r0}, \quad r = 1, ..., s, \tag{9.49}$$

$$\sum_{j=1}^{n} x_{ij}\lambda_j + \beta_i^- \leq x_{i0}, \quad i = 1, ..., m,$$

$$\sum_{j=1}^{n} \lambda_j = 1,$$

$$\beta_r^+ \geq 0, \ \beta_i^- \geq 0, \ \lambda_j \geq 0, \ r = 1, ..., s;$$
$$i = 1, ..., m; j = 1, ..., n.$$

Theorem 9.6 (Cooper, Deng, Huang, and Li, 2002). For $1 > \alpha > 0.5$.

(a) Suppose $DMU_0 \in$ E in input-output mean model (9.6), then $DMU_0 \in$ E in stochastic model (9.46) if

$$\sum_{r=1}^{s} \sigma_{r0}^o + \sum_{i=1}^{m} \sigma_{i0}^I < \left(\sum_{r=1}^{s} \theta_r^{+*} + \sum_{i=1}^{m} \theta_i^{-*} \right) / \Phi^{-1}(\alpha),$$

where $\sum_{r=1}^{s} \theta_r^{+*} + \sum_{i=1}^{m} \theta_i^{-*}$ is the optimal value of

$$\text{Min} \sum_{r=1}^{s} \theta_r^+ + \sum_{i=1}^{m} \theta_i^-$$

s.t.

$$\sum_{\substack{j \neq 0}}^{n} y_{rj}\lambda_j \geq y_{r0} - \theta_r^+, \qquad r = 1, \ldots, s, \tag{9.50}$$

$$\sum_{\substack{j \neq 0}}^{n} x_{ij}\lambda_j \leq x_{i0} + \theta_i^-, \qquad i = 1, \ldots, m,$$

$$\sum_{\substack{j \neq 0}}^{n} \lambda_j = 1,$$

$$\theta_r^+ \geq 0, \; \theta_i^- \geq 0, \; \lambda_j \geq 0 \; (j \neq 0), r = 1, \ldots, s, j = 1, \ldots, n,$$

$$i = 1, \ldots, m.$$

(b) Suppose $DMU_0 \in$ E'$\cup$F $\cup$N in input-output mean model (9.6), then $DMU_0 \in$ N in stochastic model (9.46).

5. SATISFICING DEA MODELS

In the previous section, we have discussed joint chance constrained DEA formulations and "E-model" chance constrained DEA forms. In this section, we would like to discuss another type of chance constrained DEA model. Referred to as "P-models" in the chance constrained programming literature, we can also refer to them as "satisficing" DEA models, as drawn from Cooper, Huang and Li (1996).

We start by introducing the following version of a P-Model which we use to adapt the usual definitions of "DEA efficiency" to a Chance Constrained Programming context,

$$\text{Max } P\left\{ \frac{\displaystyle\sum_{r=1}^{s} u_r \tilde{y}_{ro}}{\displaystyle\sum_{i=1}^{m} v_i \tilde{x}_{io}} \geq 1 \right\}$$

s.t.

$$P\left\{ \frac{\displaystyle\sum_{r=1}^{s} u_r \tilde{y}_{rj}}{\displaystyle\sum_{i=1}^{m} v_i \tilde{x}_{ij}} \leq 1 \right\} \geq 1 - \alpha_j \quad , \qquad j = 1, \ldots, n, \tag{9.51}$$

$$u_r, v_i \geq 0, \; \forall \; r, i.$$

Here P means "Probability" and the symbol ~ is used to identify the inputs and outputs as random variables with a known joint probability distribution. The u_r, $v_i \geq 0$ are the virtual multipliers (=weights) to be determined by solving the above problem. This model evidently builds upon the CCR model of DEA, as derived in Charnes, Cooper and Rhodes (1978), with the ratio in the objective representing output and input values for DMU_0, the DMU to be evaluated, which is also included in the $j = 1, ..., n$ DMUs with output-to-input ratios represented as chance constraints.

Evidently the constraints in (9.51) are satisfied by choosing $u_r = 0$, and $v_i > 0$ for all r and i. Hence, for continuous distributions like the ones considered in this paper, it is not vacuous to write,

$$P\left\{ \frac{\sum_{r=1}^{s} u_r^* \tilde{y}_{ro}}{\sum_{i=1}^{m} v_i^* \tilde{x}_{io}} \leq 1 \right\} + P\left\{ \frac{\sum_{r=1}^{s} u_r^* \tilde{y}_{ro}}{\sum_{i=1}^{m} v_i^* \tilde{x}_{io}} \geq 1 \right\} = 1$$

or

$$P\left\{ \frac{\sum_{r=1}^{s} u_r^* \tilde{y}_{ro}}{\sum_{i=1}^{m} v_i^* \tilde{x}_{io}} \leq 1 \right\} = 1 - \alpha^* \geq 1 - \alpha_0 .$$

Here * refers to an optimal value so α^* is the probability of achieving a value of at least unity with this choice of weights and $1 - \alpha^*$ is therefore the probability of failing to achieve this value.

To see how these formulations may be used, we note that we must have $\alpha_0 \geq \alpha^*$ since $1 - \alpha^*$ is prescribed in the constraint for $j = o$ as the chance allowed for characterizing the $\tilde{y}_{ro}$, $\tilde{x}_{io}$ values as inefficient. More formally, we introduce the following stochasticized definition of efficiency,

Definition 9.11 DMU_0 is "stochastic efficient" if and only if $\alpha^* = \alpha_0$.

This opens a variety of new directions for research and potential uses of DEA. Before indicating some of these possibilities, however, we replace (9.51) with the following

$$\text{Max } P\left\{ \frac{\sum_{r=1}^{s} u_r \tilde{y}_{ro}}{\sum_{i=1}^{m} v_i \tilde{x}_{io}} \geq 1 \right\}$$

s.t.

$$P\left\{\frac{\displaystyle\sum_{r=1}^{s}u_r\tilde{y}_{rj}}{\displaystyle\sum_{i=1}^{m}v_i\tilde{x}_{ij}}\leq1\right\}+P\left\{\frac{\displaystyle\sum_{r=1}^{s}u_r\tilde{y}_{ro}}{\displaystyle\sum_{i=1}^{m}v_i\tilde{x}_{io}}\geq1\right\}\geq1,\quad j=1,...,n,\quad(9.52)$$

$u_r, v_i \geq 0, \forall$ r, i.

This simpler model makes it easier to see what is involved in uses of these CCP/DEA formulations. It also enables us to examine potential uses in a simplified manner.

First, as is customary in CCP, it is assumed that the behavior of the random variables are governed by a known multivariate distribution. Hence we can examine the value of α^* even before the data are generated. If this value is too small then one can signal central management, say, that the situation for DMU_0 needs to be examined in advance because there is a probability of at least $1 - \alpha^* \geq 1 - \alpha_0$ that it will not perform efficiently.

Some additional uses of these concepts can be brought into view from the original work in CCP. For instance, the article by Charnes, Cooper and Symonds (1958) which introduced CCP was concerned with policies and programs involved in scheduling heating oil production for EXXON (then known as Standard Oil of New Jersey). This led to the formation of a risk evaluation committee (the first in the Company's history) to determine suitable choices of α^*. It was decided that a "policy" to supply all customers on demand would require $\alpha^* > \frac{1}{2}$ since the alternate choice of $\alpha^* \leq \frac{1}{2}$ was likely to be interpreted by customers, and others, to mean that the company was either indifferent or unlikely to be willing to supply all customers on demand. This characterization and usage of the term "policy" was important because the company was especially concerned with heating oil as a "commodity charged with a public interest" since failure to supply it to customers on demand (in cold weather) could have severe consequences. See the discussion of this "policy" in Charnes, Cooper and Symonds (1958).

If we define a "rule" as "a chance constraint which is to hold with probability one," then we can regard a "policy" as "a chance constraint which is to hold with probability $0.5 < \alpha^* < 1$." Implementation of a "policy" allows for deviations which can require managerial attention whereas a "rule" may be administered in clerical fashion since no exceptions are to be permitted. Notice, too, that a policy may be identified and evaluated by reference to ex-post data, as in an accounting or performance audit, in order to see whether the corresponding actions had been taken sufficiently frequently, or whether some "policy" other than the intended one had prevailed. See Cooper and Ijiri (1983).

We can now bring the above discussion into focus for its possible use in efficiency evaluations because the constraint for j = o in (9.52) contains complementary possibilities. Hence, accepting a value of $\alpha^* > 1/2$ means acceptance of a policy that favors efficient performance whereas a value of $\alpha^* \leq 1/2$ means that indifferent or inefficient performance is favored. This does not end the matter. The already calculated u_r^*, v_i^* remain available for use and may also be applied to the data that materialize after operations are undertaken by DMU_o. Applying the previously determined weights to the thus generated data allows us to calculate the probability that the values realized by DMU_O will occur. Using these weights, we may then determine whether the observed inputs and outputs yield a ratio that is within the allowable range of probabilities or whether a shift in the initially assumed multivariate distribution has occurred.

Further pursuit of this topic would lead into discussions of the higher order decision rules in CCP and/or the use of Bayesian procedures to modify the initially assumed probability distributions. See, e.g., R. Jaganathan (1985). We do not follow this route but prefer, instead, to move toward extensions of (9.51) that will enable us to make contact with the "satisficing" concepts of H. A. Simon (1957).

The following model represents an evident generalization of (9.51):

$$\text{Max } P \left\{ \frac{\displaystyle\sum_{r=1}^{s} u_r \tilde{y}_{ro}}{\displaystyle\sum_{i=1}^{m} v_i \tilde{x}_{io}} \geq \beta_o \right\}$$

s.t.

$$P \left\{ \frac{\displaystyle\sum_{r=1}^{s} u_r \tilde{y}_{rj}}{\displaystyle\sum_{i=1}^{m} v_i \tilde{x}_{ij}} \leq \beta_j \right\} \geq 1 - \alpha_j, \qquad j = 1, \ldots, n, \qquad (9.53)$$

$$P \left\{ \frac{\displaystyle\sum_{r=1}^{s} u_r \tilde{y}_{rj}}{\displaystyle\sum_{i=1}^{m} v_i \tilde{x}_{ij}} \geq \beta_j \right\} \geq 1 - \alpha_j, \qquad j = n+1, \ldots, n+k,$$

$u_r, v_i \geq 0, \forall$ r, i.

Here we interpret the β_0 in the objective as an "aspiration level" either imposed by an outside authority, as in the budgeting model of A. Stedry (1960), or adopted by an individual for some activity as in the satisficing concept of H. A. Simon (1957). The α_j ($0 \leq \alpha_j \leq 1$) in the constraints are predetermined scalars which represent an allowable chance (risk) of violating the constraints with which they are associated. The $u_r, v_i \geq 0$ are the virtual multipliers (=weights) to be determined by solving the above problem. This model evidently builds upon the "ratio form" for the CCR model of DEA with the ratio in the objective representing output and input values for DMU_0, the DMU to be evaluated. This ratio form is also included as one of the j = 1, ..., n $DMUs$ with output-to-input ratios represented as chance constraints. We may then think of the first j = 1, ..., n constraints as representing various conditions such as physical possibilities or the endurance limits of this individual. The added constraints j = n+1, ..., n + k may represent further refinements of the aspiration levels. This could even include a constraint $\beta_j = \beta_0$ with a prescribed level of probability for achieving this aspired level that might exceed the maximum possible value. The problem would then have no solution. In such cases, according to Simon (1957), an individual must either quit or else he must revise his aspiration level-or the risk of not achieving this level (or both). Thus, probabilistic (chance constrained programming) formulations allow for types of behavior which are not present in the customary deterministic models of satisficing.

Uses of these ideas in actual applications are yet to be made. However, we think that potential uses include possibilities for using DEA to extend the kinds of behavior that are represented in the approaches used in both economics and psychology. For instance, it is now common to contrast "satisficing" and "optimizing behavior" as though the two are mutually exclusive. Reformulation and use of DEA along lines like we have just described, however, may enable us to discover situations in which both types of behavior might be present. Indeed it is possible that behaviors which are now characterized as inefficient (with deterministic formulations) might better be interpreted as examples of satisficing behavior with associated probabilities of occurrence. This kind of characterization may, in turn, lead to further distinctions in which satisficing gives way to inefficiencies when probabilities are too low even for satisficing behavior and this, we think, provides access to sharper and better possibilities than those offered in the economics literature by G. J. Stigler's (1976) critique of H. Leibenstein's (1976) concept of "X-Efficiency."

The preceding interpretations were pointed toward individual behaviors that accord with the satisficing characterizations provided in H. A. Simon (1957). Turning now to managerial uses, we can simplify matters by eliminating the last k constraints in (9.53) from consideration. One of the DMUs of interest is then singled out for evaluating the probability that its performance will exceed the $\beta_j = \beta_0$ value assigned to (or assumed for) this entity in the constraints. We can then interpret our results as being applicable in either an <u>ex ante</u> or <u>ex post</u> manner according to whether our interest is in planning or control. In a planning mode, for instance, we can determine a maximum probability of inefficient or satisficing (tolerably inefficient) behavior that we may want to anticipate or forestall when $\alpha^* < \alpha_0$ occurs. For control purpose, we may similarly want to determine whether the observed behavior, as recorded, is too far out for us to regard it as having conformed to what should occur.

In a similar manner to the analysis of model (9.51), it is obvious that the first $j = 1, ..., n$ constraints in (9.53) are satisfied by choosing $u_r = 0$, and $v_i > 0$ for all r and i. For an optimal solution (u^*, v^*), we must have

$$P\left\{ \frac{\sum_{r=1}^{s} u_r^* \tilde{y}_{ro}}{\sum_{i=1}^{m} v_i^* \tilde{x}_{io}} \leq \beta_0 \right\} = 1 - \alpha^* \geq 1 - \alpha_0.$$

Therefore, we introduce the following stochasticized definitions of efficiency,

Definition 9.12 (Stochastic Efficiency). If $\beta_{j_0} = \beta_0 = 1$, DMU_0 is "stochastically efficient" if and only if $\alpha^* = \alpha_0$.

Definition 9.13 (Satisficing Efficiency). If $\beta_{j_0} = \beta_0 < 1$, DMU_0 is "satisficing efficient" if and only if $\alpha^* = \alpha_0$.

Following Land, Lovell and Thore (1992, 1993, and 1994), we assume that input values are deterministic so that only the outputs are to be represented as random variables with a multivariate normal distribution and known parameters. Again like Land, Lovell and Thore, we restrict attention to the class of zero order decision rules.

The class of zero order decision rules for use in chance constrained programming can be most easily explained by turning to the example of scheduling heating oil production at EXXON where the objective was to secure a best schedule for this seasonal (weather dependent) product as required to anticipate the probabilistic demands. Decisions rules were developed that allowed for changing production schedules, in conditional stochastic fashion, as sales materialized. A zero order rule, however, would have set the schedules for the entire season and use of this rule means that the vectors u and v of multipliers in (9.53) are to be treated as deterministic variables.

Choices of multivariate normal distributions and zero order decision rules are less restrictive than might at first appear to be the case. Transformations are available for bringing other types of distributions into approximately normal form--as is done in Charnes *et al.* (1968), for instance, when it was found necessary to treat the case of highly skewed (log-normal) distributions which were encountered when developing new product marketing strategies. We can also adapt our use of zero-order decision rules by interpreting them as a series of 1-period-at-a- time applications with appropriate models, to allow for changing realizations and probabilities, and regard these (in many situations) as approximations to the more complex solution procedures involved in developing higher order "conditional" decision rules to deal with full-scale treatment of the dynamics. Proceeding in this one-period-at-a-time manner also allows us to bypass additional problems such as the sample size considerations which are encountered in dealing with multiple observations.

Now we assume $\alpha_j < 0.5$ (we will relax this later). Utilizing techniques in chance constrained programming theory, a deterministic equivalent of (9.53) is then as follows

$$\text{Max } \gamma$$
$$\text{s.t.}$$
$$u^T y_0 - \beta_0 v^T x_0 \geq \Phi^{-1}(\gamma) \sqrt{u^T \sum_0 u} , \tag{9.54}$$
$$u^T y_j - \beta_j v^T x_j - \Phi^{-1}(\alpha_j)\eta_j \leq 0 , \quad j = 1, ..., n,$$
$$\eta_j^2 - u^T \sum_j u \geq 0 , \quad j = 1, ..., n,$$
$$u \geq 0, v \geq 0, \eta \geq 0,$$

where $\sum_j = (Cov(\tilde{y}_{ij}, \tilde{y}_{kj}))$.

This is a nonlinear and non-convex programming problem. However, let us consider the following quadratic programming problem:

$$\text{Max } \delta$$
s.t.
$$\mu^T y_0 - \beta_0 v^T x_0 \geq \delta, \tag{9.55}$$
$$\mu^T \sum_0 \mu \geq 1,$$
$$\mu^T y_j - \beta_j v^T x_j - \Phi^{-1}(\alpha_j)\varsigma_j \leq 0, \quad j = 1, \ldots, n,$$
$$\varsigma_j^2 - \mu^T \sum_j \mu \geq 0, \quad j = 1, \ldots, n,$$
$$\mu \geq 0, \ v \geq 0, \ \varsigma \geq 0.$$

It is easy to show that if δ^* is the optimal value of (9.55) and γ^* is the optimal value of (9.54), then we have $\Phi(\delta^*) = \gamma^*$. Therefore, we have the following result:

Theorem 9.7 (Cooper, Huang and Li, 1996). If $\beta_{j_0} = \beta_0 = 1$ ($\beta_{j_0} = \beta_0 < 1$), DMU_0 is stochastically (satisficing) efficient if and only if $\Phi(\delta^*) = \alpha_0$.

Now let us discuss the case of $\Phi(\delta^*) < \alpha_0$. In this case the risk of failing to satisfy the constraints for DMU_{j_0} falls below the level which was specified as satisfactory. To state this in a more positive manner, let us consider the fact that

$$P\left\{\frac{\sum_{r=1}^s u_r^* \tilde{y}_{ro}}{\sum_{i=1}^m v_i^* x_{io}} \leq \beta_0\right\} + P\left\{\frac{\sum_{r=1}^s u_r^* \tilde{y}_{ro}}{\sum_{i=1}^m v_i^* x_{io}} \geq \beta_0\right\} = 1.$$

Therefore,

$$P\left\{\frac{\sum_{r=1}^s u_r^* \tilde{y}_{ro}}{\sum_{i=1}^m v_i^* x_{io}} \leq \beta_0\right\} = 1 - \Phi(\delta^*) > 1 - \alpha_0.$$

This leads to the following corollary to the above theorem,

Corollary 9.1.   If $\beta_{j_0} = \beta_0 = 1$ ($\beta_{j_0} = \beta_0 < 1$), DMU_0 is stochastically (satisficing) inefficient if and only if $\Phi(\delta^*) < \alpha_0$.

Note: (1) if $\alpha_j = 0.5$ for DMU_j, the constraint $\zeta_j^2 - \mu^T \sum_j \mu \geq 0$ should be deleted from model (9.55); (2) if $\alpha_j > 0.5$ for DMU_j, the constraint $\zeta_j^2 - \mu^T \sum_j \mu \geq 0$ should be changed to be $\zeta_j^2 - \mu^T \sum_j \mu \leq 0$.

6.   CONCLUDING REMARKS

DEA with stochastic variations has recently received significant attention. Banker (1993), for example, incorporated statistical elements into DEA and developed a non-parametric approach with maximum likelihood methods to effect inferences in the presence of statistical noise. See the discussion in Chapter 11 of this handbook and an alternative which is based on "bootstrapping" in Chapter 10. Land, Lovell and Thore (1992, 1993 and 1994) utilized the chance constrained programming constraints (Charnes and Cooper (1959) and Charnes, Cooper and Symonds (1958)) which they adapted to DEA. Olesen and Petersen (1995) developed a chance constrained DEA model which uses piecewise linear envelopments of confidence regions for use with stochastic multiple inputs and multiple outputs. Cooper, Huang, and Li (1996) incorporated Simon's (1957) "Satisficing Concepts" into DEA models with chance constraints in order (1) to effect contact with theories of behavior in social psychology as well as (2) to extend the potential uses of DEA models to cases where 100% efficiency can be replaced by aspired levels of performance. Cooper et al. (1998) further developed a "joint chance constrained" DEA model to naturally generalize "Pareto-Koopmans Efficiency" to stochastic situations. Huang and Li (1996) utilized this joint chance constrained concept to discuss general dominance structures in stochastic situations. Recently, Cooper et al. (2002 & 2003) have introduced chance constrained models to deal with technical inefficiencies and congestion in stochastic situations. Additional stochastic DEA approaches can be found in, but are not limited to, Sengupta (1982, 1987, 1988, 1989, and 1990).

In the DEA literature dealing with chance constrained programming, attention has been restricted to the class of "zero order decision rules." This corresponds to a "here and now" approach to decision making in contrast to the "wait and see" approach that is more appropriate to dynamic settings in which it may be better to delay some parts of a decision until more information is available. To go further in this direction leads to the difficult problem of choosing a problem has only been addressed in any detail for the class of linear (first order) decision rules even in the general literature on chance constrained programming and even this treatment was conducted under restrictive assumptions on the nature of random variables and their statistical distributions (Cooper, Seiford, and Tone, 2000). Finally, a major assumption on probability distributions is that they are known. Hence there is a real need and interest in relaxing this assumption in future research (see Jagannathan, 1985).

REFERENCES

1. Aigner, D., C. A. K. Lovell, and P. Schmidt, 1977, Formulation and Estimation of Stochastic Frontier Production Function Models, *Journal of Econometrics*, Vol. 6, 21–37.
2. Banker, R.D., 1993, Maximum Likelihood, Consistency and DEA: Statistical Foundations, *Management Science*, Vol. 39,1265-1273.
3. Banker, R. D., A. Charnes, and W. W. Cooper, 1984, Some Models for Estimating Technical and Scale Inefficiencies in Data Envelopment Analysis, *Management Science*, Vol. 30, 1078-1092.
4. Bowlin, W. F., J. Brennan, W. W. Cooper, and T. Sueyoshi, 1984, DEA Models for Evaluating Efficiency Dominance, *Research Report*. The University of Texas, Center for Cybernetic Studies, Austin, Texas.
5. Charnes, A., R.L. Clarke and W.W. Cooper, 1989, An Approach to Testing for Organizational Slack Via Banker's Game Theoretic DEA Formulations, *Research in Governmental and Non Profit Accounting*, Vol. 5, 211–229.
6. Charnes, A. and W. W. Cooper, 1959, Chance Constrained Programming, *Management Science*, Vol. 5, 73-79.
7. Charnes, A., W.W. Cooper, J.K. DeVoe, and D.B. Learner, 1968, Demon, Mark II: An Extremal Equation Approach to New Product Marketing, *Management Science*, Vol. 14, 513-524.
8. Charnes, A., W. W. Cooper, B. Golany, L. Seiford, and J. Stutz, 1985, Foundations of Data Envelopment Analysis for Pareto-Koopmans Efficient Empirical Production Functions, *Journal of Econometrics*, Vol. 30, 91-107.
9. Charnes, A., W.W. Cooper and E. Rhodes, 1978, Measuring the Efficiency of Decision Making Units, *European Journal of Operational Research*, Vol. 2, 429–442.
10. Charnes, A., W. W. Cooper, and E. Rhodes, 1981, Evaluating Program and Managerial Efficiency: An Application of Data Envelopment Analysis to Program Follow Through in U.S. Public School Education, *Management Science*, Vol. 27, 668–697.

11. Charnes, A., W.W. Cooper and G.H. Symonds, 1958, Cost Horizons and Certainty Equivalents: An Approach to Stochastic Programming of Heating Oil, *Management Science*, Vol. 4, 235-263.
12. Charnes, A., W.W. Cooper and R.M. Thrall, 1986, Identifying and Classifying Efficiencies and Inefficiencies in Data Envelopment Analysis, *Operations Research Letter*, Vol. 5, 105-116.
13. Charnes, A., W.W. Cooper and R.M. Thrall, 1991, A Structure for Classifying Efficiencies and Inefficiencies in Data Envelopment Analysis, *Journal of Productivity Analysis*, Vol. 2, 197-237.
14. Charnes, A. S. Haag, P. Jaska, and J. Semple, 1992, Sensitivity of Efficiency Classifications in the Additive Model of Data Envelopment Analysis, *International Journal of Systems Science*, Vol. 23, 789-798.
15. Charnes, A. and L. Neralic, 1990, Sensitivity Analysis of the Additive Model in Data Envelopment Analysis, *European Journal of Operational Research*, Vol. 48, 332-341.
16. Charnes, A., J. Rousseau, and J. Semple, 1996, Sensitivity and Stability of Efficiency Classifications in Data Envelopment Analysis, *Journal of Productivity Analysis*, Vol. 7, 5-18.
17. Clarke, R.L., 1989, Effects of Repeated Applications of Data Envelopment Analysis on Efficiency of Air Force Vehicle Maintenance Units in the Tactical Air Command, *Ph.D. Thesis*, The University of Texas at Austin, Graduate School of Business. Also available from University Microfilms, Inc., in Ann Arbor, Michigan.
18. Cooper, W. W., H. Deng, Z. M. Huang, and S. X. Li, 2002, Chance Constrained Programming Approaches to Technical Efficiencies and Inefficiencies in Stochastic Data Envelopment Analysis, *Journal of the Operational Research Society*, Vol. 53, 1347-1356.
19. Cooper, W. W., H. Deng, Z. M. Huang, and S. X. Li, 2003, Chance Constrained Programming Approaches to Congestion in Stochastic Data Envelopment Analysis, *European Journal of Operational Research* (forthcoming).
20. Cooper, W.W., Z.M. Huang, V. Lelas, S.X. Li, and O.B. Olesen, 1998, Chance Constrained Programming Formulations for Stochastic Characterizations of Efficiency and Dominance in DEA, *Journal of Productivity Analysis*, Vol. 9, 53-79.
21. Cooper, W.W. and Y. Ijiri, eds., 1983, Kohler's Dictionary for Accountants, 6th edition, Englewood Cliffs, N.J., Prentice Hall, Inc.
22. Cooper, W.W., S. Li, L.M. Seiford, K. Tone, R.M. Thrall and J. Zhu, 2001, Sensitivity and Stability Analysis in DEA: Some Recent Developments, *Journal of Productivity Analysis*, Vol. 15, 217-246.

23. Cooper, W.W., Z.M. Huang and S. X. Li, 1996, Satisficing DEA Models Under Chance Constraints, *Annals of Operations Research*, Vol. 66, 279-295.

24. Cooper, W. W., L. M. Seiford, and K. Tone, 2000, Data Envelopment Analysis: A Comprehensive Text with Models, Applications, Reference and DEA-Solver Software, Kluwer Academic Publishers, Boston.

25. Huang, Z. M. and S. X. Li, 1996, Dominance Stochastic Models in Data Envelopment Analysis, *European Journal of Operational Research*, Vol. 95, 370-403.

26. Jagannathan, R., 1985, Use of Sample Information in Stochastic Recourse and Chance Constrained Programming, *Management Science*, Vol. 31, 96-108.

27. Land, K. C., C. A. K. Lovell, and S. Thore, 1992, Productivity and Efficiency Under Capitalism and State Socialism: The Chance-Constrained Programming Approach, *Proceedings of the 47th Congress of the International Institute of Public Finance*, (P. Pestieau, ed.), 109-121.

28. Land, K. C., C. A. K. Lovell, and S. Thore, 1993, Chance-Constrained Data Envelopment Analysis, *Managerial and Decision Economics*, Vol. 14, 541-554.

29. Land, K. C., C. A. K. Lovell, and S. Thore, 1994, Productivity and Efficiency Under Capitalism and State Socialism: An Empirical Inquiry Using Chance-Constrained Data Envelopment Analysis, *Technological Forecasting and Social Change*, Vol. 46, 139-152.

30. Leibenstein, H., 1976, Beyond Economic Man (Cambridge: Harvard University Press).

31. Olesen, O. B. and N. C. Petersen, 1995, Chance Constrained Efficiency Evaluation, *Management Science*, Vol. 41, 442-457.

32. Seiford, L. M. and J. Zhu, 1998, Stability Regions for Maintaining Efficiency in Data Envelopment Analysis, *European Journal of Operational Research*, Vol. 108, 127-139.

33. Sengupta, J. K., 1982, Efficiency Measurement in Stochastic Input-Output Systems, *International Journal of Systems Science*, Vol. 13, 273-287.

34. Sengupta, J. K., 1987, Data Envelopment Analysis for Efficiency Measurement in the Stochastic case, *Computers and Operations Research*, Vol.14, 117-129.

35. Sengupta, J. K., 1988, Robust Efficiency Measures in A Stochastic Efficiency, *International Journal of Systems Science*, Vol. 19, 779-791.

36. Sengupta, J. K., 1989, Data Envelopment with Maximum Correlation, *International Journal of Systems Science*, Vol. 20, 2085-2093.

37. Sengupta, J. K., 1990, Transformations in Stochastic DEA Models, *Journal of Econometrics*, Vol. 46, 109-124.
38. Simon, H. A., 1957, *Models of Man*, Wiley, New York.
39. Sinha, K.K., 1996, Moving Frontier Analysis: An Application of Data Envelopment Analysis for Competitive Analysis of a High Technology Manufacturing Plant, *Annals of Operations Research*, Vol. 66, 197-218.
40. Stedry, A.C., 1960, *Budget Control and Cost Behavior*, Englewood Cliffs, N.J., Prentice-Hall, Inc.
41. Stigler, G.J., 1976, The Xistence of X-Efficiency, *American Economic Review*, Vol. LXVI, 213-216.

Chapter 10

PERFORMANCE OF THE BOOTSTRAP FOR DEA ESTIMATORS AND ITERATING THE PRINCIPLE

Léopold Simar[1*] and Paul W.Wilson[2**]

[1] *Institut de Statistique and CORE, Université Catholique de Louvain, Voie du Roman Pays 20, Louvain-la-Neuve, Belgium email: simar@stat.ucl.ac.be*

[2] *Department of Economics, University of Texas, Austin, Texas 78712, USA*
email: wilson@eco.utexas.edu

Abstract: This paper further examines the bootstrap method proposed by Simar and Wilson (1998) for DEA efficiency estimators. Some simplifications as well as Monte Carlo evidence on the coverage probabilities of confidence intervals estimated by the method are offered. In addition, we present similar evidence for confidence intervals estimated with the so-called naive bootstrap to illustrate the fact that the naive bootstrap is inconsistent in the DEA setting. Finally, we propose an iterated version of the bootstrap which may be used to improve bootstrap estimates of confidence intervals.

Key words: Data envelopment analysis (DEA); Bootstrap; Distance function; Efficiency; Frontier models.

[*] Research support from "Projet d'Actions de Recherche Concertées" (No. 98/03-217) and from the "Inter-university Attraction Pole", Phase V (No. P5/24) from the Belgian Government is gratefully acknowledged.
[**] Research support from the Texas Advanced Computing Center is gratefully acknowledged.

1.　INTRODUCTION

Nonparametric efficiency estimators such as Data Envelopment Analysis (DEA) typically rely on linear programming (LP) techniques for computation of estimates, and are often characterized as *deterministic* (as opposed to *econometric* or *statistical*), as if to suggest that the methods lack any statistical underpinnings. Applied studies that have used these methods have typically presented point estimates of inefficiency, with no measure or even discussion of uncertainty surrounding these estimates. Indeed, many papers contain statements where efficiency is described as being *computed* or *calculated* as opposed to being *estimated,* and results are frequently referred to as *efficiencies* rather than *efficiency estimates.*

The choice of terminology in describing the nonparametric efficiency approaches and their results is perhaps understandable given the (until very recently) lack of a ``tool box" with aids for diagnostics, inference, *etc.* such as the one available to researchers using the parametric approaches. But the terminology is also unfortunate, because it has served to cloud important issues in efficiency estimation.

Today, researchers have access to a growing set of tools for statistical inference within nonparametric efficiency estimation; these tools are based on bootstrap methods. This chapter summarizes some of those tools, and shows how the bootstrap principle may be iterated to improve confidence interval estimates. Section 2 briefly reviews the microeconomic theory of the firm, and establishes our notation and an economic model. Section 3 discusses estimation of efficiency, while section 4 establishes a statistical model by augmenting the economics assumptions in section 2 with some assumptions on the data generating process (DGP). Section 5 briefly reviews existing asymptotic results, which have implications for efficiency of estimation as well as for the bootstrap. The DEA bootstrap is discussed generally in section 6, and more specifically in section 7, where we give details on its implementation. Section 8 details our Monte Carlo experiments and their results. Section 9 offers an iterated version of the bootstrap which can be used to improve estimates of confidence intervals. The final section offers some concluding remarks.

2.　EFFICIENCY AND THE THEORY OF THE FIRM

Standard microeconomics texts develop the theory of the firm by positing a production set which describes how a set of inputs may be somehow converted into outputs. To illustrate, let $x \in R_+^p$ denote a vector of p inputs and $y \in R_+^q$ denote a vector of q outputs. Then the production set may be defined as

$$P \equiv \{(x, y) | x \ can \ produce \ y\} \qquad (10.1)$$

which is merely the set of feasible combinations of x and y. The production set P is sometimes described in terms of its sections

$$Y(x) \equiv \{y | (x, y) \in P\} \qquad (10.2)$$

and

$$X(y) \equiv \{x | (x, y) \in P\} \qquad (10.3)$$

which form the output feasibility and input requirement sets, respectively. Knowledge of either $Y(x)$ for all x or $X(y)$ for all y is equivalent to knowledge of P; P implies (and is implied by) both $Y(x)$ and $X(y)$. Thus, both $Y(x)$ and $X(y)$ inherit the properties of P. Various assumptions regarding P are possible; we adopt those of Shephard (1970) and Färe (1988):

Assumption A1: P *is convex,* $Y(x)$ *is convex, bounded, and closed for all* $x \in R_+^p$, *and* $X(y)$ *is convex, bounded, and closed for all* $y \in R_+^q$.

Assumption A2: $(0, y) \notin P$ *if* $y \geq 0, y \neq 0$, *i.e., all production requires use of some inputs.*

Assumption A2 merely says that there are no free lunches[1].

Assumption A3: *for* $\tilde{x} \geq x, \tilde{y} \leq y, \tilde{x} \in R_+^p, \tilde{y} \in R_+^q$, *if* $(x, y) \in P$ *then* $(\tilde{x}, \tilde{y}) \in P$, *i.e., both inputs and outputs are strongly disposable.*

Assumption A3 is sometimes called free disposability. The presentation in this section does not depend on these particular assumptions, and the estimators discussed in Section 3 can be adapted to cases where alternative assumptions might be warranted. For example, if pollution is an inadvertent byproduct of the production process, then it might not be reasonable to assume that this particular output is strongly (freely) disposable.

The boundary of P is sometimes referred to as the *technology* or *production frontier,* and is given by the intersection of P and the closure of its complement, P^∂. Similarly, isoquants are defined by the boundary of $X(y)$,

$$X^\partial(y) = \{x \mid x \in X(y), \theta \, x \notin X(y) \, \forall \, 0 < \theta < 1\} \qquad (10.4)$$

while iso-output curves are defined by the boundary of $Y(x)$,

$$Y^\partial(x) = \{y \mid y \in Y(x), \lambda \, y \notin Y(x) \, \forall \, \lambda > 1\} \qquad (10.5)$$

---

[1] Throughout, inequalities involving vectors are defined on an element-by-element basis; *e.g.,* for, $\tilde{x}, x \in R_+^p$ means that some, but perhaps not all or none, of the corresponding elements of $\tilde{x}$ and x may be equal, while some (but perhaps not all or none) of the elements of $\tilde{x}$ may be greater than corresponding elements of x.

Firms which are *technically inefficient* operate at points in the interior of P, while those that are technically *efficient* operate somewhere along the technology defined by P^∂. Various measures of technical efficiency are possible. The Shephard (1970) output distance function provides a normalized measure of Euclidean distance from a point $(x, y) \in P$ to P^∂ in a direction orthogonal to x, and may be defined as

$$D(x, y \mid P) \equiv \inf \left\{ \theta > 0 \middle| (x, \theta^{-1} y) \in P \right\} \qquad (10.6)$$

Clearly, $D(x, y \mid P) \le 1$ for all $(x, y) \in P$. If $D(x, y \mid P) = 1$ then $(x, y) \in P^\partial$; *i.e.*, that the point (x, y) lies on the boundary of P, and the firm is technically efficient. One can similarly define the Shephard (1970) input distance function, which provides a normalized measure of Euclidean distance from a point $(x, y) \in P$ to P^∂ in a direction orthogonal to y. Alternatively, one could employ measures of hyperbolic graph efficiency defined by Färe *et al.* (1985), directional distance functions as defined by Färe *et al.* (1996), or perhaps other measures. From a purely technical viewpoint, any of these can be used to measure technical efficiency; the only real difference is in the direction in which distance to the technology is measured. However, if one wishes to take account of behavioral implications, then this might influence the choice between the various possibilities. We present our methodology only in terms of the output orientation to conserve space, but our discussion can be adapted to the other cases by straightforward changes in our notation. Note the Shephard output distance function is the reciprocal of the Farrell (1957) output efficiency measure, while the Shephard input distance function is the reciprocal of the Farrell input efficiency measure.

In addition to technical efficiency, one may consider whether a particular firm is allocatively efficient. Assuming the firm is technically efficient, its input-allocative efficiency depends on whether it operates at the optimal location along one of the isoquants $X^\partial(y)$ as determined by prevailing input prices. Alternatively, output-allocative efficiency depends on whether the firm operates at the optimal location along one of the iso-output curves $Y^\partial(x)$ as determined by prevailing output prices. Färe *et al.* (1985) combine these notions of allocative efficiency with technical efficiency to obtain measures of "overall" efficiency. See Section 5 in Chapter 1 where these relations between technical and overall inefficiency are discussed. Here, we focus only on output technical efficiency, but it is straightforward to extend our discussion to the other notions of inefficiency.

Standard microeconomic theory suggests that with perfectly competitive input and output markets, firms which are either technically or allocatively inefficient will be driven from the market. However, in the real world, even where markets may be highly competitive, there is no reason to believe that this must happen instantaneously. Indeed, due to various frictions and imperfections in real markets for both inputs and outputs, this process might take many years, and firms that are initially inefficient in one or more respects may recover and begin to operate efficiently before they are driven from the market. Wheelock and Wilson (1995, 2000) provide support for this view through empirical evidence for banks operating in the USA.

3.   ESTIMATION

Unfortunately, none of the theoretical items defined in the previous section are observed, including the production set P and the output distance function $D(x, y| P)$. Consequently, these must be *estimated.*

In the typical situation, all that is observed are inputs and outputs for a set of n firms; together, these comprise the observed sample:

$$S_n = \{(x_i, y_i)\}_{i=1}^{n} \tag{10.7}$$

These may be used to construct various estimators of P, which in turn may be used to construct estimators of $D(x, y| P)$. Charnes *et al.* (1978) used the conical hull of the free disposal hull of S_n to construct an estimator $\hat{V}$ of P in what has been called the "CCR model". This approach implicitly imposes an assumption of constant returns to scale on the production technology, P^∂. Banker *et al.* (1984) used the convex hull of the free disposal hull of S_n as an estimator $\hat{P}$ of P. This approach, which has been termed the "BCC model", allows P^∂ to exhibit increasing, constant, or decreasing returns at different locations along the frontier; hence $\hat{P}$ is said to allow variable returns to scale. Other estimators of P are possible, such as the free disposal hull (FDH) of S_n by itself (denoted $\tilde{P}$) as in Deprins *et al.* (1984), which in effect relaxes the assumption of convexity on P.

Terminology used in the DEA literature is sometimes rather confusing and misleading, and the preceding discussion offers an example. The terms "CCR model" and "BCC model" are misnomers; the conical and convex hulls of the free disposal hull of S_n are merely *different estimators* of P -- not different models. Similarly, the term "DEA model" appears far too frequently in the literature and serves only to obfuscate the fact that DEA is a class of estimators (characterized by, among other things, convexity assumptions). By contrast, a (statistical) model consists of a set of assumptions on (i) the underlying distribution from which the data are drawn and its support, and (ii) the process by which data are sampled (*e.g.,* independently or otherwise) from this distribution (see Spanos, 1986, for additional discussion). Together, these assumptions define a Data Generating Process (DGP). Given any one of these estimators of P, estimators of $D(x, y| P)$ can be constructed. For example, using $\hat{P}$, we may write

$$D(x, y \mid \hat{P}) \equiv \inf \left\{ \theta > 0 \middle| (x, \theta^{-1} y) \in \hat{P} \right\} \tag{10.8}$$

where $\hat{P}$ has replaced P on the right-hand side of (10.6). Alternative estimators of $D(x, y \mid P)$ can be obtained by defining

$$D(x, y \mid \hat{V}) \equiv \inf \left\{ \theta > 0 \middle| (x, \theta^{-1} y) \in \hat{V} \right\} \tag{10.9}$$

and

$$D(x, y \mid \tilde{P}) \equiv \inf \left\{ \theta > 0 \middle| (x, \theta^{-1} y) \in \tilde{P} \right\} \tag{10.10}$$

The estimators in (10.8) and (10.9) can be used to construct tests of returns to scale; see Simar and Wilson (2002) and the discussion in Chapter 2.

It is similarly straightforward (with the benefit of hindsight and the pioneering work of Charnes *et al.* 1978) to compute estimates of $D(x, y \mid P)$ using (10.8)-(10.10) and the data in S_n. In particular, (10.8)-(10.9) may be rewritten as linear programs:

$$\left[D(x, y \mid \hat{P}) \right]^{-1} = \max \left\{ \phi \middle| \phi y \leq Y\lambda, \ x \geq X\lambda, \ i\lambda = 1, \lambda \in R_+^n \right\} \tag{10.11}$$

and

$$\left[D(x, y \mid \hat{V}) \right]^{-1} = \max \left\{ \phi \middle| \phi y \leq Y\lambda, \ x \geq X\lambda, \ \lambda \in R_+^n \right\} \tag{10.12}$$

where $Y = [y_1 \dots y_n]$, $X = [x_1 \dots x_n]$, with each $x_i, y_i \ i = 1, \dots, n$ denoting the $(p \times 1)$ and $(q \times 1)$ vectors of observed inputs and outputs (respectively), i is a $(1 \times n)$ vector of ones, and $\lambda = [\lambda_1 \dots \lambda_n]$ is a $(n \times 1)$ vector of intensity variables. One can also rewrite (10.10) as a linear program, *e.g.*,

$$\left[D(x, y \mid \tilde{P}) \right]^{-1} = \max \left\{ \phi \middle| \phi y \leq Y\lambda, \ x \geq X\lambda, \ \lambda \in \{0, 1\} \right\} \tag{10.13}$$

although linear programming is typically not used to compute estimates for the estimator based on the free disposal hull.

Given our assumption that P is convex, by construction,

$$\tilde{P} \subseteq \hat{P} \subseteq \begin{cases} \hat{V} \\ P \end{cases} \tag{10.14}$$

If the technology P^∂ exhibits constant returns to scale everywhere, then it is also the case that $\hat{V} \subseteq P$, but otherwise, $\hat{V} \not\subset P$.

The differences between P and any of the estimators $\hat{P}$, $\hat{V}$, and $\tilde{P}$ are of utmost importance, for these differences determine the difference between $D(x, y \mid P)$ and any of the corresponding estimators $D(x, y \mid \hat{P})$, $D(x, y \mid \hat{V})$, and $D(x, y \mid \tilde{P})$. Note that P and especially $D(x, y \mid P)$ are the *true* quantities of interest. By contrast, $\hat{P}$, P, $\tilde{P}$ and are *estimators* of P, while $D(x, y \mid \hat{P})$, $D(x, y \mid \hat{V})$, and $D(x, y \mid \tilde{P})$ are each *estimators* of $D(x, y \mid P)$. The true things that we are interested in, namely P and $D(x, y \mid P)$, are fixed, but unknown. Estimators, on the other hand, are necessarily *random variables*. When the data in S_n are used together with (10.11)-(10.13) to compute numerical values for the estimators $D(x, y | \hat{P})$, $D(x, y | \hat{V})$, and $D(x, y | \tilde{P})$, the resulting real numbers are merely specific realizations of different random variables; these numbers (which are typically those reported in efficiency studies) are *estimates*.

In our view, anything we might compute from data is an estimate. The formula used for such a computation defines an estimator. The DEA setting is not an exception to this principle.

The skeptical reader may well ask, "what if I have observations on *all* the firms in the industry I am examining?" This is a perfectly reasonable question, but it begs for additional questions. What happens if an entrepreneur establishes a new firm in this industry? How well might this new firm perform? Is there any reason why, at least in principle, it cannot perform better than the original firms in the industry? In other words, is it reasonable to assume that the new firm cannot operate somewhere above the convex hull of the existing firms represented by input/output vectors in R_+^{p+q}?

We can see no reason why the general answer to this last question must be anything other than a resounding "no!" To answer otherwise would be to deny the existence of technical progress, and by extension the historical experience of the civilized world over the last several hundred years. If one accepts this answer, it seems natural then to think in terms of a conceptual infinite population of potential firms, and to view one's sample as having resulted from a draw from an infinite population.

4.   A STATISTICAL MODEL

Continuing with the reasoning at the end of the previous section, once one accepts that the world is an uncertain place and all one can hope for are reliable estimators of unobservables such as $D(x, y | P)$, some additional questions are raised:

- What are the properties of the estimators defined in (10.8)-(10.10)? Are they at least consistent? If not, then even with an infinite amount of data, one might have little hope of obtaining a useful estimate.
- How much can be learned about the true value of interest? Is inference possible?
- Can hypotheses regarding P^∂ be tested, even though P^∂ is unobservable?
- Will inferences and hypothesis tests be reliable, meaningful?

Before these questions can be answered, a statistical model must be defined. A statistical model is merely a set of assumptions on the data-generating process (DGP). While many assumptions are possible, prudence dictates that the assumptions be chosen to provide enough structure so that desirable properties of estimators can be derived, but without imposing unnecessary restrictions. In addition to Assumptions A1-A3 adopted in Section 2, we now adopt assumptions based on those of Kneip *et al.* (1998).

Assumption A4: *the sample observations in S_n are realizations of identically, independently distributed (iid) random variables with probability density function $f(x, y)$ with support over P.*

Note that a point $(x, y) \in R_+^{p+q}$ represented by Cartesian coordinates can also be represented by cylindrical coordinates (x, ω, η) where (ω, η) are the polar coordinates of y in R_+^q. Thus, the modulus $\omega = \omega(y) \in R_+^1$ of y is given by the square root of the sum of the squared elements of y, and the jth element of the corresponding angle $\eta = \eta(y) \in [0, \pi/2]^{q-1}$ of y is given by arctan (y^{j+1}/y^1) for $y^1 \neq 0$ (where y^j represents the jth element of y); if $y^1 = 0$, then all elements of $\mu(y)$ equal $\pi/2$.

Writing $f(x, y)$ in terms of the cylindrical coordinates, we can decompose the density by writing

$$f(x, \omega, \eta) = f(\omega | x, \eta) f(\eta | x) f(x) \qquad (10.15)$$

where all the conditional densities exist. In particular, $f(x)$ is defined on R_+^p, $f(\eta | x)$ is defined on $[0, \pi/2]^{q-1}$, and $f(\omega | x, \eta)$ is defined on R_+^1. Now consider a point $(x, y) \in P$, and the corresponding point $(x, y^\partial(x))$ which is the projection of (x, y) onto P^∂ in the direction orthogonal to x, i.e. $y^\partial(x) = y / D(x, by | P)$. Then the moduli of these points are related to the output distance function via

$$0 \leq D(x, y | P) = \frac{\omega(y)}{\omega(y^\partial(x))} \leq 1 \qquad (10.16)$$

Then the density $f(\omega | x, \eta)$ on $[0, \omega(y^\partial(x))]$ implies a density $f(D(x, y | P) | x, \eta)$ on the interval $[0,1]$.

In order for our estimators of P and $D(x, y | P)$ to be consistent, the probability of observing firms in a neighborhood of P^∂ must approach unity as the sample size increases:

Assumption A5: *for all $x \geq 0$ and all $\eta \in [0, \pi/2]^{q-1}$, there exist constants $\varepsilon_1 > 0$ and $\varepsilon_2 > 0$ such that for all $\omega \in [\omega(y^\partial(x)), \omega(y^\partial(x)) + \varepsilon_2]$, $f(\omega | x, \eta) \geq \varepsilon_1$.*

In addition, an assumption about the smoothness of the frontier is needed:[2]

Assumption A6: *for all* (x, y) *in the interior of* P, $D(x, y | P)$ *is differentiable in both its arguments.*

Assumptions A1-A6 define the DGP F which yields the data in S_n. These assumptions are somewhat more detailed than the set of assumptions listed in Section 2, which were motivated by microeconomic theory.

5. SOME ASYMPTOTIC RESULTS

Recent papers have provided some asymptotic results for DEA/FDH estimators. Korostelev *et al.* (1995a, 1995b) examined the convergence of estimators of P, and proved (for the special case where $p \geq 1$, $q = 1$) that both $\widetilde{P}$ and $\hat{P}$ are consistent estimators of P in the sense that

$$d(\widetilde{P}, P) = O_p\left(n^{-\frac{1}{p+1}}\right) \tag{10.17}$$

and

$$d(\hat{P}, P) = O_p\left(n^{-\frac{2}{p+2}}\right) \tag{10.18}$$

where $d(\hat{P}, P)$ is the Lebesgue measure of the difference between two sets.[3]

The result for $\widetilde{P}$ in (10.17) still holds if assumption A1 given above is dropped. However, the convexity assumption is necessary for (10.18). It would be seemingly straightforward to extend these results to obtain a convergence rate for $\hat{V}$ provided one adopted an additional assumption that P^∂ displays constant returns to scale everywhere.

These results also indicate that the *curse of dimensionality* which typically plagues nonparametric estimators is at work in the DEA/FDH setting; the convergence rates decrease as the number of inputs is increased. The practical implication of this is that researchers will need increasing amounts of data to get meaningful results as the number of inputs is increased. And, although these results were obtained with $q = 1$, presumably allowing the number of outputs to increase would have a similar effect on convergence rates and the resulting data requirements.

This intuition is confirmed by Kneip *et al.* (1998) and Park *et al.* (1999), who derive convergence rates for $D(x, y | \hat{P})$ and $D(x, y | \widetilde{P})$, respectively, for the general case where $p \geq 1$, $q \geq 1$:

---

[2] Our characterization of the smoothness condition here is stronger than required; Kneip *et al.* (1998) require only Lipschitz continuity for $D(x, y | P)$, which is implied by the simpler, but stronger requirement presented here.

[3] Banker (1993) showed, for the case $q=1$, $p \geq 1$, that $\hat{P}$ is a consistent estimator of P, but did not provide convergence rates.

$$D(x, y \mid \hat{P}) - D(x, y \mid P) = O_p\left(n^{-\frac{2}{p+q+1}}\right) \tag{10.19}$$

and

$$D(x, y \mid \tilde{P}) - D(x, y \mid P) = O_p\left(n^{-\frac{1}{p+q}}\right). \tag{10.20}$$

In both cases, the convergence rates are affected by $p + q$. The convergence rate for the FDH estimator is slower than for the convex hull estimator, and thus the convex hull estimator is preferred when P is convex. But if Assumption A1 does not hold, then there is no choice but to use the FDH estimator--the convex hull estimator as well as the convex cone estimator would be inconsistent in this case since $\tilde{P}$ and $\tilde{V}$ are convex for all $1 \le n \le \infty$, and so cannot converge to a nonconvex set as $n \to \infty$.

For the special case where $p = q = 1$, Gijbels *et al.* (1999) derived results which indicate that

$$D(x, y \mid \hat{P}) - D(x, y \mid P) \overset{asy.}{\sim} G(\cdot, \cdot) \tag{10.21}$$

where G is a known distribution function depending on unknown quantities related to the DGP F [4].

For the general case where $p \ge 1$ and $q \ge 1$, Park *et al.* (1997) prove that

$$D(x, y \mid \tilde{P}) - D(x, y \mid P) \overset{asy.}{\sim} Weibull(\cdot, \cdot), \tag{10.22}$$

where the unknown parameters of the Weibull distribution again must be estimated and depend on characteristics of the DGP.

These results are potentially very useful for applied researchers. In particular, (10.22) may be used to construct and then estimate confidence intervals for $D(x, y \mid P)$ when the FDH estimator is used. It is unfortunate that a similar result for the DEA case to date exists only for the special case where $p = q = 1$. In any case, these are asymptotic results, and the quality of confidence intervals, *etc.* estimated using these results in small samples remains an open question.

---

[4] In particular, the unknown quantities are determined by the curvature of P^∂ and the value of $f(x,y)$ at the point where (x,y) is projected onto P^∂ in the direction orthogonal to x. See Gijbels *et al.* (1999) for additional details.

6.   BOOTSTRAPPING IN DEA/FDH MODELS

The bootstrap (Efron, 1979; 1982) offers an alternative approach to classical inference and hypothesis-testing. In the case of DEA estimators with multiple inputs or outputs, the bootstrap currently offers the *only* sensible approach to inference and hypothesis-testing.

Bootstrapping is based on the analogy principle (*e.g.,* see Manksi, 1988). The presentation here is based on our earlier work in Simar and Wilson (1998, 1999c). In the *true world* we observe the data in S_n which are generated from

$$F = F(P, f(x, y)). \tag{10.23}$$

In the true world, F, P, and $D(x, y \mid P)$ are unobserved, and must be estimated using S_n.

Let $\hat{F}(S_n)$ be a consistent estimator of F:

$$\hat{F}(S_n) = F(\hat{P}, \hat{f}(x, y)). \tag{10.24}$$

One can easily simulate what happens in the true world by drawing a new dataset $S_n^* = \{x_i^*, y_i^*\}_{i=1}^n$ (analogous to (10.7)) from $\hat{F}(S_n)$, and then applying the original estimator to these new data. If the original estimator for a point (x_o, y_o) (not necessarily contained in S_n) was $D(x_0, y_0 \mid \hat{P})$, then the estimator obtained from $\hat{F}(S_n)$ would be $D(x_0, y_0 \mid \hat{P}^*)$, where $\hat{P}^*$ denotes the convex hull of the free disposal hull of S_n^*, and is obtained by solving

$$\left[D(x_0, y_0 \mid \hat{P}^*) \right]^{-1} = \max \{ \phi \mid \phi\, y_0 \leq Y^* \lambda,\, x_0 \geq X^* \lambda,\, i\lambda = 1,\, \lambda \in R_+^n \} \tag{10.25}$$

where $Y^* = [y_1^* \cdots y_n^*]$, $X^* = [x_1^* \cdots x_n^*]$, with each, (x_i^*, y_i^*), $i = 1, ..., n$, denoting observations in the pseudo dataset S_n^*. Repeating this process B times (where B is appropriately large) will result in a set of bootstrap values $\left\{ D_b(x_0, y_0 \mid \hat{P}^*) \right\}_{b=1}^B$. When the bootstrap is consistent, then

$$\left(D(x_0, y_0 \mid \hat{P}^*) - D(x_0, y_0 \mid \hat{P}) \right) \Big| F(\hat{P}, \hat{f}(x, y)) \overset{approx}{\sim}$$

$$\left(D(x_0, y_0 \mid \hat{P}) - D(x_0, y_0 \mid P) \right) \Big| F(P, f(x, y)). \tag{10.26}$$

Given the original estimate $D(x_0, y_0 \mid \hat{P})$ and the set of bootstrap values

$$\left\{ D_b(x_0, y_0 \mid \hat{P}^*) \right\}_{b=1}^B,$$

the left-hand side of (10.26) is known with arbitrary precision (determined by the choice of number of bootstrap replications, B); the approximation improves as the sample size, n, increases.

Note that we could tell the story of the previous paragraph in terms of the convex cone estimator by merely changing $\hat{P}$ to $\hat{V}$. Equation (10.26) is the essence of the bootstrap. In principle, since $F(\hat{P}, \hat{f}(x, y))$ is known, it should be possible to determine the distribution on the left-hand side analytically. In practice, however, this is intractable in most problems, including the present one. Hence Monte Carlo simulations are used to approximate this distribution. In our presentation, after the B bootstrap replications are performed, the set of bootstrap values $\left\{D_b(x_0, y_0 \mid \hat{P}^*)\right\}_{b=1}^{B}$ gives an empirical approximation to this distribution.

Once the set of bootstrap values $\left\{D_b(x_0, y_0 \mid \hat{P}^*)\right\}_{b=1}^{B}$ has been obtained, it is straightforward to estimate confidence intervals for the true distance function value $D(x_0, y_0 \mid P)$. This is accomplished by noting that *if* we knew the true distribution of

$$\left(D(x_0, y_0 \mid \hat{P}) - D(x_0, y_0 \mid P)\right),$$

then it would be trivial to find values a_∂ and b_∂ such that

$$\Pr\left(-b_\alpha \leq D(x_0, y_0 \mid \hat{P}) - D(x_0, y_0 \mid P) \leq -a_\alpha\right) = 1 - \alpha \qquad (10.27)$$

Of course, a_α and b_α are unknown, but from the empirical bootstrap distribution of the pseudo estimates $D_b(x_0, y_0 \mid \hat{P}^*)$, $b = 1, ..., B$, we can find values $\hat{a}_\alpha$ and $\hat{b}_\alpha$ such that

$$\Pr\left(-\hat{b}_\alpha \leq D(x_0, y_0 \mid \hat{P}^*) - D(x_0, y_0 \mid \hat{P}) \leq -a_\alpha \mid \hat{F}(S_n)\right) \approx 1 - \alpha \qquad (10.28)$$

Finding $\hat{a}_\alpha$ and $\hat{b}_\alpha$ involves sorting the values

$$\left(D_b(x_0, y_0 \mid \hat{P}^*) - D(x_0, y_0 \mid \hat{P})\right) \qquad b = 1, ..., B$$

in increasing order and then deleting $\left(\frac{\alpha}{2} \times 100\right)$-percent of the elements at either end of the sorted list. Then set $-\hat{b}_\alpha$ and $-\hat{a}_\alpha$ equal to the endpoints of the truncated, sorted array, with $\hat{a}_\alpha \leq \hat{b}_\alpha$.

The bootstrap approximation of (10.27) is then

$$\Pr\left(-\hat{b}_\alpha \leq D(x_0, y_0 \mid \hat{P}) - D(x_0, y_0 \mid P) \leq -\hat{a}_\alpha\right) \approx 1 - \alpha. \qquad (10.29)$$

The estimated ($1 - \alpha$)-percent confidence interval is then

$$D(x_0, y_0 \mid \hat{P}) + \hat{a}_\alpha \leq D(x_0, y_0 \mid P) \leq D(x_0, y_0 \mid \hat{P}) + \hat{b}_\alpha. \qquad (10.30)$$

This procedure can be used for any $(x_0, y_0) \in R_+^{p+q}$ for which $D(x_0, y_0 \mid \hat{P}^*)$ exists. Typically, the applied researcher is interested in the efficiency scores of the observed units themselves; in this case, the above procedure can be repeated n times, with $(x_0, y_0) = (x_i, y_i), i = 1, ..., n$, producing a set of n confidence intervals of the form (10.30), one for each firm.

The method described above for estimating confidence intervals from the set of bootstrap values $\left\{D_b(x_0, y_0 \mid \hat{P}^*)\right\}_{b=1}^{B}$ differs slightly from what we proposed in Simar and Wilson (1998); here, we avoid explicit use of a bias estimate, which adds unnecessary noise to the estimated confidence intervals. The bias estimate itself, however, remains interesting.

By definition,

$$BIAS\left(D(x_0, y_0 \mid \hat{P})\right) = E\left(D(x_0, y_0 \mid \hat{P})\right) - D(x_0, y_0 \mid P). \qquad (10.31)$$

The bootstrap bias estimate for the original estimator $D(x_0, y_0 \mid \hat{P})$ is the empirical analog of (10.31):

$$\widehat{BIAS}_B \left(D(x_0, y_0 \mid \hat{P}) \right) = B^{-1} \sum_{b=1}^{B} D_b(x_0, y_0 \mid \hat{P}^*) - D(x_0, y_0 \mid \hat{P}). \qquad (10.32)$$

It is tempting to construct a bias-corrected estimator of $D(x_0, y_0 \mid P)$ by computing

$$\hat{\hat{D}}(x_0, y_0) = D(x_0, y_0 \mid \hat{P}) - \widehat{BIAS}_B \left(D(x_0, y_0 \mid \hat{P}) \right)$$

$$= 2D(x_0, y_0 \mid \hat{P}) - B^{-1} \sum_{b=1}^{B} D_b(x_0, y_0 \mid \hat{P}^*). \qquad (10.33)$$

It is well known (*e.g.,* Efron and Tibshirani, 1993), however, that this bias correction introduces additional noise; the mean-square error of $\hat{\hat{D}}(x_0, y_0)$ may be greater than the mean-square error of $D(x_0, y_0 \mid \hat{P})$. Thus the bias-correction in (10.33) should be used with caution[5].

7.   IMPLEMENTING THE BOOTSTRAP

We have shown elsewhere (Simar and Wilson 1998, 1999a, 1999b, 2000a, 2000b) that the generation of the bootstrap pseudo data S_n^* is crucial in determining whether the bootstrap gives consistent estimates of confidence intervals, bias, *etc.* In the classical linear regression model, one may resample from the estimated residuals, or alternatively resample the original sample observations to construct the pseudo dataset S_n^*; in either case, the bootstrap will give consistent estimates. Both these approaches are variants of what has been called the *naive* bootstrap. The analog of these methods in frontier estimation would be to resample from the original distance function estimates, or alternatively from the (x, y) pairs in S_n.

In the present setting, however, there is a crucial difference from the classical linear regression model: here, the DGP F has bounded support over P, while in the classical linear regression model, the DGP has unbounded support. A related problem is that under our assumptions, the conditional density $f(D(x, y \mid P) \mid x, \eta)$ has bounded support over the interval $(0,1]$, and is right-discontinuous at 1. It is widely known that problems such as these can cause the naive bootstrap to give *inconsistent* estimates (*e.g.,* see Bickel and Friedman, 1981; Swanepoel, 1986; Beran and Ducharme, 1991; and Efron and Tibshirani, 1993), and this is true in the

---

[5] The mean-square error of the bias-corrected estimator in (10.33) could be evaluated in a second-level bootstrap along the same lines as the iterated bootstrap we propose below in section 9. See Efron and Tibshirani (1993, pp.138) for a simple example in a different context.

present setting as we have shown in Simar and Wilson (1999a, 1999b, 2000a)[6].

To address the boundary problems which doom the naive bootstrap in the present situation, we draw pseudo datasets from a smooth, consistent, nonparametric estimate of the DGP F, as represented by $f(x,\omega,\eta)$ in (10.15). In Simar and Wilson (1998), we drew values D^* from a kernel estimate $\hat{f}(D)$ of the marginal density of the original estimates $D(x_i, y_i \mid \hat{P})$, $i = 1,...,n$; given (10.16), this is tantamount to assuming $f(\omega \mid x,\eta) = f(\omega)$ in (10.15), or in other words, that the distribution of inefficiencies is homogeneous and does not depend upon location within the production set P. This assumption can be relaxed, as in Simar and Wilson (2000b), but at a cost of increased complexity and computational burden. In this paper, we will focus on the homogeneous case. In applications, one can test the homogeneity assumption using the methods surveyed by Wilson (2003), and then choose between the two bootstrap methods in Simar and Wilson (1998, 2000b).

Kernel density estimation is rather easy to apply and has been widely studied. Given values z_i, $i = 1,...,n$ on the real number line, the kernel estimate of the density $g(z)$ is given by

$$\hat{g}(z) = \frac{1}{nh} \sum_{i=1}^{n} K\left(\frac{z - z_i}{h}\right), \tag{10.34}$$

where $K(\cdot)$ is a *kernel function* and h is a *bandwidth parameter*. Both $K(\cdot)$ and h must be chosen, but there are well-established guidelines to aid with these choices. In particular, the kernel function $K(\cdot)$ must be piecewise continuous and must satisfy $\int_{-\infty}^{\infty} K(u)du = 1$ and $\int_{-\infty}^{\infty} uK(u)du = 0$. Thus any probability density function that is symmetric around zero is a valid kernel function; in addition, one could use even-ordered polynomials bounded on some interval with coefficients chosen so that the polynomial integrates to unity over this interval[7]. As Silverman (1986) shows, the choice of the kernel function is far less critical for obtaining a good estimate (in the sense of low mean integrated square error) of $f(z)$ than the choice of the bandwidth parameter h.

In order for the kernel density estimator to be consistent, the bandwidth must be chosen so that $h = O(n^{-1/5})$ for the univariate case considered here.

---

[6] Explicit descriptions of why either variation of the naive bootstrap results in inconsistent estimates are given in Simar and Wilson (1999a, 2000a). Löthgren and Tambour (1997, 1999) and Löthgren (1998, 1999) employ a bizarre, illogical variant of the naive bootstrap different from the more typical variations we have mentioned. This approach also leads to an inconsistency problem, as discussed and confirmed with Monte Carlo experiments in Simar and Wilson (2000a).

[7] Appropriately chosen high-order kernels can reduce the order of the bias in the kernel density estimator, but run the risk of producing negative density estimates at some locations.

If the data are approximately normally distributed, then one may employ the normal reference rule and set $h = 1.06\hat{\sigma} n^{-1/5}$, where $\hat{\sigma}$ is the sample standard deviation of the data whose density is being estimated (Silverman, 1986). In cases where the data are clearly non-normal, as will be the case when estimating the density $f(D)$, one can plot the density estimate for various values of h and choose the value of h that gives a reasonable estimate, as in Silverman (1978) and Simar and Wilson (1998). This approach, however, contains an element of subjectivity; a better approach is to employ least squares cross-validation, which involves choosing the value of h that minimizes an approximation to mean integrated square error; see Silverman (1986) for details. In many cases involving high dimensions (*i.e.*, large $p + q$), many of the distance function estimates will equal unity (this is especially the case when the convex hull or free disposal hull estimators are used), and this creates a discretization problem in the cross-validation procedure. Simar and Wilson (2002) propose a weighted cross-validation procedure to avoid this problem in the univariate setting; Simar and Wilson (2000b) extend this idea to the multivariate setting.

Regardless of how the bandwidth parameter is chosen, it is important to note that ordinary kernel estimators (such as the one in (10.34) above) of densities with bounded support will be biased near the boundaries of the support. Necessarily, for $(x, y) \in P$, we have $0 < D(x, y \mid P) \leq D(x, y \mid \hat{P}) \leq 1$, and so this will be a problem in estimating $f(D)$. It is easy to see why this problem arises: in (10.34), when $K(\cdot)$ is symmetric about zero, the density estimate is determined at a particular point on the real number line by adding up the value of functions $K(\cdot)$ centered over the observed data along the real number line *on either side* of the point where the density estimate is being evaluated. If, for example, $K(\cdot)$ is chosen as a standard normal probability density function, then nearby observations contribute relatively more to the density estimate at this particular point than do observations that are farther away along the real number line[8]. When (10.34) is used to evaluate the density estimate at unity in our application, then if $K(\cdot)$ is symmetric, there will necessarily be no data on the right side of the boundary to contribute to the smoothing, and this causes the bias problem. An obvious approach would be to let the kernel function become increasingly asymmetric as the boundary is approached, but this is problematic for

---

[8] This is the sense in which the kernel density estimator is a *smoother*, since it is, in effect, smoothing the empirical density function which places probability mass $1/n$ at each observed datum. Setting $h=0$ in (10.34) yields the empirical density function, while letting $h \to \infty$ yields a flat density estimate. The requirement that $h = O(n^{-1/5})$ to ensure consistency of the kernel density estimate results from the fact that as n increases, h must become smaller, but not too quickly.

several reasons as discussed by Scott (1992). A much simpler solution is to use the reflection method as in Simar and Wilson (1998).

The reflection method involves reflecting each of the n original estimates $D(x_i, y_i | \hat{P})$ about the boundary at unity (by computing $1 - D(x_i, y_i | \hat{P})$ for each $D(x_i, y_i | \hat{P})$, $i = 1, ..., n$) to obtain $2n$ points along the real number line. The output distance function estimates are also bounded on the left at 0, but typically there will be no values near zero, suggesting that the density is near zero at this boundary. Hence we ignore the effect of the left boundary, and concentrate on the boundary at unity. Viewing the reflected data (with $2n$ observations) as unbounded, we can estimate the density of these data using the estimator in (10.34) with no special problems. Then, this unbounded density estimate can be truncated on the right at unity to obtain an estimate of the density of the $D(x_i, y_i | \hat{P})$ with support on the interval $(0, 1]$.

Once the kernel function $K(\cdot)$ and the bandwidth h have been chosen, it is not necessary to evaluate the kernel density estimate in (10.34) in order to draw random values. Rather, a computational shortcut is afforded by Silverman (1986) and for cases where the kernel function $K(\cdot)$ is a regular probability density function.

We need to draw n values D^* from the estimated density of the original distance function estimates. Let $\{\varepsilon_i\}_{i=1}^{n}$ be a set of n iid draws from the probability density function used to define the kernel function; let $\{d_i\}_{i=1}^{n}$ be a set of values drawn independently, uniformly, and with replacement from the set of reflected distance function estimates $R = \{D(x_i, y_i | \hat{P}), 2 - D(x_i, y_i | \hat{P})\}$; and let $\bar{d} = n^{-1}\sum_{i=1}^{n} d_i$. Then compute

$$d_i^* = \bar{d} + (1 + h^2 / s^2)^{1/2}(d_i + h\varepsilon_i - \bar{d}), \qquad (10.35)$$

where s^2 is the sample variance of the values $v_i = d_i + h\varepsilon_i$. Using the convolution theorem, it can be shown that $v_i \sim \hat{g}(\cdot)$, where $\hat{g}(\cdot)$ is the kernel estimate of the density of the original distance function estimates and their reflections in R. As is typical with kernel density estimates, the variance of the v_i must be scaled upward as in (10.35). Straightforward manipulations reveal that

$$E(d_i^* | R) = 1 \qquad (10.36)$$

and

$$VAR(d_i^* | R) = s^2 \left(1 + \frac{h^2}{n(s^2 + h^2)}\right) \qquad (10.37)$$

so that the variance of the d_i^* is asymptotically correct. All that remains is to reflect the d_i^* about unity by computing, for each $i = 1, ..., n$,

$$D_i^* = \begin{cases} d_i^* & \text{if } d_i^* \le 1; \text{or} \\ 2 - d_i^* & \text{otherwise.} \end{cases} \tag{10.38}$$

The last step is equivalent to "folding" the right half of the symmetric (about 1) estimate $\hat{g}(\cdot)$ to the left around 1, and ensures that $d_i^* \le 1$ for all i. Putting all of this together, bootstrap estimates of confidence intervals, bias, *etc.* for the output distance function $D(x_0, y_0 \mid P)$ evaluated for a particular, arbitrary point $(x_0, y_0) \in R_+^{p+q}$ can be obtained via the following algorithm:

Algorithm #1

[1] For each $(x_i, y_i) \in S_n$, apply one of the distance function estimators in (10.11)-(10.13) to obtain estimates $D(x_i, y_i \mid \hat{P})$, $i = 1, \ldots, n$.

[2] If $(x_0, y_0) \notin S_n$, repeat step [1] for (x_0, y_0) to obtain $D(x_0, y_0 \mid \hat{P})$.

[3] Reflect the n estimates $D(x_i, y_i \mid \hat{P})$ about unity, and determine the bandwidth parameter h via least-squares cross-validation.

[4] Use the computational shortcut in (10.35) to draw n bootstrap values D_i^*, $i = 1, \ldots, n$, from the kernel density estimate of the efficiency estimates from step [1] and their reflected values from step [3].

[5] Construct a pseudo dataset S_n^* with elements (x_i^*, y_i^*) given by $y_i^* = D_i^* y_i / D(x_i, y_i \mid \hat{P})$ and $x_i^* = x_i$.

[6] Use (10.25) (or an analog of (10.25) if the convex cone estimators were used in step [1]) to compute the bootstrap estimate $D^*(x_0, y_0 \mid \hat{P}^*)$, where $\hat{P}^*$ denotes the convex hull of the free disposal hull of the bootstrap sample S_n^*.

[7] Repeat steps [4]-[6] B times to obtain a set of B bootstrap estimates

$$\left\{ D_b(x_0, y_0 \mid \hat{P}^*) \right\}_{b=1}^{B}.$$

[8] Use (10.28) to determine $\hat{a}_\partial$ and $\hat{b}_\partial$, and then use these in (10.30) together with the original estimate $D^*(x_0, y_0 \mid \hat{P}^*)$ obtained in step [2] (or step [1] if $(x_i, y_i) \in S_n$) to obtain an estimated confidence interval for $D(x_0, y_0 \mid \hat{P}^*)$. In addition, the bootstrap estimates can be used to obtain an estimate of the bias of $D(x_0, y_0 \mid P)$ from (10.32), and to obtain the bias-corrected estimator in (10.33) if desired.

Note that the arbitrary point (x_0, y_0) might or might not be in P. Typically, however, researchers are interested in the efficiency of firms represented in the observed sample, S_n. In that case, (x_0, y_0) will correspond in turn with each of the $(x_i, y_i) \in S_n$. Rather than repeatedly apply Algorithm #1 n times, considerable computational cost can be saved by noting that step [2] is not applicable in this case, and performing step [6] n times (for each element of S_n) inside each of the bootstrap loops in step [7]. This will result in n distinct sets of bootstrap estimates when step [8] is reached; each can then be used to estimate a confidence interval for its corresponding true distance function.

It is rather straightforward to extend Algorithm #1 to estimation of confidence intervals for Malmquist indices (Simar and Wilson, 1999c) or to the problem of testing hypotheses regarding returns to scale as in Simar and Wilson (2002). In choosing between the convex hull and convex cone estimators defined in (10.11)-(10.12), it is important to have some idea of whether the true technology P^∂ exhibits constant returns to scale everywhere; if it does not, the convex cone estimator in (10.12) will necessarily be inconsistent. It would be trivial to extend the results in Simar and Wilson (2002) to test for concavity of the production set P using the convex hull estimator and the FDH estimator defined in (10.13). Tests of other hypotheses regarding the shape of the technology should also be possible.

8.   MONTE CARLO EVIDENCE

We conducted a series of Monte Carlo experiments to examine the coverage probabilities of estimated confidence intervals for output distance functions. We simulated two different DGPs for these Monte Carlo experiments, but to minimize computational costs, we set $p = q = 1$ in both cases so that we examine only a single output produced from a single input. To consider a true technology with constant returns to scale everywhere, we used the model

$$y = xe^{-|v|} \tag{10.39}$$

where $v \sim N(0,1)$ and $x \sim Uniform$ on $(1,9)$. To allow for variable returns to scale, we used the model

$$y = (x-2)^{2/3} e^{-|v|}; \tag{10.40}$$

in experiments where this model was used, $v \sim N(0,1)$ as in the previous case, but $x \sim Uniform$ on $(3,12)$. Pseudo random uniform deviates were generated from a multiplicative congruential pseudo random number generator with modulus $2^{31} - 1$ and multiplier 7^5 (see Lewis *et al.*, 1969). Pseudo random normal deviates were generated via the Box-Muller method (see Press *et al.*, 1986, pp.202-203). In each Monte Carlo trial, the fixed point is given by $x_0 = 7.5$, $y_0 = 2$.

Each Monte Carlo experiment involved 1000 Monte Carlo trials; on each trial, we performed $B = 2000$ bootstrap replications. For each trial, we estimated confidence intervals for $D(x_0, y_0 | P)$ and then checked whether the estimated confidence intervals included the true value $D(x_0, y_0 | P)$. After 1000 trials, we divided the number of cases where $D(x_0, y_0 | P)$ was included in the estimated confidence intervals by 1000 to produce the results shown in Table 10-1.

Table 10-1. Monte Carlo Estimates of Confidence Interval Coverages (p = q =1; *i.e.,* one input, one output)

n	Significance Level	(1) Smooth	(2) x,y	(3) $D(x, y \mid \hat{P})$	(4) Smooth	(5) x,y	(6) $D(x, y \mid \hat{P})$
10	.80	0.745	0.757	0.757	0.626	0.757	0.450
	.90	0.858	0.789	0.789	0.754	0.830	0.609
	.95	0.916	0.899	0.899	0.852	0.917	0.740
	.975	0.948	0.936	0.936	0.895	0.938	0.826
	.99	0.976	0.957	0.957	0.941	0.969	0.894
25	.80	0.750	0.770	0.771	0.646	0.683	0.393
	.90	0.861	0.776	0.775	0.777	0.808	0.562
	.95	0.932	0.894	0.890	0.843	0.889	0.689
	.975	0.965	0.928	0.922	0.905	0.926	0.807
	.99	0.980	0.967	0.950	0.944	0.963	0.885
50	.80	0.752	0.769	0.765	0.645	0.683	0.369
	.90	0.863	0.783	0.778	0.770	0.808	0.515
	.95	0.920	0.896	0.891	0.842	0.889	0.640
	.975	0.950	0.934	0.941	0.901	0.926	0.751
	.99	0.972	0.962	0.960	0.947	0.963	0.850
100	.80	0.766	0.767	0.749	0.620	0.580	0.285
	.90	0.864	0.797	0.800	0.749	0.724	0.438
	.95	0.921	0.889	0.891	0.830	0.810	0.565
	.975	0.956	0.929	0.947	0.882	0.869	0.676
	.99	0.980	0.955	0.970	0.925	0.927	0.786
200	.80	0.780	0.757	0.736	0.600	0.531	0.280
	.90	0.877	0.795	0.803	0.728	0.674	0.406
	.95	0.937	0.879	0.888	0.811	0.774	0.514
	.975	0.963	0.939	0.939	0.867	0.825	0.607
	.99	0.984	0.961	0.963	0.902	0.886	0.726

Table 10-1 continues

n	Significance Level	(1) Smooth	(2) x,y	(3) $D(x,y \mid \hat{P})$	(4) Smooth	(5) x,y	(6) $D(x,y \mid \hat{P})$
400	.80	0.785	0.759	0.772	0.632	0.540	0.306
	.90	0.878	0.809	0.811	0.758	0.665	0.430
	.95	0.936	0.883	0.889	0.845	0.754	0.529
	.975	0.970	0.937	0.936	0.891	0.828	0.626
	.99	0.990	0.963	0.961	0.926	0.896	0.720
800	.80	0.767	0.769	0.733	0.614	0.487	0.281
	.90	0.884	0.811	0.791	0.730	0.616	0.380
	.95	0.950	0.886	0.871	0.810	0.708	0.478
	.975	0.971	0.938	0.930	0.858	0.790	0.571
	.99	0.984	0.966	0.955	0.907	0.853	0.667
1600	.80	0.786	0.752	0.735	0.663	0.522	0.302
	.90	0.897	0.799	0.793	0.777	0.657	0.423
	.95	0.957	0.876	0.868	0.846	0.739	0.517
	.975	0.981	0.945	0.929	0.896	0.815	0.629
	.99	0.990	0.970	0.960	0.944	0.887	0.730
3200	.80	0.801	0.758	0.765	0.653	0.500	0.303
	.90	0.904	0.822	0.813	0.775	0.639	0.430
	.95	0.951	0.897	0.864	0.860	0.717	0.527
	.975	0.976	0.951	0.924	0.899	0.785	0.611
	.99	0.992	0.973	0.951	0.938	0.865	0.715
6400	.80	0.839	0.765	0.743	0.666	0.537	0.337
	.90	0.917	0.816	0.811	0.789	0.656	0.462
	.95	0.960	0.878	0.868	0.868	0.737	0.588
	.975	0.977	0.943	0.937	0.909	0.806	0.671
	.99	0.987	0.964	0.962	0.942	0.873	0.744

Table 10-1 contains 6 numbered columns; the first 3 correspond to the constant returns to scale model in (10.39) where the convex cone estimator defined in (10.12) was employed, while the last 3 correspond to the variable returns to scale model in (10.40) where the convex hull estimator defined in (10.11) was used. The columns labeled ``Smooth'' correspond to Algorithm #1 given earlier, using the kernel smoothing described in section 7. Results in the columns labeled "(x,y)" were obtained using the naive bootstrap where resampling is from the elements of S_n. In terms of Algorithm #1, this amounts to deleting step [3] and replacing steps [4]-[5] with a single step where we draw n observations from S_n uniformly, independently, and with replacement to form the pseudo sample S_n^*. Results in the columns labeled "$D(x, y | \hat{P})$" were obtained with the other variant of the naive bootstrap mentioned previously, i.e., where the original distance function estimates were resampled. Again in terms of Algorithm #1, this involved deleting step [3] and replacing step [4] with a step where we draw n times uniformly, independently, and with replacement, from the set $\{D(x_i, y_i | \hat{P})\}_{i=1}^n$ to obtain the bootstrap values D_i^*.

Under constant returns to scale, the smooth bootstrap performs reasonably well with only 10 observations in the single input, single output case. The minor fluctuations in the reported coverages in column (1) that are seen as the sample size increases beyond 25 are to be expected due to sampling variation in the Monte Carlo experiment, and due to the fact that a finite number of bootstrap replications are being used. Also, the two variants of the naive bootstrap appear to perform, for the particular model in (10.39), reasonably well in terms of coverages, although not as well as the smooth bootstrap. One should not conclude from this, however, that it is safe to use the naive bootstrap--since it is inconsistent, there is no reason to think that coverages would be reasonable in any other setting than the one considered here. Moreover, the case of $p = q = 1$ with constant return to scale, although not typical of actual applications, is the most favorable case for the naive bootstrap since typically the estimated frontier will intersect only one sample observation. In higher dimensions, or where the convex hull or FDH estimators are used, a far higher portion of the original observations will have corresponding efficiency estimates equal to unity, which will be problematic for either version of the naive bootstrap.

Differences between the smooth bootstrap and the variants of the naive bootstrap become more pronounced with variable returns to scale, as columns (4)-(6) in Table 10-1 reveal. The smooth bootstrap does not perform quite as well as in the constant returns to scale case, but this is to be expected given the greater degree of curvature of the frontier in (10.40) (see Gijbels *et al.,* 1999, for details on the relation between the curvature of the frontier and the convergence rate of the output distance function estimator used here). Also, the coverages obtained with the smooth bootstrap appear to fluctuate somewhat as the sample size increases, but eventually seem to be on a path toward the nominal significance levels. This merely reflects that fact that consistency does not imply that convergence must be monotonic. With the two variants of the naive bootstrap, however, even with 6400 observations, coverages at all significance levels are worse than with the smooth bootstrap at any of the sample sizes we examined. Again, the problem is that with variable returns to scale, many sample observations will typically have efficiency estimates equal to unity. Between the two variants of the naive bootstrap, the one based on resampling from the original distance function estimates seems to give poorer coverage than the one based on resampling from S_n ; but this distinction is irrelevant, since both methods yield inconsistent estimates.

It is important to note that the performance of the bootstrap will vary not only as sample size increases, but also depending on features of the true model, particularly the curvature of the frontier and the shape of the density $f(x,y)$, The data generating process represented by (10.40) is somewhat unfavorable to the smooth bootstrap, in that the range of the inputs is considerably larger than the range of the outputs. To illustrate this point, we performed additional Monte Carlo experiments for $n = 200$, generating data for each trial using (10.40), but varying the range of the input variable. Results from these experiments are shown in Table 10-2, where we report estimated confidence interval coverages as in Table 10-1. We considered four cases, where x is alternately distributed uniformly on the (6.5, 8.5), (5.5, 9.5), (4.5, 10.5), and (3.5, 11.5) intervals (other features of the experiments remained unchanged). The results show that as the distribution of the input becomes more disperse, the coverages of confidence intervals estimated with each method worsen. As before, however, the smooth bootstrap dominates both variants of the naive bootstrap in every instance.

Table 10-2. Monte Carlo Estimates of Confidence Interval Coverages Showing the Influence of Range of the Input for the Variable Returns to Scale Case (p=q=1, n=200)

Significance	$x \sim U(6.5, 8.5)$	$x \sim U(5.5, 9.5)$	$x \sim U(4.5, 10.5)$	$x \sim U(3.5, 11.5)$
		Smooth Bootstrap		
.80	0.734	0.727	0.659	0.633
.90	0.846	0.829	0.790	0.748
.95	0.906	0.888	0.858	0.830
.975	0.946	0.930	0.902	0.886
.99	0.967	0.963	0.946	0.920
		Resample (x, y)		
.80	0.686	0.620	0.565	0.540
.90	0.813	0.763	0.719	0.689
.95	0.897	0.849	0.808	0.769
.975	0.941	0.913	0.872	0.839
.99	0.969	0.955	0.932	0.902
		Resample $D(x, y \mid \hat{P})$		
.80	0.326	0.276	0.274	0.279
.90	0.489	0.442	0.406	0.398
.95	0.640	0.568	0.526	0.516
.975	0.744	0.686	0.635	0.615
.99	0.858	0.801	0.753	0.725

Table 10-3. Widths of Estimated Confidence Intervals for Constant Returns to Scale Case

n	Mean	Std. Dev.	Range
		Smooth Bootstrap	
10	0.1384	0.0335	0.0485—0.2838
25	0.0551	0.0099	0.0271—0.0935
50	0.0283	0.0044	0.0183—0.0477
100	0.0146	0.0019	0.0094—0.0255
200	0.0076	0.0008	0.0055—0.0110
400	0.0039	0.0004	0.0029—0.0053
800	0.0019	0.0002	0.0016—0.0026
1600	0.0010	0.0001	0.0008—0.0012
3200	0.0005	0.0001	0.0004—0.0007
6400	0.0003	0.0001	0.0002—0. 0003
		Resample (x, y)	
10	0.2018	0.1404	0.0079—0.9945
25	0.0664	0.0384	0.0048—0.2556
50	0.0320	0.0186	0.0018—0.1391
100	0.0154	0.0087	0.0011—0.0527
200	0.0078	0.0046	0.0001—0.0306
400	0.0037	0.0021	0.0001—0.0141
800	0.0019	0.0011	0.0002—0.0066
1600	0.0009	0.0005	0.0 —0.0040
3200	0.0005	0.0003	0.0 —0.0018
6400	0.0002	0.0001	0.0 —0.0009
		Resample $D(x, y \mid \hat{P})$	
10	0.2018	0.1404	0.0079—0.9945
25	0.0693	0.0431	0.0038—0.3975
50	0.0315	0.0175	0.0018—0.1218
100	0.0157	0.0089	0.0011—0.0580
200	0.0074	0.0040	0.0003—0.0295
400	0.0038	0.0022	0.0002—0.0141
800	0.0019	0.0010	0.0001—0.0072
1600	0.0009	0.0006	0.0001—0.0034
3200	0.0005	0.0003	0.0 —0.0017
6400	0.0002	0.0001	0.0 —0.0009

Table 10-4. Widths of Estimated Confidence Intervals for Constant Returns to Scale Case

n	Mean	Std. Dev.	Range
		Smooth Bootstrap	
10	0.1384	0.0335	0.0485—0.2838
25	0.0551	0.0099	0.0271—0.0935
50	0.0283	0.0044	0.0183—0.0477
100	0.0146	0.0019	0.0094—0.0255
200	0.0076	0.0008	0.0055—0.0110
400	0.0039	0.0004	0.0029—0.0053
800	0.0019	0.0002	0.0016—0.0026
1600	0.0010	0.0001	0.0008—0.0012
3200	0.0005	0.0001	0.0004—0.0007
6400	0.0003	0.0001	0.0002—0. 0003
		Resample (x, y)	
10	0.2018	0.1404	0.0079—0.9945
25	0.0664	0.0384	0.0048—0.2556
50	0.0320	0.0186	0.0018—0.1391
100	0.0154	0.0087	0.0011—0.0527
200	0.0078	0.0046	0.0001—0.0306
400	0.0037	0.0021	0.0001—0.0141
800	0.0019	0.0011	0.0002—0.0066
1600	0.0009	0.0005	0.0 —0.0040
3200	0.0005	0.0003	0.0 —0.0018
6400	0.0002	0.0001	0.0 —0.0009
		Resample $D(x, y \mid \hat{P})$	
10	0.2018	0.1404	0.0079—0.9945
25	0.0693	0.0431	0.0038—0.3975
50	0.0315	0.0175	0.0018—0.1218
100	0.0157	0.0089	0.0011—0.0580
200	0.0074	0.0040	0.0003—0.0295
400	0.0038	0.0022	0.0002—0.0141
800	0.0019	0.0010	0.0001—0.0072
1600	0.0009	0.0006	0.0001—0.0034
3200	0.0005	0.0003	0.0 —0.0017
6400	0.0002	0.0001	0.0 —0.0009

Table 10-5. Widths of Estimated Confidence Intervals for Variable Returns to Scale Case

n	Mean	Std. Dev.	Range
		Smooth Bootstrap	
10	0.3020	0.1520	0.0499—2.3832
25	0.1331	0.0407	0.0426—0.3941
50	0.0741	0.0195	0.0208—0.1619
100	0.0426	0.0103	0.0181—0.0819
200	0.0254	0.0053	0.0133—0.0460
400	0.0161	0.0030	0.0081—0.0271
800	0.0102	0.0018	0.0057—0.0167
1600	0.0066	0.0011	0.0036—0.0105
3200	0.0042	0.0007	0.0022—0.0062
6400	0.0028	0.0004	0.0016—0.0043
	Resample	(x, y)	
10	0.4911	0.2834	0.0603—2.9616
25	0.1667	0.0741	0.0237—0.4875
50	0.1667	0.0741	0.0237—0.4875
100	0.0447	0.0185	0.0101—0.1252
200	0.0245	0.0098	0.0046—0.0723
400	0.0147	0.0060	0.0035—0.0417
800	0.0089	0.0034	0.0021—0.0262
1600	0.0057	0.0022	0.0012—0.0153
3200	0.0035	0.0013	0.0007—0.0088
6400	0.0022	0.0009	0.0002—0.0059
	Resample $D(x, y \mid \hat{P})$		
10	0.2701	0.1746	0.0264—1.6783
25	0.1067	0.0529	0.0169—0.4355
50	0.0529	0.0241	0.0039—0.1742
100	0.0270	0.0125	0.0057—0.0800
200	0.0147	0.0064	0.0014—0.0392
400	0.0090	0.0037	0.0015—0.0229
800	0.0054	0.0021	0.0011—0.0123
1600	0.0036	0.0013	0.0003—0.0092
3200	0.0023	0.0008	0.0005—0.0053
6400	0.0015	0.0005	0.0003—0.0031

Table 10-6. Bootstrap Bias Estimates Constant Returns to Scale Case

n	"True"	Mean	Std. Dev.	Range
		Smooth Bootstrap		
10	0.0517	0.0362	0.0085	0.0133—0.0756
25	0.0203	0.0147	0.0025	0.0077—0.0241
50	0.0101	0.0076	0.0011	0.0050—0.0123
100	0.0048	0.0039	0.0005	0.0025—0.0067
200	0.0024	0.0020	0.0002	0.0015—0.0029
400	0.0012	0.0010	0.0001	0.0008—0.0014
800	0.0006	0.0005	0.0001	0.0004—0.0007
1600	0.0003	0.0003	0.0000	0.0002—0.0004
3200	0.0002	0.0001	0.0000	0.0001—0.0002
6400	0.0001	0.0001	0.0	0.0001—0.0001
		Resample (x, y)		
10	0.0517	0.0324	0.0250	0.0025—0.2503
25	0.0203	0.0117	0.0082	0.0008—0.0585
50	0.0101	0.0058	0.0039	0.0003—0.0278
100	0.0048	0.0028	0.0019	0.0003—0.0152
200	0.0024	0.0014	0.0010	0.0001—0.0066
400	0.0012	0.0007	0.0005	0.0 —0.0034
800	0.0006	0.0004	0.0003	0.0 —0.0017
1600	0.0003	0.0002	0.0001	0.0 —0.0009
3200	0.0002	0.0001	0.0001	0.0 —0.0005
6400	0.0001	0.0000	0.0001	0.0 —0.0003
		Resample $D(x, y \mid \hat{P})$		
10	0.0517	0.0324	0.0250	0.0025—0.2503
25	0.0203	0.0121	0.0085	0.0008—0.0650
50	0.0101	0.0057	0.0037	0.0003—0.0302
100	0.0048	0.0030	0.0020	0.0003—0.0135
200	0.0024	0.0014	0.0009	0.0001—0.0071
400	0.0012	0.0007	0.0005	0.0001—0.0036
800	0.0006	0.0004	0.0002	0.0 —0.0016
1600	0.0003	0.0002	0.0001	0.0 —0.0009
3200	0.0002	0.0001	0.0001	0.0 —0.0006
6400	0.0001	0.0000	0.0000	0.0 —0.0002

Table 10-7. Bootstrap Bias Estimates Variable Returns to Scale Case

n	"True"	Mean	Std. Dev.	Range
		Smooth Bootstrap		
10	0.1615	0.1017	0.0487	0.0149—0.7556
25	0.0735	0.0511	0.0171	0.0140—0.1452
50	0.0403	0.0298	0.0090	0.0074—0.0745
100	0.0247	0.0178	0.0050	0.0066—0.0403
200	0.0150	0.0110	0.0029	0.0049—0.0216
400	0.0088	0.0071	0.0017	0.0031—0.0132
800	0.0060	0.0045	0.0010	0.0022—0.0090
1600	0.0035	0.0029	0.0007	0.0013—0.0054
3200	0.0023	0.0019	0.0004	0.0009—0.0032
6400	0.0014	0.0013	0.0003	0.0007—0.0021
		Resample (x, y)		
10	0.1615	0.2263	0.5805	-3.3019—0.3282
25	0.0735	0.0337	0.0320	-0.5655—0.1450
50	0.0735	0.0337	0.0320	-0.5655—0.1450
100	0.0247	0.0100	0.0048	0.0015—0.0346
200	0.0150	0.0057	0.0028	0.0013—0.0183
400	0.0088	0.0034	0.0016	0.0006—0.0123
800	0.0060	0.0020	0.0010	0.0004—0.0064
1600	0.0035	0.0013	0.0006	0.0002—0.0037
3200	0.0023	0.0008	0.0004	0.0001—0.0021
6400	0.0013	0.0005	0.0002	0.0001—0.0017
		Resample $D(x, y \mid \hat{P})$		
10	0.1615	0.0506	0.0412	0.0030—0.6932
25	0.0735	0.0207	0.0142	0.0017—0.1226
50	0.0403	0.0101	0.0069	0.0005—0.0642
100	0.0247	0.0052	0.0035	0.0004—0.0224
200	0.0150	0.0029	0.0020	0.0002—0.0126
400	0.0088	0.0018	0.0012	0.0001—0.0076
800	0.0060	0.0011	0.0007	0.0001—0.0044
1600	0.0035	0.0008	0.0005	0.0 —0.0030
3200	0.0023	0.0005	0.0003	0.0001—0.0016
6400	0.0013	0.0003	0.0002	0.0 —0.0010

Coverage probabilities of estimated confidence intervals tell only part of the story. Consequently, we also examined the widths of the confidence intervals estimated in the experiments reported in Table 10-1. Table 10-3 shows, for the case of constant returns to scale, the mean width of the estimated confidence intervals over 1000 Monte Carlo trials, the standard deviation of these widths, and the range of the widths of the estimated confidence intervals. Up to about 100 observations, the smooth bootstrap produces narrower confidence interval estimates than either of the naive bootstrap methods. Differences beyond $n = 100$ appear inconsequential. Table 10-4 shows similar results for the case of variable returns to scale. Here, the differences between the smooth bootstrap and the naive variants are more pronounced at large sample sizes. The smooth bootstrap yields narrower intervals than the naive methods at smaller sample sizes, but wider intervals at very large sample sizes. With either constant or variable returns to scale, the widths of the confidence interval estimates from the smooth bootstrap have smaller standard deviations than those from the naive methods.

In Tables 10-5 and 10-6, we give results on the bias estimates (based on (10.32)) for the constant returns to scale case (Table 10-5) and the variable returns to scale case (Table 10-6). In both tables, the first column gives the sample size, while the second column (labeled "True") gives the Monte Carlo estimate of the true bias, *i.e.*, the mean of the original efficiency estimates $D(x_0, y_0 | \hat{P})$ produced in step [1] of Algorithm #1 over the 1000 Monte Carlo Trials, minus the known, true value $D(x_0, y_0 | P) = 0.6419$. The remaining columns in the tables give the mean and standard deviation of the bias estimates (10.32) over 1000 Monte Carlo trials, as well as the range of the bias estimates.

In both the constant and variable returns to scale cases, the naive methods yield, on average, smaller and less accurate bias estimates than the smooth bootstrap, with the differences becoming more pronounced in the variable returns to scale case. With variable returns to scale, the naive methods typically give bias estimates whose averages are equal to about half the average of the corresponding bias estimates from the smooth bootstrap. The naive methods also produce bias estimates with larger standard deviation than in the case of the smooth bootstrap for sample sizes up to 100, but these differences become minimal for larger sample sizes. The naive method based on resampling (x, y)-pairs yields negative bias estimates in some cases for $n = 10$, 25, and 50 with variable returns to scale--even though bias is necessarily positive for our simulated DGP.

9.   ENHANCING THE PERFORMANCE OF THE BOOTSTRAP

The results of the preceding section show that in less favorable situations, even if the bootstrap is consistent, the coverage probabilities could be poorly approximated in finite samples.

Let I_α denote the $(1-\alpha)$-level confidence interval for $D(x_0, y_0 \,|\, P)$ given by (10.30). We have:

$$I_\alpha = \left[D(x_0, y_0 \,\big|\, \hat{P}) + \hat{a}_\alpha, D(x_0, y_0 \,\big|\, \hat{P}) + \hat{b}_\alpha \right]$$

where $\hat{a}_\alpha$ and $\hat{b}_\alpha$ are solutions to (10.28). Our Monte-Carlo experiments indicate that, in finite samples, the error $\Pr(D(x_0, y_0 \,|\, P) \in I_{\hat{a}}) - (1-\alpha)$ may sometimes be substantial. Of course, in practice, with real data, we cannot evaluate this error as in Monte-Carlo experiments.

We show in the present section that a method based on the iterated bootstrap may be used to estimate the true coverage,

$$\pi(\alpha) = \Pr(D(x_0, y_0 \,|\, P) \in I_\alpha), \tag{10.41}$$

and also to estimate the value of α for which the true coverage is equal to a predetermined nominal level $(1 - \alpha_0)$, such as 0.95. If $\hat{\pi}(\alpha)$ is our estimator of $\pi(\alpha)$, then we can search for a solution $\hat{\alpha}$ in the equation

$$\hat{\pi}(\hat{\alpha}) = 1 - \alpha_0 \tag{10.42}$$

and then re-calibrate our initial confidence interval I_α, choosing instead $I_{\hat{a}}$ as our final, corrected confidence interval for $D(x_0, y_0 \,|\, P)$.

The idea of iterating the bootstrap is discussed in detail by Hall (1991), and has been used in parametric frontier models by Hall *et al.* (1995). Using the notation introduced above, the method may be defined as follows. After steps [5]-[6] of Algorithm #1, where S_n^* and $D(x_0, y_0 \,|\, \hat{P}^*)$ have been computed, we generate a second-level bootstrap sample S_n^{**} from S_n^* along the same lines by which we generated S_n^* from S_n.

In particular, the second level bootstrap estimator $D(x_0, y_0 \,|\, \hat{P}^{**})$ corresponds to S_n^{**}; *i.e.*, $\hat{P}^{**}$ is the convex hull of the free disposal hull of the elements of S_n^{**}. So, from the bootstrap distribution of $D(x_0, y_0 \,|\, \hat{P}^{**})$, we can find values $\hat{a}_\alpha^*$ and $\hat{b}_\alpha^*$ such that

$$\Pr\!\left(-\hat{b}_\alpha^* \le D(x_0, y_0 \,\big|\, \hat{P}^{**}) - D(x_0, y_0 \,\big|\, \hat{P}^*) \le -\hat{a}_\alpha^* \,\Big|\, \hat{F}(S_n^*)\right) \approx 1 - \alpha \tag{10.43}$$

where $D(x_0, y_0 \,|\, \hat{P}^*)$ and $\hat{F}(S_n^*)$ have replaced $D(x_0, y_0 \,|\, \hat{P})$ and $\hat{F}(S_n)$ in (10.28).

This will provide a confidence interval for $D(x_0, y_0 \,|\, \hat{P})$ (in the second-level bootstrap) denoted I_α^*, just as I_α was the confidence interval for $D(x_0, y_0 \,|\, P)$ in the first-level bootstrap:

$$I_\alpha^* = \left[D(x_0, y_0 \,\big|\, \hat{P}^*) + \hat{a}_\alpha^*, D(x_0, y_0 \,\big|\, \hat{P}^*) + \hat{b}_\alpha^* \right]. \tag{10.44}$$

However, here $D(x_0, y_0 \,|\, \hat{P})$ is known; so by repeating the first-level bootstrap many times, we can record $\hat{\pi}(\alpha)$, the proportion of times that I_α^* covers $D(x_0, y_0 \,|\, \hat{P})$:

$$\hat{\pi}(\alpha) = \Pr\!\left(D(x_0, y_0 \,\big|\, \hat{P}) \in I_\alpha^* \,\Big|\, \hat{F}(S_n)\right). \tag{10.45}$$

This quantity is the estimator of the function $\pi(\alpha)$ defined in (10.41); by solving (10.42), we obtain the iterated bootstrap confidence interval, $I_{\hat{a}}$.

The algorithm can be implemented as follows:

Algorithm #2

[1] For each $(x_i, y_i) \in S_n$, apply one of the distance function estimators in (10.11)-(10.13) to obtain estimates $D(x_i, y_i | \hat{P})$, $i = 1, ..., n$. Here the reference set is S_n.

[2] Repeat step [1] for (x_0, y_0) to obtain $D(x_0, y_0 | \hat{P})$.

[3] Reflect the n estimates $D(x_i, y_i | \hat{P})$ about unity, and determine the bandwidth parameter h via least-squares cross-validation.

[4] Use the computational shortcut in (10.35) to draw n bootstrap values D_i^*, $i = 1, ..., n$, from the kernel estimate of the density of the $D(x_i, y_i | \hat{P})$.

[5] Construct a pseudo dataset S_n^* with elements (x_i^*, y_i^*) given by $y_i^* = D_i^* y_i / D(x_i, y_i | \hat{P})$ and $x_i^* = x_i$.

[6] Use (10.25) (or an analog of (10.25) if the convex cone estimators were used in step [1]) to compute the bootstrap estimate $D(x_0, y_0 | \hat{P}^*)$; here the reference set is S_n^*. [6.1] For each $(x_i^*, y_i^*) \in S_n^*$, apply the same distance function estimators chosen in [1] to obtain estimates $D(x_i^*, y_i^* | \hat{P}^*)$, $i = 1, ..., n$, where the reference set is S_n^*. [6.2] Reflect the n estimates $D(x_i^*, y_i^* | \hat{P}^*)$ about unity. [6.3] Use the computational shortcut in (10.35) to draw n bootstrap values D_i^{**}, $i = 1, ..., n$, from the kernel estimate of the density of the $D(x_i^*, y_i^* | \hat{P}^*)$. [6.4] Construct a pseudo dataset S_n^{**} with elements (x_i^{**}, y_i^{**}) given by $y_i^{**} = D_i^{**} y_i^* / D(x_i^*, y_i^* | \hat{P}^*)$ and $x_i^{**} = x_i^*$. [6.5] Use (10.25) (or an analog of (10.25) if the convex cone estimators were used in step [1]) to compute the bootstrap estimate $D(x_0, y_0 | \hat{P}^{**})$; here the reference set S_n^{**}. [6.6] Repeat steps [6.3]-[6.5] B_2 times to obtain a set of B_2 bootstrap values

$$\left\{ D_{b_2}(x_0, y_0 | \hat{P}^{**}) \right\}_{b_2 = 1}^{B_2}.$$

[6.7] Use (10.43) to determine $\hat{a}_\alpha^*$ and $\hat{b}_\alpha^*$, and then use these in (10.44) together with the estimate $D(x_0, y_0 | \hat{P}^*)$ obtained in step [6] to obtain I_α^*, an estimated confidence interval for $D(x_0, y_0 | \hat{P})$.

[7] Repeat steps [4]-[6.7] B_1 times to obtain a set of B_1 bootstrap estimates

$$\left\{ D_{b_1}(x_0, y_0 | \hat{P}^*) \right\}_{b_1 = 1}^{B_1}$$

and a set of confidence intervals $\left\{ I_{\alpha, b_1}^* \right\}_{b_1 = 1}^{B_1}$.

[8] Compute the proportion of cases where $I_{\hat{a}, b_1}^*$ covers $D(x_0, y_0 | \hat{P})$:

$$\hat{\pi} B_1 B_2(\alpha) = \frac{1}{B_1} \sum I(D(x_0, y_0 | \hat{P}) \in I_{\alpha, b_1}^*)$$

where $I(\cdot)$ is the indicator function.

[9] Solve the equation $\hat{\pi}B_1B_2(\alpha)=1-\alpha_0$ for α, where $1-\alpha_0$ is the desired, nominal coverage. Denote the solution by $\hat{\alpha}$. [10] Use (10.28) to determine $\hat{a}_{\hat{a}}$ and $\hat{b}_{\hat{a}}$, and then use these in (10.30) together with the original estimate $D(x_0,y_0|\hat{P})$ obtained in step [2] to obtain $I_{\hat{a}}$, the estimated confidence interval for $D(x_0,y_0|P)$. In addition, the bootstrap estimates can be used with (10.32) to obtain an estimate of the bias of $D(x_0,y_0|\hat{P})$, and with (10.33) to obtain a bias-corrected estimator if the condition in (6.13) is satisfied.

10.   CONCLUSIONS

As noted in the introduction, DEA methods are still quite young compared to many existing parametric statistical techniques. Because of their nonparametric nature and the resulting lack of structure, obtaining asymptotic results is difficult, and even when they can be obtained, their practical use remains to be demonstrated. The bootstrap is a natural tool to apply in such cases. Our Monte Carlo evidence confirms, however, that the structure of the underlying, true model plays a crucial role in determining how well the bootstrap will perform in a given applied setting. In those cases, where the researcher does not have access to the true model as we did in our Monte Carlo experiments, the iterated bootstrap offers a convenient, analogous approach for evaluating the performance of the bootstrap and providing corrections if needed.

REFERENCES

1. Banker, R.D., 1993, Maximum likelihood, consistency and data envelopment analysis: a statistical foundation, Management Science 39(10), 1265-1273.
2. Banker, R.D., A. Charnes, and W.W. Cooper, 1984, Some models for estimating technical and scale inefficiencies in data envelopment analysis, Management Science 30, 1078-1092.
3. Beran, R., and G. Ducharme, 1991, Asymptotic Theory for Bootstrap Methods in Statistics, Montreal: Centre de Reserches Mathematiques, University of Montreal.
4. Bickel, P.J., and D.A. Freedman, 1981, Some asymptotic theory for the bootstrap, Annals of Statistics 9, 1196-1217.
5. Charnes, A., W.W. Cooper, and E. Rhodes, 1978, Measuring the inefficiency of decision making units, European Journal of Operational Research 2(6), 429-444.

6. Deprins, D., L. Simar, and H. Tulkens, 1984, Measuring Labor Inefficiency in Post Offices, in The Performance of Public Enterprises: Concepts and Measurements, ed. by M. Marchand, P. Pestieau and H. Tulkens. Amsterdam: North-Holland, 243-267.

7. Efron, B., 1979, Bootstrap methods: another look at the jackknife, Annals of Statistics 7, 1-16.

8. Efron, B., 1982, The Jackknife, the Bootstrap and Other Resampling Plans, CBMS-NSF Regional Conference Series in Applied Mathematics, #38. Philadelphia: SIAM.

9. Efron, B., and R.J. Tibshirani, 1993, An Introduction to the Bootstrap. London: Chapman and Hall.

10. Färe, R., 1988, Fundamentals of Production Theory, Berlin: Springer-Verlag.

11. Färe, R., and C.A.K. Lovell (1978), Measuring the technical efficiency of production, Journal of Economic Theory 19, 150-162.

12. Färe, R., S. Grosskopf, and C.A.K. Lovell (1985), The Measurement of Efficiency of Production. Boston: Kluwer-Nijhoff Publishing.

13. Farrell, M.J., 1957, The measurement of productive efficiency, Journal of the Royal Statistical Society A 120, 253-281.

14. Gijbels, I., E. Mammen, B.U. Park, and L. Simar, 1999, On estimation of monotone and concave frontier functions, Journal of the American Statistical Association 94, 220-228.

15. Hall, P., 1992, The Bootstrap and Edgeworth Expansion, New York: Springer-Verlag.

16. Hall, P., W. Härdle, and L. Simar, 1995, Iterated bootstrap with application to frontier models, The Journal of Productivity Analysis 6, 63-76.

17. Kneip, A., B.U. Park, and L. Simar, 1998, A note on the convergence of nonparametric DEA estimators for production efficiency scores, Econometric Theory, 14, 783-793.

18. Korostelev, A., L. Simar, and A.B. Tsybakov, 1995a, Efficient estimation of monotone boundaries, The Annals of Statistics 23, 476-489.

19. Korostelev, A., L. Simar, and A.B. Tsybakov, 1995b, On estimation of monotone and convex boundaries, Publications de l'Institut de Statistique des Universités de Paris XXXIX 1, 3-18.

20. Lewis, P.A., A.S. Goodman, and J.M. Miller, 1969, A pseudo-random number generator for the System/360, IBM Systems Journal 8, 136-146.

21. Löthgren, M., 1998, How to Bootstrap DEA Estimators: A Monte Carlo Comparison (contributed paper presented at the Third Biennal Georgia Productivity Workshop, University of Georgia, Athens, GA,

October 1998), Working paper series in economics and Finance #223, Department of Economic Statistics, Stockhold School of Economics, Sweden.

22. Löthgren, M., 1999, Bootstrapping the Malmquist productivity index--A simulation study, Applied Economics Letters 6, 707-710.

23. Löthgren, M., and M. Tambour, 1997, Bootstrapping the DEA-based Malmquist Productivity Index, in Essays on Performance Measurement in Health Care, Ph.D. dissertation by Magnus Tambour, Stockholm School of Economics, Stockholm, Sweden.

24. Löthgren, M., and M. Tambour, 1999, Testing scale efficiency in DEA models: A bootstrapping approach, Applied Economics 31, 1231-1237.

25. Manski, C.F., 1988, Analog Estimation Methods in Econometrics, New York: Chapman and Hall.

26. Park, B., L. Simar, and C. Weiner, 1999, The FDH estimator for productivity efficiency scores: Asymptotic Properties, Econometric Theory 16, 855-877.

27. Press, W.H., B.P. Flannery, S.A. Teukolsky, and W.T. Vetterling, 1986, Numerical Recipes, Cambridge University Press, Cambridge.

28. Scott, D.W., 1992, Multivariate Density Estimation. John Wiley and Sons, Inc., New-York.

29. Shephard, R.W., 1970, Theory of Cost and Production Function. Princeton: Princeton University Press.

30. Silverman, B.W., 1978, Choosing the window width when estimating a density, Biometrika 65, 1-11.

31. Silverman, B.W. (1986), Density Estimation for Statistics and Data Analysis, Chapman and Hall Ltd., London.

32. Simar, L., 1992, Estimating efficiencies from frontier models with panel data: a comparison of parametric, non-parametric and semi-parametric methods with bootstrapping, Journal of Productivity Analysis 3, 167-203.

33. Simar, L., 1996, Aspects of statistical analysis in DEA-type frontier models, Journal of Productivity Analysis 7, 177-185.

34. Simar, L., and P.W. Wilson, 1998, Sensitivity analysis of efficiency scores: How to bootstrap in nonparametric frontier models, Management Science 44(11), 49-61.

35. Simar, L., and P.W. Wilson, 1999a, Some problems with the Ferrier/Hirschberg bootstrap idea, Journal of Productivity Analysis 11, 67-80.

36. Simar, L., and P.W. Wilson, 1999b, Of course we can bootstrap DEA scores! But does it mean anything? Logic trumps wishful thinking, Journal of Productivity Analysis 11, 93-97.

37. Simar, L., and P.W. Wilson, 1999c, Estimating and bootstrapping Malmquist indices, European Journal of Operations Research 115, 459-471.
38. Simar, L., and P.W. Wilson, 2000a, Statistical inference in nonparametric frontier models: The state of the art, Journal of Productivity Analysis 13, 49-78.
39. Simar, L., and P.W. Wilson, 2000b, A general methodology for bootstrapping in nonparametric frontier models, Journal of Applied Statistics 27, 779-802.
40. Simar, L. and Wilson, P.W., 2001, Testing restrictions in nonparametric efficiency models, Communications in Statistics 30, 159-184.
41. Simar, L. and Wilson, P.W., 2002, Nonparametric tests of returns to scale, European Journal of Operational Research 139, 115-132.
42. Spanos, A., 1986, Statistical Foundations of Econometric Modelling, Cambridge: Cambridge University Press.
43. Swanepoel, J.W.H., 1986, A note on proving that the (modified) bootstrap works, Communications in Statistics: Theory and Methods 15, 3193-3203.
44. Wheelock, D.C., and P.W. Wilson, 1995, Explaining bank failures: deposit insurance, regulation, and efficiency, Review of Economics and Statistics 77, 689-700.
45. Wheelock, D.C., and P.W. Wilson, 2000, Why do banks disappear? The determinants of US bank failures and acquisitions, Review of Economics and Statistics 82, 127-138.
46. Wilson, P.W., 2003, Testing independence in models of productive efficiency, Journal of Productivity Analysis, forthcoming.

Chapter 11

STATISTICAL TESTS BASED ON DEA EFFICIENCY SCORES

Rajiv D. Banker[1] and Ram Natarajan[2]
School of Management, The University of Texas at Dallas, Richardson, TX 75083-0688 USA
email: [1] *rbanker@utdallas.edu* [2]*nataraj@utdallas.edu*

Abstract: This chapter is written for analysts and researchers who may use Data Envelopment Analysis (DEA) to statistically evaluate hypotheses about characteristics of production correspondences and factors affecting productivity. Contrary to some characterizations, it is shown that DEA is a full-fledged statistical methodology, based on the characterization of DMU efficiency as a stochastic variable. The DEA estimator of the production frontier has desirable statistical properties, and provides a basis for the construction of a wide range of formal statistical tests (Banker 1993). Specific tests described here address issues such as comparisons of efficiency of groups of DMUs, existence of scale economies, existence of allocative inefficiency, separability and substitutability of inputs in production systems, analysis of technical change and productivity change, impact of contextual variables on productivity and the adequacy of parametric functional forms in estimating monotone and concave production functions.

Key words: Data envelopment analysis (DEA); Statistical tests

1.   INTRODUCTION

Data Envelopment Analysis (DEA) continues to be used extensively in many settings to analyze factors influencing the efficiency of organizations. The DEA approach specifies the production set only in terms of desirable properties such as convexity and monotonicity, without imposing any parametric structure on it (Banker, Charnes and Cooper, 1984). Despite its widespread use, many persons continue to classify DEA as a non-statistical approach. However, many recent advances have established the statistical properties of the DEA efficiency estimators. Based on statistical representations of DEA, rigorous statistical tests of various hypotheses have also been developed.

To start with, Banker (1993) provided a formal statistical foundation for DEA by identifying conditions under which DEA estimators are statistically consistent and maximize likelihood. He also developed hypothesis tests for efficiency comparison when a group of DMUs (decision making units) is compared with another. Since the publication of Banker (1993), a number of significant advances have been made in developing DEA-based hypothesis tests to address a wide spectrum of issues of relevance to users of DEA. These include issues such as efficiency comparison of groups, existence of scale inefficiency, impact of contextual variables on productivity, adequacy of parametric functional forms in estimating monotone and concave production functions, examination of input separability and substitutability in production systems, existence of allocative inefficiency and evaluation of technical change and productivity change. In the rest of this chapter, we describe different DEA-based tests of hypotheses in the form of a reference list for researchers and analysts interested in applying DEA to efficiency measurement and production frontier estimation[1].

The paper proceeds as follows. In the next section, section 2, we present statistical tests that are relevant for applications of DEA in environments where the deviation from the frontier is caused by a single one-sided stochastic random variable representing DMU inefficiency. We describe salient aspects of the statistical foundation provided in Banker (1993), discuss hypothesis tests for efficiency comparison of groups of DMUs, for the existence of scale inefficiency or allocative inefficiency and for input substitutability. In section 3, we address situations where the production frontier shifts over time. We describe DEA-based techniques and statistical tests to evaluate and test for existence of productivity and technical change over time. In section 4, we discuss the application of DEA in environments where the deviation from the production frontier arises as a result of two stochastic variables, one representing inefficiency and the other random noise. We explain how efficiency comparisons across groups of DMUs can be carried out, and describe DEA-based methods to determine the impact of contextual or environmental variables on inefficiency. We also suggest methods to evaluate the adequacy of an assumed parametric form for situations where prior guidance specifies only monotonicity and concavity for a functional relationship. Finally, we summarize and conclude in section 5.

---

[1] Bootstrapping, discussed in chapter 10, offers an alternative approach for statistical inference within nonparametric efficiency estimation.

2. HYPOTHESIS TESTS WHEN INEFFICIENCY IS THE ONLY STOCHASTIC VARIABLE

The tests described in this section build on the work in Banker (1993) that provides a formal statistical basis for DEA estimation techniques. We briefly describe the salient aspects of Banker (1993) before discussing the hypothesis tests.

2.1 Statistical Foundation for DEA

Consider observations on $j = 1,....N$ decision making units (DMUs), each observation comprising a vector of outputs $\mathbf{y}_j \equiv (y_{1j}, ...y_{Rj}) \geq 0$ and a vector of inputs $\mathbf{x}_j \equiv (x_{1j}, ...x_{Ij}) \geq 0$, where $y \in Y$ and $x \in X$, and Y and X are convex subsets of $\mathfrak{R}^R$ and $\mathfrak{R}^I$, respectively. Input quantities and output mix proportion variables are random variables[2]. The production correspondence between the frontier output vector $\mathbf{y}^0$ and the input vector $\mathbf{x}^0$ is represented by the production frontier $g(\mathbf{y}^0, \mathbf{x}^0) = 0$. The support set of the frontier is a monotonically increasing and convex set $T \equiv \{(\mathbf{y}, \mathbf{x}) \mid \mathbf{y}$ can be produced from $\mathbf{x}\}$. The inefficiency of a specific DMU j is, $\theta_j \equiv \max \{\theta \mid (\theta\mathbf{y}_j, \mathbf{x}_j) \in T\}$. It is modeled as a scalar random variable that takes values in the range $[1, \infty)$ and is distributed with the probability density function $f(\theta)$ in this range. Banker (1993) imposes additional structure on the distribution of the inefficiency variable requiring that there is a non-zero likelihood of nearly efficient performance i.e., $\int_1^{1+\delta} f(\theta)d\theta > 0$ for all $\delta > 0$.

In empirical applications of DEA, the inefficiency variable θ is not observed and needs to be estimated from output and input data. The following Banker, Charnes and Cooper (BCC 1984) linear program is used to estimate the inefficiency:

$$\hat{\theta}_j = \arg\max \left\{ \theta \left|
\begin{array}{l}
\sum_{k=1}^{N} \lambda_k y_{rk} \geq \theta y_{rj}, \forall r = 1,..R; \sum_{k=1}^{N} \lambda_k x_{ik} \leq x_{ij}, \forall i = 1,...I; \\
\sum_{k=1}^{N} \lambda_k = 1, \lambda_k \geq 0, \forall k = 1,..N
\end{array}
\right. \right\} \quad (11.1)$$

where $\hat{\theta}_j$ is the DEA estimator of θ_j.

Modeling the inefficiency deviation as a stochastic variable that is distributed independently of the inputs, enabled Banker (1993) to derive several results that provide a statistical foundation for hypothesis tests using

---

[2] Alternative specifications that specify output quantities and input mix proportion as random variables or endogenous input mix decisions based on input prices modeled as random variables can also be used.

DEA. He demonstrated that the DEA estimators of the true inefficiency values maximize likelihood provided the density function of the inefficiency random variable $f(\theta)$ is monotone decreasing. He also pointed out that a broad class of probability distributions, including the exponential and half-normal distributions, possesses monotone decreasing density functions. Banker (1993) also shows that the DEA estimator of the inefficiency underestimates the true inefficiency in finite samples. More importantly, he shows that asymptotically this bias reduces to zero; that is the DEA estimators are consistent if the probability of observing nearly efficient DMUs is strictly positive[3].

While consistency is a desirable property of an estimator, it does not by itself guide the construction of hypothesis tests. However, Banker (1993) exploits the consistency result to prove that for "large" samples the DEA estimators of inefficiency for any given subset of DMUs follow the same probability distribution as the true inefficiency random variable. This is, perhaps, the most important result in the Banker (1993) paper since it implies that, for large samples, distributional assumptions imposed for the true inefficiency variable can be carried over to the empirical distribution of the DEA estimator of inefficiency and test-statistics based on the DEA estimators of inefficiency can be evaluated against the assumed distribution of the true inefficiency.

2.2   Efficiency Comparison of Two Groups of DMUs

The asymptotic properties of the DEA inefficiency estimator are used by Banker (1993) to construct statistical tests enabling a comparison of two groups of DMUs to assess whether one group is more efficient than the other. Banker (1993) proposes parametric as well as nonparametric tests to evaluate the null hypothesis of no difference in the inefficiency distributions of two sub-samples, G_1 and G_2, that are part of the sample of N DMUs when the sample size, N, is large. For N_1 and N_2 DMUs in subgroups G_1 and G_2, respectively, the null hypothesis of no difference in inefficiency between the two subgroups can be tested using the following procedures:

---

[3] The consistency result does not require that the probability density function $f(\theta)$ be monotone decreasing. It only requires that there is a positive probability of observing nearly efficient DMUs, which is a much weaker condition than the monotonicity condition required for the DEA estimators to be maximum likelihood.

(i) If the logarithm[4] of the true inefficiency θ_j is distributed as exponential over $[0, \infty)$ for the two subgroups, then under the null hypothesis that there is no difference between the two groups, the test statistic is calculated as

$$\left[\sum_{j\in G_1} \ln(\hat{\theta}_j)/N_1\right]\Big/\left[\sum_{j\in G_2} \ln(\hat{\theta}_j)/N_2\right]$$ and evaluated relative to the critical value of

the F distribution with $(2N_1, 2N_2)$ degrees of freedom.

(ii) If logarithm of the true inefficiency θ_j is distributed as half-normal over the range $[0, \infty)$ for the two subgroups, then under the null hypothesis that there is no difference between the two groups, the test statistic is calculated as

$$\left[\sum_{j\in G_1} \left\{\ln(\hat{\theta}_j)\right\}^2/N_1\right]\Big/\left[\sum_{j\in G_2} \left\{\ln(\hat{\theta}_j)\right\}^2/N_2\right]$$ and evaluated relative to the critical

value of the F distribution with (N_1, N_2) degrees of freedom.

(iii) If no such assumptions are maintained about the probability distribution of inefficiency, a non-parametric Kolmogorov-Smirnov's test statistic given by the maximum vertical distance between $F^{G_1}(\ln(\hat{\theta}_j))$ and $F^{G_2}(\ln(\hat{\theta}_j))$, the empirical distributions of $\ln(\hat{\theta}_j)$ for the groups G_1 and G_2, respectively, is used. This statistic, by construction, takes values between 0 and 1 and a high value for this statistic is indicative of significant differences in inefficiency between the two groups.

2.3   Tests of Returns to Scale

Examining the existence of increasing or decreasing returns to scale is an issue of interest in many DEA studies. We provide a number of DEA-based tests to evaluate returns to scale using DEA inefficiency scores[5]. Consider the inefficiency $\hat{\theta}_j^C$ estimated using the CCR (Charnes, Cooper and Rhodes 1978) model obtained from the BCC linear program in (11.1) by deleting the constraint $\sum_{k=1}^{N}\lambda_k = 1$ i.e.,

$$\hat{\theta}_j^C = \mathrm{argmax}\left\{\theta \middle| \begin{array}{l} \sum_{k=1}^{N}\lambda_k y_{rk} \geq \theta y_{rj}, \forall r = 1,...R;\ \sum_{k=1}^{N}\lambda_k x_{ik} \leq x_{ij}, \forall i = 1,...I; \\ \lambda_k \geq 0, \forall k = 1,...N \end{array}\right\} \quad (11.2)$$

---

[4] Alternatively, the assumption may be maintained that $t(\theta_j)$ is distributed as exponential over $[0, \infty)$ where $t(.)$ is some specified transformation function. Then the test statistic is

given by $$\left[\sum_{j\in G_1} t(\hat{\theta}_j)/N_1\right]\Big/\left[\sum_{j\in G_2} t(\hat{\theta}_j)/N_2\right].$$

[5] Chapter 2 of this handbook provides a detailed discussion of qualitative and quantitative aspects of returns to scale in DEA.

By construction, $\hat{\theta}_j^C \geq \hat{\theta}_j$. Scale inefficiency is then estimated as $\hat{\theta}_j^S = \hat{\theta}_j^C / \hat{\theta}_j$. Values of scale inefficiency significantly greater than 1 indicate the presence of scale inefficiency to the extent operations deviate from the *most productive scale size (MPSS)* (Banker 1984, Banker, Charnes and Cooper 1984). All observations in the sample are scale efficient if and only if the sample data can be rationalized by a production set exhibiting constant returns to scale.

Under the null hypothesis of no scale inefficiency (or equivalently, under the null hypothesis of constant returns to scale), $\hat{\theta}_j^C$ is also a consistent estimator of θ_j (Banker 1993, 1996). The null hypothesis of no scale inefficiency in the sample can be evaluated by constructing the following test statistics:

(i) If the logarithm of the true inefficiency θ_j is distributed as exponential over $[0, \infty)$, then under the null hypothesis of constant returns to scale, the test statistic is calculated as $\sum \ln(\hat{\theta}_j^C) / \sum \ln(\hat{\theta}_j)$. This test statistic is evaluated relative to the half-F distribution $|F_{2N,2N}|$ with 2N,2N degrees of freedom over the range $[1,\infty)$, since by construction the test statistic is never less than 1. The half-F distribution is the F distribution truncated below at 1, the median of the F distribution when the two degrees of freedom are equal.

(ii) If the logarithm of the true inefficiency θ_j is distributed as half-normal over the range $[0, \infty)$, then under the null hypothesis of constant returns to scale, the test statistic is calculated as $\sum \left\{ \ln(\hat{\theta}_j^C) \right\}^2 / \sum \left\{ \ln(\hat{\theta}_j) \right\}^2$ and evaluated relative to the half-F distribution $|F_{N,N}|$ with N,N degrees of freedom over the range $[1,\infty)$.

(iii) If no such assumptions are maintained about the probability distribution of inefficiency, a non-parametric Kolmogorov-Smirnov's test statistic given by the maximum vertical distance between $F^C(\ln(\hat{\theta}_j^C))$ and $F(\ln(\hat{\theta}_j))$, the empirical distributions of $\ln(\hat{\theta}_j^C)$ and $\ln(\hat{\theta}_j)$, respectively, is used. This statistic, by construction, takes values between 0 and 1 and a high value for this statistic is indicative of the existence of significant scale inefficiency in the sample.

The above tests evaluate the null hypothesis of constant returns to scale against the alternative of variable returns to scale. In addition, it is also possible to test the null hypothesis of non-decreasing returns to scale against the alternative of decreasing returns to scale and the null hypothesis of non-increasing returns to scale against the alternative of increasing returns to scale. Two additional inefficiency estimators $\hat{\theta}_j^D$ and $\hat{\theta}_j^E$ required for these tests are calculated by solving the program in (1) after changing the constraint $\sum_{k=1}^{N} \lambda_k = 1$ to $\sum_k \lambda_k \leq 1$ for $\hat{\theta}_j^D$ and to $\sum_k \lambda_k \geq 1$ for $\hat{\theta}_j^E$. By construction, $\hat{\theta}_j^C \geq \hat{\theta}_j^D \geq \hat{\theta}_j$ and $\hat{\theta}_j^C \geq \hat{\theta}_j^E \geq \hat{\theta}_j$.

The following is a test of the null hypothesis of non-decreasing returns to scale against the alternative of decreasing returns to scale:

(i) If the logarithm of the true inefficiency θ_j is distributed as exponential over $[0, \infty)$, the test statistic is calculated as $\sum \ln(\hat{\theta}_j^E)/\sum \ln(\hat{\theta}_j)$ or $\sum \ln(\hat{\theta}_j^C)/\sum \ln(\hat{\theta}_j^D)$. Each of these statistics is evaluated relative to the half-F distribution $|F_{2N,2N}|$ with 2N,2N degrees of freedom over the range $[1,\infty)$.

(ii) If the logarithm of the true inefficiency θ_j is distributed as half-normal over the range $[0, \infty)$, the test statistic is calculated as either $\sum \left\{\ln(\hat{\theta}_j^E)\right\}^2 / \sum \left\{\ln(\hat{\theta}_j)\right\}^2$ or $\sum \left\{\ln(\hat{\theta}_j^C)\right\}^2 / \sum \left\{\ln(\hat{\theta}_j^D)\right\}^2$ and evaluated relative to the half-F distribution $|F_{N,N}|$ with N,N degrees of freedom over the range $[1,\infty)$.

(iii) If no such assumptions are maintained about the probability distribution of inefficiency, a non-parametric Kolmogorov-Smirnov's test statistic given by either the maximum vertical distance between $F^E(\ln(\hat{\theta}_j^E))$ and $F(\ln(\hat{\theta}_j))$, or that between $F^C(\ln(\hat{\theta}_j^C))$ and $F^D(\ln(\hat{\theta}_j^D))$ is used.

The test statistics for testing the null of non-increasing returns to scale against the alternative of increasing returns to scale can be developed in a similar fashion by interchanging $\hat{\theta}_j^E$ and $\hat{\theta}_j^D$ in the statistics above.

2.4   Tests of Allocative Efficiency

In this section, we describe DEA-based tests that can be used to examine the existence of allocative inefficiencies associated with input utilization. In many DEA studies that have examined inefficiency associated with input utilization, inefficiency is often estimated using aggregate cost expenditure information. Banker et al. (2003) address the situation when information about input prices is not available, except for the knowledge that the firms procure the inputs in the same competitive market place. They employ the result that the DEA technical inefficiency measure using a single aggregate cost variable, constructed from multiple inputs weighted by their unit prices, reflects the aggregate technical and allocative inefficiency. This result is then used to develop statistical tests of the null hypothesis of no allocative inefficiency analogous to those of the null hypothesis of no scale inefficiency described earlier.

For the purposes of this section, consider observations on $j = 1,....N$ DMUs, each observation comprising an output vector $\mathbf{y}_j \equiv (y_{1j}, ...y_{Rj}) \geq 0$ and a vector of input costs $\mathbf{c}_j \equiv (c_{1j}, ...c_{Ij}) \geq 0$ for $i=1,...I$ inputs. Each input i, $i=1,...I$, is bought by all firms in the same competitive market at a price p_i. Let $\mathbf{p} = (p_1,....., p_I)$ be the vector of input prices. The cost of input i for DMU j is then $c_{ij} = p_i x_{ij}$. The total cost of inputs for DMU j is $c_j = \sum_i p_i x_{ij} = \sum_i c_{ij}$. The input quantities x_{ij} and the price vector $\mathbf{p}$ are not observable by the researcher[6]. Only the output and cost information are observed.

The aggregate technical and allocative inefficiency estimator, $\hat{\theta}_j^Z \geq 1$, is estimated using the following linear program that utilizes output and aggregate cost data:

$$\hat{\theta}_j^Z = \operatorname{argmax} \left\{ \theta \left| \begin{array}{l} \sum_{k=1}^N \lambda_k y_{rk} \geq y_{rj}, \forall r = 1,...R; \sum_{k=1}^N \lambda_k c_k \leq c_j / \theta; \\ \sum_{k=1}^N \lambda_k = 1, \lambda_k \geq 0, \forall k = 1,...N \end{array} \right. \right\} \quad (11.3)$$

The technical inefficiency estimator $\hat{\theta}_j^B \geq 1$ is estimated as

$$\hat{\theta}_j^B = \operatorname{argmax} \left\{ \theta \left| \begin{array}{l} \sum_{k=1}^N \lambda_k y_{rk} \geq y_{rj}, \forall r = 1,...R; \sum_{k=1}^N \lambda_k c_{ik} \leq c_{ij} / \theta \ \forall i = 1,...I; \\ \sum_{k=1}^N \lambda_k = 1, \lambda_k \geq 0, \forall k = 1,...N \end{array} \right. \right\} \quad (11.4)$$

$\hat{\theta}_j^B$ is a consistent estimator of the true technical inefficiency θ_j^B (Banker 1993). Further the estimator for the allocative inefficiency, $\hat{\theta}_j^V$, can be calculated as $\hat{\theta}_j^Z / \hat{\theta}_j^B$. Under the null hypothesis that the sample data does not exhibit any allocative inefficiency, $\hat{\theta}_j^Z$ is also a consistent estimator of the true technical inefficiency θ_j^B. This leads to the following tests of the null hypothesis of no allocative inefficiency in the utilization of the inputs as opposed to the alternative of existence of allocative inefficiency:

(i) If $\ln(\theta_j^B)$ is distributed as exponential over $[0, \infty)$, then under the null hypothesis of no allocative inefficiency, the test statistic is calculated as $\sum_{j=1}^N \ln(\hat{\theta}_j^Z) / \sum_{j=1}^N \ln(\hat{\theta}_j^B)$ and evaluated relative to the critical value of the half-F distribution with $(2N, 2N)$ degrees of freedom.

(ii) If $\ln(\theta_j^B)$ is distributed as half-normal over the range $[0, \infty)$, then under the null hypothesis of no allocative inefficiency, the test statistic is

---

[6] Sections 5 and 6 of chapter 1 discuss efficiency estimation for situations in which unit prices and unit costs are available.

calculated as $\sum\limits_{j=1}^{N}(\ln(\hat{\theta}_j^Z))^2 \Big/ \sum\limits_{j=1}^{N}(\ln(\hat{\theta}_j^B))^2$ and evaluated relative to the critical

value of the half-F distribution with (N, N) degrees of freedom.

(iii) If no such assumptions are maintained about the probability distribution of inefficiency, a non-parametric Kolmogorov-Smirnov's test statistic given by the maximum vertical distance between $F(\ln(\hat{\theta}_j^Z))$ and $F(\ln(\hat{\theta}_j^B))$, the empirical distributions of $\ln(\hat{\theta}_j^Z)$ and $\ln(\hat{\theta}_j^B)$, respectively, is used. This statistic, by construction, takes values between 0 and 1 and a high value is indicative of the existence of allocative inefficiency.

There could also be situations involving multiple outputs and multiple inputs where output quantity information may not be available but monetary value of the individual outputs along with input quantity information may be available. In such situations, output based allocative inefficiency can be estimated and tested using procedures similar to those outlined above for input-based allocative efficiency (Banker et al. 1999).

2.5 Tests of Input Separability

In this section we describe DEA-based tests that can be used to evaluate the null hypothesis of input separability, i.e., the influence of each of the inputs on the output is independent of other inputs, against the alternative hypothesis that the inputs are substitutable. The DEA-based tests proposed in this section evaluate the null hypothesis of input separability over the entire sample data in contrast to the parametric (Berndt and Wood 1975) tests which are operationalized only at the sample mean.

Once again, consider observations on j = 1,....N DMUs, each observation comprising an output vector $\mathbf{y}_j \equiv (y_{1j}, ...y_{Rj}) \geq 0$, a vector of inputs $\mathbf{x}_j \equiv (x_{1j}, ...x_{Ij}) \geq 0$ and a production technology characterized by a monotone increasing and convex production possibility set $T \equiv \{(\mathbf{y}, \mathbf{x}).| \mathbf{y}$ can be produced from $\mathbf{x}\}$. The input-oriented inefficiency measure for this technology is estimated using the following BCC -- Banker, Charnes and Cooper (1984) -- linear program:

$$\hat{\theta}_j^{SUB} = \arg\max \left\{ \theta \left| \begin{array}{l} \sum\limits_{k=1}^{N}\lambda_k y_{rk} \geq y_{rj}, \forall r = 1,...R; \sum\limits_{k=1}^{N}\lambda_k x_{ik} \leq x_{ij}/\theta \; \forall i = 1,...I; \\ \sum\limits_{k=1}^{N}\lambda_k = 1, \lambda_k \geq 0, \forall k = 1,...N \end{array} \right. \right\} \quad (11.5)$$

When the inputs are separable, input inefficiency is first estimated considering only one input at a time, resulting in I different inefficiency measures corresponding to the I inputs. The overall DMU inefficiency is then estimated as the minimum of these I inefficiency measures. Specifically, the inefficiency corresponding to input i is measured as:

$$\hat{\theta}^i_j = \text{argmax} \left\{ \theta_i \left| \begin{array}{l} \sum_{k=1}^{N} \lambda_k y_{rk} \geq y_{rj}, \forall r = 1,...R; \; \sum_{k=1}^{N} \lambda_k x_{ik} \leq x_{ij}/\theta_i; \\ \sum_{k=1}^{N} \lambda_k = 1, \lambda_k \geq 0, \forall k = 1,...N \end{array} \right. \right\} \tag{11.6}$$

The inefficiency measure under the input separability assumption is then estimated as $\hat{\theta}^{SEP}_j = \text{Min} \left\{ \hat{\theta}^i_j \mid i = 1,.....,I \right\}$. Since $\hat{\theta}^{SEP}_j$ is estimated from a less constrained program, $\hat{\theta}^{SEP}_j \geq \hat{\theta}^{SUB}_j$. Under the null hypothesis of input separability, the asymptotic empirical distributions of $\hat{\theta}^{SEP}_j$ and $\hat{\theta}^{SUB}_j$ are identical, with each retrieving the distribution of the true input inefficiency θ.

The above discussion leads to the following tests of the null hypothesis of separability in the utilization of the inputs as opposed to the alternative of substitutability of inputs:

(i) If $\ln(\theta_j)$ is distributed as exponential over $[0, \infty)$, then under the null hypothesis of separability of inputs, the test statistic is calculated as $\sum_{j=1}^{N} \ln(\hat{\theta}^{SEP}_j) / \sum_{j=1}^{N} \ln(\hat{\theta}^{SUB}_j)$ and evaluated relative to the critical value of the half-F distribution with (2N, 2N) degrees of freedom.

(ii) If $\ln(\theta_j)$ is distributed as half-normal over the range $[0, \infty)$, then under the null hypothesis of input separability, the test statistic is calculated as $\sum_{j=1}^{N} (\ln(\hat{\theta}^{SEP}_j))^2 / \sum_{j=1}^{N} (\ln(\hat{\theta}^{SUB}_j))^2$ and evaluated relative to the critical value of the half-F distribution with (N, N) degrees of freedom.

(iii) If no such assumptions are maintained about the probability distribution of inefficiency, a non-parametric Kolomogorov-Smirnov's test statistic given by the maximum vertical distance between $F(\ln(\hat{\theta}^{SEP}_j))$ and $F(\ln(\hat{\theta}^{SUB}_j))$, the empirical distributions of $\ln(\hat{\theta}^{SEP}_j)$ and $\ln(\hat{\theta}^{SUB}_j)$, respectively, is used. This statistic, by construction, takes values between 0 and 1 and a high value is indicative of the rejection of input separability in the production technology.

3.   HYPOTHESIS TESTS FOR SITUATIONS CHARACTERIZED BY SHIFTS IN FRONTIER

In the previous section, we presented DEA tests that are useful in situations where cross-sectional data on DMUs is used for efficiency analysis. In this section, we describe estimation procedures and statistical tests when both longitudinal and cross-sectional data on DMUs are available and where the object of interest is change in productivity over time[7]. Productivity researchers have advocated both parametric and nonparametric approaches to estimate and analyze the impact of technical change and efficiency change on productivity change. The nonparametric literature (e.g. Färe et al 1997, Ray and Desli 1997, Førsund and Kittelsen 1998) has focused exclusively on the measurement of productivity change using Data Envelopment Analysis (DEA) without attempting to provide a statistical basis to justify those methods. Recently, Banker, Chang and Natarajan (2002) have developed DEA-based estimation methods and tests of productivity change and technical change. This section summarizes salient aspects of the Banker et al. (2002) study. For ease of exposition, we focus on a single output rather than a multiple output vector to illustrate the techniques and tests in this section.

Let $\mathbf{x}_t = (x_{1t},...x_{it},...x_{It}) \in X$, $x_{it} > 0$, $t=0,1$ be the I-dimensional input vector in period t. Consider two possible values of the time subscript t corresponding to a base period ($t = 0$) and another period ($t = 1$). The production correspondence in time t between the frontier output y_t^* and the input $\mathbf{x}_t$, is represented as

$$y_t^* = \phi^t(\mathbf{x}_t), \; t = 0,1 \tag{11.7}$$

The function $\phi^t(.): X{\rightarrow}\Re^+$ is constrained to be monotone increasing and concave in the input $\mathbf{x}_t$ but not required to follow any specific parametric functional form. The input vector $\mathbf{x}_t$, the production function $\phi^t(\mathbf{x}_t)$, and a relative efficiency random variable, α_t, that takes values between $-\infty$ and 0, together determine the realized output, y_t, in period t[8]. Specifically,

$$y_t \equiv e^{\alpha_t} \, \phi^t(\mathbf{x}_t) \tag{11.8}$$

For the above setup, estimators of technical change, relative efficiency change and productivity change, respectively, $\hat{b}_j^{(N)}$, $\hat{r}_j^{(N)}$ and $\hat{g}_j^{(N)}$, for the jth DMU can be estimated from observed output values and *estimators* of frontier outputs using the following expressions:

$$\hat{b}_j^{(N)} = \left\{ \ln\left(\hat{\phi}^1(\mathbf{x}_{j1})\right) - \ln\left(\hat{\phi}^0(\mathbf{x}_{j1})\right) \right\}$$

$$\hat{r}_j^{(N)} = \left\{ \ln\left(y_{j1} / \hat{\phi}^1(\mathbf{x}_{j1})\right) - \ln\left(y_{j0} / \hat{\phi}^0(\mathbf{x}_{j0})\right) \right\}$$

$$\hat{g}_j^{(N)} = \left\{ \ln\left(y_{j1} / \hat{\phi}^0(\mathbf{x}_{j1})\right) - \ln\left(y_{j0} / \hat{\phi}^0(\mathbf{x}_{j0})\right) \right\} \tag{11.9}$$

---

[7] The treatment of productivity change in this section provides an alternative to treatments using the Malmquist index described in chapter 8 and additionally has the advantage of providing explicit statistical characterizations.

[8] The relative efficiency random variable α_t and the inefficiency measure θ_t are linked by the relationship $\alpha_t = -\ln(\theta_t)$.

These estimators satisfy the fundamental relationship that productivity change is the sum of technical change and relative efficiency change i.e., $\hat{g}_j^{(N)} = \hat{b}_j^{(N)} + \hat{r}_j^{(N)}$. The frontier outputs required for the estimation of the various change measures in (11.9) are estimated using linear programs. The estimation of $\hat{\phi}^0(x_{j0})$ and $\hat{\phi}^1(x_{j1})$, is done using only the input-output observations from the base period or period 1, as the case may be. The linear program for estimating $\hat{\phi}^0(x_{j0})$ is the following Banker, Charnes and Cooper (BCC 1984) model:

$$\hat{\phi}^0(x_{j0}) = \text{argmax} \left\{ \hat{\phi} \left| \begin{array}{c} \sum_{k=1}^{N} \lambda_{k0}^0 y_{k0} \geq \hat{\phi}; \sum_{k=1}^{N} \lambda_{k0}^0 x_{ik0} \leq x_{ij0}, \forall i = 1,...I; \\ \sum_{k=1}^{N} \lambda_{k0}^0 = 1, \lambda_{k0}^0 \geq 0, \forall k = 1,...N \end{array} \right. \right\} \quad (11.10)$$

A similar program is used to estimate $\hat{\phi}^1(x_{j1})$ from input and output data from period 1. $\hat{\phi}^0(x_{j0})$ and $\hat{\phi}^1(x_{j1})$ are consistent estimators of $\phi^0(x_{j0})$ and $\phi^1(x_{j1})$, respectively, (Banker 1993).

The base period frontier value, $\phi^0(x_{j1})$, corresponding to *period 1* input, x_{j1}, is estimated based on the following linear program:

$$\hat{\phi}^0(x_{j1}) = \text{argmax} \left\{ \hat{\phi} \left| \begin{array}{c} \sum_{k=1}^{N} \lambda_{k0}^0 y_{k0} \geq \hat{\phi}; \sum_{k=1}^{N} \lambda_{k0}^0 x_{ik0} \leq x_{ij1}, \forall i = 1,...I; \\ \sum_{k=1}^{N} \lambda_{k0}^0 = 1, \lambda_{k0}^0 \geq 0, \forall k = 1,...N \end{array} \right. \right\} \quad (11.11)$$

or $= y_{j1}$ when the above linear program is not feasible

Note that the difference between the above model and the traditional BCC model is that the observation under evaluation is not included in the reference set for the constraints in (11.11) as in the super-efficiency model described first in Banker, Das and Datar (1989) and Anderson and Petersen (1993). It is the case that $\hat{\phi}^0(x_{j1})$ is a consistent estimator of $\phi^0(x_{j1})$.

Given the N values of technical, relative efficiency and productivity changes estimated using (11.9), the estimators of the medians of these performance measures are

$$\hat{b}^{MD(N)} = \text{argmin} \frac{1}{N} \sum_{j=1}^{N} |\hat{b}_j^{(N)} - b|$$

$$\hat{r}^{MD(N)} = \text{argmin} \frac{1}{N} \sum_{j=1}^{N} |\hat{r}_j^{(N)} - r|$$

$$\hat{g}^{MD(N)} = \text{argmin} \frac{1}{N} \sum_{j=1}^{N} |\hat{g}_j^{(N)} - g| \quad (11.12)$$

Banker et al. (2002) show that $\hat{b}^{MD(N)}$, $\hat{r}_j^{MD(N)}$ and $\hat{g}_j^{MD(N)}$ are consistent estimators of the population median technical change β^{MD}, population median relative efficiency change ρ^{MD} and population median productivity change γ^{MD}, respectively.

Consider the number of observations, $\hat{p}_\beta^{(N)}$, $\hat{p}_\rho^{(N)}$ and $\hat{p}_\gamma^{(N)}$ (out of the sample of N observations) for which $\hat{b}_j^{(N)}$, $\hat{r}_j^{(N)}$ and $\hat{g}_j^{(N)}$, respectively, is strictly positive. Tests for the median of the various performance measures being equal to zero are conducted as follows:

(a) Under the null hypothesis of zero median technical change between the base period and period t i.e., $\beta^{MD} = 0$, the statistic $\hat{p}_\beta^{(N)}$ is asymptotically distributed as a binomial variate with parameters N and 0.5 i.e., $\hat{p}_\beta^{(N)} \sim$ b(N,0.5).

(b) Under the null hypothesis of zero median relative efficiency change between the base period and period t i.e., $\rho^{MD} = 0$, the statistic $\hat{p}_\rho^{(N)}$ is asymptotically distributed as a binomial variate with parameters N and 0.5 i.e., $\hat{p}_\rho^{(N)} \sim$ b(N,0.5).

(c) Under the null hypothesis of zero median technical change between the base period and period t i.e., $\gamma^{MD} = 0$, the statistic $\hat{p}_\gamma^{(N)}$ is asymptotically distributed as a binomial variate with parameters N and 0.5 i.e., $\hat{p}_\gamma^{(N)} \sim$ b(N,0.5).

Banker et al. (2002) also provide methods for estimating and testing the location of the population mean of the various performance measures. The mean technical change $\overset{\triangle(N)}{b}$, mean relative efficiency change $\overset{\triangle(N)}{r}$ and mean productivity change $\overset{\triangle(N)}{g}$ are estimated as

$$\overset{\triangle(N)}{b} = \frac{1}{N}\sum_{j=1}^{N}(\ln(\hat{\phi}^1(\mathbf{x}_{j1})) - \ln(\hat{\phi}^0(\mathbf{x}_{j1})))$$

$$\hat{\bar{r}}^{(N)} = \frac{1}{N}\left\{\ln\left(y_{j1}/\hat{\phi}^1(\mathbf{x}_{j1})\right) - \ln\left(y_{j0}/\hat{\phi}^0(\mathbf{x}_{j0})\right)\right\}$$

$$\hat{\bar{g}}^{(N)} = \frac{1}{N}\left\{\ln\left(y_{j1}/\hat{\phi}^0(\mathbf{x}_{j1})\right) - \ln\left(y_{j0}/\hat{\phi}^0(\mathbf{x}_{j0})\right)\right\} \qquad (11.13)$$

$\overset{\triangle(N)}{b}$, $\hat{\bar{r}}^{(N)}$ and $\hat{\bar{g}}^{(N)}$ are consistent estimators of the population mean technical change, $\bar{\beta}$, population mean relative efficiency change, $\bar{\rho}$, and population productivity change, $\bar{\gamma}$, respectively.

Next consider the statistics $\hat{t}^{(N)}(\beta) = \sqrt{N}\overline{\hat{b}}^{(N)}\Big/\hat{s}(\beta)$, $\hat{t}^{(N)}(\rho) = \sqrt{N}\overline{\hat{r}}^{(N)}\Big/\hat{s}(\rho)$

and $\hat{t}^{(N)}(\gamma) = \sqrt{N}\overline{\hat{g}}^{(N)}\Big/\hat{s}(\gamma)$ where

$$\hat{s}^2(\beta) = \left(\sum_{j=1}^{N}\left(\ln(\hat{\phi}^1(\mathbf{x}_{j1})) - \ln(\hat{\phi}^0(\mathbf{x}_{j1}))\right)^2 - N\left(\overline{\hat{b}}^{(N)}\right)^2\right)\Big/(N-1)$$

$$\hat{s}^2(\rho) = \left(\sum_{j=1}^{N}\left(\ln(y_{j1}/\hat{\phi}^1(\mathbf{x}_{j1})) - \ln(y_{j0}/\hat{\phi}^0(\mathbf{x}_{j0}))\right)^2 - N\left(\overline{\hat{r}}^{(N)}\right)^2\right)\Big/(N-1)$$

$$\hat{s}^2(\gamma) = \left(\sum_{j=1}^{N}\left(\ln\left(y_{j1}/\hat{\phi}^0(\mathbf{x}_{j1})\right) - \ln\left(y_{j0}/\hat{\phi}^0(\mathbf{x}_{j0})\right)\right)^2 - N\left(\overline{\hat{g}}^{(N)}\right)^2\right)\Big/(N-1) \quad (11.14)$$

For large samples, the distribution of each of $\hat{t}^{(N)}(\beta)$, $\hat{t}^{(N)}(\rho)$ and $\hat{t}^{(N)}(\gamma)$ approaches that of a Student's T variate with N-1 degrees of freedom. Therefore, a simple t-test for the mean of the DMU-specific estimators for the various performance measures estimated using (11.9) is appropriate when the sample size is large.

4.   HYPOTHESIS TESTS FOR COMPOSED ERROR SITUATIONS

The tests described in the previous sections are conditioned on a data generating process that characterizes the deviation of the actual output from the production frontier as arising only from a stochastic inefficiency term. In this section, we describe the application of DEA-based tests for composed error situations where the data generating process involves not only the one-sided inefficiency term but also a noise term that is independent of the inefficiency. Recently, Gsatch (1998) and Banker and Natarajan (2001) have developed DEA-based estimation procedures for environments characterized by both inefficiency and noise. The tests developed by Banker (1993) can be adapted to these environments through an appropriate transformation of the inefficiency term. We describe these tests below:

4.1   Tests for Efficiency Comparison

Consider observations on $j = 1,....N$ DMUs, each observation comprising a single output $y_j \geq 0$ and a vector of inputs $\mathbf{x}_j \equiv (x_{1j}, ...x_{1j}) \geq 0$. The production correspondence between the frontier output y^0 and the I inputs is represented as $y^0 = g(\mathbf{x})$ subject to the assumption that $g(.)$ is monotonically increasing and concave in $\mathbf{x}$. The deviation from the frontier for the jth DMU could be positive or negative and is represented as $\varepsilon_j = u_j - v_j$

$= g(x_j) - y_j$. Thus, the deviation is modeled as the sum of two components, a one-sided inefficiency term, u_j, and a two-sided random noise term v_j bounded above at V^M, analogous to composed error formulations in parametric stochastic frontier models (Aigner et al. 1977, Meussen and van den Broeck 1977, Banker and Natarajan 2001). In this stochastic framework, Banker and Natarajan (2003) propose two statistical tests to compare the efficiency of two groups of DMUs.

As before, consider two sub-samples, G_1 and G_2, that are part of the sample of N DMUs when the sample size, N, is large. Let the true inefficiency, u_j, be distributed with means $\bar{u}_1$ and $\bar{u}_2$ in the two groups. Further assume that the variance of the inefficiency is the same in both groups. Define $\tilde{u}_j = V^M - v_j + u_j$. We can estimate $\hat{\tilde{u}}_j$, a consistent estimator of $\tilde{u}_j$, by applying DEA on input and output data from the full sample of N DMUs. For N_1 and N_2 DMUs in subgroups G_1 and G_2, respectively, the null hypothesis of no difference in mean inefficiency between the two subgroups can be tested using the following procedures:

(i) Consider the OLS regression $\hat{\tilde{u}}_j = a_0 + a_1 z_j + e_j$ estimated using a total of $N_1 + N_2$ DEA inefficiency scores. z_j is a dummy variable that takes a value of 0 if a particular DMU belongs to group G_1 and 1 if it belongs to G_2 and e_j is an i.i.d error term. The regression coefficient $\hat{a}_1$ is a consistent estimator of $\bar{u}_2 - \bar{u}_1$, the difference in mean inefficiency between groups G_2 and G_1. The t-statistic associated with this regression coefficient can be used to evaluate whether two groups are significantly different in terms of mean inefficiency.

(ii) Assume that the probability distributions of the inefficiency random variable u_j and the noise random variable v_j are such that $\tilde{u}_j = V^M - v_j + u_j$ is distributed as a log-normal variable in the two groups. Under the null hypothesis that the mean inefficiencies are equal i.e., $\bar{u}_1 = \bar{u}_2$ and assuming that the variance of u_j is the same in the two groups, the Student-t statistic $\hat{t}$

$$= \left(\overline{\widetilde{lu}}_1 - \overline{\widetilde{lu}}_2 \right) \Big/ \hat{S} \sqrt{\frac{1}{N_1} + \frac{1}{N_2}} \quad \text{distributed with } (N_1 + N_2 - 2) \text{ degrees of freedom}$$

can be used to evaluate the null hypothesis of no difference in mean inefficiency across the two groups. $\overline{\widetilde{lu}}_1$ is $\dfrac{1}{N_1} \sum_{j=1}^{N_1} \ln(\hat{\tilde{u}}_{j1})$, $\overline{\widetilde{lu}}_2 = \dfrac{1}{N_2} \sum_{j=1}^{N_2} \ln(\hat{\tilde{u}}_{j2})$

and $\hat{S} = \left(\dfrac{1}{N_1 + N_2 - 2} \left\{ \sum_{j=1}^{N_1} \left\{ \ln(\hat{\tilde{u}}_{j1}) - \overline{\widetilde{lu}}_1 \right\}^2 + \sum_{j=1}^{N_2} \left\{ \ln(\hat{\tilde{u}}_{j2}) - \overline{\widetilde{lu}}_2 \right\}^2 \right\} \right)^{0.5}$.

4.2 Tests for Evaluating the Impact of Contextual Variables on Efficiency

Analysis of factors contributing to efficiency differences has been an important area of research in DEA. Ray (1991), for instance, regresses DEA scores on a variety of socio-economic factors to identify key performance drivers in school districts. The two-stage approach of first calculating productivity scores and then seeking to correlate these scores with various explanatory variables has been in use for over twenty years but explanations of productivity differences using DEA are still dominated by ad hoc speculations (Førsund 1999). Banker and Natarajan (2001) provide a general framework for the evaluation of contextual variables affecting productivity by considering a variety of Data Generating Processes (DGPs) and present appropriate estimation methods and statistical tests under each DGP. In this section, we describe the DEA-based tests developed in Banker and Natarajan (2001) that can be used to determine the impact of contextual or environmental variables on efficiency.

Consider observations on $j = 1,....N$ decision making units (DMUs), each observation comprising a single output $y_j \geq 0$, a vector of inputs $x_j \equiv (x_{1j}, ...x_{Ij}) \geq 0$, and a vector of contextual variables $z_j \equiv (z_{1j}, ... z_{Sj})$ that may influence the overall efficiency in transforming the inputs into the outputs. The production function $g(.)$ is monotone increasing and concave in x, and relates the inputs and contextual variables to the output as specified by the equation

$$y_j = g(x_j) + v_j - h(z_j) - u_j \qquad (11.15)$$

where v_j is a two-sided random noise term bounded above at V^M, $h(z_j)$ is a non-negative monotone increasing function, convex in z and u_j is a one-sided inefficiency term. The inputs, contextual variables, noise and inefficiency are all independently distributed of each other. Defining $\tilde{g}(x_j) = g(x_j) + V^M$ and $\tilde{\delta}_j = (V^M - v_j) + h(z_j) + u_j \geq 0$, (11.15) can be expressed as

$$y_j = \tilde{g}(x_j) - \tilde{\delta}_j \qquad (11.16)$$

Since $\tilde{g}(.)$ is derived from $g(.)$ by multiplication with a positive constant, $\tilde{g}(.)$ is also monotone increasing and concave. Therefore, the DEA inefficiency estimator, $\hat{\tilde{\delta}}_j$, obtained by performing DEA on the inputs-output observations (y_j, x_j), $j = 1, . . .N$, is a consistent estimator of $\tilde{\delta}_j$ (Banker 1993). This consistency result is used by Banker and Natarajan (2001) to develop the following DEA-based tests corresponding to different specifications of $h(.)$, the function linking contextual variables to inefficiency.

Consider the case where $h(\mathbf{z}) = h(\mathbf{z};\beta)$ and $h(\mathbf{z};\beta)$ is a non-negative function, monotone increasing in $\mathbf{z}$, linear in β. In this case, the impact of contextual variables can be consistently estimated by regressing the first stage DEA estimate $\hat{\hat{\delta}}_j$ on the various contextual variables associated with the various components of the β vector. This procedure yields consistent estimators of the parameter vector β. In the special case where $h(\mathbf{z};\beta) =$

$$\mathbf{z}'\beta = \sum_{j=1}^{N} z_j \beta_j,$$ the independent variables in the regression are the same as the S contextual variables.

Now, suppose no additional structure can be placed on $h(\mathbf{z}_j)$ except that it is a non-negative monotone increasing function, convex in $\mathbf{z}$. Let $\tilde{\varepsilon}_j = (V^M - v_j) + u_j$. Then $\hat{\hat{\delta}}_j = h(\mathbf{z}_j) + \tilde{\varepsilon}_j$ and a second stage DEA estimation on the pseudo "input-outputs" observations $(\hat{\hat{\delta}}_j, \mathbf{z}_j)$ yields a consistent estimator $\hat{\tilde{\varepsilon}}_j$ for $\tilde{\varepsilon}_j$ (Banker 1993). This estimator is obtained by solving the following linear programming formulation analogous to the BCC model (Banker, Charnes and Cooper 1984) in DEA individually for each observation in the sample:

$$\hat{\psi}_j = \operatorname{argmin}\left\{ \psi \left| \begin{array}{l} \sum\limits_{k=1}^{N} \lambda_k \hat{\hat{\delta}}_k = \psi; \sum\limits_{k=1}^{N} \lambda_k z_{ik} \geq z_{ij}, \forall i = 1,...S; \\ \sum\limits_{k=1}^{N} \lambda_k = 1, \lambda_k \geq 0, \forall k = 1,...N \end{array} \right. \right\} \quad (11.17)$$

A consistent estimator for $\tilde{\varepsilon}_j$ for each observation in the sample is obtained as $\hat{\tilde{\varepsilon}}_j = \hat{\hat{\delta}}_j - \hat{\psi}_j$.

To evaluate the statistical significance of individual z_s, a third stage DEA estimation is first performed on the pseudo observations $(\hat{\hat{\delta}}_j, \mathbf{z}_j^{-s})$ where $\mathbf{z}^{-s}$ is the original $\mathbf{z}$ vector without the z_s variable. The following modified version of the program in (11.17) is used for this purpose.

$$\hat{\psi}_j^{-s} = \operatorname{argmin}\left\{ \psi \left| \begin{array}{l} \sum\limits_{k=1}^{N} \lambda_k \hat{\hat{\delta}}_k = \psi; \sum\limits_{k=1}^{N} \lambda_k z_{ik} \geq z_{ij}, \forall i = 1,.,s-1,s+1,..S; \\ \sum\limits_{k=1}^{N} \lambda_k = 1, \lambda_k \geq 0, \forall k = 1,...N \end{array} \right. \right\} \quad (11.18)$$

Let the resulting estimator of $\tilde{\varepsilon}_j$ be $\hat{\tilde{\varepsilon}}_j^{-s} = \hat{\hat{\delta}}_j - \hat{\psi}_j^{-s}$. Since (11.18) is a less constrained program than (11.17), $\hat{\tilde{\varepsilon}}_j^{-s} \geq \hat{\tilde{\varepsilon}}_j$ for all observations j = 1,...N.

Under the null hypothesis that the marginal impact of z_s (i.e. $\partial h(\mathbf{z})/\partial z_s$ if $h(.)$ is differentiable) is zero, the asymptotic distributions of $\tilde{\varepsilon}$ and $\tilde{\varepsilon}^{-s}$ are identical (Banker 1993). If the asymptotic distribution of $\tilde{\varepsilon}_j$ is assumed to be exponential or half-normal, the null hypothesis of no impact of z_s is tested by comparing the ratios $\sum_{j=1}^{N} \hat{\tilde{\varepsilon}}_j^{-s} \Big/ \sum_{j=1}^{N} \hat{\tilde{\varepsilon}}_j$ or $\sum_{j=1}^{N}\left[\hat{\tilde{\varepsilon}}_j^{-s}\right]^2 \Big/ \sum_{j=1}^{N}\left[\hat{\tilde{\varepsilon}}_j\right]^2$ against critical values obtained from half-F distributions with (2N,2N) or (N,N) degrees of freedom, respectively. If $\tilde{\varepsilon}_j$ is exponentially distributed, the test statistic is evaluated relative to the half-F distribution $|F_{2N,2N}|$ with 2N,2N degrees of freedom over the range $[1,\infty)$, since by construction the test statistic is never less than 1. If $\tilde{\varepsilon}_j$ is a half-Normal variate then the test statistic is evaluated relative to the half-F distribution $|F_{N,N}|$ with N,N degrees of freedom over the range $[1,\infty)$. Recall that $\tilde{\varepsilon} = u + (V^M - v)$. Therefore, statistics based on Banker (1993) may not be valid unless the variance of the noise term v is considerably smaller than the variance of the inefficiency term, u, and u is distributed with a mode at its lower support.

The Kolmogorov-Smirnov statistic, which is based on the maximum vertical distance between the empirical cumulative distribution of $\hat{\tilde{\varepsilon}}_j^{-s}$ and $\hat{\tilde{\varepsilon}}_j$, can also be used to check whether the empirical distributions of $\hat{\tilde{\varepsilon}}_j^{-s}$ and $\hat{\tilde{\varepsilon}}_j$ are significantly different. If they are significantly different, then it can be established that z_s has a significant impact on productivity.

4.3 Tests for Evaluating the Adequacy of Parametric Functional Forms

While DEA provides a theoretically correct way to estimate monotone and concave (or convex) functional relationships, it is often useful to represent the relationship in a more parsimonious functional form that is afforded by a parametric specification. Specific parametric functional forms, such as the Cobb-Douglas, are useful if they provide a good approximation to the general monotone and concave (or convex) function as evidenced by sample data. In this section, we present methods developed in Banker, Janakiraman and Natarajan (2002) to evaluate the adequacy of a parametric functional form to represent the functional relationship between an endogenous variable and a set of exogenous variables given the minimal maintained assumption of monotonicity and concavity.

Consider sample data on an endogenous variable and I exogenous variables for N observations. For the jth observation, denote the endogenous variable as y_j and the vector of exogenous variables as $X_j \equiv (x_{1j}, x_{2j}, \dots x_{lj})$. The relationship between y_j and X_j is specified as:

$$y_j = g(X_j) \ e^{\varepsilon_j} \tag{11.19}$$

where $g(.)$ is a monotone increasing and concave function. It is assumed further that ε_j is independent of X_j and i.i.d. with a probability density function $f(\varepsilon)$ over the range $[-S_L, \ S_U] \subseteq \Re$, where $S_L \geq 0$, and $S_U \geq 0$ are unknown parameters that describe the lower and upper supports of the distribution. Define $\tilde{g}(X) = g(X)e^{S_U}$ and $\tilde{\varepsilon}_j = S_U - \varepsilon_j \geq 0$ such that (11.19) can be rewritten as $y_j = \tilde{g}(X_j) \ e^{-\tilde{\varepsilon}_j}$. The DEA estimator of $\tilde{g}(X_j)$ can be estimated using the following linear program:

$$\hat{\tilde{g}}^{DEA}(X_j) = \text{argmax} \left\{ y \left| \begin{array}{l} \sum_{k=1}^{N} \lambda_k y_k = y; \ \sum_{k=1}^{N} \lambda_k x_{ik} \leq x_{ij}, \forall i = 1, \dots I; \\ \sum_{k=1}^{N} \lambda_k = 1, \lambda_k \geq 0, \forall k = 1, \dots N \end{array} \right. \right\} \tag{11.20}$$

An estimator for $\tilde{\varepsilon}_j$ for each observation in the sample can then be obtained as $\hat{\tilde{\varepsilon}}_j^{DEA} = \ln(\hat{\tilde{g}}^{DEA}(X_j)) - \ln(y_j)$. Further, $\hat{\tilde{g}}^{DEA}(X_j)$ and $\hat{\tilde{\varepsilon}}_j^{DEA}$ are consistent estimators of $\tilde{g}(X_j)$ and $\tilde{\varepsilon}_j$ respectively (Banker 1993).

The parametric estimation is carried out by specifying a parametric form $g(X; \beta)$ and regressing $\ln(y)$ on the exogenous variables in $\ln(g(X; \beta))$. The residuals from the regression are used to obtain $\hat{S}_U = \max\{\ln(y) - \ln(\hat{g}(X_j; \hat{\beta}))\}$ which is a consistent estimator of S_U (Greene 1980). The estimated deviation from the parametric frontier is then calculated as $\hat{\tilde{\varepsilon}}_j^{PARAM} = \ln(\hat{g}(X_j; \hat{\beta})) + \hat{S}_U - \ln(y_j)$. In addition to $\hat{\tilde{\varepsilon}}^{DEA}$, $\hat{\tilde{\varepsilon}}_j^{PARAM}$ also is a consistent estimator of $\tilde{\varepsilon}_j$ under the null hypothesis that $g(X; \beta) = g(X)$ for all X. Banker et al. (2002) prove that the asymptotic distribution of $\hat{\tilde{\varepsilon}}^{PARAM}$ retrieves the true distribution of $\tilde{\varepsilon}$ if the parametric specification is, in fact, the true specification of the production function (i.e., $g(X; \beta) = g(X)$ for all X). Further, they also show that if the parametric specification is, in fact, the true specification of the production function then as N $\rightarrow \infty$, (a) the asymptotic distribution of $\hat{\tilde{\varepsilon}}^{PARAM}$ converges to that of $\hat{\tilde{\varepsilon}}^{DEA}$ and (b) both $\hat{\tilde{\varepsilon}}_j^{PARAM}$ and $\hat{\tilde{\varepsilon}}_j^{DEA}$ converge asymptotically to $\tilde{\varepsilon}_j$ for all $j \in J$, where J is a given set of observations. Based on these results, they suggest the following four tests for testing the adequacy of the parametric functional form:

(a) The first test uses the Kolmogorov-Smirnov test statistic given by the maximum vertical distance between $\hat{F}(\hat{\varepsilon}_j^{DEA})$ and $\hat{F}(\hat{\varepsilon}_j^{PARAM})$, where $\hat{F}(\hat{\varepsilon}_j^{DEA})$ and $\hat{F}(\hat{\varepsilon}_j^{PARAM})$ denote the empirical distributions of $\hat{\varepsilon}_j^{DEA}$ and $\hat{\varepsilon}_j^{PARAM}$ respectively. A low value is indicative of support for the null hypothesis that $g(X;\beta)$ adequately represents $g(X)$.

(b) The second procedure is based on the regression of rank of $\hat{\varepsilon}_j^{DEA}$ on the rank of $\hat{\varepsilon}_j^{PARAM}$ (Iman and Conover 1979). Under the null hypothesis that the parametric form is adequate, the expected value of the coefficient on $\hat{\varepsilon}_j^{PARAM}$ in the rank regression is asymptotically equal to 1. The null hypothesis is evaluated against the alternative hypothesis that the regression coefficient has a value less than 1.

(c) The third test procedure employs the Wilcoxon rank-sum test to evaluate whether the empirical distributions $\hat{F}(\hat{\varepsilon}_j^{DEA})$ and $\hat{F}(\hat{\varepsilon}_j^{PARAM})$ are different. If the test shows these distributions to be different, the adequacy of the parametric form is rejected.

(d) The fourth procedure is based on Theil's (1950) distribution-free test. This test evaluates the null hypothesis that $\mu_1 = 1$ against the alternative $\mu_1 \neq 1$ in the relation $\hat{\varepsilon}^{DEA} = \mu_0 + \mu_1\hat{\varepsilon}^{PARAM}$. To compute Theil's statistic, the difference $D_j = \hat{\varepsilon}_j^{DEA} - \hat{\varepsilon}_j^{PARAM}$ is calculated and then the data are sorted by $\hat{\varepsilon}_j^{PARAM}$. Next, a score $c_{ji} = 1$, 0 or -1 is assigned for each $i < j$ depending on whether $D_j - D_i > 0$, $= 0$ or < 0 respectively. Theil's test statistic C is defined as $C = \sum_{j=1}^{N} c_{ji}$. Theil's test statistic is distributed as a standard normal variate for large samples and a high absolute value of C rejects the adequacy of the parametric functional form.

Banker et al. (2002) also propose an alternative approach to evaluating the adequacy of a parametric functional form. The approach suggested by them relies on Wooldridge's (1992) Davidson-Mackinnon type test to evaluate a linear null model against a nonparametric alternative. Banker et al. (2002) propose the use of a sieve DEA estimator as the nonparametric alternative for the purposes of Wooldridge's test since the test cannot be applied directly when the nonparametric alternative is based on the traditional DEA estimator. Interested readers are referred to section 2.3 of Banker et al. (2002) for additional details on how Wooldridge's test can be applied to examine the adequacy of a specified parametric form for a monotone and concave function.

5. CONCLUDING REMARKS

We have described here several statistical tests that can be used to test hypotheses of interest and relevance to applied users of Data Envelopment Analysis. A common underlying theme of these tests is that the deviation from the DEA frontier can be viewed as a stochastic variable. While the DEA estimator is biased in finite samples, the expected value of the DEA estimator is almost certainly the true parameter value in large samples. The tests described in this paper rely on this asymptotic property of the DEA estimator.

An important caveat is that the tests described in this paper are designed for large samples. Results of simulation studies conducted on many of the tests proposed in this study suggest that these tests perform very well for sample sizes similar to those used in many typical applications of DEA[9]. These tests need to be used with caution in small samples. We believe additional simulation studies are warranted to provide evidence on small sample performance of the tests described here. Clearly, this is an important area for future research.

We believe that there are many more avenues and areas where DEA-based statistical tests can be applied. This is because the flexible structure of DEA facilitates application in a large number of situations where insufficient information or guidance may preclude the use of parametric methods. Statistical tests developed during the past 10 years have contributed significantly to the reliability of managerial and policy implications of DEA studies and we believe that they will continue to enrich future applications of DEA.

REFERENCES

1. Aigner, D.J., Lovell, C.A.K. and Schmidt, P., 1977. Formulation and Estimation of Stochastic Frontier Production Function Models. *Journal of Econometrics*, 6, 21-37.
2. Anderson, P. and Petersen, N.C., 1993, A Procedure for Ranking Efficient Units in Data Envelopment Analysis. *Management Science* 39, 1261-1264.
3. Banker, R.D., 1984. Estimating Most Productive Scale Size Using Data Envelopment Analysis. *European Journal of Operations Research*, July, 35-44.

---

[9] Banker (1996) provides details on some of these simulation studies. Based on the results of these studies, it appears that the tests described in this paper perform well in sample sizes of the order of 50 or more.

4. Banker, R.D., 1993. Maximum Likelihood, Consistency and Data Envelopment Analysis: A Statistical Foundation. *Management Science,* October, 1265-1273.
5. Banker, R.D., 1996. Hypothesis Tests Using Data Envelopment Analysis. *Journal of Productivity Analysis,* 7, 139-159.
6. Banker, R.D., Charnes, A. and Cooper, W.W., 1984. Models for the Estimation of Technical and Scale Inefficiencies in Data Envelopment Analysis. *Management Science.* 30, 1078-1092.
7. Banker, R.D., Chang, H. and Natarajan, R., 1999. Statistical Tests of Allocative Efficiency Using DEA: An Application to the U.S. Public Accounting Industry. Working Paper, The University of Texas at Dallas.
8. Banker, R.D., Chang, H. and Natarajan, R., 2002. Nonparametric Analysis of Productivity Change and Technical Progress. Working Paper, The University of Texas at Dallas.
9. Banker, R.D., Das, S. and Datar S., 1989. Analysis of Cost Variances for Management Control in Hospitals. *Research in Governmental and Nonprofit Accounting,* 5, 269-291.
10. Banker, R.D., Janakiraman, S. and Natarajan, R., 2002. Evaluating the Adequacy of Parametric Functional Forms in Estimating Monotone and Concave Production Functions. *Journal of Productivity Analysis.* Volume 17, 111-132.
11. Banker, R.D., Janakiraman, S. and Natarajan, R., 2003. Analysis of Trends in Technical and Allocative Efficiency: An Application to Texas Public School Districts. Forthcoming *European Journal of Operations Research.*
12. Banker, R. D., and Natarajan, R., 2001. Evaluating Contextual Variables Affecting Productivity using Data Envelopment Analysis. Working paper, The University of Texas at Dallas.
13. Banker, R.D., and Natarajan, R., 2003. DEA-based Hypothesis Tests for Comparison of Two Groups of Decision Making Units. Working paper, The University of Texas at Dallas.
14. Berndt, E., and Wood, D., 1975. Technology, Prices and the Derived Demand for Energy. *Review of Economics and Statistics.* 259-268.
15. Charnes, A., Cooper, W.W., and Rhodes, E., 1978. Measuring the Efficiency of Decision Making Units. *European Journal of Operational Research,* 429-444.
16. Färe, R., Griffel-tatje, E., Grosskopf, S., Lovell, C.A.K., 1997. Biased Technical Change and the Malmquist Productivity Index. *Scandinavian Journal of Economics,* Volume 99: 119-127.
17. Førsund, F.R., 1999. The Evolution of DEA – The Economics Perspective. *Working Paper,* University of Oslo, Norway.

18. Førsund, F., and Kittelsen, S., 1998. Productivity Development of Norwegian Electricity Distribution Utilities. *Resource and Energy Economics*. 20:207-224.
19. Greene, W.H., 1980. Maximum Likelihood Estimation of Econometric Frontier Production Functions. *Journal of Econometrics* 13, 27-56.
20. Gstach, D., 1998. Another Approach to Data Envelopment Analysis in Noisy Environments: DEA+. *Journal of Productivity Analysis*, 9, 161-176.
21. Iman, R. L. and Conover, W. J., 1979. The Use of Rank Transform in Regression. *Technometrics*. 499-509.
22. Meeusen, W. and van den Broeck, J., 1977. Efficiency Estimation from Cobb-Douglas Production Functions with Composed Error. *International Economic Review*. June, 435-444.
23. Ray, S., 1991. Resource-use Efficiency in Public Schools: A Study of Connecticut Data. *Management Science*. 1620-1628.
24. Ray, S., and Desli, E., 1997. Productivity Growth, Technical Progress, and Efficiency Change in Industrialized Countries: Comment. *American Economic Review* 87: 1033-1039.
25. Theil, H., 1950. A Rank-invariant Method of Linear and Polynomial Regression Analysis, *I. Proc. Kon. Ned. Akad. V. Wetensch.* A 53, 386-392.
26. Wooldridge, J. M., 1992. A Test for Functional Form Against Nonparametric Alternatives. *Econometric Theory* 8, 452-475.

Chapter 12

PERFORMANCE EVALUATION IN EDUCATION
Modeling Educational Production

John Ruggiero
Department of Economics and Finance, University of Dayton, Dayton, Ohio 45469-2251
email: John.Ruggiero@notes.udayton.edu

Abstract: The performance of decision making units (DMUs) that provide educational services has been a concern in the academic and policy arenas. The topic is of the utmost importance considering the amount of money that is devoted to this public sector activity. Recent reform initiatives have moved away from equity towards accountability and efficiency. Application of data envelopment analysis (DEA) to the education sector is natural given these recent concerns. Standard DEA models fail, however, due to the heterogeneity of DMUs arising from socioeconomic differences. Models suitable for application to education have been developed and will be discussed in this chapter.

Key words: Education, DEA, Heterogeneity, Efficiency

1.   INTRODUCTION

Recent court rulings regarding the provision of public education in the United States has moved away from traditional arguments of equity towards achieving educational goals efficiently. There is widespread concern that public education is inefficient. In a recent ruling, the Ohio Supreme Court ruled that the system of financing public education was unconstitutional because Ohio had not lived up to the promise of providing public education in a thorough and efficient manner. In the last 15 years, similar challenges have appeared in Kentucky, Montana, New Jersey and Texas. For useful discussions, see Odden and Picus (1992), Reschovsky (1994), Clotfelter and Ladd (1996), and Duncombe, Ruggiero and Yinger (1996). The understanding of efficiency of educational production is timely given the recent litigation and important considering the amount of resources used in providing public education. In Ohio for example, it was estimated that over $1 billion in additional funding was necessary to overhaul the school finance system.

Recent reform in education was fueled by the damning criticism of public education that appeared in the 1983 report *A Nation at Risk*. Some of the indicators included poor performance by United States' students relative to other industrial countries. Business leaders complained that additional resources were required to provide remedial education and training in basic skill. Further, at the time, 23 million American adults were functionally illiterate; 13 percent of all 17-year olds were also found to be functionally illiterate. Given these stylized facts, there can be no question about the failure of public education.

While most of the DEA articles have analyzed technical efficiency of educational DMUs, it is important to note that there is a vast literature on educational production in the economics literature and theories explaining inefficiency in the political science and economics literature. Useful references for educational production include Hanushek (1979, 1986), Bridge, Judd and Moock (1979), and Cohn and Geske (1990). Leibenstein (1966, 1978), Niskanken (1971, 1975) and Wyckoff (1990) provide useful theories of bureaucratic and X-inefficiency and Chubb and Moe (1990) provide a thorough analysis of politics and education.

There have been numerous articles and dissertations devoted to the evaluation of education units beginning with Rhodes (1978), who analyzed U.S. educational production (Program Follow Through) in his dissertation. Charnes, Cooper and Rhodes (1978) provide a brief discussion and Charnes, Cooper and Rhodes (1981) provide a detailed analysis, separating out program and managerial efficiency. Since this early work, many papers have appeared in the academic literature. Due to the volume, I am sure that some of the papers will be accidentally omitted. Due to space consideration, I will only consider articles that contain DEA applications, recognizing a large literature exists using the stochastic frontier approach.

Perhaps a useful starting point for listing papers is the level of analysis. The education process exists at the student level. However, these data are not easily available. For an exception, see Thanassoulis (1999). Most DEA analyses of educational production in the United States focus on the school or the school district as the decision making unit. Data are aggregated to these levels and are readily available in many states. The appendix contains a table of select DEA education applications that have appeared in the literature.

2. PRODUCTION OF EDUCATION

A useful framework for analyzing educational production was provided by Bradford, Malt and Oates (1969). This model considered public sector production as a two stage model where intermediate output is determined in the first stage by the levels of discretionary inputs. Final outcomes are determined by the intermediate outputs and by environmental (exogenous) variables. The model is well-supported in the empirical education production literature; see Hanushek (1979, 1986) for examples.

The provision of educational services provides a useful application of the Bradford, Malt and Oates model. Schools (or classrooms or school districts) use labor (teachers, administrators, etc.) and capital (computers, facilities, etc.) to provide instruction in mathematics, reading, writing and other content areas. These outputs are generally not of primary interest to the community; rather, the final outcomes (for example, test scores, graduation rates, drop-out rates) are of more concern. Additionally, there is strong evidence that non-discretionary socio-economic factors play a large and an important role in the educational process. Parental wealth and education and the socio-economic background of classmates will influence a child's ability to learn. A useful summary of the education process is presented in Figure 12-1.

Based on the model, and assuming that the socioeconomic characteristics are exogenous and hence, non-discretionary for the decision making unit, it is necessary to control for these factors in the evaluation of educational decision making units. The effect that the environment has on the production frontier is shown in Figure 12-2. It is assumed that 7 DMUs producing one output y_l with one input x_l. It is also assumed that an index of environmental harshness is available. In section 3 an approach is provided to construct such an index from multiple exogenous factors. It is assumed that DMUs $A - C$ have a favorable socioeconomic environment relative to DMUs $D - G$. For example, DMUs $A - C$ might have less students with limited English proficiency.

Time Period 1

Educational
DMU

Time Period 2

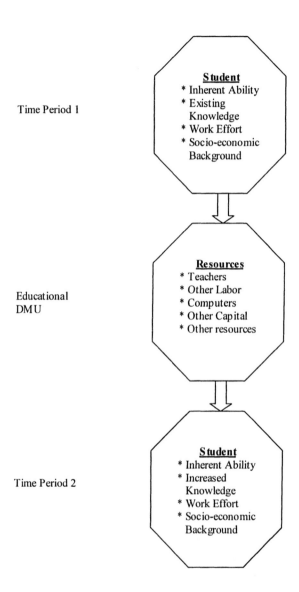

Figure 12-1. Educational Production

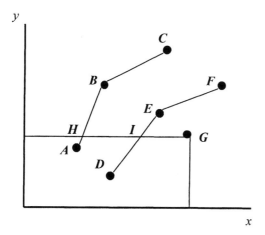

Figure 12-2. Frontier Production with Non-discretionary Inputs

The presence of non-discretionary factors has direct implications for evaluating efficiency. In order to evaluate the efficiency of production, a given DMU must be projected to the appropriate frontier. As shown in Figure 12-2, and remembering the assumption that DMU G faces an environment similar to DMUs $D - F$, a proper projection of G would be to production possibility I and not H. Evaluating G relative to H indicates frontier production if G had a more favorable environment. As a result, information about how the harshness of the environment as it effects G's production is revealed by the distance between I and H.

2.1   Production without Exogenous Inputs

Following Lovell (1993), the production technology transforming inputs $x = (x_1,...,x_M)$ into outputs $y = (y_1,...,y_S)$ for $j = 1,...,n$ firms can be represented with an input set
$$L(y) = \{x : (y,x) \text{ is feasible}\}. \tag{12.1}$$
For every output vector y, $L(y)$ has isoquant
$$IsoqL(y) = \{x : x \in L(y), \theta x \notin L(y), \theta \in [0,1)\} \tag{12.2}$$
and efficient subset
$$Eff L(y) = \{x : x \in L(y), x' \notin L(y), x' \le x\} \tag{12.3}$$
We adopt the notation DMU_i to represent production possibility (y_i, x_i). Technical efficiency can be evaluated relative to the efficiency subset using

a Farrell (1957) input measure of efficiency for production possibility (y_i, x_i), defined as $F(y_i, x_i) = \min \{\theta : \theta x_i \in L(y_i)\}$.

Allowing variable returns to scale, technical efficiency for DMU$_i$ can be evaluated using the Banker, Charnes and Cooper (1984) model:

$$F(y_i, x_i) = \underset{\theta, \lambda_1, \ldots, \lambda_n}{Min} \theta$$

$$\text{s.t.} \sum_{j=1}^{n} \lambda_j y_{sj} \geq y_{si} \quad s = 1, \ldots, S$$

$$\sum_{j=1}^{n} \lambda_j x_{mj} \leq \theta x_{mo} \quad m = 1, \ldots, M \quad (12.4)$$

$$\sum_{j=1}^{n} \lambda_j = 1$$

$$\lambda_j \geq 0 \qquad j = 1, \ldots, n$$

2.2 Production with Exogenous Inputs

The original DEA models assumed that all inputs were discretionary. Assume instead that each decision making unit (DMU) uses a vector x of inputs to produces a vector y of outputs given a vector of non-discretionary inputs $z = (z_1, \ldots, z_R)$. These non-discretionary inputs affect the transformation of discretionary inputs into outputs. For convenience, the vector z is defined so that increases in any component leads to a more favorable environment, *ceteris paribus*. The production technology transforming input vector x into output vector y can be represented by the conditional input set

$$L(y, z) = \{x : (y, x) \text{ is feasible given } z\}. \quad (12.5)$$

For every output vector y, $L(y, z)$ has isoquant

$$IsoqL(y, z) = \{x : x \in L(y, z), \theta x \notin L(y, z), \theta \in [0,1)\} \quad (12.6)$$

and efficient subset

$$Eff L(y, z) = \{x : x \in L(y, z), x' \notin L(y, z), x' \leq x\} \quad (12.7)$$

It is assumed that $L(y,z) \subseteq L(y,z')$ implies that z' is a more favorable environment than z. Given multiple non-discretionary inputs, it is necessary to identify the importance of each exogenous factor in the production process. As we will see later in this chapter, this can be achieved via a three-stage model.

3. DATA ENVELOPMENT ANALYSIS IN EDUCATION

3.1 Measuring Best Practice

A model that could be used to analyze educational production is a one-stage model that incorporates only discretionary inputs and outputs. This approach was adopted by Bessent *et al.* (1983) to evaluate the efficiency of educational programs. In this case, model 12.4 is used to project each school district to the outermost envelope. If non-discretionary factors exist in the production process, the resulting index will capture not only inefficiency, but also the effect that the environment has on production. This can be seen in Figure 12.2 where DMU G is projected to H instead of I. In this context, we will consider best practice to be consistent with efficiency under the most favorable environment possible while recognizing that best practice may not be feasible without improvements in the environment. We turn now to models that attempt to control for the exogenous inputs.

3.2 Models with Exogenous Inputs

Due to the importance of non-discretionary factors in certain production processes, there have been a number of approaches developed to control for these factors. One of the important applications of these models is to education, where empirical studies have found evidence that socio-economic influences are largely responsible for educational outcomes. In this section, we present some of the models. There are other models that have been applied to these situations but current research has not evaluated the approaches. Two of these approaches involve using a handicapping function (see Paradi, Vela, and Yang, 2003 for further discussion) and Muñiz (2002) who applies an alternative three-stage model to assess education in Spain. Comparing the approaches provides a useful direction for future research.

3.2.1   One-Stage Models

The first DEA model to allow continuous exogenous variables was developed by Banker and Morey (1986). Recognizing the inappropriateness of treating fixed factors as discretionary, the authors modified the constraints on the fixed inputs. Assuming an input-orientation with variable returns to scale, the model is presented as follows:

$$BM(y_i, x_i) = \min_{\theta, \lambda_1, ..., \lambda_n} \theta$$

$$\text{s.t.} \sum_{j=1}^{n} \lambda_j y_{sj} \geq y_{si} \quad s = 1, ..., S$$

$$\sum_{j=1}^{n} \lambda_j x_{mj} \leq \theta x_{mo} \quad m = 1, ..., M \tag{12.8}$$

$$\sum_{j=1}^{n} \lambda_j z_{rj} \leq x_{mo} \quad r = 1, ..., R$$

$$\sum_{j=1}^{n} \lambda^j = 1$$

$$\lambda_j \geq 0 \qquad j = 1, ..., n$$

Model (12.8) is similar to the standard DEA model (12.4) with one major difference; an additional constraint is included for each non-discretionary input. These additional constraints are similar to the constraints for the discretionary inputs except the efficiency component is not included on the right-hand side. The important insight of the Banker and Morey model is to define efficiency with respect to the discretionary inputs only. Effectively, these constraints force the referent "frontier" production possibility to have a level of each non-discretionary factor that is less than or equal to the DMU under analysis. Ruggiero (1996a) shows however that this formulation leads to referent points that are not feasible.

Suppose that there are three DMUs *A*, *B* and *C* that produce one output *y* using one discretionary input *x* with a non-discretionary input *z*. We will assume that all DMUs are efficient with production data shown in Table 12-1.

Table 12-1. Efficient Production with Non-discretionary Inputs

DMU	*y*	*x*	*z*
A	10	10	20
B	10	8	30
C	10	1	40

The production data revealed in Table 12-1 are consistent with efficient production. All DMUs produce the same output but use different amounts of the discretionary factor depending on the level of the non-discretionary factor. DMUs A and B have to use more discretionary inputs to compensate for the relatively harsh environment each faces. Applying model 12.8 under constant returns to scale we have:

$$BM(y_B, x_B) = \underset{\theta, \lambda_A, \lambda_B, \lambda_C}{Min} \; \theta$$
$$\text{s.t. } 10\lambda_A + 10\lambda_B + 10\lambda_C \geq 10$$
$$10\lambda_A + 8\lambda_B + 1\lambda_C \leq 10\theta$$
$$20\lambda_A + 30\lambda_B + 40\lambda_C \leq 30$$
$$\lambda_A, \lambda_B, \lambda_C \geq 0$$

with solution $\lambda_A^* = \lambda_B^* = 0.5, \lambda_C^* = 0 \; and \; \theta^* = 0.6875.$ Since the sum of the weights is unity, the results are the same without or without the convexity constraint. According to the Banker and Morey model, DMU B could have produced the same output using 5 units of the discretionary input. This contradicts the assumption that all were operating efficiently.

Ruggiero (1996) provided a one-stage model to correct this problem by excluding all DMUs with a more favorable environment from the analysis of each DMU. This is achieved with the following linear programming model:

$$R(y_i, x_i) = \underset{\theta, \lambda_1, ..., \lambda_n}{Min} \; \theta$$
$$\text{s.t. } \sum_{j=1}^{n} \lambda_j y_{sj} \geq y_{si} \quad s = 1, ..., S$$
$$\sum_{j=1}^{n} \lambda_j x_{mj} \leq \theta x_{mi} \quad m = 1, ..., M \tag{12.9}$$
$$\lambda_j = 0 \quad if \; z_{rj} > z_{ri} \quad r = 1, ..., R$$
$$\sum_{j=1}^{n} \lambda_j = 1$$
$$\lambda_j \geq 0 \qquad j = 1, ..., n$$

For DMU B, the following model results:

$$R(y_B, x_B) = \underset{\theta, \lambda_A, \lambda_B, \lambda_C}{Min} \theta$$

$$\text{s.t. } 10\lambda_A + 10\lambda_B + 10\lambda_C \geq 10$$

$$10\lambda_A + 8\lambda_B + 1\lambda_C \leq 10\theta$$

$$\lambda_A, \lambda_B \geq 0$$

$$\lambda_C = 0$$

with solution $\lambda_A^* = \lambda_C^* = 0, \lambda_B^* = 0 \text{ and } \theta^* = 1$. This model provides the correct measure of efficiency in this case.

Similar to the Banker and Morey (1986) model, it is necessary to specify the continuous non-discretionary inputs and the direction of influence. In some cases, this may not be possible. A potentially more serious problem is the inability to handle many non-discretionary factors. As the number of non-discretionary inputs increases, the probability of overestimating efficiency increases. As a result, inefficient DMUs could be identified as efficient by default. This model does not recognize tradeoffs that exist between the non-discretionary variables; a given DMU under analysis could have a favorable environment because it has favorable levels of most non-discretionary factors but have a limited referent set only because it has an unfavorable level of at least one non-discretionary input. In this case, the one-stage models are insufficient.

3.2.2   Two-Stage Models

The Banker and Morey and Ruggiero models discussed include the non-discretionary variables directly into the DEA model. The Banker and Morey model, in general, does not generate comparisons to the efficient frontier and the Ruggiero model breaks down as the number of non-discretionary inputs increases for a given number of observations and discretionary variables. Ray (1991) provided an alternative two-stage model that overcomes these weaknesses. See also Ray (1988). In the first-stage, DEA model 12.4 is performed using outputs and discretionary inputs. As mentioned above, the resulting index of best practice does not measure efficiency because the index also captures the non-discretionary factors.

After obtaining the first-stage index using 12.4, Ray (1991) used a second stage OLS model that regressed the first-stage index on the non-discretionary factors. The second-stage regression is specified as:

$$F = \alpha + \beta_1 z_1 + ... + \beta_R z_R + \varepsilon. \tag{12.10}$$

Based on the regression results, Ray's two-stage measure of efficiency can be calculated as:

$$RAY = F - \alpha - \beta_1 z_1 - ... - \beta_R z_R, \qquad (12.11)$$

which will have mean zero. This is easily remedied by adjusting the intercept similar in a similar way as used in the Corrected OLS models.

The distortions introduced in the original DEA model from the exclusion of the non-discretionary variables are factored out and the primary advantage is computational. The two-stage approach requires solving the original DEA model once. This introduces considerable flexibility and allows sensitivity analysis in the second stage. As a result, different sets of non-discretionary inputs can be tested. The added flexibility also overcomes weaknesses in model 12.9. If there are a large number of non-discretionary variables, the regression analysis provides the means to explicitly weight the contribution each non-discretionary factor has on the first-stage estimate.

There are some potential drawbacks to this approach. Firstly, the upper limit on the efficiency index obtained in the first-stage is 1, leading to a censored sample. OLS produces biased estimates. A solution to this potential problem is to use Tobit. McCarty and Yaisawarng (1993) use a second-stage Tobit model in an efficiency analysis of New Jersey school districts. Secondly, the model requires *a priori* specification of functional form. One could use a flexible functional form to address this problem. As will be revealed below, linear and log-linear results produced similar results using simulated data. Thirdly, the model can overestimate inefficiency because of the adjustments. Finally, if the non-discretionary variables are correlated with efficiency, endogeneity results. The latter case was addressed in Ruggiero (2003).

3.2.3 Three-Stage Models

An alternative approach involves extending Ray's two-stage model using a third stage. The first two stages are identical to the two-stage model. However, rather than using the error term from the regression to measure efficiency, the parameters from the regression are used to construct an index for the non-discretionary factors. In particular, the regression equation provides the means to weight the influence that each non-discretionary factor has on the first-stage scores. From (12.10), we consider the following index z:

$$z = \sum_{i=1}^{R} B_i z_i . \tag{12.12}$$

Given this index, we can return to model 12.9 and measure efficiency in the third stage:

$$R3S(y_i, x_i) = \underset{\theta, \lambda_1, \dots, \lambda_n}{Min} \theta$$

$$\text{s.t. } \sum_{j=1}^{n} \lambda_j y_{sj} \geq y_{si} \quad s = 1, \dots, S$$

$$\sum_{j=1}^{n} \lambda_j x_{mj} \leq \theta x_{mi} \quad m = 1, \dots, M \tag{12.13}$$

$$\lambda_j = 0 \quad if \ z_j > z_i$$

$$\sum_{j=1}^{n} \lambda^j = 1$$

$$\lambda_j \geq 0 \quad\quad\quad j = 1, \dots, n$$

There are a few advantages to using this model. Firstly, assuming that the second-stage regression produces unbiased estimates of the parameter weights, the model maintains the desirable properties of (12.9). Secondly, like the two-stage model, the identified weakness in (12.9) of identifying DMUs as efficient by default is corrected. Further, this new model (unlike Ray's models) maintains other desirable properties of the original DEA models. In particular, this model controls for non-discretionary factors and uncovers the efficient referent set. As a result, the nature of scale economies can be revealed and benchmarking capabilities are not lost.

3.3   Simulating Production with Exogenous Inputs

3.3.1   Simulation Design

Ruggiero (1998) provided an analysis of the models discussed above using simulated data. This simulation and results are presented in this section to facilitate comparisons between the various approaches. We assume that there are 200 DMUs that produce one output y using two discretionary inputs x_1 and x_2 in the presence of non-discretionary inputs z_1, z_2 and z_3. Three production models are considered depending on the number of non-discretionary factors. This effectively shows problems that arise in the one-stage models and the importance of the multiple-stage models. It is further assumed that constant returns to scale prevail with respect to the discretionary inputs. Importantly, these variables are not discretionary even in the long run. Consequently, scale cannot be defined relative to these exogenous variables as is normally the case with fixed factors becoming variable in the long run. The three models we consider are:

$$Model\ 1: \quad y = z_1 x_1^{0.5} x_2^{0.5};$$

$$Model\ 2: \quad y = z_1 z_2^{0.75} x_1^{0.5} x_2^{0.5}; \quad and$$

$$Model\ 3: \quad y = z_1 z_2^{0.75} z_3^{0.5} x_1^{0.5} x_2^{0.5}.$$

All inputs are randomly generated from a uniform distribution according to the following intervals:

$$x_i : (30, 50) \quad for\ i = 1,\ 2;\ and$$

$$z_j : (1, 2) \quad for\ j = 1,\ 2,\ 3.$$

With the generated inputs, observed output is calculated for models (1) – (3). Efficiency is calculated as $\gamma = \exp(-|u|) \le 1$, where u is normally distributed with mean 0 and standard deviation 0.3. For the first 25 observations the value of γ is set to unity to ensure frontier producers. Observed input usage was obtained by scaling efficient input usage up by the reciprocal of γ and only constant returns to scale models are considered.

3.3.2   Simulation Results

The results of the simulation are reported in Table 12-2. Four criteria were selected. The primary measures used are the correlation and rank correlation coefficient between the true and measured efficiency. A higher (rank) correlation coefficient suggests better performance in measuring efficiency. Also considered is the mean absolute deviation (MAD) between true and measured efficiency and the percent of the DMUs for which measured efficiency was less than the true efficiency. A lower MAD for a given measure indicates that the measure is closer in proximity to the true measure. Finally, the percent of DMUs for which measured efficiency is less than the true measure indicates the extent to which inefficiency is overstated. In the absence of non-discretionary factors, one purported advantage of the original DEA model was the generation of a lower bound on efficiency; as a result, the DEA measure did not overstate inefficiency.

Table 12-2. Efficient Production with Non-discretionary Inputs

Model	Efficiency Measure	Correlation with True Efficiency*	MAD**	Measured Efficiency < True Efficiency (% of firms)
1	F	0.65 (0.63)	0.16	83.5
	BM	0.64 (0.63)	0.16	79.5
	R	0.94 (0.92)	0.04	0.0
	RAY			
	Linear	0.97 (0.98)	0.05	87.5
	Log-Linear	0.99 (0.98)	0.04	94.5
	R3S			
	Linear	0.94 (0.92)	0.04	0.0
	Log-Linear	0.94 (0.92)	0.04	0.0
2	F	0.56 (0.56)	0.23	93.0
	BM	0.49 (0.22)	0.22	84.5
	R	0.84 (0.80)	0.09	0.0
	RAY			
	Linear	0.92 (0.92)	0.11	95.5
	Log-Linear	0.95 (0.94)	0.10	97.0
	R3S			
	Linear	0.93 (0.91)	0.04	9.5
	Log-Linear	0.93 (0.91)	0.04	4.0
3	F	0.53 (0.54)	0.25	91.5
	BM	0.47 (0.25)	0.25	88.0
	R	0.68 (0.63)	0.13	0.0
	RAY			
	Linear	0.94 (0.93)	0.07	89.0
	Log-Linear	0.97 (0.96)	0.08	96.5
	R3S			
	Linear	0.93 (0.91)	0.04	10.0
	Log-Linear	0.93 (0.91)	0.04	8.5

*Pearson's correlation coefficient is reported; Spearman's rank correlation coefficient is shown in parentheses. **MAD is the mean absolute deviation between measured and true efficiency.

Some interesting but expected results emerge from the simulation analysis. The original DEA measure (F) does poorly in the presence of non-discretionary inputs. With only one non-discretionary input, the correlation (rank correlation) between F and the true level of efficiency is only 0.65 (0.63). Also, the MAD is a relatively high 0.16. As expected, the performance declines as the number of non-discretionary factors increases. The rank correlation decreases to 0.54 when there are three non-discretionary inputs while the MAD increases to 0.25. In addition, the percent of DMUs where inefficiency is overstated varies from 83.5 to over 90 with multiple non-discretionary inputs. This result confirms Banker and Morey's important insight that non-discretionary inputs must be controlled for before efficiency can be measured.

The Banker and Morey model performs worse than the original DEA model. The rank correlation between *BM* and the true level of efficiency varies from 0.63 with one non-discretionary factor to under 0.25 with multiple environmental factors. In addition, the MAD increases from 0.16 to 0.25 and the percent of DMUs for which inefficiency is overstated varies from 79.5 to 88.

Ruggiero's measure (*R*) performs well when there is only one non-discretionary input. However, as the number of non-discretionary factors increases, the performance declines dramatically. The correlation (rank correlation) coefficient decreases from 0.94 (0.92) to 0.68 (0.63) and the MAD increases from 0.04 to 0.13. Notably, this measure does not overstate efficiency. These results confirm the discussion above: as the number of non-discretionary inputs increase, *R* will tend to 1.0 as the measured efficiency of DMUs increases by default. Interestingly, however, *R* outperforms *F* and *BM* in this analysis.

The *RAY* measure performs better than all other models based on the correlation and rank correlation criteria. This holds true for both the linear and log-linear second-stage regression specifications. The minimum correlation and rank correlation coefficient achieved by *RAY* was 0.92. Unlike all of the other models discussed, the performance of *RAY* did not decrease as the number of non-discretionary inputs increased. The MAD achieved across specifications was relatively low, varying from 0.04 to 0.11. Of note, however, is the percent of DMUs for which inefficiency was overstated. In all situations, about 90 percent of the efficiency DMUs was under the true level; in fact, *RAY* identifies almost all efficient DMUs as inefficient. Hence, while *RAY* generates an efficiency index that is highly correlated with the true level of efficiency, it does not properly distinguish between efficient and inefficient DMUs. Further, it overstates inefficiency for about 90 percent of the DMUs.

Finally, the three-stage model (*R3S*) performs relatively well. Like the *RAY* measures, the *R3S* measure achieved high correlation and rank correlation coefficients with the true level of efficiency. The correlation (rank correlation) was approximately 0.935 (0.915) in all model situations. Hence, as the number of non-discretionary inputs increased, the performance of *R3S* did not decline. Thus, as discussed above, *R3S* overcomes the weakness of overstating efficiency inherent in the *R* measure. Further, the MAD was the lowest for the *R3S* measure in all model situations. Hence, by the MAD criterion, the new measure performed the best. Also, the new measure over-measured inefficiency in the fewest number of cases and identified all efficient DMUs as efficient.

3.4   Other Issues

There are other issues that must be addressed to provide more reliable efficiency estimates. In addition to the presence of non-discretionary variables, the standard problems with DEA arise. In particular, the Farrell measure doesn't account for excess slack that exists after weak efficiency is achieved. Also, the problem of measurement error is present. The cross-sectional stochastic frontier model does not improve efficiency estimation in the presence of statistical noise and cannot be relied upon. For a further discussion, see Ondrich and Ruggiero (2001). Current theoretical research is bridging the gap with regard to the problems with efficiency measurement; however, there are other concerns as it relates to educational production.

The education process as envisioned in Figure 12-1 is student based; the education production function is modeled at the micro level using students as in Hanushek (1979). With few exceptions (see Thanassoulis, 1999 and Silva Portela and Thanassoulis, 2001) studies use aggregated data. In the United States, data are readily available at the school or school district level. With aggregation, however, information is lost and can lead to average results that are not reliable at the aggregated level. A potentially important research topic would be to analyze how aggregation effects efficiency estimation as it pertains to education.

More important is the selection of outcomes. As Hanushek (1979) points out, most production analyses measure output using standardized test scores. The use of test scores is appealing but does not necessarily reflect change in quality of students. Notably, these tests lack external validation. Hanushek (1986) reports that the evidence that test scores relate to future achievement is inconclusive. Regardless, standardized test performance is valued by parents and is the focus of state policies seeking to establish accountability. For useful discussions, see Hanushek (1979) and Hanushek (1986).

4.   APPLICATION TO OHIO SCHOOL DISTRICTS

For illustrative purposes, we include an efficiency analysis of 607 school districts in Ohio School Districts for school year 2000 using Data Envelopment Analysis. This application was chosen given the recent court decision that found that the system of financing education in Ohio was unconstitutional. Recently, school districts have been classified according to the number of standards met from a list of 27 outcome indicators. The outcomes consist of pass rates on standardized tests, the attendance rate and the graduation rate. Districts meeting at least 26 of the 27 standards are considered excellent. The other categories are effective (21 to 25 standards met), continuous improvement (13 to 20), academic watch (8 to 12), and academic emergency (less than 8 standards met.) Report cards are issued to facilitate comparisons with other school districts. However, the state does not currently identify the efficiency of these school districts. As a result, the classification could arise from inefficiency, low resource usage, or from a harsh socio-economic environment.

We include five outcomes: the percent of students passing math and reading tests in fourth and sixth grades and the graduation rate. Correlation analysis revealed that these five measures are highly correlated with the other measures. We also include a median district income to control for the socio-economic environment. Finally, we employ a cost model discussed in Ruggiero (1999) and use expenditures per pupil as the input variable. The results of the analysis are summarized in Table 12-3.

Interesting results emerge from the analysis. On average, the "Excellent Districts" tend to produce higher outcomes but do so by spending over $8,000 per pupil. In addition, these top rated schools tend to have a higher income level, representing higher costs due to higher resource prices. Even though they have been selected as excellent, there still tends to be a high level of inefficiency; on average, the excellent districts waste over $1,400 per pupil. The stylized facts suggest that any reform using these Ohio benchmarks could be bad policy. Inefficient districts should not be used as benchmarks; and the state needs to take care in not comparing other districts to some of these wealthier districts.

The districts that are identified as "Academic Emergency" are, on average, the most inefficient, spending in excess of $2,400 per pupil. Outcome-based reforms in education typically include additional spending. However, as this analysis shows, many of the poor districts suffer from mis-management. Additional spending financed through higher taxes is clearly not the answer because there is no guarantee that additional revenue will be spent to increase outcomes. Future research should consider the important question of how to design a finance system to encourage efficient behavior.

Table 12-3. Efficiency in Ohio School Districts

District Type	Variable	Average	Std. Dev.
Academic **Emergency** (N = 12)	Expenditures ($/pupil)	$8,410	$1,272
	Minimum Cost	$5,988	$416
	Inefficiency	$2,423	$1,217
	Math 4	56.49	18.45
	Reading 4	56.26	16.72
	Math 6	59.41	18.19
	Reading 6	59.13	15.74
	Graduation Rate	83.46	8.98
	FY95 Income ($)	$24,127	$6,107
Academic **Watch** (N = 38)	Expenditures ($/pupil)	$7,560	$1,200
	Minimum Cost	$5,847	$421
	Inefficiency	$1,713	$1,348
	Math 4	55.76	18.47
	Reading 4	52.71	15.90
	Math 6	59.27	19.41
	Reading 6	56.78	17.97
	Graduation Rate	80.44	14.46
	FY95 Income ($)	$23,759	$4,673
Continuous **Improvement** (N = 350)	Expenditures ($/pupil)	$6,896	$921
	Minimum Cost	$5,723	$461
	Inefficiency	$1,173	$999
	Math 4	65.11	14.17
	Reading 4	61.15	12.68
	Math 6	66.10	14.69
	Reading 6	63.82	13.17
	Graduation Rate	87.22	8.58
	FY95 Income ($)	$25,386	$4,979
Effective (N = 136)	Expenditures ($/pupil)	$7,112	$1,074
	Minimum Cost	$5,891	$474
	Inefficiency	$1,221	$1,166
	Math 4	65.73	15.84
	Reading 4	60.49	12.83
	Math 6	66.60	14.67
	Reading 6	62.86	13.06
	Graduation Rate	89.23	8.04
	FY95 Income ($)	$24,842	$3,664
Excellent (N = 71)	Expenditures ($/pupil)	$8,024	$2,073
	Minimum Cost	$6,582	$1,081
	Inefficiency	$1,442	$2,067
	Math 4	67.88	11.80
	Reading 4	63.76	12.15
	Math 6	69.30	11.65
	Reading 6	66.31	10.13
	Graduation Rate	89.86	6.00
	FY95 Income ($)	$26,011	$4,965

5.   POLICY IMPLICATIONS/CONCLUSIONS

School finance equity remains at the forefront of educational reform more than 25 years after the *Serrano v. Priest* (1971) decision by the California Supreme Court sparked intense debate. The court challenges which commenced with this case continue to the present day. A new round of court cases has broadened traditional equity standards in calling for remedies that go beyond equalization of per-pupil spending (e.g., Ohio, Kentucky and Texas) to standards for equity based on adequacy of resources and outcomes; see Reschovsky (1994). Most of the school finance debate has focused on differences in fiscal capacities of school districts and the efficacy of aid formulas intended to compensate (i.e., "equalize") for low-capacity. Equity across districts and hence, the effectiveness of compensatory aid generally is measured in relation to expenditure per pupil This standard, however, is severely flawed because it omits differences in district costs and efficiency, both of which have major impacts on the levels of per pupil educational services provided across districts and on the educational outcomes that result.

6.   POLICY IMPLICATIONS/CONCLUSIONS

School finance equity remains at the forefront of educational reform more than 25 years after the *Serrano v. Priest* (1971) decision by the California Supreme Court sparked intense debate. The court challenges which commenced with this case continue to the present day. A new round of court cases has broadened traditional equity standards in calling for remedies that go beyond equalization of per-pupil spending (e.g., Ohio, Kentucky and Texas) to standards for equity based on adequacy of resources and outcomes; see Reschovsky (1994). Most of the school finance debate has focused on differences in fiscal capacities of school districts and the efficacy of aid formulas intended to compensate (i.e., "equalize") for low-capacity. Equity across districts and hence, the effectiveness of compensatory aid generally is measured in relation to expenditure per pupil This standard, however, is severely flawed because it omits differences in district costs and efficiency, both of which have major impacts on the levels of per pupil educational services provided across districts and on the educational outcomes that result.

Even if fully adjusted for cost differences, however, expenditure will not be indicative of output and equity so long as there is variation in the efficiency of service provision. Studies of educational production and costs have found little systematic relationship between inputs or expenditures and outcomes, suggesting that school districts vary in the efficiency of service provision; see Hanushek (1993). With such variation, two districts will not provide the same level of student outcomes even if they spend equivalent amounts per pupil and face identical cost environments. The conclusion is clear: cost-adjusted spending remains a distorted measure of output and hence, of equity unless, ultimately, it is adjusted for districts' relative efficiency. Ruggiero, Miner and Blanchard (2002) found that nearly one-half of the perceived inequity that existed in New York State was attributable to the cost environment arising from non-discretionary inputs and the efficiency of production. Serious reform must account for inefficiency; the operations research and economics literature can contribute to the debate with advancements in Data Envelopment Analysis.

In addition to the question of reform, evaluating the performance of educational DMUs is extremely important. True to the spirit of the Niskanen models of bureaucratic supply (see Niskanen, 1971 and Niskanen, 1975) there is pressure on local governments to increase spending in spite of the overwhelming evidence of inefficiency in educational production. As a result, taxpayers have to bear a heavy burden to finance the waste that exists in local governments. Recently, it was suggested that Ohio would need to increase spending by $1 billion to overhaul the current system. However, as discussed in Ruggiero (2001) the estimate was derived assuming all districts were efficient. Application of DEA showed that there exists over $800 million in inefficiency. Hence, the proposed increase in resources purportedly needed to overhaul the system did not need to be passed on to the taxpayers.

REFERENCES

1. Banker, R.D., Charnes, A., and Cooper, W.W., 1984, Some Models for Estimating Technical and Scale Inefficiencies in Data Envelopment Analysis, *Management Science* 30, 1078-1092.
2. Banker, R.D., and Morey, R., 1986, Efficiency Analysis for Exogenously Fixed Inputs and Outputs, Operations Research, 34, 513-521.
3. Bessent, A., Bessent, E.W., Charnes, A., Cooper, W.W., and Thorogood, N.C. (1983), Evaluation of Educational Program Proposals by Means of DEA, Educational Administration Quarterly, 19, 82-107.
4. Bessent, A., Bessent, E.W., Kennington, J. and Reagan, B., 1982, An Application of Mathematical Programming to Assess Productivity in the Houston Independent School District, Management Science, 28, 1335-1366.

5. Bradford, D., Malt, R. and Oates, W., 1969, The Rising Cost of Local Public Services: Some Evidence and Reflections, National Tax Journal 22, 185-202.
6. Bradley, S., Johnes, G. and Millington, J., 2001, The Effect of Competition on the Efficiency of Secondary Schools in England, European Journal of Operational Research, 135, 545-568.
7. Bridge, R., Judd, C., and Moock, P., 1979, The Determinants of Educational Outcomes, Ballinger Publishing, Cambridge, MA.+
8. Chalos, P., and Cherian, J., 1995, An Application of Data Envelopment Analysis to Public Sector Performance Measurement and Accountability, Journal of Accounting and Public Policy, 14, 143-160.
9. Charnes, A., Cooper, W.W., and Rhodes, E., 1978, Measuring the Efficiency of Decision Making Units, European Journal of Operational Research, 2, 429-444.
10. Charnes, A., Cooper, W.W., and Rhodes, E., 1981, Evaluating Program and Managerial Efficiency: An Application of Data Envelopment Analysis to Program Follow Through, Management Science, 27, 668-697.
11. Chubb, J., and Moe, T., 1990, Politics, Markets and America's Schools, Brookings Institute, Washington, DC.
12. Clotfelter, C., and Ladd, H., 1996, Recognizing and Rewarding Success in Public Schools, in Ladd, H. (Ed.), Holding Schools Accountable: Performance Based Reform in Education, Brookings Institute, Washington, DC, 23-64.
13. Cohn, E. and Geske, T., 1990, The Economics of Education, Pergamon Press, Oxford, England.
14. Duncombe, W., Miner, J., and J. Ruggiero, 1997, Empirical Evaluation of Bureaucratic Models of Inefficiency, Public Choice, 93, 1-18.
15. Duncombe, W., Ruggiero, J., and Yinger, J., 1996, Alternative Approaches to Measuring the Cost of Education, in Ladd, H. (Ed.), Holding Schools Accountable: Performance Based Reform in Education, Brookings Institute, Washington, DC, 327-356.
16. Färe, R., Grosskopf, S., and Weber, W., 1989, Measuring School District Performance, Public Finance Quarterly, 17, 409-428.
17. Farrell, M.J., 1957, The Measurement of Productive Efficiency, Journal of the Royal Statistical Society, Series A, 120, 253-281.
18. Farren, D., 2002, The Technical Efficiency of Schools in Chile, Applied Economics, 34, 1533-1542.
19. Fukuyama, H., and Weber, W., 2002, Evaluating Public School District Performance via DEA Gain Functions, Journal of the Operational Research Society, 53, 992-1003.

20. Grosskopf, S., Hayes, K., Taylor, L., and Weber, W., 1999, Anticipating the Consequences of School Reform, A New Use of DEA, Management Science, 45, 608-620.
21. Grosskopf, S. and Moutray, C., 2001, Evaluating Performance in Chicago Public High Schools in the Wake of Decentralization, Economics of Education Review, 20, 1-14.
22. Grosskopf, S. and Yaisawarng, S., 1990, Economies of Scope in the Provision of Local Public Services, National Tax Journal, 43, 61-74.
23. Hanushek, E., 1979, Conceptual and Empirical Issues in the Estimation of Educational Production Functions, Journal of Human Resources, 14, 351-388.
24. Hanushek, E., 1981, Throwing Money at Schools, Journal of Policy Analysis and Management, 1, 19-41.
25. Hanushek, E., 1986, The Economics of Schooling, Production and Efficiency in Public Schools, Journal of Economic Literature, 24, 1141-1177.
26. Hanushek, E., 1993, Can Equity be Separated from Efficiency in School Finance Debates, in Emily, H. (Ed.), Essays on the Economics of Education, Upjohn Institute, 35-74.

27. Leibenstein, H., 1966, Allocative Efficiency vs. X-Efficiency, American Economic Review, 56, 392-415.
28. Leibenstein, H., 1978, On the Basic Propositions of X-Efficiency Theory, American Economic Review, 68, 328-332.
29. Lovell, C.A.K., 1993, Production frontiers and productive efficiency, in H.O. Fried, C.A.K. Lovell and S.S. Schmidt (eds.), *The Measurement of Productive Efficiency*, Oxford University Press, New York, 3-67.
30. McCarty, T. and Yaisawarng, S., 1993, Technical Efficiency in New Jersey School Districts, in H.O. Fried, C.A.K. Lovell and S.S. Schmidt (eds.), *The Measurement of Productive Efficiency*, Oxford University Press, New York, 271-287.
31. McMillan, M., and Datta, D., 1998, The Relative Efficiencies of Canadian Universities: A DEA Perspective, 24, 485-513.
32. Muñiz , M.A., 2002, Separating Managerial Inefficiency and External Conditions in Data Envelopment Analysis, European Journal of Operational Research, 143, 625-643.
33. National Commission on Excellence in Education, 1983, A Nation at Risk, Washington, DC.
34. Niskanen, W.A., 1971, Bureaucracy and Representative Government, Aldine-Atherton, Chicago.
35. Niskanen, W.A., 1975, Bureaucrats and Politicians, The Journal of Law and Economics, 18, 617-643.

36. Odden, A., and Picus, L., 1992, School Finance: A Policy Perspective, McGraw-Hill, New York.
37. Ondrich, J. and Ruggiero, J., 2001, Efficiency Measurement in the Stochastic Frontier Model, European Journal of Operational Research, 129, 434-442.
38. Paradi, J., Vela, S., and Yang, Z., 2003, Assessing Bank and Bank Branch Performance: Modeling Considerations and Approaches, This Volume.
39. Ray, S., 1988, Data Envelopment Analysis, Nondiscretionary Inputs and Efficiency: An Alternative Interpretation, Socio-Economic Planning Sciences, 22, 167-176.
40. Ray, S., 1991, Resource-Use Efficiency in Public Schools: A Study of Connecticut Data, Management Science, 37, 1620-1628.
41. Reschovsky, A., 1994, Fiscal Equalization and School Finance, National Tax Journal, 47, 211-224.
42. Ruggiero, J., 1996a, On the Measurement of Technical Efficiency in the Public Sector, European Journal of Operational Research, 90, 553-565.
43. Ruggiero, J., 1996b, Measuring Technical Efficiency in the Public Sector: An Analysis of Educational Production, Review of Economics and Statistics, 78, 499-509.
44. Ruggiero, J., 1998, Non-discretionary Inputs in Data Envelopment Analysis, European Journal of Operational Research, 111, 461-468.
45. Ruggiero, J., 1999, Nonparametric Analysis of Educational Costs, European Journal of Operational Research, 119, 605-612.
46. Ruggiero, J., 2000, Nonparametric Estimation of Returns to Scale in the Public Sector with an Application to the Provision of Educational Services, Journal of the Operational Research Society, 51, 906-917.
47. Ruggiero, J., 2001, Determining the Base Cost of Education: An Analysis of Ohio School Districts, Contemporary Economic Policy, 19, 268-279.
48. Ruggiero, J., 2003, Performance Evaluation When Non-Discretionary Inputs Correlate With Technical Efficiency, European Journal of Operational Research (forthcoming.)
49. Ruggiero, J., Miner, J. and Blanchard, L., 2002, Equity vs. Efficiency in the Provision of Educational Services, European Journal of Operational Research 142, 642-652.
50. Ruggiero, J., and Vitaliano, D., 1999, Assessing the Efficiency of Public Schools Using Data Envelopment Analysis and Frontier Regression, Contemporary Economic Policy, 17, 321-331.
51. Sarrico, C.S., and Dyson, R., 2000, Using DEA for planning in UK Universities – An Institutional Perspective, Journal of the Operational Research Society, 51, 789-790.

52. Serrano V. Priest, 1971, 487 P.2d 1241.
53. Sexton, T., Sleeper, S., and Taggart, R., 1994, Improving Pupil Transportation in North Carolina, Interfaces, 24, 87-103.
54. Silva Portela, M.C., and Thanassoulis, E., 2001, Decomposing School and School-Type Efficiency, European Journal of Operational Research, 132, 357-373.
55. Sinuany-Stern, Z., and Mehrez, A., 1994, Academic Departments Efficiency via DEA, Computers and Operations Research, 21, 543-556.
56. Smith, P., and Mayston, D., 1987, Measuring Efficiency in the Public Sector, Omega: International Journal of Management Science, 15, 181-189.
57. Thanassoulis, E., 1999, Setting Achievement Targets for School Children, Education Economics, 7, 101-118.
58. Thanassoulis, E., and Dunstan, P., 1994, Guiding Schools to Improved Performance Using Data Envelopment Analysis: An Illustration with Data from a Local Education Authority, Journal of the Operational Research Society, 45, 1247-1262.
59. Tompkins, C., and Green, R., 1988, An Experiment in the Use of Data Envelopment Analysis for Evaluating the Efficiency of UK University Departments of Accounting, Financial Accountability & Management, 4, 147-164.
60. Wyckoff, P., 1990, The Simple Analytics of Slack-Maximizing Bureaucracy, Public Choice, 67, 35-67.

Acknowledgements
The author would like to extend his sincere gratitude to Joe Zhu for providing helpful comments throughout the process and to William W. Cooper for providing a careful reading of the manuscript and numerous suggestions.

Part of the material in this chapter is adapted from Ruggiero, J., 1996, On the Measurement of Technical Efficiency in the Public Sector, European Journal of Operational Research 90, 553-565, Ruggiero, J., 1998, Non-discretionary Inputs in Data Envelopment Analysis, European Journal of Operational Research 111, 461-469 and Ruggiero, J., 1999, Nonparametric Analysis of Educational Costs, European Journal of Operational Research, 119, 605-612, with permission from Elsevier Science.

Appendix Selected DEA Education Applications

Author(s)	Sample
Bessent, Bessent, Charnes, Cooper, and Thorogood, 1983	San Antonio College Occupational-Technical Programs
Bessent, Bessent, Kennington and Reagan, 1982	Elementary Schools, Houston Independent School District
Bradley, Johnes and Millington, 2001	British Secondary Schools
Chalos and Cherian, 1995	Illinois School Districts
Charnes, Cooper, Rhodes 1978, 1981	Program Follow Through
Duncombe, Miner and Ruggiero, 1997	New York School Districts
Duncombe, Ruggiero and Yinger, 1996	New York School Districts
Färe, Grosskopf and Weber, 1989	Missouri School Districts
Farren, 2002	Chilean Schools
Fukuyama and Weber, 2002	Texas Public School Districts
Grosskopf and Moutray, 2001	Chicago High Schools
McCarty and Yaisawarng, 1993	New Jersey School Districts
McMillan and Datta, 1998	Canadian Universities
Muñiz, 2002	Spanish Public High Schools
Ray, 1991	Connecticut High Schools
Ruggiero, 1996a, 1996b, 1999, 2000	New York School Districts
Ruggiero, 2001	Ohio School Districts
Ruggiero, Miner, and Blanchard, 2002	New York School Districts
Ruggiero and Vitaliano, 1999	New York School Districts
Sarrico and Dyson, 2000	UK Universities
Sexton, Sleeper and Taggart, 1994	North Carolina Pupil Transportation
Silva Portela and Thanassoulis, 2001	British Secondary School Pupils
Sinuany-Stern and Mehrez, 1994	Ben-Gurion University Academic Departments

Smith and Mayston, 1987	Local Education Authorities, London Boroughs
Thanassoulis, 1999	British Pupils
Thanassoulis and Dunstan, 1994	Great Britain Schools
Tomkins and Green, 1988	UK University Accounting Departments

Chapter 13

ASSESSING BANK AND BANK BRANCH PERFORMANCE
Modeling Considerations and Approaches

Joseph C. Paradi[1], Sandra Vela[2] and Zijiang Yang[3]
[1]Centre for Management of Technology and Entrepreneurship, University of Toronto, Toronto, Ontario, M5S 3E5, Canada.

[2]TD Canada Trust, 55 King Street, TD Tower, 18th floor, Toronto, Ontario, M5K 1A2, Canada

[3]Department of Mathematics and Statistics, York University, 4700 Keele Street, Toronto, ON M3J 1P3, Canada

Abstract: The Banking industry has been the object of DEA analyses by a significant number of researchers and probably is the most heavily studied of all business sectors. The Financial Services Industry is a great subject because both the need by management is there (due to competition) and the data is available in great detail affording the analyst much scope for study. Moreover, there are interesting challenges presented because there are many influences on this industry that must be accounted for in the models. Many new theoretical innovations in DEA were spawned by the need that arose because of the unusual situation at hand.

Key words: Banking performance, DEA models, Performance, Efficiency

1.   INTRODUCTION

As the principal sources of financial intermediation and means of making payments, banks play a vital role in the economic development and growth of a country. In addition to their large economic significance, the existence of a very competitive market structure highlights the importance of evaluating the banks' performance and monitoring their financial condition. There are many uses for performance analyses by bank regulators who need to determine how the industry will respond to the introduction of new regulations, non-traditional entrants and distribution systems, world-wide competition and thereafter, in directing future government policies. Equally important use of these methods is made by bank management who is concerned about the impact of ongoing structural changes on bank operations, and their ability to realign their businesses with the current and most profitable trends. The challenge still remains in selecting the methodology most suitable for these types of assessments.

In a rapidly changing world with an increasingly global business environment, continuous improvement is vital for any successful organization. Historically, banks have been in the forefront of trying to improve their operating efficiency, but typically have not had at their disposal a sound performance analysis system to evaluate their branch networks. Capturing the essential aspects of the process, such a system should yield a relevant and trustworthy measure, suitable to establish benchmarks. Indicative of the unit's ability to use its resources to generate desirable outcomes, this measure should lead to a better understanding of the process in terms of what is achieved and how it is achieved. It should allow a meaningful investigation of hypotheses concerning the sources of inefficiency, in order to separate the effects of the environment, as well as to isolate the impact of differences in production technology. Finally, it should provide management with a control mechanism with which to monitor performance.

2.   PERFORMANCE MEASUREMENT APPROACHES IN BANKING

Long before DEA was introduced, banks have been engaged in performance measurement techniques and most large banks have from one person to whole departments engaged in continuous performance monitoring activities. However, the arsenal of tools and the variety of approaches are typically limited to a few relatively simple methods.

2.1   Ratio Analysis

Ratio analysis has historically been the standard technique used by regulators, industry analysts and management to examine banking performance. Ratios measure the relationship between two variables chosen to provide insights into different aspects of the banks' multifaceted operations, such as liquidity, profitability, capital adequacy, asset quality, risk management, and many others. Any number of ratios can be designed depending on the objective of the analysis, generally for comparisons within the same bank over different time periods, as well as for benchmarking with reference to other banks.

All major investment houses and market participants continually review banking as an investment opportunity. Their analyses are based on a large number of Key Performance Indicators (KPIs), usually in the form of ratios or dollar amounts. Various market tools have also been created for supervisory, examination and management purposes to monitor the financial condition of the banks. Compustat Research Insight, for instance, is a corporate and financial information research database of publicly traded corporations compiled by Standard and Poors (S&P). This CD-ROM database contains up to twenty years of financial ratios, stock market prices, geographic segment data, and S&P's ratings and rankings.

Although the traditional ratio measures are attractive to analysts due to their simplicity and ease of understanding, there are several limitations that must be considered. For example, the analysis assumes comparable units, which implies constant returns-to-scale (Smith 1990). Each of the indicators yields a one-dimensional measure by examining only a part of the organization's activities, or combining the multiple dimensions into a single, unsatisfactory number. Moreover, the seemingly unlimited number of ratios that can be created from financial statement data are often contradictory and confusing, thus ineffective for the assessment of overall performance. This overly simplistic analytical approach offers no objective means of identifying inefficient units and requires a biased separation of the inefficient and efficient levels. Failure to account for multidimensional input and output processes, combined with its inability to pinpoint the best performers in any culturally homogeneous group, makes ratio analysis inadequate for efficiency evaluations.

2.2   Frontier Efficiency Methodologies

The limitations associated with ratio analysis have led to the development of more sophisticated tools for assessing corporate performance. Frontier analyses, either parametric or non-parametric, measure the relative efficiency of production units based on the distance from the "best-practice" frontier, which must be empirically estimated from the data set since the true theoretical frontier is unknown. In utilizing these methods, management is provided with a framework that supports the planning, decision-making and control processes. For one, frontier techniques can objectively identify areas of best performance within complex business operations and further separate the efficient production units from their inefficient counterparts. Sources and magnitude of inefficiency can be determined that may ultimately lead to a reduction in the cost of operations or an increase in the services provided without expending additional resources. Achievable targets for inefficient units and the effects of environmental variables can be determined in order to provide additional insights and improve the overall understanding of production systems.

Several forms of parametric and non-parametric frontier techniques have been developed to evaluate a firm's performance relative to an empirically defined best-practice standard. The parametric, or econometric, methods for measuring efficiency include the Stochastic Frontier Approach (SFA), Distribution-Free Approach (DFA), and Thick Frontier Approach (TFA). Among the most commonly used non-parametric, or mathematical programming, techniques is Data Envelopment Analysis (DEA), including the Free Disposal Hull (FDH) approach which can be viewed as a special case of the DEA model. The parametric and non-parametric methods differ in the restrictions imposed on the functional form of the efficient frontier, the existence of random error, and the distributional assumptions on the inefficiencies and random error (Bauer 1990). Econometric analyses require an *a priori* specification of the form of the production function, and typically include two error components: an error term that captures inefficiency and a random error. While mathematical, non-parametric methods require few assumptions when specifying the best-practice frontier, they generally do not account for random errors.

2.2.1   Stochastic Frontier Approach (SFA)

The most commonly employed econometric method is the Stochastic Frontier Analysis (SFA) as in Kaparakis et al. (1994), Berger and Humphrey (1997), Hao et al. (2001), which specifies frontier functions for one input–multiple output or one output–multiple input scenarios. The methodology separates random errors from the inefficiency component and also varies from other parametric approaches as it imposes different distributional assumptions to accomplish this disentanglement. The SFA approach ranks the firm with lower costs for a given set of input prices (but the same output quantities) as more efficient than another firm.

2.2.2   Distribution-Free Approach (DFA)

The Distribution-Free Approach (DFA) as in Berger, Hancock and Humphrey (1993), Akhavein et al. (1997) and DeYoung (1997) estimates efficiency by specifying a functional form for the frontier, much like the SFA, but does not impose a parametric structure on the random errors or inefficiencies. DFA makes the assumption that random errors average out to zero, while the average efficiency of each firm remains constant over time. Shifts in the efficiency frontier over time, caused by any number of external forces, result in DFA scores that reflect the average deviation of each firm from the best average-practice frontier instead of the efficiency at any one point in time (Berger and Humphrey 1997).

2.2.3 Thick Frontier Approach (TFA)

The Thick Frontier Approach (TFA) as in Berger and Humphrey (1991), Clark (1996) and DeYoung (1998) is the least frequently used parametric technique. TFA measures the general level of overall efficiency, rather than point estimates of efficiency for individual firms.

2.2.4 Data Envelopment Analysis (DEA)

Theoretical aspects of Data Envelopment Analysis (DEA) are discussed in Chapter 1, so there will be no repetition of that here. But in contrast to DEA's models, FDH (Tulkens (1993) and Chang (1999)) does not use convex technologies and assumes that no substitution is possible between observed input combinations on a piecewise linear frontier. Instead, the FDH frontier is represented by a step function, which is formed by the intersection of lines drawn from the input combinations.

3. DATA ENVELOPMENT ANALYSIS IN BANKING

Sherman and Gold (1985) wrote the first significant DEA bank analysis paper and started what turned out to be a long list of DEA applications to banking from several different angles:
- Country-wide bank (companies) analysis
- Bank branch analysis within one banking organization
- Cross national banking analysis
- Bank merger efficiencies
- Branch deployment strategies

Unparalleled growth in the volumes of theoretical and empirical investigations of banking efficiency has resulted in the main Operational Research journals devoting special issues just to this research area (i.e., Journal of Productivity Analysis (1993), Journal of Banking and Finance (1993), European Journal of Operational Research (1997) and Interfaces

(1999)). A considerable number of papers have been published on banking (both banks as entities and branches) using DEA since the technology was introduced. Berger (1997) summarized these works and listed 41 DEA studies in a number of countries with U.S. banking research predominant with 23 papers. Since then, a further 17 banking studies appeared up to early 2003 and of these 5 involved U.S. banks. Clearly, the work keeps going on and will likely to continue in the future.

3.1   Banking Corporations

There have been a considerably larger number of studies that focus on the efficiency of banking institutions at the industry level compared to those that measure performance at the branch level (Berger and Humphrey 1997). Lack of publicly available branch data is perhaps the reason behind this trend.

3.1.1   In-Country

In any country that has a sufficient number of banks an in-country study using DEA is quite feasible. A number of such studies have been done, many of them involving the U.S. banking system while other researchers focused on Denmark, Finland, India, Italy, Norway and Spain. Bhattacharyya et al. (1997) examined the productive efficiency of 70 Indian banks during 1986-1991, which was a period of liberalization. They found the publicly owned Indian banks to have been the most efficient, followed by foreign-owned banks and privately owned Indian banks. Favero and Papi (1995) derived measures of technical and scale efficiencies in Italy examining a cross-section of 174 Italian banks in 1991. Fukuyama (1993) employed DEA to compute and assess technical efficiency and scale efficiency for 143 Japanese banks. He compared statistically the relationships between organizational status (as well as bank size) and the various efficiency scores. Among the three organizational types, the performance of city banks were, on average, the best in every dimension. Relative to both assets and revenues, scale efficiency was weakly associated with bank size. Thompson et al. (1997) applied classification, sensitivity, uniqueness, linked cones and profits ratios to a bank panel of the 100 largest U.S. banks from 1986 to 1991.

3.1.2   Cross Country Studies

These are quite difficult studies because allowances must be made for a host of natural differences between the nations involved. These include real cultural differences (as opposed to DEA cultural issues), regulatory environments, presence in a common market structure (EU, NAFTA) and the "sophistication" of the banking institutions[1]. For example, Berg et al. (1993) examined the banking community in Norway, Sweden and Finland. They created individual and then combined frontiers to examine the similarities and differences between the banks. Later, in a follow-up study, Bukh et al. (1995), added Denmark to the group to cover all the Scandinavian countries. The study by Pastor (1997) covered 429 banks in 8 developed countries, and pooled cross-country data to define a common frontier.

3.2   Bank Branches

Much of the published literature on bank branch performance utilizes methodologies based on non-parametric frontier estimations, primarily DEA and FDH. Appendix 13.1 provides a comprehensive reference of DEA-based *bank branch* analyses up to 2002 (Yang and Paradi 2003), containing information on the country and number of branches analyzed, the focus of the study and specification of the model variables and characteristics.

These studies examine branch performance by assuming various behavioral objectives.

- In the *production* approach, branches are regarded as using labor and capital to produce deposits and loans
- The *intermediation* approach captures the process in which deposits are being converted into loans.
- *Profitability* is the measure of how well branches generate profits from their use of labor, assets and capital.

Depending upon behavioral goals, the analysts would choose between input minimization and output maximization models and CCR and BCC approaches to achieve the desired answers for managers who will have to implement the findings.

3.2.1   Small Number of Branches

The early studies on branch performance are fairly simple and evaluate efficiency using small sample sizes. Sherman and Gold (1985) used a basic CCR model to analyze the production (operational) efficiency of 14 branches belonging to a U.S. savings bank. This then set the direction for subsequent studies by demonstrating the practical and diverse opportunities associated with the methodology. Parkan (1987) used the CCR model to

---

[1] For example, in the Czech Republic, only in the past few years have credit cards been issued by banks.

benchmark 35 branches of a Canadian bank, but he failed to consider the possible effects of returns-to-scale on operations.

Based on the work by Vassiloglou and Giokas (1990), Giokas (1991) estimated relative efficiencies of 17 branches in Greece using both DEA and log-linear methodologies. Oral and Yolalan (1990) introduced a DEA measurement model that forced each of the twenty branches in a Turkish bank to compare itself with the global leader. This was the first study to include time-based outputs (activities were given standard times) in the production model, instead of the *number of transactions* that are used in traditional DEA applications. An investigation on whether there exists a relationship between operating and profitability levels of bank branches was further pursued. Oral et al. (1992) developed two banking models on the same dataset for analyzing both the operational efficiency and profitability of those Turkish bank branches.

Al-Faraj et al. (1993) applied the basic formulations of DEA to assess the performance of 15 bank branches in Saudi Arabia. They used eight inputs and seven outputs, and subsequently identified all but three branches as relatively efficient. They inadvertently illustrated one of the limitations associated with the DEA approach, its inability to effectively discriminate between efficient and inefficient units when a limited number of observations, relative to the number of input/output variables, are used. The efficiency scores obtained by Haag and Jaska (1995) were achieved using an additive model on 14 branches operating in the U.S. Sherman and Ladino (1995) reported that the implementation of their DEA results in the restructuring process of the 33 branches belonging to a U.S. bank led to actual annual savings of over $6 million. But both studies were potentially biased due to an insufficient number of branches. Similarly, Athanassopoulos and Giokas (2000) reported on the successful adoption of DEA by the management in a Greek bank with 47 branches. Cook et al. (2000), using only 20 bank branches of a large Canadian bank, offered a methodology for splitting shared resources among branch activities in order to maximize the aggregate efficiency scores.

3.2.2   Large Number of Branches

Tulkens (1993) examined 773 branches of a public bank and 911 branches of a private bank in Belgium using the FDH model for the first time in banking. Drake and Howcroft (1994) reported the efficiencies of 190 branches in the United Kingdom under both the CCR and BCC assumptions. The analysis was taken further to reveal an optimum branch size in terms of the number of staff employed. Schaffnit et al. (1997) specifically focused on personnel performance using a 291 branch segment of a large Canadian bank. Several DEA models with output and input multiplier constraints were developed to measure the operating efficiency in providing both transactions and maintenance services at the branch.

Lovell and Pastor (1997) evaluated the operating efficiency of 545 branch offices based on the target setting procedure employed in a large bank in Spain. Athanassopoulos (1998) conducted a performance assessment of 580 branches of a commercial bank in the UK using cost efficiency, then a new dimension in bank branch performance.

Kantor and Maital (1999) used the activity-based costing (ABC) system to generate data on cost drivers for two separate activities in 250 Israeli branches using a basic CCR model. Camanho and Dyson (1999) estimated operational efficiencies of 168 branches of a large bank in Portugal. The efficiency-profitability matrix was then employed for a more comprehensive characterization of their performance profile.

Golany and Storbeck (1999) conducted a multi-period efficiency analysis of 182 selected branches in the U.S. using a DEA model that allowed for the separation of inputs into discretionary and non-discretionary (i.e., not controllable by management) factors. A resource-allocation model was introduced to estimate the maximal output enhancement that could be expected from the branch given its level of inputs. Zenios et al. (1999) extended the DEA framework to capture the effects of the environment on efficiency measures of 144 branches in Cyprus. They examined whether superior efficiency was the result of the technology used or managerial practices.

3.2.3 Branch Studies Incorporating Service Quality

Only a few researchers incorporated service quality dimensions into the analyses. A study by Soteriou and Stavrinides (1997) provided benchmarks of employee service quality perception of the branches. Soteriou and Zenios (1999) built on this work to provide a more complete assessment of the overall performance of a larger network consisting of 144 branches in Cyprus. The paper attempted to link service quality, operations and profitability in a common framework of efficiency benchmarks. Athanassopoulos, (1997) and (2000), also dealt with the notion of service quality and its interaction with operating efficiency.

3.2.4 Unusual Banking Applications of DEA

Nash and Sterna-Karwat (1996) measured the cross-selling effectiveness of financial products for 75 bank branches in Australia. Soteriou and Zenios (1999) presented a novel approach to efficient bank product costing at the branch level. The focus was on allocating total branch cost to the product mix offered by the branch and obtaining a reliable set of cost estimates for these products.

In summary, it can be seen that there are many reasons for stimulating the analysts' ingenuity in formulating the appropriate DEA models. Most of the work referred to here address real life issues that are usually less than ideal (from an analytical or theoretical point of view), hence the analyst must innovate in order to use what data is available to come up with the answers required by management.

4.   MODEL BUILDING CONSIDERATIONS

There are two crucial considerations the analyst must keep in mind when preparing a study in a banking environment – selecting inputs and outputs and the choice of DEA technology. These issues will be pointed out and the reasons for making the choices as the chapter progresses.

4.1   Approaching the Problem

Analysts and managers must come to a common ground about what each model is to accomplish. The researcher must satisfy the needs for specific information that is likely to be useful to the bank (branch) manager and to show the way to better performance. Before bank (branch) efficiency can be measured, a definition of what its business processes are is required. Countless studies have been done to determine accurate ways of measuring bank (branch) efficiency – as detailed above. Kinsella (1980) discussed some of the reasons banks are difficult to measure: they offer multiple products; have complex services (many of which are interdependent); provide some services that are not directly paid for; and have complex government regulations that may affect the way in which services are offered or priced. Given these issues, it becomes obvious that there is no one way of accurately measuring branch efficiency and that clearly a combined set of metrics is required.

Moreover, for the study to be successful from an operational and implementation point of view, it is necessary that management be involved right from the start. Decision makers must understand all model designs and the selection of measures to use in the models and must see the outcomes as *fair* and *equitable*.

4.2   Input or Output?

When defining models, often the question arises about what to do with an input or output that should not be minimized or maximized respectively. One example of these are bad loans, which is obviously an output, but it is not desirable to reward the DMU for having more bad loans than its peers have. So how can the problem be solved? Two different approaches have been used in the literature: the first is to leave bad loans as an output but use the inverse value. The other method is to move it to the input side where the lower this value the better. This seems like a fairly simple decision to make. But for the sake of managerial understanding the inverse value on the output side may be preferred.

Now how about deposits? Here it could be argued that the higher the value the better because that shows efficiency in attracting depositors. On the other hand, one could make a case that the lower the deposit value, the better because the bank is doing more lending with less deposits. This, of course, implies that the bank has sources of funds that are cheaper than deposits. Fortunately the analyst can have it either way, but also both ways if there are several models being built - really depending on what the model is intended to achieve.

4.3   Too Few DMUs / Too Many Variables

DEA, as many other methods, require that there be enough observations to allow good separation and discrimination between DMUs. Several methods can be used to address this problem. One is to increase the number of DMUs by using the Windows Analysis approach, which allows different years' observations to be compared to each other.

Another possibility is to decrease the number of inputs and outputs in the models by creating more than one model where each has fewer total (inputs + outputs) number of variables. In short, the number of DMUs should be at least two to three times the total number of inputs plus outputs used in the models. So, for example, if there are 3 inputs and 5 outputs, the minimum number of DMUs should be 16 – 24. This is a rule of thumb decision but from a practical point of view, works reasonably well (see Cooper et al (2000) Section 9.2.1 pg. 252 for comments).

Too few variables of course reduce the model almost to a ratio measure – but this may still be usable for some BCC type studies.

4.4   Relationships and Proxies

Data may be available on measures that really represent the same variable, although often expressed in different units. An example of this may be staff in a branch where data on both salaries and FTEs are available. Of course, these represent the same variable, so typically only one is used – but which one? The decision is dictated by the objectives of the study. More often the FTE measure is used because this eliminates the dispute over pay scales that may be different depending on the local economic realities (large city vs. small community). However, if the manager has the flexibility of using staff in different capacities – less costly worker assisting a more costly one (really good sales people receive more support) the salary costs may be a better measure to bring out the efficiency gained by more effective management of the resource.

But there are other situations where different variables prove to be very highly correlated, even if they are not related logically. We may consider one as the proxy of the other. In this case, it is appropriate to use only one of the measures because the highly correlated other(s) only decrease the discriminatory power of the model without adding useful information. Hence, it is customary to run correlation analyses on all inputs and outputs selected for the model to see if one or more are highly correlated and then decide which may be dropped from the model. One might also take into consideration suspected relationships between certain input(s) and output(s) and test them to ensure that those that do correlate are left out of the analysis as appropriate.

4.5   Outliers

Essentially there will always be problems with empirical data either because some data elements are wrong or missing or because some DMUs are outliers and really do not belong to the dataset. Before meaningful DEA results can be obtained, these outliers must be dealt with.

The obvious first step is to double check the data of those DMUs that appear to be performing too well, or too poorly. One way to find the former is to check the number of peers that use them as an efficient reference – it will be easy to see if this is much larger than for the other efficient DMUs. DMUs scoring very low, for example 0.2 or less, in a bank branch analysis are also suspicious. After all, in a typical bank, branches are closely controlled, have numerous policies, procedures and rules of operation, so it is not possible to have branches that are 20% or less efficient in real life. Hence, checking data integrity is the first step, followed by either correcting data errors or removing problematic DMUs.

Removing outliers is justified only if they can be identified as having erroneous or missing data, or even when the data is correct, these DMUs really are in a different business than the others. For example, a "real-estate" branch will always have significantly higher loans and assets than other branches. It is, therefore, unrealistic or unfair to expect otherwise similar sized or located branches to emulate this activity.

4.6   Size Does Matter

At first glance, the size of the bank or branch appears to have significant effects on the operations of the unit. Surprisingly, while this indeed is the case for banking corporations, it does not hold for branches. Banks are best modeled using BCC models to allow for their size differences. Branches on the other hand, tend to operate as CCR entities. Most banks first segment their branch networks into various groups based on business type and then evaluate them separately within their own group. Typically, large commercial branches are taken out of the analysis altogether because they need different models than retail branches.

Retail branches are often segmented into four groups:
- Small Rural (small towns and villages)
- Small Urban (local residential areas in large towns and cities)
- Large Rural or Regional (located in larger towns and some branches serve local businesses as well)
- Major Urban (Large cities, sophisticated clientele, investment and business orientation).

This segmentation is often employed in DEA studies of larger banks where there are a sufficient number of branches.

4.7   Too Many DMUs on the Frontier

Banks tend to be very well managed, or at least well controlled, as they have to follow policies and rules laid out by the Head Office and the Government regulator. In many DEA analyses this results in a substantial portion of the branches being on the frontier – typically 25-50%. While this is not a problem with the technique per se, it is a problem if management wishes to improve operations across the branch network because frontier resident branch managers see themselves as already being the best they can be. Sowlati and Paradi (2003) addressed this issue by developing a management opinion based technique that created a "Practical Frontier" that enveloped the empirical one, thus offering targets to the empirically efficient units.

4.8   Cross Cultural Comparisons

When one compares branches across banks, differences in the basic banking culture are inherent, as top management tends to determine what segment of the banking business they want to focus on.   Hence, some methodology must be introduced to allow for the systemic differences caused by the managerial direction the branches are getting. A later example will show how this might be addressed.

4.9   Environmental Factors

Fairness and equitable treatment are both key components if people are to accept the outcomes from DEA (or any other) studies. The main pushback by those being measured is usually based on factors not typically included in whatever measuring technique is being used. These factors can be classified as environmental factors.   Examples of these may be the economic environment in a geographic area being different than that in another (a disadvantaged State/Province/District vs. an advantaged one); different opening hours; demographics etc.

Another such factor, which is more difficult to quantify but very important is the competitive environment the bank or branch operates in. Clearly, if the branch is the only one in town, it gets almost all the business, but if it is located across the street from three other bank branches on the same intersection and have two or three others within a block, they have to fight for their market share. An effective way to include this factor is to develop a "competitive index" as was done by Vance (2000). She incorporated in her index the type and number of competitors in the reasonable geographical drawing area of a branch.

4.10   Service Quality

Service quality is an important issue but it is difficult to measure and even more difficult to incorporate into the model. Typically, there are two subjective measures used when estimating such data: Customer Satisfaction and Employee Satisfaction. As data on both of these are acquired through questionnaires, quality control of the data collection process is a crucial issue. Often, the questions offer choices from a balanced 5 step Likert scale response set. Banks tend to count, as the percentage of all responses, only the top two scores (Excellent, Very Good or if applicable the most negative two) and ignore the rest. Not a very satisfactory process, but typically, it is the best available data.

4.11 Validating Results

For most managers, DEA is an unknown "black-box" and without meaningful validation, they will not use it or believe in the recommendations offered on the basis of such results. Hence, validation of the results is important, although not at all easy. One method is to compare DEA results with the bank's own performance measures. Another technique is to use statistical methods to establish the credibility of the approach. But care must be taken with these because the need is to show how much better DEA is, and not leave the impression that it is no better than what they have already.

5. BANKS AS DMUS

One of the characteristics of DEA is that it requires that all DMUs under examination be comparable and therefore have the same "cultural" background or that adjustments can be made for the differences between, for instance, banks in different countries. Clearly, before the DEA study can be made some method must be adopted to adjust the variables in order to place all the DMUs on a level playing field. Different approaches have been developed to address this issue.

5.1 Cross Country/Bank Comparisons

The example used here is a cross bank comparison of branches from three large banks with branches in a mid sized Canadian city of about 500,000 inhabitants. When comparing the branch efficiencies of different banks with each other using DEA, the main problem is that the performance of the branch can be materially affected by upper management's decisions that the branch can not control. The typical DEA methodology does performance analysis for units assumed to be operating in similar environments, or "culture". An approach that works quite well for comparing branches across banks depends on a handicapping function that includes various environmental or cultural factors to account for corporate differences.

Compared to the normal DEA methodology, handicapped DEA models (Yang and Paradi 2003) include a handicapping function denoted by "h" that can be used to handicap either inputs or outputs. Assume h_j is the handicapping measure to be applied to input variables and $\hat{h}_j$ handicaps output variables. Now we have $h_j x_{ij}$ and $\hat{h}_j y_{rj}$ as the adjusted inputs and outputs respectively. Hence, the DMUs with the advantaged environment are penalized more by a higher input-handicapping measure and/or a lower output-handicapping measure, either or both of which makes the DMUs less efficient. On the other hand assigning a lower input-handicapping measure or a higher output-handicapping measure, which enables these DMUs to reach the efficient frontier sooner, compensates those DMUs with a disadvantaged environment. The brief mathematical formulation is presented as follows:

Input oriented BCC model:

$$\min_{\theta,\lambda,s^+,s^-} z_0 = \theta - \varepsilon \left[\sum_{i=1}^{m} s_i^+ + \sum_{r=1}^{s} s_r^- \right]$$

$$s.t. \qquad \sum_{j=1}^{n} (\hat{h}_j y_{rj})\lambda_j - s_r^- = (\hat{h}_0 y_{r0}) \qquad r = 1,\dots, s$$

$$\theta(h_0 x_{i0}) - \sum_{j=1}^{n} (h_j x_{ij})\lambda_j - s_i^+ = 0 \qquad i = 1,\dots, m \qquad (13.1)$$

$$\sum_{j=1}^{n} \lambda_j = 1$$

$$\lambda_j, s_i^+, s_r^- \geq 0$$

The handicapped DEA model directly incorporates cultural differences into its formulation rather than introducing non-discretionary (sometimes referred to as "exogenously fixed") variables that require a natural nesting of categories (Banker and Morey 1986). Oftentimes, real world situations do not have such a hierarchy. In addition, the handicapping approach will have wider application since it reduces multiple uncontrollable environmental variables into a single handicapping measure without complicating the analysis, and this substantially improves the discriminatory power of DEA. Finally, this methodology ensures, given a proper handicapping approach, that the deviation from the frontier comes from pure operational inefficiency, not a combination of inefficiency and the cultural disadvantage imposed by the Head Office created environment. However, analysts need to find their own handicapping system for each application domain, which obviously depends on the specific situation.

One type of handicapping function incorporates the index number approach that has been applied to the banking industry by Fixler and Zieschang (1993). Their approach views the technological difference between different entities as the source of inefficiency and therefore, the relative productivity is the ratio of the strategy differences between the banks. Consequently, the handicapping factor can be defined as the ratio of the output index over the input index for each bank.

The specific index function used in the following example will not be described in detail here, but the appropriate handicapping factor would be usable in this manner. These factors are constructed for all the banks in the sample. Such factors represent a "difference in strategy" parameter among the banks in the comparison and then applied to the branches for each bank.

To illustrate the process a real example of 70 bank branches operated by three large Canadian banks were used (Yang and Paradi 2003). All these branches focus on personal banking and all potential outliers were removed from the analysis. The annual data for the fiscal year 2000 are shown in Table 13-1.

Table 13-1. Annual Data for Fiscal 2000

	Max	Median	Min	Average	Standard Deviation
Inputs (in dollars)					
FTE	1,394,461	468,553	155,093	545,279	260,498
Premises/ IT Expenses	1,303,433	271,467	42,075	319,073	209,762
Other Expenses	1,600,860	385,192	22,674	397,228	281,884
Outputs (in dollars)					
Deposit Balances ($'000s)	140,464	47,208	300	55,120	34,537
Loan Balances ($'000s)	356,847	78,751	23,100	96,915	66,312
Security Balances ($'000s)	96,796	18,748	400	23,702	19,484
Gross Revenue	6,642,668	2,175,473	44,427	2,381,524	1,798,402

The DEA model used by Yang and Paradi (2003) was:

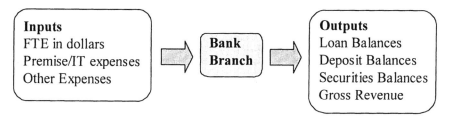

Inputs		Outputs
FTE in dollars	**Bank Branch**	Loan Balances
Premise/IT expenses		Deposit Balances
Other Expenses		Securities Balances
		Gross Revenue

Figure 13-1. DEA Model

Input orientation was chosen for this model since the branches have little direct control over the services required by their customers. Hence, the objective for bank managers is to improve their staff performance while maintaining the current level of service. The results of this analysis are shown in Table 13-2 (Yang and Paradi, 2003).

The handicapped DEA results suggest that branch performance from the three banks were very similar. This is in accordance with Schaffnit et al. (1997), who state that the large Canadian banks are quite sophisticated and technologically intensive and that all branches have similar performances.

The handicapping function for Bank B and C compensate their branches as they operate under less favorable environments; therefore, their branches' efficiency scores increase. For Bank A, the handicapping function penalizes its branches since they operate under advantaged environments, which results in a decrease in its efficiency scores.

Table 13-2. Handicapped Results Compared to Conventional DEA

	Conventional Approach	Handicapped Approach
Whole sample		
% technically efficient branches	50%	50%
average BCC score	0.912	0.923
% technically and scale efficient branches	24%	29%
average CCR score	0.841	0.865
standard deviation	0.148	0.144
Bank A		
% technically and scale efficient branches	40%	40%
average CCR score	0.880	0.869
Bank B		
% technically and scale efficient branches	15%	22%
average CCR score	0.810	0.866
Bank C		
% technically and scale efficient branches	25%	29%
Average CCR score	0.851	0.862

It should be noted that the handicapped DEA approach is not banking specific, indeed, it could be used in any situation where normal DEA analysis could be used if cultural differences did not exist. This approach relaxes the requirement of similar operating circumstances (cultures) and allows the comparison of dissimilar units if a deterministic or stochastic function can be used to quantify the differences between DMUs.

5.2   Bank Mergers

One of the hot topics in the Financial Services Industry during the 1990s was the significant merger activity U.S. and European Banks engaged in (Vela and Paradi, 2002). Some of the major players were provided in Table 13-3.

Much work has been done on the efficiency effects arising from bank mergers using various methodologies, ranging from simple ratio analyses to more complex frontier approaches; both econometric and mathematical methods were used (Berger (1998), DeYoung (1993), Peristiani (1993, 1997), Grabowski *et al.* (1995), Vander Vennet (1996), Resti (1998), Avkiran (1999), Haynes and Thompson (1999)). In addition to measuring post-merger gains, the importance of estimating the *potential* for achieving efficiencies through the merger process may be even more interesting.

Drawing on the concept of organizational cultures' effects on DMUs introduced in the previous section, the proposed framework can be further extended to account for the impact of the banks' cultures on merger efficiencies. The 'culture' for this purpose could be defined as the management style and operating infrastructure adopted by the *merged* bank, from each of the merger partners for implementation in its combined branch network.

To illustrate the process, two banks of different cultural backgrounds wish to assess the potential impact on efficiency from merging their branches (Vela and Paradi 2002). One bank is positioned as a market leader in service quality, while the other employs very effective corporate/investment strategies. Through simulating merger outcomes and taking the best approaches from the merger partners, a general approach to estimating possible efficiency gains can be developed in a step-by-step process.

Table 13-3. Selected Recent Bank Merger Activity

Year	Merging Banks	Country of Origin
2002	Industrial Bank of Japan and Dai Ichi Kangyo and Fuji Bank	Japan
2002	Sanwa Bank and Tokai Bank	Japan
2002	Bank of Taiwan and Land Bank of Taiwan and Central Trust of China	Taiwan
2001	Sumitomo Bank and Sakura Bank	Japan
2001	Bank of Tokyo and Mitsubishi Bank and Mitsubishi Trust and Bank	Japan
2000	Chase Manhattan and J.P. Morgan	US
2000	Deutsche Bank and Dresdner Bank	Germany
1999	Banque Nationale de Paris and Paribas	France
1999	Fleet Financial and Bank of Boston	US
1998	Deutsche Bank and Bankers Trust	Germany/US
1998	Wells Fargo and Norwest	US
1998	NationsBank and BankAmerica	US
1998	Banc One and First Chicago NDB	US
1998	Hypo-Bank and Bayerische Vereinsbank	Germany
1998	United Bank of Switzerland and Swiss Bank	Switzerland
1997	NationsBank and Barnett	US
1996	Mitsubishi Bank and Bank of Tokyo	Japan
1996	Chemical Bank and Chase Manhattan	US
1996	Wells Fargo and First Interstate Bancorp	US
1990	ABN and AMRO	Netherlands

5.2.1   Selecting pairs of branch units for merger evaluation

This step involves the strategic selection of consolidating branch units based on intuitive criteria, such as: branches in close physical proximity; similar business natures and objectives (i.e., distinguishing between general branches and processing cost centers); etc.

5.2.2   Defining a strategy for hypothetically merging two units

Simulation of efficiency gains involves constructing a hypothetical unit that represents the combination of inputs and outputs of the two merged branches, less the expected reductions in costs specified by management. It is assumed assets are not re-optimized for further cost savings, but rather staffing, back-office operations and other inputs are rationalized to realize cost efficiencies. In measuring the potential output production of the combined entity, additivity in the output levels is used since the business of the branch to be closed is assumed to be transferred to its merger partner – less some unavoidable attrition.

5.2.3   Developing models for evaluating the overall performance of merged units through the selection of appropriate input and output variables

An example of a model to evaluate the performance of a merged branch is depicted in Figure 13-2. It includes salaries, rent and other general expenses as inputs and three variables as outputs: deposits, loans and revenues.

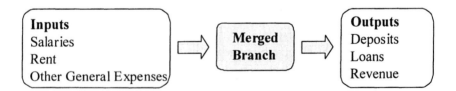

Figure 13-2. Model for measuring efficiency gains from mergers

5.2.4   Calculating potential efficiency gains

Performance gains are estimated by comparing the technical efficiency of the post-merger unit (hypothetically combined using the merging strategy defined above) to the corresponding pre-merger efficiencies of the merging branches (Vela and Paradi 2002). More specifically, the change in technical efficiency (ΔTE) is calculated as the difference between the technical efficiency of the hypothetically merged unit (TE_M) and the weighted-sum of the pre-merger technical efficiencies of the two branches (TE_1 and TE_2) as shown in Equation 13.2 below:

$$\Delta TE = TE_M - \left[w_1(TE_1) + w_2(TE_2) \right] \qquad (13.2)$$

where w_1 and w_2 are the weights for the two branches before the merger, whose sum equals unity or $w_1 + w_2 = 1$. The weights are based on total assets (*TA*), such that $w_i = TA_i / (TA_1 + TA_2)$, where *i* represents Branch 1 or Branch 2 involved in the merger.

Efficiencies of consolidating branches, as well as the hypothetically combined unit are measured using standard DEA models. A sample of non-merging units is also included in the analysis for comparison purposes.

5.2.5 Identifying differences in cultural environments between the merging banks.

Differences in culture between the two banks are quantified using existing methodologies that provide relative index measures of cultural advantages or disadvantages. For the given example, the culture represents corporate strategies (corporate index or *CI*), along with organizational core competencies that affect service quality (service index or *SI*). The cultural framework can be represented mathematically as follows, where *F* indicates *as a function of*:
- *Corporate Strategy* = *F* (Resources, Cost of Input, Product Portfolio, Price of Output)
- *Service Quality* = *F* (Human Capital, Technology, Operating Processes, Size of Activity)
- *Overall Performance* = *F* (Corporate Strategy, Service Quality, Branch Size, Cost, Products, Revenue)

Over time, the two branches merged into a single entity are assumed to adopt each other's more favorable or stronger cultural elements, i.e., more efficient investment strategies or higher level of customer service. This cultural adoption is reflected in adjustments of the level of inputs consumed and/or outputs produced.

5.2.6 Calculating potential synergies

A merged organization is assumed to eventually increase the level of revenues generated with the implementation of the corporate strategies from the culturally favorable branch. The overall increase is calculated by multiplying the revenues of the branch employing less efficient corporate strategies, i.e., relatively unfavorable, with the ratio difference of CI's associated with favorable and unfavorable units, i.e., CI_F / CI_{UF}. Also, the merged branch is assumed to produce higher levels of customer service, by embracing the methods of the favorable merger partner. Similarly, increases are calculated using the ratio difference of SI's associated with favorable and unfavorable units, i.e., SI_F / SI_{UF}. The final values for the merged branch's inputs, deposits and loans, along with revenues adjusted for culture (X_C, $Y_{D\&L,C}$, $Y_{R,C}$ respectively) are calculated as follows (Vela and Paradi 2002):

$$\left(X_C, Y_{D\&L,C}, Y_{R,C} \right) =$$
$$\left[X_F + X_{UF}\left(\frac{SI_F}{SI_{UF}} \right), Y_{D\&L,F} + Y_{D\&L,UF}\left(\frac{SI_F}{SI_{UF}} \right), Y_{D\&L,F} + Y_{D\&L,UF}\left(\frac{CI_F}{CI_{UF}} \right) \right] \tag{13.3}$$

where X_F and X_{UF} are the original input values of the branches that operate in favorable and unfavorable cultural environments, respectively. The same notation is used for Y_U and Y_{UF}.

Potential synergies are estimated as the difference in efficiencies between the hypothetically merged branch and the same branch that has undergone cultural adoption, which are also calculated using standard DEA models. The final step involves testing whether additional gains in efficiency are possible when the impact of cultural environments is incorporated into the analysis.

5.3   Temporal Studies

In many instances progress over time becomes the main objective for measuring the efficacy of management strategies. There are two interesting techniques one can employ to deal with time series data and gain valuable insights from the pattern and direction of changes during the study period. Here are two methods (Windows Analysis and Malmquist Index) which can shed light on both what is happening and why.

5.3.1   The Models

The first model (Production Model) used here is given in Figure 13-3 (Aggarwall 1996).

Other Non-Interest Earning Assets such as accrued interest, deferred and recoverable income taxes, goodwill, accounts receivables and prepayments, deferred items and some other assets are not considered in the model. It is neither a management objective to increase these assets, nor do they qualify as a service provided by the banks. There is also no interest earned on this category of assets. The results from this model will be referred to as the Technical Efficiency results.

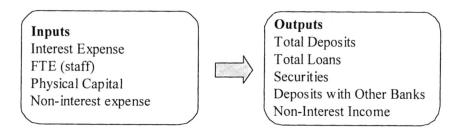

Figure 13-3. Variables in the Production Model

The above model can be though of as a model that shows how well the banks performed their production of tasks, dealing with customers and making transactions as required.

For the sake of brevity the development of another DEA model where both input and output constraints were used on a different set of inputs and outputs will not be shown here, except for the model variables in the following figure.

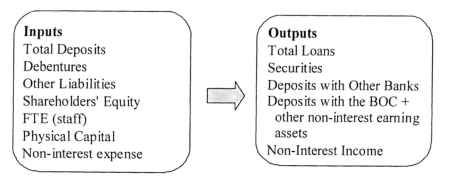

Figure 13-4. Variables in the Intermediation Model (Aggarwall 1996)

The above model can be considered as the model that explains how well the banks did in their efforts to make a profit by dealing with money and securities in a risky environment. The results from this model are referred to later as the Overall Efficiency results.

The weights in both models were constrained using real prices and management provided information as appropriate (Aggarwall 1996). But again, these developments are not shown here.

Table 13-4 gives the statistics on the data used:

Table 13-4. Data Statistics

	Min	Max	Average
Inputs [1]			
Interest Expense	1,790,898	16,358,317	6,391,627
FTE (staff) [1]	10,877	52,745	30,280
Physical Capital	124,941	2,057,000	862,912
Other Non-interest Expenses	216,881	1,837,556	779,608
Total Deposits [2]	23,166,413	135,815,000	78,237,069
Debentures	25,157	282,182	133,098
Other Liabilities	996,507	27,748,000	5,664,598
Shareholders' Equity	825,683	8,589,000	4,419,406
Outputs [1]			
Total Deposits [2]	23,166,413	135,815,000	78,237,069
Total Loans	19,626,193	117,013,807	65,867,548
Securities	2,202,023	28,753,000	10,781,395
Deposits with Other Banks	630,112	22,345,822	8,396,201
Deposit with BOC + other non-int earning assets	762,946	8,791,198	3,707,324
Non-interest income	195,071	2,697,000	961,179

1. All figures except FTE are in thousands of Canadian Dollars
2. Total Deposits appear as input in one model and as output in the other

5.3.2   Window Analysis

One of the challenges in DEA modeling is the situation when there is an insufficient number of DMUs in comparison to the number of relevant inputs and outputs in the model. One of the approaches is to collect a time series panel data and then use the DEA Window Analysis approach. This technique works on the principle of *moving averages* (Charnes *et al.* 1994B, Yue 1992), and is useful to detect performance trends of a unit over time. Each unit in a different year is treated as if it were a "different" unit. In doing so, the performance of a unit in a particular year is compared with its performance in other periods in addition to the performance of the other units. This means that the units of the same DMU in different years are treated as if they were independent of each other – but comparable.

A notable feature of this technique is that there are *nk* units (DMUs) in each window where *n* is the number of units in a given time period (and it is the same in all time periods), and *k* is the width of each window (equal for all windows). This feature is extremely important in the case of a small number of DMUs and a large number of inputs and outputs since it increases the discriminatory power of the results. This is accomplished by dividing the total number of time periods, *T*, into a series of overlapping periods or *windows*, each of width k $(k<T)$ and thus having nk units. Hence the first window has nk DMUs for the time periods $\{1,....,k\}$, the second one has nk DMUs and the time periods $\{2,......,k+1\}$, and so on and the last window consists of nk DMUs and the time periods $\{T-k+1,......,T\}$. In all, there are $T-k+1$ separate analyses where each analysis examines nk DMUs.

An important factor is the determination of the window size. If the window is too narrow, there may not be enough DMUs in the analysis and thus not enough discrimination in the results (which also depends on the number of DMUs and variables in the model). On the other hand, too wide a window may give misleading results because of significant changes that occur over periods covered by each window. The best window size is usually determined by experimentation.

Panel data, from the six largest Canadian banks during a 14 year period (1981 to 1994) was used in the analysis (Aggarwall 1996). However, during the 14 year period a substantial difference can be expected in the production possibilities of the banks due to changes in the laws governing banks, acquisitions, technology growth and even leadership changes. Therefore, it is not reasonable to compare DMUs over such a long and disparate period in one analysis. Hence, the window analysis technique is used to obtain efficiency results for a few (just 6 DMUs) banks over time and to identify performance trends.

The first step of a window analysis is to select a *window size*. For this model (with output multipliers constraints), not much interpretation (discrimination) is possible using three year windows ($k=3$, $n=6$). This is due to the insufficient number of units in each window (only 18) compared to the number of variables (nine). Four year windows are not discriminatory enough either, hence, *five year windows* are selected as the narrowest window that still enables a reasonable discrimination for the analysis.

Table 13-5. Technical efficiency scores from the output constrained model with 5 year windows for one bank

Window	1981	1982	1983	1984	1985	1986	1987	1988	1989	1990	1991	1992	1993	1994
1981-85	0.638	0.603	0.898	0.918	1									
1982-86		0.553	0.830	0.848	0.931	1								
1983-87			0.702	0.715	0.777	0.829	1							
1984-88				0.715	0.777	0.829	1	0.951						
1985-89					0.777	0.829	1	0.951	0.771					
1986-90						0.829	1	0.951	0.771	0.779				
1987-91							1	0.951	0.771	0.779	0.894			
1988-92								0.773	0.606	0.607	0.706	1		
1989-93									0.526	0.521	0.610	0.870	0.988	
1990-94										0.493	0.579	0.829	0.928	0.993
Average	0.638	0.578	0.810	0.799	0.852	0.863	1.000	0.915	0.689	0.636	0.697	0.900	0.958	0.993

Using the DEA model described above with 5 year windows, it is evident that the efficiency of the banks reflect the economic and managerial activities that took place during the time period analyzed. Table 13-5 shows the results for one of the banks. From all the tables such as these a comparison of the various bank performances can be made.

The results of both models are presented here and the Overall Efficiency results only are shown (not how they were produced). When these results are combined with the Technical Efficiency detailed above, a Graph is constructed for all six banks as a group to show the industry trends as can be seen in the Figure below.

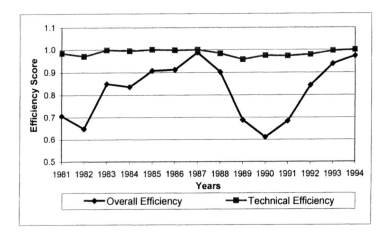

Figure 13-5. Technical and Overall Efficiency for all banks for each year

The striking observation here is how well the overall efficiency reflects the economic events of the 1980s and 1990s in Canada. A boom through the latter part of the 1980's followed the recession in the early part of the decade. The performance again declines with the next economic down cycle as the 1990s arrive and then rises as expected when the economy rebounded later in the decade.

Technical efficiency on the other hand shows a very consistent record as the banks are in a lock step with each other in a very highly controlled and extremely competitive environment where producing transactions is not effected much by the economic environment.

This section illustrated a way in which a small number of DMUs can be analyzed when a temporal database is available.

5.3.3 Malmquist Productivity Index

The banking industry is continually being shaped by market forces, which include world-wide competition, technological advances in production processes and non-banking firms competing for business. The continuous disintermediation is being accelerated by technology that enables competitors to enter markets merely by using the Internet and the World Wide Web. Domestic banks in countries that traditionally were almost oligopolies also face erosion of their market share by new competitors, such as "white label" ATM machines, large retailers that provide a "cash back" service to customers, foreign banks doing business strictly electronically, and via the Internet. Hence, productivity gains for banks will partially be caused by technological advancements and partially by more efficient management. How does one separate these effects? One method is the Malmquist index (for the technical development of this approach see the book by Coelli, Rao and Battese, 1998).

Productivity growth measurement involves separating changes in *pure productivity* (level of inputs necessary to produce a level of outputs) from changes in the *relative efficiency* of the DMUs with time. The Malmquist Productivity Index can be used to measure the *productivity growth* of DMUs between any two time periods t_1 and t_2. In this method, the relative distances of the DMUs from the frontier are used in conjunction with the relative changes in the position of the frontier from one period to the next to reveal the total changes in the productivity of a DMU.

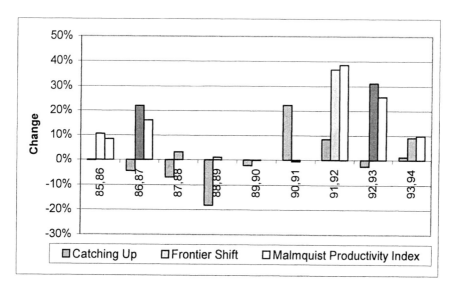

Figure 13-6. Malmquist Productivity Index components for adjacent periods

In the Malmquist Productivity Index approach a DMU is studied at two different points in time with respect to a common reference technology. Hence, *productivity* change is measured as the ratio of distances to the common technology, while *technical efficiency* change can be expressed as the relative distance from an observation to the frontier, keeping constant observed proportions between outputs and inputs. The path to the frontier can be accomplished either by input reduction or by output augmentation. The Malmquist Index can then be expressed as a ratio of corresponding efficiency measures for the two observations of the DMU.

The mathematical formulation of the Malmquist Productivity Index can also be found in a number of papers (c.f. Malmquist 1953, Cave et al. 1982, Färe et al. 1994). In this section the results from the 14 year panel data (Aggarwall 1996) are shown applying the Malmquist Productivity Index to DEA window analysis scores. The following Figure shows the catching up, frontier shift and Malmquist Productivity Index change for the average of the 6 banks studied. The technology is fixed for 1985.

As can be seen from Figure 13-6, the banks, on average, are not keeping pace with the frontier's movement as their catching up component is negative for 5 out of the nine periods examined here. When the changes are evaluated over longer periods of time (1985-94, 1985–89 and 1989–94) a similar picture emerges as seen in Figure 13-7.

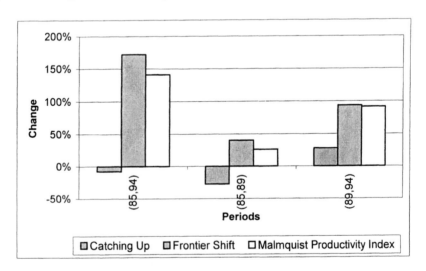

Figure 13-7. Average productivity change for the banks over longer periods

These results indicate that the Canadian banks increased their productivity quite significantly during the 10 year period presented here, in fact, they more than doubled the services delivered per unit resource consumed from 1985 to 1994. However, it is evident that such growth is mainly due to the frontier shift that resulted from the massive technology deployment that occurred during the period. The catching up component is actually negative for the first 4 years and for the entire 10 year period as shown by the bars in Figure 13-7. The conclusion from this is that the banks have actually been falling behind in management induced productivity rather than catching up. The lesson for management is that there are opportunities to improve productivity in the catching up dimension and that this could make a significant contribution to increased efficiency.

6. BANK BRANCHES AS DMUS

From many aspects, bank branch studies are more popular because from a managerial point of view, improvements can be achieved directly by the branch managers and so results from the analyses affecting the bottom line are close at hand. In cross bank analysis the identified benchmarks may be hard to use as role models because the difficulties in relating to what others are doing gives rise to criticisms of unfair and inequitable outcomes based on what demands are made on people. Therefore, there are more practical analyses where bank branches of the same bank were used as DMUs.

Traditional productivity measures were first devised in manufacturing concerns where absolute standards were available for comparison. All participants knew what a "good job" is and the standard was not debatable. Moreover, when productivity fell, the reasons were relatively easy to find and remedied. But services industries are fundamentally different organizations and while the inputs are no more difficult to identify (i.e. labor and other operational expenses), the outputs are much more troublesome to properly define and more importantly, measure. Some of the outputs may be such hard to measure variables as: how well is the customer served? How many transactions per period are possible per staff member? What types of transactions are appropriate to measure, each with varying degrees of difficulty and time? What are the metrics for these different types of transactions? And then there are no theoretical maximums for these measures. Hence, reaching 100% productivity can not be observed either.

To illustrate the methodology of how to build bank branch models data from a large Canadian bank's with a branch network was used (Rouatt 2003). The dataset has 816 branches with very extensive and detailed data items for each branch.

The decision was to build three different models which could be considered as three different "camera angles" on productivity – each yielding different improvement opportunities for the branch. Furthermore, by introducing three models, there is a natural separation of performance dimensions and any branch can perform well in one or more dimensions while lagging in another. Managers like the idea of "looking good" in some sense and they may more readily accept the need to improve in areas where they are not excelling in if they feel that they get credit for at least some aspects of their work.

6.1   The Production Model

In branch bank analyses, the first model is usually a production model to show how well the branch is able to handle the transactions they are required to do. These transactions may be face-to-face with the customer in the branch, carried out in the back office or even delivered at customer premises. In all cases, however, the efficiency measures are aimed at finding the best performers among the branches so that they can be both rewarded and used as benchmarks for others. The branch is a service "factory" and customer satisfaction is a key outcome of a good effort.

When creating the Production Model, transactions were divided based on the three main customer types - Corporate, Commercial, and Retail (Rouatt 2003). Individuals were considered retail customers, SME (small and medium sized) businesses were considered commercial, and large businesses (i.e. Wal-Mart, Microsoft, and Air Canada, BP Petroleum) were considered corporate.

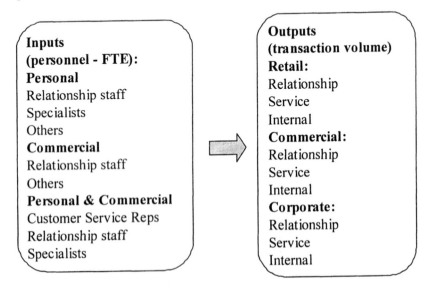

Figure 13-8. Production Model Inputs and Outputs

This model is designed to measure staff performance in a Branch in terms of producing transactions. Schaffnit et al (1997) carried out a similar study on another Canadian bank's branches examining performance with the purpose of estimating how a branch would be effected if a substantial portion of the internal (back office) transactions were moved to centralized facilities. The study showed that for a significant portion of the branches the time freed up this way could not be used for other purposes. This tends to be the problem for small branches where they are already at minimal staffing levels.

6.2 Profitability Model

The profitability model is designed to examine the process of how well a branch uses its inputs (expenses) to produce revenues (Rouatt 2003). This is fairly straightforward because it treats the branch as the producer of a product as opposed to offering a service. While this does simplify the identification of inputs and outputs there are still some complexities to address. There is the issue of separating some revenues from their products. For example, a bank provides a below prime interest loan to a customer but requires that a certain percentage of the funds lent be held in the bank account (which pays minimal or no interest to the customer). This results in certain products appearing more profitable because the customer is paying interest on money that the bank has not released in practice. In this case, there is a lower lending revenue (because of the below prime interest) but higher commercial banking revenue from the interest earned on the portion left on deposit at the bank. The model then looks at those measures that show how well the branch is doing in producing profits as seen in the Figure below.

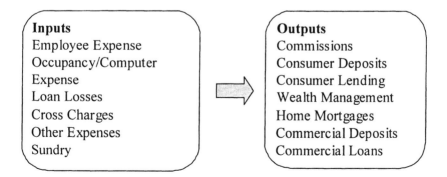

Figure 13-9. Profitability Model Inputs and Outputs

Expenses included were those that branch management was able to directly influence, others such as depreciation and capital expenditures were excluded.

Loan losses, while clearly an output measure are included as an input because the need is for minimizing this outcome, thus penalizing those branches with higher losses.

6.3   Intermediation Model

The idea behind the intermediary model is to examine the bank's ability to take in deposits and lend the money to worthy and profitable borrowers. This approach was used in some of the earlier banking studies such as Colwell and Davis (1992). Deposits, liabilities, and assets are the "raw materials of investible funds" (Berger and Humphrey 1990). The assumption is that the branches, in order to maximize income, should attempt to lend or invest as much money as possible, thus minimizing the funds on hand.

As mentioned earlier, there has been much debate on whether to include deposits as an input or output. Characterizing deposits as an input unfairly penalizes branches for seeking new depositors – an important sales function – and banks generate a significant amount of non-interest revenue from handling deposits. However, deposits are consistently positioned as an output under the user cost approach (Fixler and Zieschang 1999). Colwell and Davis (1992) also found that using just assets (loans plus investments) that earn an interest, the model excluded other assets, thus inflating the unit costs of larger banks. Their model uses deposits as an input to illustrate this alternative.

Loan quality and losses are critical factors in a bank's health. Berger and DeYoung (1996) point out that the majority of research on the causes of bank and thrift failures has found that failing institutions have large proportions of non-performing loans prior to failure. Thus, loan losses were included in the model, but on the input side to appropriately reward those with small loan losses. On the other hand, too low loan losses may also be a problem because that implies that the bank is also passing up good business by being overly restrictive in its credit criteria.

The Intermediary Model involves measures that most closely reflect the branch's ability to lend to and invest on behalf of, their customers. The following Figure shows the selection of measures and the placement into input or output.

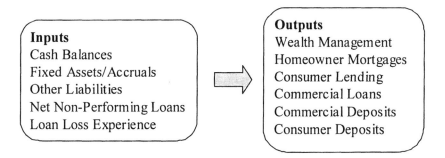

Figure 13-10. Intermediary Model Inputs and Outputs

6.4   Model Results

It is easy to see that a considerable effort must be made to determine which variables to include and where (Rouatt 2003). Frequent management consultation is a must to ensure that all items are considered and approved by those who will eventually have to execute the suggested changes. Buy-in is essential for successful DEA result implementation. So the above three models, all constructed during the same study and using the same data source, illustrates an approach that is likely to provide considerable amounts of useful information to the branch manager while being regarded as both fair and equitable.

From a practical point of view, branch managers can see what aspects of their branch need improvement and where they are doing well. It is not at all unusual for a branch to show efficient performance in one or two models while they could improve in the third. Moreover, they can contact other branch managers with whom they can exchange information on how to improve, and often this will be a mutual exchange of information in their respective areas of excellence. A spillover effect of this process is the morale level, which should not be negatively effected since they are exchanging information on an equal basis, and neither has to feel inferior to the other. It is not a "competition" after all.

Most models can be processed either as input oriented or output oriented and sometime both are carried out for the same model. Table 13-6 shows the possibilities in this set of three models. The X represents the orientation and technology actually used in this work (Rouatt 2003) and the O represents orientation and technology that could have been used but was not. Only the Production Model is not a typical candidate to be used in an output oriented analysis because the branch can not do much, if anything, about getting customers into the branch to do business.

Table 13-6. Model orientation possibilities

Model	CCR-I	BCC-I	CCR-O	BCC-O
Production	X	X		
Profitability	X	X	O	O
Intermediation	X	X	O	O

Table 13-7 shows the results from different models for the 816 branches of this large bank.

This is a typical situation where 25-40% of the branches are DEA efficient, whatever the model. Understandably, this makes it difficult for senior managers to motivate those who are already on the frontier. The advantage of using three different models is that it is seldom the case that the same branch is efficient in all the models. Hence opportunities for improvement will be present.

However, this does not permit us to examine the individual effects on the whole system and the trade-offs involved in focusing on certain areas. Managers may choose to focus on only one or two aspects of their branch as part of a competitive or purely reactive strategy. A branch attempting to improve its lending results (or intermediation efficiency) could potentially lower their production efficiency. For example, to improve their average loan quality and reduce loan defaults, a branch may choose to increase the number of loan officers and the amount of time they spend with each client. However, this may well have a negative impact on the branch's production efficiency, as there is an increase in staffing (additional loan officers) combined with a decrease in the number of transactions performed (due to more time being spent with each client). Additionally, there could also be adverse effects on profitability (due to the increased staffing costs) if there was not a subsequent reduction in loan losses to offset the cost increase.

Table 13-7. DEA Results Summary

Model	Efficiency Scores	Production Model		Profitability Model		Intermediation Model	
		BCC	CCR	BCC	CCR	BCC	CCR
Global	% Efficient	33%	21%	38%	26%	29%	20%
	Average	0.77	0.71	0.87	0.82	0.81	0.76
	Minimum	0.28	0.25	0.32	0.26	0.34	0.29
	Std. Dev.	0.21	0.20	0.15	0.16	0.17	0.18

6.5   Senior Management

Unfortunately, most DEA studies of banks and financial services firms limit themselves to only one point of view, typically that of the individual they work with or for. However, banks tend to be large and complex organizations in a managerial sense. Hence, models, even as focused as those shown thus far, might not meet the needs of the banks' top managers when you consider that they have the task of managing progress in all aspects of the firm. Therefore, they tend to use some ranking mechanism that will sort their branches from best to worst using either a single ratio, or most often a combination of "important" ratios. This results in rankings from 1 to n but the branches in the last half of the rankings may not be properly motivated to improve themselves in the standings. After all what difference does it make whether you are 659th or 597th?

Clearly, DEA can be a very useful tool to offer managers valid and achievable goals as shown above. But the typical DEA study shows a substantial number of branches as efficient and that often means that there is little differentiation for far too many branches and this ends up in some top managers rejecting DEA as a valid process for their firms. In any case, top management needs tools appropriate to their management processes.

6.6 A Two Stage Process

A way to address top management needs is to conduct a two stage study (Rouatt 2003). For example, one can carry out the three separate studies shown above and then devise a second stage where the outputs from the three models are used as variables.

While gauging branch performance is important, being able to rank them is beneficial as well – mainly for assurance that the methodology validates the bank's own. A ranking scheme enables senior management to target branches in most need of assistance and to reward those branch managers who perform exceptionally well. With an approach that encompasses all three performance measures outlined above, management is able to view a much more comprehensive ranking of their branches, but this ranking is one where targets for improvement are available.

A more holistic approach was the creation of a second stage that measures the combined effects of the three models shown above. This second stage (or combined) model is formed by using the input oriented CCR DEA results from the initial three models as outputs in the new model and a dummy variable as the input.

The Slacks Based Measure (SBM) Additive model was used to provide a more representative measure of efficiency. This resulted in the efficiency scores being derived purely from the slacks. It does however give the same number of efficient DMUs as the BCC model. The SBM Additive model can be stated as shown in Equation (13.4) from Cooper et al. (2000).

$$\rho = \left(\frac{1}{m} \sum_{i=1}^{m} \frac{x_{i0} - s_i^-}{x_{i0}} \right) \left(\frac{1}{s} \sum_{r=1}^{s} \frac{y_{r0} + s_r^+}{y_{r0}} \right)^{-1}$$

(13.4)

As the input was a dummy variable of 1 this eliminated the first half of the equation (the x_{i0} terms are all 1 and the input slacks, s_i^-, are all 0). A DMU is considered efficient if $\rho = 1$; this condition is met only when all slacks (s^- and s^+) are equal to zero (see Cooper et al. 2000). Due to the elimination of the first half of the equation, the SBM Additive model is simply the average of the inverted scores. Using these scores as opposed to just the average of the three scores makes sense. The inverted scores penalize branches for being *overly* bad in one area. For example, a branch with inverted output scores of 0.9, 0.9 and 0.9 would have a combined score of 1.11 (or 90% efficient). Conversely, another branch with scores of 1.0, 1.0, and 0.7 would have a combined score of 1.14 (or 88% efficient). The simple mean of both branches is the same (0.90) but because the inverses were used, the branch with the 0.7 output score is penalized for performing poorly in one area. Here are the results:

Table 13-8. Model Results

	SBM Additive Model
Average	0.77
Std. Deviation	0.13
Median	0.78
Min.	0.41
% Efficient	7%

As can be seen, the second stage SBM model had yielded significant benefits for senior managers over any single stage model, even if there are three as shown here. First, the model reduced the portion of efficient units from a range of 29% - 38% for the BCC model and 20% - 26% for the CCR model to just 7%. Managers can gain substantial new knowledge about their organization's overall performance and, if they implement the DEA suggestions, improve in a meaningful way.

6.7 Targeted Analysis

With all the tools available to the DEA analyst today, targeted work can be planned and executed with excellent results. The following is an example where management was planning to change the way they operate back office branch work. The idea was to concentrate the back office work to a number of regional centers where specialized processes with trained staff could do the work more efficiently than in the branch. If this worked, they could either reduce the workforce in the branch or free them up for more person-to-person customer work. The question was: Should this be done?

A dataset of 291 branches of a large Canadian bank was used for the study (see Schaffnit et al. 1997). A personnel or staffing model was constructed with management's assistance. Six types of employees were identified and 6 business transactions (involving customer contact) plus 3 maintenance or back office transactions (not involving direct customer contact) were agreed upon. The transactions were used as outputs while the staff (counted as FTEs) were the inputs. The interaction between staff and services performed are shown in Table 13-9.

Notice, however, that not all staff were able (trained) to perform all required transactions.

Various input oriented models can be applied here, but performance can only be fairly measured if appropriate constraints are placed on the multipliers. A BCC input oriented model was chosen and applied to the whole model and then a reduced model where the last 3 outputs were omitted.

The findings indicated that if the bank removed all the back office work from all the branches, many branches with minimum staff configuration would end up losing work without an opportunity to do something else. So if the bank did implement these plans across the branch network, it would result in a significant productivity loss as some people would have had nothing to do once the back office work was removed from the branch.

Table 13-9. Connections for work being done

Staff type / Transactions	Teller	Ledger handler	Accounting	Supervision	Typing	Credit
Business						
Counter interaction	X	X		X		
Counter sales		X	X	X		
Securities			X	X		
Deposit sales			X	X		
Commercial Loans			X	X	X	
Personal Loans			X	X	X	
Maintenance						
Term			X	X	X	
Commercial Loan				X	X	X
Personal Loan				X	X	X

6.8 Environmental Effects

In many real world situations DEA could be applied with considerable success but yet it is found that the typical DEA results are not satisfactory because management (the "measured") push back citing unfairness due to a lack of consideration of "external effects". Much work has been done to address these concerns and the following is an example where economic conditions are incorporated in the models along with a measure of the branch's business risks on its portfolio of loans. The dataset is of 90 commercial branches of a large Canadian bank, across 8 economic districts in Canada (Paradi and Schaffnit 2003).

Several models were built and both input and output oriented BCC models were utilized. The regional economic data was obtained from the Bank's economics department and represented the average rate of change of the real district domestic product. This factor was used to avoid unfair comparisons of branches operating, for example in the economically disadvantaged Atlantic region of Canada to their West Coast counterparts where the economic conditions are a lot more favorable.

The following table shows the results of the study district by district, restricted by the environmental variable - the district domestic product (growth factor).

Table 13-10. Overall results by District (Paradi and Schaffnit 2003)

District	D1	D2	D3	D4	D5	D6	D7	D8
Growth factor	2	2.5	3.3	3.3	2.7	2.5	2.9	3.3
# Branches	10	12	27	13	6	2	8	12
# Effective	7	1	1	1	4	0	1	2
% Effective	70	8	4	8	67	0	12	17
Average score	0.91	0.4	0.38	0.36	0.83	0.36	0.62	0.6

Clearly, District D1 has the highest percentage of effective branches. However, this is the undesirable result of using the economic constraints on peer selection because this District having the lowest economic growth rate selects peers only from among others in the District. Hence, unrestricted analysis must also be considered before final managerial decisions are taken. This is a good illustration of the problem one can encounter when both relatively few DMUs are available and environmental factors are used to restrict full comparison.

7. VALIDATION

Managers are interested in improving their operations and whatever measurement methods are used to evaluate performance needs to be validated against other methods or even just against the managers' experience or expectations of how their world works. Formal validation is a critical success factor in having the results of an analysis accepted by those who are being measured. There are various approaches to validating results and this is not the topic of this chapter. But the following example illustrates one of these approaches and may inspire the reader to devise their own strategies.

7.1   Validating a method statistically

A useful approach to validating DEA work, when appropriate, is to use statistical methods, such as the Monte Carlo simulation. Recall from section 5.1 that a handicapping function was used to adjust for cultural differences across three different banks (Yang and Paradi 2003). The goal was to eliminate corporation specific influences on the branches to be compared. The results for the branches within each bank after handicapping proved to be quite similar and much closer than the unadjusted results. The question naturally comes to mind: how valid is this approach?

To address this issue 100 sets of random numbers were generated and used as handicapping factors and compared to the specific handicapping factors defined in the study referred to above. Since 1.0 represents the average performance and the standard deviation (σ) is 0.138 in the handicapping system, the random numbers were generated to have the population mean of 1.0 and standard deviation of 0.138 to keep them within the range of $1.0 \pm 2\sigma$ as required (Yang and Paradi 2003).

The results show that most of the random numbers increased average performance differences between the banks and their standard deviations. Some of the random numbers did not have very much effect on the average efficiency and the standard deviation, but the differences between the banks were still much larger than when using the real handicapping function. Only one set of random numbers produced results nearly identical to those of the handicapping system. However, this specific set was almost exactly the same as the handicapping factors calculated, hence the very close results. Therefore the assertion that the handicapping system developed is valid to adjust for cultural differences between banks is supported by this test.

8.   CONCLUSIONS

Banks offer an almost ideal study subject because they are ubiquitous, operate both domestically in every country in the world as well as internationally, either directly (large banks from the developed Countries) or indirectly through corresponding relationships. They have many branches, are typically well managed and collect copious amounts of data on very detailed operational activities. Moreover, they are well accustomed to measuring performance and controlling operations by the very nature of what they do.

Nevertheless, considerable caution is advisable when collecting data because notwithstanding the large amounts available, there are many small and not so small errors in every dataset. So outliers may, in fact, be benefiting from data errors as easily as grossly inefficient DMUs may well be suffering from bad data. Also, banks tend not to define data in exactly the same manner, so "assets" may mean generally the same thing, but in fact, there are different components included depending on the banks' own decision or even by legislation, which complicates cross-bank or cross-country analyses.

Very different approaches must be taken when studying banks as the DMUs as opposed to the cases where the bank branches are the DMUs. This is naturally the case because the availability of data and the measures that matter to a bank, as an entity, or to a branch, as a unit, are very different. Much discussion also arises when defining the models with respect to which variables to include as inputs and which ones as output. Consideration must be given to both the technical issues involved and to the perception by management when results are delivered.

Much has been written on bank performance in many different domains of study and much more will be done in the future as this is an interesting and challenging field to explore. DEA is a significant tool in this arena because it is non-parametric, does not need preconceived models, can be adapted to many views, is units independent (except for the Additive model) and allows for large models where the data is available. But it also has shortcomings, so the best approach is to use it in conjunction with other methodologies and validate the results against accepted processes in the financial services industry.

REFERENCES

1. Aggarwall, V., 1996, Performance Analysis of Large Canadian Banks over Time using DEA", MASc Dissertation, Centre for Management of Technology and Entrepreneurship.
2. Akhavein, J.D., Berger, A.N., and Humphrey, D.B., 1997, The Effects of Megamergers on Efficiency and Prices: Evidence from A Bank Profit Function, Review of Industrial Organizations 12, 95-139.
3. Al-Faraj, T.N., Alidi, A.S., and Bu-Bshait, K.A., 1993, Evaluation of Bank Branches By Means of Data Envelopment Analysis, International Journal of Operations and Production Management 13, No. 9, 45-53.

4. Athanassopoulos, A.D., 1997, Service Quality and Operating Efficiency Synergies for Management in The Provision of Financial Services: Evidence from Greek Bank Branches, European Journal of Operational Research 98, 300-313.
5. Athanassopoulos, A.D., 1998, Non-Parametric Frontier Models for Assessing the Market and Cost Efficiency of Large-Scale Bank Branch Networks, Journal of Money, Credit, and Banking 30, No. 2, 172-192.
6. Athanassopoulos, A.D., 2000, An Optimization Framework of the Triad: Service Capabilities, Customer Satisfaction and Performance, In: P.T. Harker, and S.A. Zenios (Eds.), Performance of Financial Institutions, Cambridge University Press, Cambridge, England, 312-335.
7. Athanassopoulos, A.D., and Giokas, D., 2000, The Use of Data Envelopment Analysis in Banking Institutions: Evidence from the Commercial Bank of Greece, Interfaces 30, No. 2, 81-95.
8. Avkiran, N.K., 1999, The Evidence on Efficiency Gains: The Role of Mergers and the Benefits to the Public, Journal of Banking and Finance 23, 991-1013.
9. Banker, R.D., and Morey, R., 1986, The Use of Categorical Variables in Data Envelopment Analysis", Management Science, 32(12), 1613-1627.
10. Bauer, P.W., 1990, Recent Developments in the Econometric Estimation of Frontiers, Journal of Econometrics 46, 39-56.
11. Berg, S.A., Førsund, F.R. and Jansen, E.S., 1992, Malmquist Indices of Productivity Growth During the Deregulation of Norwegian Banking, 1980-89, *Scandinavian Journal of Economics*, vol. 94 (Supplement)
12. Berger, A.N., and DeYoung, R., 1996, Problem Loans and Cost Efficiency in Commercial Banks, Working Paper, 96-01, Wharton Financial Institutions Paper.
13. Berger, A.N, Hancock, D. and Humphrey, D.B., 1993, Bank Efficiency Derived from the Profit Function, Journal of Banking and Finance 17, No. 2/3, 317-348.
14. Berger, A.N., Hunter, W.C., and Timme, S.G., 1993, The Efficiency of Financial Institutions: A Review and Preview of Research Past, Present, And Future, Journal of Banking and Finance 17, 221-249.
15. Berger, A.N., and Humphrey, D.B., 1990, Measurement and Efficiency Issues in Commercial Banking, Finance and Economic Discussion Series, Working Paper #151, Board of Governors of the Federal Reserve System.
16. Berger, A.N. and Humphrey, D.B., 1991, The Dominance of Inefficiencies over Scale and Product Mix Economies in Banking, Journal of Monetary Economics 28, No. 1, 117-148.

17. Berger, A.N. and Humphrey, D.B., 1997, Efficiency of Financial Institutions: International Survey and Directions For Future Research, European Journal of Operational Research 98, 175-212.
18. Berger, A.N., 1998, The Efficiency Effects of Bank Mergers and Acquisitions: A Preliminary Look at The 1990s Data, In: Amihud, Y., And Miller, G. (Eds.), Bank Mergers and Acquisitions, Kluwer Academic Publishers, Dordrecht, 79-111.
19. Bhattacharyya, A., Lovell, C.A.K., and Sahay, P., 1997, The Impact of Liberalization on the Productive Efficiency of Indian Commercial Banks, European Journal of Operations Research 98, 333-346
20. Bukh, P.N.D., Berg, S.A., and Førsund, F.R., 1995, Banking Efficiency in the Nordic Countries: A Four-country Malmquist Index Analysis, Working paper, University of Aarhus, Denmark
21. Camanho, A.S. and Dyson, R.G., 1999, Efficiency, Size, Benchmarks and Targets for Bank Branches: an Application of Data Envelopment Analysis, Journal of The Operational Research Society 50, 903-915.
22. Caves, D. W., Christensen, L. R. and Diewert, W. E., 1982, The Economic Theory of Index Numbers and the Measurement of Input, Output and Productivity, Econometrica, 50, 1393-1414.
23. Chang, K.P., 1999, "Measuring Efficiency with Quasiconcave Production Frontiers", European Journal of Operational Research, 115(3), 497-506.
24. Charnes, A., W.W. Cooper, A.Y. Lewin and L.M. Seiford (eds.), 1994, Data Envelopment Analysis: Theory, Methodology and Applications. Kluwer Academic Publishers.
25. Clark, J.A., 1996, Economic Cost, Scale Efficiency, and Competitive Viability in Banking, Journal of Money, Banking and Credit 28, No. 3, 342-364.
26. Coelli, T., Rao, D.S.P. and Battese, G. E. 1998 An Introduction to Efficiency and Productivity Analysis, Kluwer Academic Publishers, Boston.
27. Colwell, R.J., Davis, E.P., 1992, Output and Productivity in Banking, Scandinavian Journal of Economics.
28. Cook, W.D., Hababou, M. and Tuenter, H.J., 2000, Multicomponent Efficiency Measurement and Shared Inputs in Data Envelopment Analysis: an Application to Sales and Service Performance in Bank Branches, Journal of Productivity Analysis 14, 209-224.
29. Cooper, W.W., Seiford, L.M. and Tone, K. (eds.), 2000, Data Envelopment Analysis: A Comprehensive Text with Models, Applications, References, and DEA-Solver Software, Kluwer Academic Publishers, Boston.

30. DeYoung, R., 1993, Determinants of Cost Efficiencies in Bank Mergers, Economic and Policy Analysis, Working Paper 93-01, Washington: Office of the Comptroller of the Currency.
31. DeYoung, R., 1997, A Diagnostic Test for the Distribution-Free Efficiency Estimator: An Example Using U.S. Commercial Bank Data, European Journal of Operational Research 98, No. 2, 243-249.
32. DeYoung, R., 1998, Management Quality and X-Inefficiency in National Banks, Journal of Financial Services Research 13, No. 1, 5-22.
33. Drake, L. and Howcroft, B., 1994, Relative Efficiency in The Branch Network of a UK Bank: An Empirical Study, Omega 22, No. 1, 83-91.
34. Favero, C., and Papi, L., 1995, Technical Efficiency and Scale Efficiency in the Italian Banking Sector: a Non-Parametric Approach, Applied Economics, 27, 385-395
35. Färe, R., Grosskopf, S., Norris, M. and Zhang, Z., 1994, Productivity Growth, Technical Progress, and Efficiency Changes in Industrialized Countries, American Economic Review, 84, 66-83.
36. Fixler, D.J. and Zieschang, K.D., 1993, An Index Number Approach to Measuring Bank Efficiency: An Application to Mergers, Journal of Banking and Finance 17, 437-450.
37. Fixler, D., and Zieschang, K., 1999, The Productivity of the Banking Sector: Integrating Financial and Production Approaches to Measuring Financial Service Output, Canadian Journal of Economics, 32, No. 2.
38. Fukuyama, H., 1993, Technical and Scale Efficiency of Japanese Commercial Banks: A Non-Parametric Approach, Applied Economics 25, 1101-1112
39. Giokas, D., 1991, Bank Branch Operating Efficiency: A Comparative Application of DEA and the Log-Linear Model, Omega 19, No. 6, 549-557.
40. Golany, B. and Storbeck, J.E., 1999, A Data Envelopment Analysis of the Operation Efficiency of Bank Branches, Interfaces 29, No. 3, 14-26.
41. Grabowski, R., Mathur, I. and Rangan, N., 1995, The Role of Takeovers in Increasing Efficiency, Managerial and Decision Economics 16, No. 3, 211-224.
42. Haag, S.E. and Jaska, P.V., 1995, Interpreting Inefficiency Ratings: An Application of Bank Branch Operating Efficiencies, Managerial and Decision Economics 16, No. 1, 7-15.
43. Hao, J., Hunter, W.C. and Yang, W.K., 2001, Deregulation and Efficiency: The Case of Private Korean Banks, Journal of Economics and Business 53, No. 2/3, 237-254.
44. Haynes, M. and Thompson, S., 1999, The Productivity Effects of Bank Mergers: Evidence from the UK Building Societies, Journal of Banking and Finance 23, 825-846.

45. Kantor, J. and Maital, S., 1999, Measuring Efficiency by Product Group: Integrating DEA with Activity-Based Accounting in A Large Mideast Bank, Interfaces 29, No. 3, 27-36.
46. Kaparakis, E.I., Miller, S.M. and Noulas, A.G., 1994, Short-Run Cost Inefficiency of Commercial Banks: A Flexible Stochastic Frontier Approach, Journal of Money, Credit And Banking 26, No. 4, 875-894.
47. Kinsella, R.P., 1980, The Measurement of Bank Output, Journal of the Institute of Bankers in Ireland, 82, pp. 173-183.
48. Lovell, C.A.K. and Pastor, J.T., 1997, Target Setting: An Application to a Bank Branch Network, European Journal of Operational Research 98, 290-299.
49. Malmquist, S. 1953, Index Numbers and Indifference Surfaces, Trabajos de Estatistica, 4,209-42.
50. Nash, D. and Sterna-Karwat, A., 1996, An Application of DEA To Measure Branch Cross Selling Efficiency, Computers Operations Research 23, No. 4, 385-392.
51. Oral, M., Kettani, O. and Yolalan, R., 1992, An Empirical Study on Analyzing the Productivity of Bank Branches, IIE Transactions 24 No 5.
52. Oral, M. and Yolalan, R., 1990, An Empirical Study on Measuring Operating Efficiency and Profitability of Bank Branches, European Journal of Operational Research 46, 282-294.
53. Paradi, J. C. and Schaffnit, C., 2003, Commercial Branch Performance Evaluation and Results Communication in a Canadian Bank - a DEA Application. Accepted for publication in the European Journal of Operational Research.
54. Parkan, C., 1987, Measuring The Efficiency of Service Operations: An Application to Bank Branches, Engineering Costs and Production Economics 12, 237-242.
55. Pastor, J., Perez, F., and Quesada, J. 1997, Efficiency Analysis in Banking Firms: An International Comparison, European Journal of Operations Research 98, 396-408
56. Peristiani, S., 1993, Evaluating The Postmerger X-Efficiency And Scale Efficiency of U.S. Banks, Federal Reserve Bank Of New York.
57. Peristiani, S., 1997, Do Mergers Improve the X-Efficiency and Scale Efficiency of U.S. Banks? Evidence from the 1980s, Journal of Money, Credit and Banking 29, No. 3, 326-337.
58. Resti, A., 1998, Regulation Can Foster Mergers, Can Mergers Foster Efficiency? The Italian Case, Journal of Economics and Business 50, 157-169.
59. Rouatt, S., 2003, Two Stage Evaluation of Bank Branch Efficiency Using Data Envelopment Analysis, MASc Dissertation, Centre for Management of Technology and Entrepreneurship.

60. Schaffnit, C., Rosen, D. and Paradi, J.C., 1997, Best Practice Analysis of Bank Branches: An Application of DEA in a Large Canadian Bank, European Journal of Operational Research 98, No. 2, 269-289.
61. Sherman, H.D. and Ladino, G., 1995, Managing Bank Productivity Using Data Envelopment Analysis (DEA), Interfaces 25, No. 2, 60-73.
62. Sherman, H.D. and Gold, F., 1985, Bank Branch Operating Efficiency: Evaluation with Data Envelopment Analysis, Journal of Banking and Finance 9, No. 2, 297-316.
63. Smith, P., 1990, Data Envelopment Analysis Applied to Financial Statements, OMEGA International Journal of Management Science, 18(2), pp. 131-138.
64. Soteriou, A.C. and Stavrinides, Y., 1997, An Internal Customer Service Quality Data Envelopment Analysis Model for Bank Branches, International Journal of Operations and Production Management 17, No. 8, 780-789.
65. Soteriou, A, and Zenios, S.A., 1999, Operations, Quality, and Profitability in the Provision of Banking Services, Management Science 45, No. 9, 1221-1238.
66. Sowlati, T. and Paradi, J. C. 2003, Establishing the "Practical Frontier" in Data Envelopment Analysis, Working Paper.
67. Thompson, R. G., Brinkmann, E. J., Dharmapala, P. S., Gonzales-Lima. M. D., and Thrall, R. M., 1997, DEA/AR Profit Ratios and Sensitivity of 100 Large U. S. Banks, European Journal of Operations Research 98, 213-229
68. Tulkens, H., 1993, On FDH Efficiency Analysis: Some Methodological Issues and Applications to Retail Banking, Courts and Urban Transit, Journal of Productivity Analysis 4, No. 1/2, 183-210.
69. Vance, H., 2000, Opportunity Index Development for Bank Branch Networks, MASc Dissertation, Centre for Management of Technology and Entrepreneurship.
70. Vander Vennet, R., 1996, The Effect of Mergers and Acquisitions on the Efficiency and Profitability of EC Credit Institutions, Journal of Banking and Finance 20, 1531-1558.
71. Vassiloglou, M. and Giokas, D., 1990, A Study of The Relative Efficiency of Bank Branches: An Application of Data Envelopment Analysis, Journal of The Operational Research Society 41, No. 7, 591-597.
72. Vela, S. and Paradi, J. C., 2002, Measuring the Effects of Cultural Differences in Bank Branch Performance Utilizing DEA, Working paper, Centre for Management of Technology and Entrepreneurship.

73. Yang, Z. and Paradi, J. C., 2003, Evaluating Competitive Banking Units Using "Handicapped" Data Envelopment Analysis to Adjust for Systemic Differences. Working Paper.

74. Yue, P., 1992, Data Envelopment Analysis and Commercial Bank Performance: A Primer with Applications to Missouri Banks, Federal Reserve Bank of St. Louis Review, Vol. 74-1.

75. Zenios, C.V., Zenios, S.A., Agathocleous, K. and Soteriou, A.C., 1999, Benchmarks of the Efficiency of Bank Branches, Interfaces 29, No. 3, 37-51.

Part of the material in this chapter is adapted from Schaffnit, C., Rosen, D. and Paradi, J.C., 1997, "Best Practice Analysis of Bank Branches: An Application of DEA in a Large Canadian Bank", European Journal of Operational Research 98, No. 2, 269-289. and Paradi, J. C. and Schaffnit, C., 2003, "Commercial Branch Performance Evaluation and Results Communication in a Canadian Bank - a DEA Application", accepted for publication in the European Journal of Operational Research, with permission from Elsevier Science.

We are indebted to Dr. Asmild for proof reading the chapter, the graduate students in the CMTE who had offered some of their work, especially Steve Rouatt whose data and results were used here to illustrate different bank branch models and the two stage evaluation process.

AppendixBank Branch Performance DEA Models in Literature (Yang and Paradi 2003)

Author(s)	Variables	Sample	RTS	Orientation

Al-Faraj, Alidi and Bu-Bshait 1993	Inputs:
	No. of employees,
	% Employees with college degree,
	Years of experiences, Index for:
	  Location,
	  Expenditure on decoration
	  Average monthly salaries
	  Other operational expenses
	Outputs:
	Average monthly:
	  Net profit
	  Balance of current accounts
	  Balance of savings accounts
	  Balance of other accounts
	  Value of mortgages
	Index for loans
	No. of current accounts
		15 branches in Saudi Arabia	CCR	Input-Oriented Production Model

Athanassopoulos 1997	Inputs: No. of Employees, On-line and ATM, No. of computers Outputs: Deposit accounts, Credit transaction, Debit transactions Loan applications, Transactions involving commissions	68 branches in Greece	CCR	Input-Oriented Production Model
Athanassopoulos 1997	Inputs: Total non-interest costs, Total interest costs Outputs: Non-interest income, Total volume of loans, Time deposit accounts, Saving deposit accounts, Current deposit accounts	68 branches in Greece	BCC	Non-Radial Intermediation Model
Athanassopoulos 1998	Inputs: No. of transactions, Potential market, Sales representatives, Internal automatic facilities, No. of branch outlets in the surrounding area Outputs: Liabilities sales, Loans and mortgages, Insurance and securities, Number of credit cards	580 branches in UK	BCC	Output-Oriented Production Model

	sold			
Athanassopoulos 1998	Inputs: Direct labor costs, Technology facilities Outputs: No. of transactions, Liabilities sakes, Loans and mortgages, Insurance and securities, Number of credit cards sold	580 branches in UK	BCC	Input-Oriented Production Model

Athanassopoulos and Giokas 2000	Inputs:
	Labor hours, Branch size, Computer terminals, Operating expenditure
	Outputs:
	Number of transactions:
	  Easiest
	  Medium-easy
	  Most-difficult
	Credit transactions
	Deposit transactions
	Foreign receipts
		47 branches in Greece	CCR	Input-Oriented Production Model

Athanassopoulos and Giokas 2000	Inputs: Labor, Operating cost, Running costs of the building Outputs: Income from commissions Volume of loans Accounts: Time deposit Savings deposit Current deposit Demand deposit	47 branches in Greece	CCR	Output-Oriented Production Model
Camanho and Dyson 1999	Input: No. of employees Square meters of floor space Operational costs No. of external ATMs Outputs: No. of general service transactions No. of transactions in external ATMs No. of all types of accounts Value of savings Value of loans	168 branches in Portugal	CCR BCC	Input-Oriented Output-Oriented Production Model
Cook, Hababou and Tuenter 2000	Inputs: No. of staff:	20 branches in Canada	CCR	Input-Oriented Production Model

	Service, sales, support, other Outputs: No. of counter level deposits, No. of transfers between accounts, No. of retirement savings plan openings, No. of mortgage accounts opened.			
Drake and Howcroft 1994	Inputs: No. of interview rooms, No. of ATMs, Square meters of space, Management grades, Clerical grades, Stationary expenses Outputs: Teller transactions, lending products, Deposit products, Automated transfers, Clearing items, Ancillary business, Insurance business	190 branches in UK	CCR BCC	Input-Oriented Production Model
Giokas 1991	Inputs: No. of employees hours worked, Square meters of space, Operating costs excluding labor Outputs: Weighted no. of transactions: Deposit Credit Foreign receipts	17 branches in Greece	CCR BCC	Input-Oriented Production Model
Golany and Storbeck 1999	Inputs: Personnel: Officers Tellers Salaries employees Size of branch, Unemployment Outputs: Loans: Direct Indirect Commercial Equity Deposits: Checking Savings CDs Ave. No. of accounts per customer, Customer satisfaction	182 branches in US	BCC	Output-Oriented Production Model
Haag and Jaska 1995	Inputs: FTE personnel, Rent, Supplies Outputs: Number of transactions: Least time consuming	14 branches in US	BCC	Additive Production Model

	Medium-low time consuming Medium-high time consuming Most time consuming			
Kantor and Maital 1999	Inputs: Labor costs, Transactions; area for transactions only Outputs: Credit card; Weighted output index- transactions, Commissions on import and export, Commercial accounts, Saving accounts, Activity	250 branches in Mideast	CCR	Input-Oriented Production Model
Lovell and Pastor 1997	No Inputs Outputs: 17 performance targets including several deposit types, several loan types and several miscellaneous services	545 branches in Spain	BCC	Output-Oriented Production Model
Nash and Sterna-Karwat 1996	No inputs Outputs: Sales of four products associated with housing loans	75 branches in Australia	BCC	Additive Production Model
Oral and Yolalan 1990	Inputs: Labor, Terminals, Number of accounts, Credit applications Outputs: Transactions	20 branches in Turkey	CCR	Input-Oriented Production Model
Oral and Yolalan 1990	Inputs: Labor, Expenses, Interest Outputs: Income	20 branches in Turkey	CCR	Input-Oriented Mixed Model
Paradi and Schaffnit 2003	Inputs: Staff (5 types) IT expenses Rent Other non-interest expenses Outputs: Deposits Loans Fee income Maintenance	90 Com'l Branches in Canada	CCR BCC	Input-Oriented Production Model
Paradi and Schaffnit 2003	Inputs: Staff (5 types) IT expenses Rent Other non-interest expenses Non-accrual loans Outputs:	90 Com'l Branches in Canada	CCR BCC	Input and Output Oriented Strategic Model

	Deposits Loans Fee income Maintenance Deposit spread Loan Spread			
Parkan 1987	Inputs: Labor, Expenses, Space (rent), Space quality, Terminals, Marketing Outputs: Transactions, Customer survey, Error corrections	35 branches in Canada	CCR	Input-Oriented Production Model
Schaffnit, Rosen and Paradi 1997	Inputs Teller, Accounting, Supervisor, Typing, Credit Outputs: Counter transactions, Counter sales, Security transaction, Deposit sales, Commercial loan sales, Personal loan sales, Term accounts, Commercial loan accounts, Personal loan accounts	291 branches in Canada	CCR	Input-Oriented Production Model
Sherman and Gold 1985	Inputs: Labor, Expenses, Space (rent) Outputs: Transactions	14 branches in US	CCR	Input-Oriented Production Model
Sherman and Ladino 1995	Inputs: Labor, Expenses, Space (rent) Outputs: Transactions	33 branches in US	CCR	Input-Oriented Production Model
Soteriou and Zenios 1999	Inputs: Managerial personnel, Clerical personnel, Computer terminals, Space Outputs: Total amount of work	144 branches in Cyprus	CCR	Input-Oriented Production Model
Soteriou and Zenios 1999	Inputs: Personnel, Computer, Space Outputs: Incident duration, Waiting time, Reliability, Responsiveness, Assurance, Tangibles, Empathy	144 branches in Cyprus	CCR	Input-Oriented Production Model
Soteriou and Zenios 1999	Inputs: Managerial personnel, Clerical personnel, Computer terminals,	144 branches in Cyprus	CCR	Input-Oriented Production Model

	Space Outputs: Profits			
Tulkens 1993	Inputs: Labor, Space (rent), ATM Outputs: Transactions	773 branches in Belgium	CCR BCC	Input-Oriented FDH Production Model
Vassiloglou and Giokas 1990	Inputs: Labor, Expenses, Space (rent), ATM Outputs: Transactions	20 branches in Greece	CCR	Input-Oriented Production Model
Yang and Paradi 2003	Inputs: FTE in dollars Premise/IT expenses Other Expenses Outputs: Loan Balances Deposit Balances Securities Balances Gross Revenue	70 branches of 3 Canadian banks in one city	CCR	Input-Oriented Production Model

Chapter 14

ENGINEERING APPLICATIONS OF DATA ENVELOPMENT ANALYSIS
Issues and Opportunities

Konstantinos P. Triantis
Virginia Tech, System Performance Laboratory, Grado Department of Industrial and Systems Engineering, Northern Virginia Center, 7053 Haycock Road, Falls Church, VA 22043-2311 USA; E mail: triantis@vt.edu.

Abstract: Engineering is concerned with the design of products, services, processes, or in general with the design of systems. These design activities are managed and improved by the organization's decision-makers. Therefore, the performance evaluation of the production function where engineering plays a fundamental role is an integral part of managerial decision-making. In the last twenty years, there has been limited research that uses Data Envelopment Analysis (DEA) in Engineering. One can attribute this to a number of issues that include but are not limited to: the lack of understanding of the role of DEA in assessing and improving design decisions, the inability to open the input/output process transformation box, and the unavailability of production and engineering data. Nevertheless, the existing DEA applications in Engineering have focused on the evaluation of alternative design configurations, have proposed performance improvement interventions for production processes at the disaggregated level, assessed the performance of hierarchical manufacturing organizations, studied the dynamical behavior of production systems, and have dealt with data imprecision issues. This chapter discusses the issues that the researcher faces when applying DEA to engineering problems, proposes an approach for the design an integrated DEA based performance measurement system, summarizes studies that have focused on engineering applications of DEA, and suggests some systems thinking concepts that are appropriate for future DEA research in Engineering.

Key words: data envelopment analysis (DEA), engineering design and decision-making, disaggregated process definition and improvement, hierarchical manufacturing system performance, dynamical production systems, data imprecision, integrated performance measurement systems, systems thinking.

1.   BACKGROUND AND CONTEXT

Investments are made by the engineering organization to enhance its ability to provide quality products/services to its customers in the most efficient and effective manner. Investments are typically made in facilities and equipment and for process improvement programs. Consequently, the engineering organization's investment and production functions are not only inter-linked but also inseparable. A part of the domain of the investment function is the provision of capital for projects at the benchmark or better rate of return. On the other hand, the production function is primarily involved with the design of the production processes or systems that provide the organization's products and services. As part of the production function, design engineers define and introduce product and service concepts based on customer requirements and process engineers design and improve production processes.

Consequently, the organization needs to be successful on both fronts, i.e., in managing the investment and production function activities concurrently. Well-known criteria such as net present worth, internal rate of return, benefit/cost ratios, etc. are used to assess the organization's success in carrying out its investment activities. Additionally, operational performance is typically assessed using a different set of criteria such as efficiency, productivity, quality, outcome performance, etc. Nevertheless, investments in equipment and products typically have not been evaluated using efficiency-based operational criteria. For notable exceptions see Bulla, Cooper, Wilson, and Park (2000) and Sun (2002).One can make the assertion that Engineering is primarily concerned with the design of products, services, processes or in general with the design of systems. Furthermore, these design activities are managed and improved by the organization's decision-makers. Therefore, the performance evaluation of the production function where engineering plays a fundamental role is an integral part of managerial decision-making.Nevertheless, even though engineers have always been involved with the design and improvement of the production function, in academia, the study of the production function performance has been primarily the domain of economists and operations research analysts. Surprisingly, very few engineering curricula have courses and/or research programs dedicated to the teaching and study of productive performance.

On the other hand, it should be noted that the measurement and evaluation of the efficiency performance of production processes is what initially provided the impetus for the theoretical developments of Debreu (1951), Koopmans (1951), and Shephard (1953, 1970). However, many of the first and subsequent analyses of efficiency performance have focused on entire economies or sectors of the economy.

Therefore, the empirical evaluation of the disaggregated efficiency performance of production processes has not been pursued as rigorously as the efficiency evaluations that focus on the economy, industries, or firms. This is not necessarily surprising given that traditional cost and financial accounting systems rarely define and collect data that would be useful for process-related studies. On the other hand, engineers typically track alternative performance measures in inventory management, product quality, and customer order management. Nevertheless, the relationship of these engineering-based measures to the Farrell (1957), Koopmans (1951), and Charnes, Cooper and Rhodes (1978) representations of efficiency performance have not been extensively analyzed nor documented.

Also, within engineering, the linkage of performance measurement to process performance improvement is of great concern. In practice, performance improvement requires that the engineer identify the causal factors that are associated with efficiency performance. However, Färe, Grosskopf, and Lovell (1994) argue that the measurement of efficiency performance is by itself an appropriate and necessary research endeavor. They continue by stating that the investigation of the causes of efficiency or inefficiency is important but that the measurement of efficiency by itself can provide some important insights, i.e., discovering the patterns of efficiency performance without hypothesizing about the causal factors.

Causality and process improvement have been critical points of departure from what engineers are practically involved with on a daily basis and at least initially what researchers in efficiency and productivity performance literature have not shown a great deal of interest in. This perhaps is one of the reasons why there has been such a gap between the research in efficiency and productivity and the practice of engineering. However, it is encouraging to observe that recently the efficiency literature is actively pursuing the issues of causality and process improvement.

Consequently, even though efficiency measurement has been a fruitful area of scholarly research, decision-makers and engineers have not extensively implemented efficiency measurement concepts to improve the design performance of products and the transformation performance of production processes. The challenge remains to focus on the measurement of product design alternatives and disaggregated process performance and evaluate the appropriateness of the non-parametric and parametric approaches (Fried, Lovell, and Schmidt (1993)) that have been proposed in the literature insofar as to their leading to improved decision making resulting in product and process improvements.

Given that Data Envelopment Analysis (DEA) (Charnes, *et al.* (1978)) research does not have a well-defined role in the various engineering disciplines, the National Science Foundation provided funding for a workshop that was held at Union College in December of 1999 to explore issues and applications of DEA in Engineering. A number of engineers from Union College attended the workshop where they were exposed to efficiency measurement and DEA (Seiford (1999)) and to some existing DEA engineering applications. These applications involved road maintenance (Cook, Kazakov, and Roll (1990, 1994)), turbofan jet engines (Bulla, Cooper, Wilson, and Park (2000)), electric power systems (Criswell and Thompson (1996)), and circuit board manufacturing (Otis (1999), Hoopes and Triantis (2000, 2001), Triantis and Otis (2003)). During the workshop small break out sessions explored research issues and attempted to define interesting engineering applications of DEA for further study. One of the key issues that arose during the workshop was how well established engineering design methodologies such as design of experiments (DOE) could effectively interface with the DEA methodology. One of the fundamental conclusions of the workshop is the need to explore these methodological interfaces and to continue to pursue innovative engineering DEA applications.

The objectives of this chapter is to present some of the fundamental research issues and opportunities when exploring engineering applications of DEA (Section 2), to propose an approach that can be used to design an integrated DEA based performance measurement system (Section 3), to summarize selected engineering DEA applications (Section 4), to discuss some systems thinking concepts that can guide future DEA research in Engineering (Section 5) and to provide an extensive bibliography (Section 6) of the engineering applications of DEA. This reference list includes refereed papers, Ph.D. dissertations, and M.S. theses. It should be noted that even though the coverage in this chapter is intended to be as comprehensive as possible, there are references that have been inadvertently omitted. The author apologizes to the authors of the references that have not been included and would appreciate if they sent him an e-mail stating that a reference has been omitted so as to include these references in a possible subsequent edition of this handbook.

2.   RESEARCH ISSUES AND OPPORTUNITIES

The intent of this section is to concisely discuss some of the issues and opportunities faced by researchers when studying the engineering applications of DEA. These issues are by no means the only ones that researchers face when engaged in this research domain. However, they can be viewed as a means to establish a background on which future discussions on these and other issues can occur. Finally, the sequence of the issues presented in this Section is by no means an indication of their relative importance.

2.1   Evaluating Design Alternatives

As stated earlier, the primary concern of engineering is design. The fundamental question is how can DEA provide assistance for engineers that choose among different design alternatives? Within the context of this question lie three other fundamental questions.

Can DEA be used effectively with other methodologies such as Design of Experiments (DOE) to assist in establishing better product design configurations? On the surface, there is no reason why this research direction should not to be pursued. Cooper (1999) states that DEA has been used in conjunction with other disciplines and/or approaches. The fundamental provision is that DEA provides important additional insights for the analyst, engineer, and decision-maker. This remains a conjecture that needs to be ascertained for each DEA research study that combines different methodologies.

Can DEA be used to contribute to the design of effective process improvement interventions? The answer to this question is positive as long as the specification of the input/output variables accurately represents the underlying production processes. Farrell (1957) discussed that although there are many possibilities in evaluating productive performance, two at once suggest themselves - a theoretical function specified by engineers[1] and an empirical function based on the best results observed in practice.

The former would be a very natural concept to choose since a theoretical function represents the best that is theoretically obtainable. Certainly it is a concept used by engineers themselves when they discuss the efficiency of a machine or process. However, although the theoretical function is a reasonable and perhaps the best concept for the efficiency of a single production process, there are considerable objections to its application to anything so complex as a typical manufacturing firm, let alone an industry. Nevertheless, one can argue that in order to make any reasonable suggestion for a process improvement intervention there is a need for the input/output variables of the empirical function to effectively represent the underlying production processes, i.e., accurately denote how inputs are transformed into outputs. Cook, Kazakov, and Roll (1990, 1994) and Hoopes and Triantis (2001) document examples where this is accomplished.

---

[1] There have been attempts in the literature to define the theoretical production function. However, the focus of this chapter is to build on the notion of the empirical production function and how it has been used in engineering applications.

Is the DEA model valid? If it is not, then the researcher will be hard pressed to recommend the DEA results as a means to choose among alternative product and process design configurations. There have been attempts in the literature to investigate the issue of pitfalls/protocols (Dyson, Allen, Camanho, Podinovskim Sarrico, and Shale (2001)) and quality (Pedraja-Chaparro, Salinas-Jiménez and Smith (1999)) of the DEA analysis. However, researchers need to investigate further what is meant by model validity in DEA. This should be pursued in conjunction with an understanding of the relationship between the virtual representation of the DEA model and the "real" world (see Section 5.3 of this Chapter).

2.2   Disaggregated Process Evaluation and Improvement: Opening the "Input/Output Transformation Box"

As stated earlier, even though efficiency measurement has been a fruitful area of scholarly research, decision-makers have not used the DEA framework (Charnes, *et al.* (1978)) to evaluate and improve the performance of production/manufacturing processes. A notable exception is provided by the work of Cooper, Sinha and Sullivan (1996).

Consequently, the challenge remains on how to focus on disaggregated production processes, i.e., how to study in detail the "input/output transformation box". This entails understanding the technologies used, the processes employed, the organizational structures, and the decision-making rules and procedures. An important modeling and research issue for one who decides to pursue this line of research is to decide the appropriate level of aggregation (or the necessary level of detail complexity) of the DEA analysis.

One of the main reasons for the lack of efficiency studies at the disaggregated production process level is that one needs to invest a considerable amount of time within organizations studying their transformation processes, interviewing both decision-makers and employees, and collecting useful data and information. This means that the researcher is interested in acquiring a deep appreciation of the physical, organizational, and decision-making structures within the organization. By understanding these structures the researcher will also understand the performance behavior of the organization and consequently gain an in-depth appreciation of all performance measurement issues. To gain access to organizations, decision-makers need to accept that the DEA approach can potentially lead to significant organizational learning and important improvements in performance.

Once within the organization, the researcher may not have the data to support a complete model where all the input and output variables are included. Nevertheless, formulations that attempt to capture performance purely from the input or the output perspective can lead to potentially useful insights. This is important when there is an effort to incorporate alternative concepts of "inputs" and "outputs" in performance evaluation such as process and product characteristics. This notion of focusing on performance evaluation models solely from the input or the output perspective is consistent with the unified framework for classifying Data Envelopment Analysis models presented by Adolphson, Cornia and Walters (1990). Nevertheless, there is a need for academic research to emphasize innovative input/output specifications that represent the underlying processes and structures whose performance is being evaluated.

One of the DEA modeling approaches that has been proposed to measure and improve disaggregated process performance is the network model (Färe and Grosskopf (2000)). The fundamental concept behind this approach is for the researcher to open the "input/output transformation box" represented by the DMU meaning, as stated earlier, that the researcher is concerned with how the transformation of inputs into outputs is accomplished. For example, intermediate outputs of one stage may be used as inputs into another stage. One can proceed to evaluate the efficiency performance of the specific production stages (represented by nodes) as well as the overall performance of the DMU. In addition to evaluating the relationship of different production stages, the analyst can also evaluate the performance of a production node and the performance of a customer node (Löthgren and Tambour (1999)) that are interrelated. This approach facilitates the allocation of the resources between the customer oriented activities and the traditional production activities. In this way, one can also study the satisfaction of different goals associated with each node.

2.3 Hierarchical Manufacturing System Performance[2]

In addition to the study of disaggregated efficiency performance, DEA research can focus performance management decision-making that coordinates activities among different levels in a manufacturing organization. Nijkamp and Rietveld (1991) present three principal problems of policy making in multi-level environments: Interdependencies among the components of the system; conflicts among various priorities, objectives and

---

[2] Part of the material of this Section is adopted from Hoopes, B., Triantis, K., and N. Partangel, 2000, The Relationship Between Process and Manufacturing Plant Performance: A Goal Programming Approach, *International Journal of Operations and Quantitative Management*, 6(4), 287-310.

targets *within* individual components of the system; conflicts among priorities, objectives and targets *among* the various components of the system.

In the context of manufacturing environments analogies can be found for each of these three principal problems. Clearly, technological and administrative interdependencies exist among the various production processes. Within a single production process, conflicts exist with respect to objectives; for example, quality assurance and throughput objectives are usually at odds, especially in the short term. Finally, conflicts with respect to manufacturing performance targets in terms of throughput, cost, efficiency, allocation of resources, etc. exist among the various manufacturing departments.

Furthermore, the achievements of the production-planning goals at different time periods in the planning cycle are complementary. The achievement of the overall plant-level manufacturing targets usually depends on the performance of individual processes that provide the intermediate products. Consequently, one would expect that the improvement of the operational performance of the individual manufacturing processes at specific time periods should favorably impact the overall plant performance of the manufacturing facility for the entire planning cycle. This constitutes a fundamental research hypothesis that needs to be explored further.

Based on the three previously mentioned problems associated with multi-level organizations, Nijkamp and Rietveld (1991) suggest the usefulness of multi-objective analytical techniques such as goal programming for addressing planning problems. However, within the efficiency measurement field, the use of goal programming in conjunction with data envelopment analysis (DEA) has not been studied extensively. Efficiency performance has been linked with other performance dimensions such as quality and effectiveness but only rarely with respect to accomplishment of conflicting and/or complementary performance goals. Goal programming is an analytical approach that allows for the explicit incorporation of these types of goals. DEA provides performance assessments with respect to resource utilization and/or output/outcome achievement. Therefore, the combination of goal programming with DEA allows the analysis to focus on goal achievement within an organizational hierarchy and at the same time address performance measurement with respect to resource utilization and output production. Given the nature of the production planning function in most manufacturing organizations, a goal programming data envelopment analysis approach is appropriate when assessing organizational performance.

Thanassoulis and Dyson (1992) and Athanassopoulos (1995) have introduced goal programming as means of assessing whether organizational input/output targets are met. Thanassoulis and Dyson's (1992) formulation provides a method to estimate the input/output targets for each individual decision making unit (DMU) in an organization. However, it does not address the planning and resource allocation issues at the global organizational level while considering all the decision-making units (DMUs) simultaneously. This means that their formulation does not include global organizational targets or global organizational constraints. Athanassopoulos (1995) provided these enhancements with his formulation of a Goal Programming and Data Envelopment Analysis (GODEA) approach that combines conflicting objectives of efficiency, effectiveness and equity in resource allocation for service organizations.

It should be noted, that the extension of the GODEA approach to manufacturing environments requires a fundamental re-thinking of how to incorporate the decisions made at the different levels within the manufacturing hierarchy. Process supervisors are primarily concerned with attaining the required throughput that is defined by the master production plan. Process engineers are concerned with well-balanced production processes that minimize bottlenecks and reduce idle time. Plant managers require that plant level output requirements be met so as to ensure that demand requirements are satisfied. Above all, management is interested in the overall performance of the manufacturing facility and the extent to which all of these previously defined goals are aligned.

Additionally, within a manufacturing context, the goal programming analytical approach needs to define what constitutes an appropriate unit of analysis, labeled as a decision-making unit (DMU) in DEA. At the very minimum the goal programming approach should provide information as to how realistic the process and plant level targets are, should identify processes that are under-performing, and should point to serious bottlenecks. This is considerably more information than one would obtain through the completion of a typical DEA evaluation of each production process and of the plant as a whole.

Note that a fundamental assumption of these modeling approaches is that each of the processes in a manufacturing facility is uniquely different from a technological point of view. This implies that the operational performance of a single process cannot be used to benchmark other processes in the manufacturing facility. Consequently, each process is compared to itself over the planning horizon, and one must evaluate the performance of each production process independently, while recognizing the process interdependencies by linking adjacent processes through their respective output levels.

Based on the need to align overall plant and production process goals, Hoopes, Triantis and Partangel (2000) assess the contribution of the operational performance of individual production processes at specific time periods for the achievement of the overall plant production-planning targets for the entire production planning cycle. They present a goal programming data envelopment analysis formulation that evaluates the achievement of plant and production process targets for a manufacturing technology that involves serial production stages. They then illustrate their approach by evaluating the process and plant effectiveness and serial production performance of a circuit board manufacturing facility for which plant and process level data have been accumulated over a two-year time horizon. The results obtained from the proposed goal programming approach are compared to the radial variable returns to scale efficiency scores (Banker, Charnes and Cooper (1984)) in terms of evaluating the overall performance of the manufacturing facility.

In this research, the concepts of effectiveness and balance are addressed in a manufacturing context. The production-planning function predetermines the goals or targets both at the plant and production process levels. Process effectiveness is the degree to which a production process meets its resource (input) and throughput (output) targets for a specific time period. Plant-level effectiveness is the degree to which the manufacturing facility meets its plant-level production-planning targets over the entire planning period. Finally, the concept of production line balance takes into account the serial nature of the production processes so as to evaluate the technology's production bottlenecks. Note that by substituting actual input and output levels, the model proposed in this research can be used to also evaluate the performance of the production processes in terms of their resource utilization and output production process efficiency.

2.4   Data Measurement Imprecision in Production Systems[3]

Whether one uses a frontier (DEA, stochastic frontier (Aigner, Lovell and Schmidt (1977), Battese and Corra (1977), Meeusen and Van Den Broeck (1977)) or a non-frontier approach (Free Disposal Hull (Deprins, Simar and Tulkens (1984), pair-wise dominance (Koopmans (1951)), data measurement imprecision is more the norm rather than the exception for production and manufacturing systems.

Production data are usually not gathered for the sole purpose of conducting production analyses. The problem is further compounded by the fact that there are multiple measurement systems in place that account for different segments of the production process (Otis (1999)). These

---

[3] Part of the material of this section is adopted from Triantis, K., Sarangi, S. and D. Kuchta, 2003, Fuzzy Pair-Wise Dominance and Fuzzy Indices: An Evaluation of Productive Performance, *European Journal of Operational Research*, 144, 412-428.

measurement systems are rarely consistent in the way they collect information and in their definitions of the accounts associated with production processes.

Furthermore, in most of the real-world decision-making problems, the data are not always known *precisely* or the information regarding certain parameters that are part of a mathematical model is not readily available. For example when evaluating the performance of urban transit alternatives, imprecision exists when measuring the headway between vehicles.

The "classical" DEA formulation lacks the flexibility to deal with *imprecise* data or data expressed in *linguistic form*. This problem can be approached by expanding the current DEA technology to allow imprecision in the ordinal rankings and the knowledge only of bounds of certain parameters (Cooper, Park, and Gang (2001)) or with the use of fuzzy set theory (Zadeh (1965) Triantis and Girod (1998), Kao and Liu (2000)).

This is not to say that other approaches such as a standard probability approach cannot address the problem of imprecision associated with the production data. Nor is it necessary to delve into the endless debate of probability versus fuzzy set theory. Others have done so extensively (see the August 1995 issue of Technometics (volume 37, no.3) and the February 1994 issue of IEEE Transactions on Fuzzy Systems (volume 2, no. 1) for more details). Nevertheless, one can agree with the final statement of Laviolette, *et al.* (1995, p. 260) that states, "we maintain that development of equitable comparative approaches is essential" and with the statement made by Laviolette and Seaman (1994. p. 38) "we have not found no evidence that fuzzy set theory is ever *exclusively* useful in solving problems."

For example, fuzzy set theory has provided an alternative modeling avenue to stochastic data envelopment analysis (Sengupta (1987), Land, Lovell and Thore (1993), Olesen and Petersen (1995)). While both of these modeling approaches can be used to address the measurement imprecision associated with the input and output variables, the data requirements are quite different. Stochastic data envelopment analysis and specifically the chance-constrained programming methodology requires that the analyst "supply information on expected values of all variables, variance-covariance matrices for all variables, and probability levels at which feasibility constraints are to be satisfied" (Fried, Lovell and Schmidt (1993), p. 35).

On the other hand, the fuzzy mathematical programming approach requires the definition and measurement of the most and least plausible production bounds for all fuzzy input/output variables and an assumption on the form of their membership functions (Triantis and Girod (1998)). Furthermore, Almond (1995, p. 268) points outs that "being precise requires more data, and it requires more work."

It should be noted that it is also more expensive for organizations to obtain more precise data. In context of the manufacturing technology used as an illustration of the fuzzy DEA approach Girod and Triantis (1999) were faced with the following constraint. Data necessary for the fuzzy approach, stochastic approach, and the crisp data envelopment analysis approach could be obtained at a cost ranging in the hundreds, tens of thousands, and millions of dollars respectively. This fact provides strong impetus for developing cost effective methodologies to capture imprecision in productivity and efficiency analysis in general.

Depending on the nature of the imprecision, fuzzy numbers can represent coefficients in the objective function and/or set of constraints. Furthermore, uncertainty may also appear in the formulation of the constraints and the objective function, i.e., the extent to which the objective function is optimized and the constraints hold. During the decision-making process in a fuzzy environment, fuzzy objectives, fuzzy constraints and fuzzy decisions, represented through fuzzy sets with corresponding membership functions, can be considered. In this case, the decision is simply defined as a selection of the sets of feasible solutions or merely one solution, which simultaneously satisfies the fuzzy objective and fuzzy constraints.

However, not all data are imprecise. Some data are known with precision whereas other data are approximately known or are described with linguistic information. Methodologies need to take into account all types of available data, i.e., crisp, approximately known, and/or data expressed in a linguistic form.

There are many examples where DEA mathematical programming formulations were expanded using fuzzy set theory to incorporate the following: imprecision in the data as in Triantis and Girod (1998) and Girod and Triantis (1999), missing data and imprecise linguistic data as in Kao and Liu (2000), or fuzziness in the objective function and constraints as in Sengupta (1992), Sheth (1999), Kabnurkar (2001). Furthermore, Hougaard (1999) extends DEA technical efficiency scores to efficiency scores defined over fuzzy intervals. The fuzzy scores allow the decision-maker to use technical efficiency scores in combination with other sources of available performance information such as expert opinions, key figures, etc.

2.5   Dynamical Production Systems[4]

One of the important characteristics of production and engineering systems is that they are dynamic. This characteristic needs to be captured in

---

[4] Part of the material in this Section is adopted from Vaneman, W. and K. Triantis, 2003, The Dynamic Production Axioms and System Dynamics Behaviors: The Foundation for Future Integration, *Journal of Productivity Analysis*, 19 (1), 93-113.

efficiency analyses. Samuelson (1947, p. 314) states a *"system is dynamical if its behavior over time is determined by functional equations in which 'variables at different points of time' are involved in an 'essential' way."*

Dynamical systems can be classified in three distinct ways: (1) dynamic and historical (Samuelson (1947)); (2) dynamic and causal (Samuelson (1947)); and (3) dynamic, causal, and closed (Vaneman and Triantis (2003)). Systems that are dynamic and historical exhibit a high degree of correlation between the variables at the initial time t_0 with the variables at the final time t. Neither the passage of time nor the structure of the system are considered, thus variables that become active during the interval between the initial time and final time, (t_0, t), are not considered when determining the final state of the system.

Dynamic and historical systems correlate the initial conditions of the system to the final conditions, and do not contain information about the structure or behavior of the system. Dynamic and historical systems can be expressed (Samuelson (1947)) as:

$$y_{jt} = f\{t_0; t; x_{it_0}\} \tag{14.1}$$

where x_{it_0} is the ith input at the initial time t_0, $i=1,2,3,...n$ y_{jt} is the jth output at time t, $j=1,2,3,...m$.

Dynamic and causal systems consider the initial system inputs, x_{it_0}, along with the passage of time. Dynamic and causal systems can be expressed as:

$$y_{jt} = f\{t - t_0; x_{it_0}; x_{it_d}\} \tag{14.2}$$

This type of system allows inputs to be added at some intermediate time t_d, where $t_0 < t_d < t$, and for the behavior of the system to be evaluated at any given time. However, the inputs added during the interval (t_0, t) are not a result of feedback mechanism from within the system.

In many applications, the systems discussed could be classified as open systems. An open system converts inputs into outputs, but the outputs are isolated from the system so that they cannot influence the inputs further. Conversely outputs are not isolated in a closed system, thus are aware of and can influence the system behavior by providing information to modify the inputs via a feedback mechanism.

Dynamic, causal, and closed systems can be expressed as:

$$y_{jt} = f\{t - t_0; x_{it_0}; x_{it_d}; y_{j(t_d - t_0)}\} \tag{14.3}$$

Where $y_{j(t_d - t_0)}$ is the jth output resulting from action during the interval $[t_0, t_d]$. By adding the output variable $y_{j(t_d - t_0)}$ to (14.3), this relationship is

defined in an "essential way" such that the system brings results from past
actions to influence or control future actions via a feedback mechanism

In the efficiency literature, Färe and Grosskopf (1996) were the first to
study dynamical and historical systems. They defined and developed the
dynamic data envelopment analysis (DDEA) by adding the element of time
to the DEA model. This is accomplished by extending the static DEA model
into an infinite sequence of static equations (Färe and Grosskopf (1996)).
While this methodology does evaluate organizational performance over time,
it is important to understand that this methodology is static at each discrete
point in time, and provides a linear solution.

Figure 14-1 (adapted from Färe and Grosskopf (1996)) illustrates the
concept of DDEA. Each production cycle (P) has three types of inputs:
fixed ($x(f)$), variable ($x(v)$), and inputs that were intermediate outputs to the
last production cycle ($y(i)$). The outputs for each production cycle are the
final outputs ($y(f)$) that represent the products that go to the customer, and
intermediate outputs that are used as inputs to subsequent production cycles
($y(i)$).

As systems become more complex and time dependent, alternative
methods for evaluating production and engineering system performance in a
dynamic environment must be explored. The efficiency literature to date has
primarily been interested in system performance measurement rather than
causation, because it was thought that: (i) uncovering the pattern of efficient
and inefficient practices should be paramount; and (ii) that the comparative
advantage is with performance measurement and not determining the causal
factors associated with system performance (Färe, Grosskopf, and Lovell
(1994)).

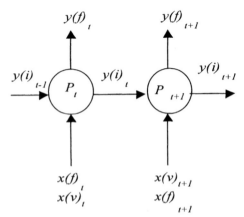

Figure 14-1. Dynamic Data Envelopment Analysis Approach (Färe and Grosskopf (1996))

Nevertheless, Färe and Grosskopf (1996) also suggest the need to explore
what is inside the black box of production technologies to determine how

inputs are converted into outputs, so that their efficiency could be better understood. To this end, as stated in Section 2.2, they develop a network technology model. In their model they evaluate how multiple inputs placed in the production process, at multiple time periods, can produce multiple outputs. While this approach is evolutionary, it fails to consider the causal relationships that exist within the network.

System performance is inherent within the system's structure and policies[5]. Thus if the system structure (that includes physical processes with a representation of input and outputs, decisions, and organizational structure to incorporate the dissemination of information) is understood, the sources of good system performance can be replicated for future system design, and the causes of poor system performance can be corrected. Since policies are deep-seated within the system's structure, determining and evaluating causal relationships within a system will provide an understanding as to how system policies affect system performance.

In order to study causal relationships, the impact of system structure and of policies, Vaneman (2002) uses system dynamics (Forrester (1961), Sterman (2000)) to explore productive efficiency. System dynamics is a mathematical approach that employs a series of interrelated equations to define the system that is dynamic, closed and causal and considers time over a continuous spectrum. System dynamics differs from traditional methods of studying productive efficiency in two fundamental ways. First, the system dynamics approach allows for problems with both dynamic and combinatorial complexity to be studied concurrently, within the same framework. Second, system dynamics is more concerned with the behavior of the system over time versus the measure of central tendency and historic trends. Differences between the Dynamic Data Envelopment Analysis Approach (Färe and Grosskopf (1996)) and Efficiency Evaluation using Systems Dynamics are summarized in Table 14-1.

Furthermore, in the context of evaluating the efficiency performance of dynamic, causal, and closed systems, one must also investigate their equilibrium and stability. Systems can be categorized as being in in equilibrium or in disequilibrium, stable or unstable. Equilibrium can be further categorized as either static or dynamic. Static equilibrium is defined as the condition that exists when there is no flow within the system (Sterman (2000)). Two conditions must be satisfied for a system to be in static

---

[5] The term policy is used to describe how a decision process converts information into action to show a change in the system (Forrester (1961)). Forrester (1968) further identifies four concepts found within any policy statement:
  1. A goal.
  2. An observed condition of the system.
  3. A method to express any discrepancy between the goal and the observed condition.
  4. Guidelines of which actions to take based on the discrepancy.

equilibrium: (1) All first order derivatives x'_{it}, y'_{jt} are zero at the time considered; and (2) all higher order derivatives are also zero. A system in which only condition (1) is satisfied is said to be momentarily at rest (Frisch (1935-36)).

Table 14-1. Differences Between the DDEA (Färe and Grosskopf (1996)) and System Dynamics Approaches when Measuring Productive Efficiency (adopted from Vaneman (2002))

Attribute	Dynamic Data Envelopment Analysis (DDEA)	System Dynamics Approach for Measuring Productive Efficiency
Primary Authors	Färe and Grosskopf (1996)	Coyle (1996), Wolstenholme (1990), Vaneman (2002)
Goal	Decision-making based upon efficiency measurement, and estimation of the effects of policy change over time.	Decision-making based upon the behavior of the endogenous elements of the system with respect to efficiency measurement, and the simulated effects of policy changes over time
Technical Approach	Optimization through linear programming	Optimization through system dynamics. (Heuristic approach).
Characteristics	-Dynamic -Deterministic or probabilistic -Linear -Discrete Time	-Dynamic -Deterministic or probabilistic -Linear or non-linear -Continuous Time
Advantages	-Linear programming approach guaranteed to find optimal solution. -Optimal solution is easily interpreted.	-Model represents causal relationships well. -Suggested policy changes are calculated and simulated over time. -Model allows for non-linear relationships. -Model allows for feedback within the structure. -Model can represent information flows.
Disadvantages	-Model does not allow for causal relationships to be defined -Policy changes and their effects are estimated (not simulated) and observed over time. -The model only accommodates linear relationships. -Does not allow for feedback within networks. -Does not allow for information flows to be modeled.	-Heuristic approach not guaranteed to find optimal solution. -Optimal solution is not always readily apparent.

A system in dynamic equilibrium is a system where there is a constant flow going through the system. Viewing the system from a macro level, dynamic equilibrium gives the appearance that nothing within the system changes over time. A closer look reveals that there is a constant flow of inputs into the system, and a constant flow of outputs from the system (Sterman (2000)). All derivatives will have non-zero values for dynamic equilibrium.

System stability refers to how a system that was previously in equilibrium behaves when a disturbance is introduced. Consider a small disturbance introduced to the system at time t_d. If the system returns to its original (or closely related) state of equilibrium after being disturbed, the system is considered stable (Firsch (1935-36), Sterman (2000)). If the small disturbance forces the system further away from equilibrium with the passage of time, the system is said to be in unstable equilibrium (Sterman (2000)).

2.6 Visualization of the DEA Results: Influential Data Identification

On of the important aspects of making effective decisions for engineering applications is the need to improve upon the existing visual representation of the DEA results. The question that needs to be addressed and answered is how to incorporate in DEA existing computer science, statistical, and graphical based visualization techniques so that decision-makers can more effectively use the information generated by the DEA analytical framework. These results primarily include, the DEA scores, the peers and the performance targets.

However, any discussion on data visualization inevitably leads to a subsequent discussion on outliers and influential observations. This is because what become apparent from any visualization approach are the extremes in efficiency performance, i.e., the efficient and extremely inefficient observations along with the main mass of observations.

In general, influential data points fall into three categories: (1) Observations that are outliers in the space of the independent variables (referred to as leverage influence). (2) Points that deviate from the linear pattern of the majority of the data (referred to as residual influence). (3) Points that have both leverage and residual influence (referred to as interaction influence).

There are three main issues that arise from the research with respect to influential observation in efficiency analysis. The first has to do with identification, i.e., how to identify the influential observations given the masking effect that exists in most datasets (Seaver and Triantis (1989, 1992, 1995), Wilson (1993, 1995)). The second issue has to do with the impact that each influential observation has on the efficiency scores. The third and often misguided issue is what to do with influential observations once they are identified. The temptation is to remove extreme observations from the data set. However, in most cases extreme observation are the observations that have the most information as they represent extreme production occurrences. Therefore, the identification of influential observations allows the decision-maker and engineer to isolate important observations that represent unique occurrences that can be subsequently used for benchmarking purposes and process improvement interventions.

Nevertheless, irrespective of the methods used for outlier and influential observation identification and the assessment of their impact on the efficiency scores, there is a need to study existing data visualization approaches and their appropriateness in effectively representing DEA results to decision-makers.

3. A DEA BASED APPROACH USED FOR THE DESIGN OF AN INTEGRATED PERFORMANCE MEASUREMENT SYSTEM

Based on the implementation of DEA in manufacturing (Triantis, Coleman, Kibler, and Sheth (1998), Girod and Triantis (1999)) and service environments (Triantis and Medina Borja (1996)) as part of the funded research activities of the System Performance Laboratory at Virginia Tech, an approach has been developed that allows the researcher not only to define appropriate efficiency based performance measures but to link these performance measures to performance improvement actions.

Initially, one needs to communicate to decision-makers that organizational performance is difficult to measure and improve. The complexity of organizations no longer allows performance measurement to be addressed with "off-the -cuff" solutions. Furthermore, one needs a language to represent production complexity, i.e., to understand how the production structure of the organization creates the behavior that is observed. One can represent this organizational production complexity with multiple key variables that directly or indirectly account for the fashion in which inputs are transformed into outputs. Additionally, it is necessary to evaluate the organization's current key performance measures or indicators in terms of how they account for the different organizational performance dimensions such as cost, quality, availability, safety, capacity utilization, etc.

This leads to the requirement to design and implement a performance evaluation system that is based on Data Envelopment Analysis. This system should integrate existing key performance indicators or measures, key variables that represent the organization's production structure as well as the data used to represent both the measures and variables. The integrated performance measurement system should provide performance benchmarks, identify best practices, define performance targets and provide input to the operational planning process of the organization that will plan for process and organizational performance improvements.

The proposed system can be used to address key questions such as:

1. Given the plethora of performance measures collected by the organization, which ones should management take into account when focusing on key performance dimensions (such as cost, quality, etc.)?
2. How should the organization integrate its key performance measures given the fact that different sources of data are currently used?
3. Are current performance measurement systems effective in terms of meeting its current goals and affecting change?
4. How do lessons learned from the performance improvement interventions feedback into the definition of performance objectives?
5. How does one make a fair comparison among different DMUs within the organization using the key performance measures?
6. How does one identify achievable performance improvement interventions for each DMU?

Figure 14-2 provides a DEA based approach that can be used to design the integrated performance measurement system.

1) *Task 1*: The first task involves two critical activities. First, a measurement team should be established. The team will include representatives from the organization and the researchers/analysts. The measurement team should interact in team meetings with the decision-makers and should provide key information for the analytical model. Second, the performance measurement problem should be defined and the purpose and use of the model should be established.

2) *Task 2*: Once the problem is identified in Task 1, theories about the causes of the current organizational performance levels can be developed. These theories, also known as hypotheses, are believed to account for the behavior demonstrated by the organization. The fundamental hypothesis is that the variables designated as inputs determine the response of the output variables that will be selected to represent the behavior of the organization. Additionally, it will be hypothesized that specific infrastructure and environmental variables also impact performance. Once the most appropriate variables representing the behavior of the organization are selected, their units

of measurement should be defined, and an assessment will be made as to how to best obtain data for these variables.

3) *Task 3*: The third task is the system conceptualization. This task constitutes the beginning of model development. At this point there are a number of activities that are required. First, the input and output variables selected in Task two will require further classification. These variables will be classified as representing good or bad inputs/outputs, whether they are controllable (discretionary) or not, whether they represent quality or quantity performance, whether they are fixed or variable, whether they represent environmental conditions or not, whether they represent infrastructure etc. Second, uncertainty or imprecision associated with the measurement of specific variables should be determined. Third, various appropriate input/output specifications should be identified. This implies that there may be more than one input/output model that is appropriate for the evaluation of the organization.

4) *Task 4*: The DEA model formulation is the next task. During this task, the DEA input/output specifications generated during the system conceptualization phase are transformed into mathematical models so they can be "solved." At this point, solution algorithms will be proposed and the DEA formulations will be programmed in the appropriate software platform.

5) *Task 5*: The DEA model behavior and validation task will be conducted during this task under the premise that all models of the real world are inaccurate to some extent. These inaccuracies occur because it is difficult to model every variable that affects the performance of the organization. The results of the DEA formulations will be evaluated considering the identification of the best and worst performing DMUs, the benchmark DMUs, and the performance targets.

6) *Task 6:* At this point, the policy analysis can begin. The DEA results will point to different operational interventions that decision-makers can make.

7) *Task 7*: The last task is to plan for the implementation and deployment of the policies for a real world system. By this time policies have been evaluated and tested under numerous conditions to ensure that they represent the direction best for the real world system.

The first six tasks outlined in this section will be completed in an iterative fashion. The first iteration will require extraction of a very small number, but important, factors (variables) associated with the performance of the organization, collect data, formulate and solve the model, test and validate it, and then evaluate alternative performance improvement interventions. During the second iteration additional factors (variables) will be added and the first six tasks will be repeated. The process will be repeated until a point is reached where all the important elements of the problem are included. Finally, completing task seven will finalize the process.

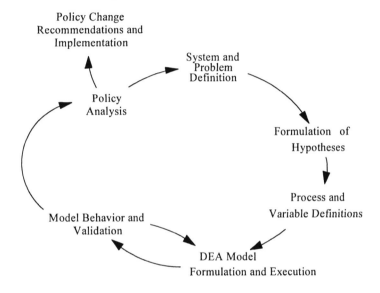

Figure 14-2. A DEA Based Approach Used to Develop an Integrated Performance Measurement System

4. SELECTED DEA ENGINEERING APPLICATIONS

This section provides an overview of some of the existing work on DEA engineering applications. Not all applications are described in this Section. The remaining applications are found in the Bibliography at the end of this Chapter.

In the first part of this Section (Section 4.1) the four applications presented at the 1999 NSF Workshop on Engineering Applications of DEA at Union College are summarized. In the remaining parts of this Section, existing applications that consider the role of environmental issues in production (Section 4.2), that evaluate the performance of transportation systems (Section 4.3) and that present studies in other areas of engineering (Section 4.4) are briefly presented.

4.1   Four Applications

4.1.1   Evaluating Efficiency of Turbofan Jet Engines (Bulla, Cooper, Wilson and Park (2000))

DEA is explained and illustrated with an application to data on turbofan jet engines. The purpose of this study is to augment, not replace, traditional methods for measuring efficiency. What is interesting is that the results of this study are compared with standard engineering methods for measuring efficiency. This is something that more researchers should focus on in the future. It should be noted that the input and output variables are defined based on the turbofan engine characteristics. In contrast to the engineering methods for measuring efficiency, DEA handles data on multiple inputs (fuel consumption, engine weight, drag) and multiple outputs (airflow and cruise thrust) that can be used to characterize the performance of each engine. Furthermore, the DEA analysis is accomplished without any use of prearranged engineering based weights. DEA also provides information on the sources and amounts of inefficiencies in each input and output of each engine without requiring knowledge of the functional relations among the inputs and outputs.

4.1.2   Measurement and Monitoring of Relative Efficiency of Highway Maintenance Patrols (Cook, Kazakov, and Roll (1990, 1994))

This study was motivated in part from the fact that the Ontario Ministry of Transportation in Canada needed a management tool by which to assess quantitatively the relative efficiency of the different patrols. Additionally, a desire was voiced to evaluate the impact of different "external" (regional) factors as well as maintenance policy options (e.g., extent on privatization) on patrol efficiency. Comparisons were made both region wide and province wide.

The innovation of this research stems from the definition of inputs (maintenance expenditure, capital expenditure, and climate factor) and outputs (assignment size factor, average traffic served, rating change factor, and accident prevention factor). More specifically, the area served factor captures the extent of the workload for which the patrol has responsibility. The average traffic served is a measure of the overall benefit to the users of the highway system in a patrol. The pavement rating change factor measures the actual change in the various road sections. The accident prevention factor presents the accident prevention goal of maintenance. The maintenance and capital expenditures are straightforward. However, the climatic factor takes into account snowfall, major/minor temperature cycles, and rainfall.

The authors conclude by emphasizing the extreme importance to choose the correct input/output factors that will enter the DEA analysis. Furthermore, the unbounded version of DEA does not satisfactorily capture real-life situations. However, there is no "mechanical" process for choosing bounds on the inputs and outputs. What is required is a thorough knowledge of the process in which the DMUs are engaged, a clear vision for the purposes for which efficiency is measured, and sound managerial considerations.

4.1.3 Data Envelopment Analysis of Space and Terrestrially-Based Large Scale Commercial Power Systems for Earth (Criswell and Thompson (1996))

This research addresses an important energy problem that society will face in the 21st century. The Lunar Solar Power system (LSP) is a new option that is independent of the biosphere. LSP captures sunlight on the moon, converts the solar power to microwaves, and beams the power to receivers on Earth the output electricity. The collimated microwave beams are low in intensity, safe, and environmentally benign. LSP is compared to fossil, fission, thermal, and nuclear technologies using DEA. The comparison suggests that the efficiencies that are gained from LSP are large. In terms of a normalized cost/benefit ratio, DEA reveals that LSP is much more efficient than the other technologies. What is noteworthy is that the gains remain even if the resources needed for LSP are tenfold greater than what is estimated from governmental studies.

4.1.4 The Relationship of DEA and Control Charts (Hoopes and Triantis (2001))

It is the contention of Hoopes and Triantis (2001) that measurement of efficiency performance in conjunction with statistical process control charts can be the starting point from which the causal factors can be identified and process improvement interventions can be initiated. Consequently, given this challenge, the following objective is identified for this study, i.e., to provide a conceptual linkage between efficiency performance assessment and the traditional control charts (Shewhart (1980)). In order to create the conceptual linkage to traditional control charts, the specification of the production function in this study uses the concepts of process and product characteristics. Control charts evaluate the stochastic behavior of the production process by studying process and/or product characteristics one at time. On the other hand, efficiency measurement approaches include as part of their evaluation the entire set of critical product and/or process characteristics simultaneously.

This research shows that these two approaches can be used in a complementary manner to identify unusual or extreme production instances, benchmark production occurrences, and evaluate the contribution of individual process and product characteristics to the overall performance of the production process. The identification of extreme production instances in conjunction with the evaluation of their technological and managerial characteristics are used to identify potential root causes. Decision-makers can use this information to make process improvements. These issues are illustrated by studying the inner layer production process of a circuit board manufacturing facility. The inner layer process is one of the first stages in a manufacturing system that produces multi-layer printed circuit boards.

4.2 The Effect of Environmental Controls on Productive Efficiency[6]

The impact of pollution prevention methods on process efficiency performance has been a concern of a number of studies. This issue is extremely important for chemical process manufacturing organizations. In the literature, the generation of pollution has been taken into account using a number of different analytical methods. Two fundamental assumptions underlying these methods are that: a) Pollution controls are after-the-fact, end-of-pipe and do not explicitly take into account process and other changes that can reduce pollution (i.e., pollution prevention) and can potentially improve efficiency performance. b) Interactions with other production systems represented by variations in input and output mixes are not taken into account.

In measuring productive efficiency, waste products are sometimes explicitly considered and sometimes not. Both econometric and Data Envelopment Analysis based techniques are used to evaluate the effect of pollution controls on productive performance. There are three approaches

---

[6] Part of the material in this Subsection is adopted from Triantis, K. and P. Otis, 2003, A Dominance Based Definition of Productive Efficiency for Manufacturing Taking into Account Pollution Prevention and Recycling, forthcoming, *European Journal of Operational Research*.

that have been used in the literature. One is to apply standard DEA or econometric analysis to decision-making units with and without pollution controls. The results obtained by considering and not considering pollution controls provide estimates of the cost of lost production associated with pollution controls. The other two approaches explicitly consider pollution or undesirable outputs and determine the cost of pollution control based on loss of disposability or assign shadow prices to the undesirable outputs.

Tyteca (1995, 1996) compares several methods of evaluating environmental performance of power plants. The methods are those developed by Färe, *et al.* (1989) where both output oriented and input oriented efficiencies are considered and an approach developed by Hayes, *et al.* (1993) that maximizes ratios of weighted sums of desirable outputs to weighted sums of pollutants where pollutants are treated as inputs. In addition, an index method is used to calculate performance metrics as a means of comparison. Depending on the approach chosen, significant variations were found in the relative efficiencies of the power plants evaluated.

Recently, Wang, Zhang, and Wang, (2002) propose a DEA model where desirable and undesirable outputs are treated synchronously. A sample of efficiency measures for ten paper mills is given. Ramanathan (2002) considers several variables simultaneously when comparing carbon emissions of countries. He illustrates the use of the DEA methodology with four variables that include CO_2 emissions, energy consumption and economic activity. Soloveitchik, Ben-Aderet, Grinman, and Lotov (2002), use DEA in conjunction with multiple objective optimization models to examine the long-run capacity expansion problem of power generation systems as a base for defining the marginal abatement cost. Sarkis, and Weinrach (2001) analyze a decision-making case study concerning the investment and adoption of environmentally conscious waste treatment technology in a government-supported agency. Sarkis (1999) provides a DEA based methodological framework for evaluating environmentally conscious manufacturing programs. Finally, Otis (1999) expands the Full-Disposal Hull (Deprins, *et al.* (1984)) to evaluate the impact of pollution prevention on efficiency performance of circuit board manufacturing facilities.

4.3 The Performance of Transit Systems

DEA has also been used to evaluate the performance of public transit systems. For example, Husain, Abdullah, and Kuman (2000) presents a study that evaluates the performance of services of a public transit organization in Malaysia. Chu *et al.* (1992) compare transit agencies operating in the U.S whereas Boile (2001) evaluates public transit agencies. Tone and Sawada (1991) evaluate bus enterprises under public management and explore the notions of service efficiency, cost efficiency, income efficiency and public service efficiency separately. Nolan (1996) applies DEA to evaluate the efficiency of mid-sized transit agencies in the U.S. using the data from Section 15 US DOT 1989-1993. Carotenuto *et al.* (2001) stress the fact that in recent years, it has become imperative for public transit organizations to rationalize the operating costs and to improve the quality of the services offered. They obtain measures of pure technical, scale and overall efficiency of both public and private agencies. Kerstens (1996) investigates the performance of the French Urban Transit sector by evaluating the single mode bus operating companies and Dervaux, Kerstens and Vanden Eeckaut (1998) compute radial and non-radial measures of efficiency and compute congestion. Nakanishi, and Norsworthy (2000) estimate the relative efficiency of transit agencies providing motorbus service. Odeck (2000) assesses the relative efficiency and productivity growth of Norwegian motor vehicle inspection services for the period 1989-91 using DEA and Malmquist indices. Finally, Cowie and Asenova (1999) examine the British bus industry in light of fundamental reform in ownership and regulation.

As seen by the utilization of DEA for the purpose of evaluating transit systems, the research primarily focuses of the assessment of companies and/or agencies. Very little research has focused on the evaluation of performance at the transportation process level. Notable exceptions include the evaluation of two-way intersections by Kumar (2002), the prioritization of highway accident sites by Cook, Kazakov and Persaud (2001).

4.4   Other Engineering Applications of DEA

Some of the most recent engineering applications of DEA focus on two primary issues, i.e., evaluating the investment in equipment and assessing alternative organizational structures (such as worker teams).

Sun (2002) reports on an application of data envelopment analysis (DEA) to evaluate computer numerical control (CNC) machines in terms of system specification and cost. The evaluation contributed to a study of advanced manufacturing system investments carried out in 2000 by the Taiwanese Combined Service Forces. The methodology proposed for the evaluation of the twenty-one CNC machines is based on the combination of the Banker, Charnes, and Cooper (BCC) model (Banker *et al.* (1984)) and cross-efficiency evaluation (Doyle and Green (1994)). Both Karsak (1999) and Braglia and Petroni (1999) use DEA to select industrial robots. Karsak (1999) argues that a robust robot selection procedure necessitates the consideration of both quantitative criteria such as cost and engineering attributes, and qualitative criteria, e.g. vendor-related attributes, in the decision process. The qualitative attributes are modeled using linguistic variables represented by fuzzy numbers. Braglia and Petroni (1999) adopt a methodology that is based on a sequential dual use of DEA with restricted weights. This approach increases the discriminatory power of standard DEA and makes it possible to achieve a better evaluation of robot performance. Finally, Sarkis and Talluri (1999) present a model for evaluating alternative Flexible Manufacturing Systems by considering both quantitative and qualitative factors. The evaluation process utilizes Data Envelopment Analysis (DEA) model, which incorporates both ordinal and cardinal measures.

Paradi, Smith, and Schaffnit-Chatterjee (2002) propose DEA as an approach to evaluate knowledge worker productivity. Data envelopment analysis (DEA) is used to examine the productivity, efficiency, and effectiveness of one such knowledge worker group - the Engineering Design Teams (EDT) at Bell Canada, the largest telecommunications carrier in Canada. Two functional models of the EDT's were developed and analyzed using input oriented constant returns to scale (CRS) and variable returns to scale (VRS) DEA models. Finally, Miyashita and Yamakawa (2002) study collaborative product design teams. This research stems from the need to reduce the time for product development where engineers in each discipline have to develop and improve their objectives collaboratively. Sometimes, they have to cooperate with those who have no knowledge at all for their own discipline. Collaborative design teams are proposed to solve these kinds of the problems and consequently their effectiveness needs to be assessed.

5. SYSTEMS THINKING CONCEPTS AND FUTURE DEA RESEARCH IN ENGINEERING

One of the skills that engineers are taught and practice is that of systems thinking and modeling (O'Connor and McDermott (1997), Richmond (2000), Sterman (2000)). Furthermore, systems engineering as a discipline is gaining more and more acceptance. Growth in this discipline is expected in the future as it brings together not only all aspects of engineering but also the management of engineering practices, functions and processes. Perhaps, the future success of implementing the DEA approach in engineering is to communicate through a systems-based framework in terms of defining the performance measurement problem, stating and testing hypotheses and implementing performance improvement strategies. This framework would not be inconsistent with the firm/organizational/process view that the DEA research has held over the years. Nevertheless, there are three, if not more, systems related issues briefly described in this Section that need to be considered when completing DEA research in engineering. In fact, future researchers should explore other systems thinking issues that are pertinent for DEA research.

5.1   The Need for Operational Thinking

In order to open the "input/output transformation box" as discussed in Section 2.2, the researcher needs to understand the operational structure of the production process that is being represented by the DEA model. Operational thinking addresses the question about how performance is actually being generated. One can say that operational thinking is concerned with the "physics" of how systems use scarce resources to add value and generate outputs. The researcher attempts to answer the question: "How does this actually work?"

Operational thinking as indicated in Section 2.2 requires access to and investment in time with engineering/manufacturing organizations. The difficulty in getting access to these organizations is that they do not want to necessarily invest time in researchers who would like to "learn" about their processes. Decision-makers do not immediately see the long-term benefits that may accrue from the analytical production studies. However, both organizations and the researchers have to meet half way. The researchers need to be willing to make problem-solving contributions to the organizations during their visits within organizations and decision-makers need to learn about and appreciate the benefits that they will accrue from the analytical studies.

5.2   Contribution to Performance Measurement Science

It is perhaps not stated often enough that one of the fundamental contributions of DEA research is its impact on the science of performance measurement and improvement. As such, a discussion needs to be initiated by the research community that will further define the scope of this scientific domain and will more specifically identify the impacts of the DEA modeling framework.

Fundamentally, as Figure 14-3 indicates, the researcher starts with a problem (perhaps some type of performance behavior) formulates a theory/hypothesis that is primarily represented by the DEA model and to the extent that the model can generate the problem (the observed performance behavior), the researcher has a reasonable hypothesis/theory and model. If not, one must modify the model and retest it and/or modify the hypothesis/theory.   Once the researcher has arrived at a reasonable hypothesis/theory, one can then communicate the newfound clarity to others and begin to implement process changes.

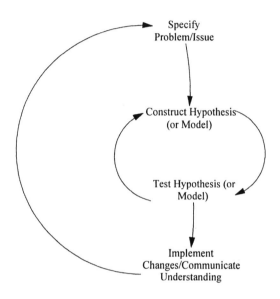

Figure 14-3. Contribution of Modeling to the Scientific Process

Scientific thinking is most applicable *after* one has constructed the model since it helps build a better, shared understanding of a system that in our case is a production process.   Furthermore, discarding current paradigms marks progress in science. However, it is not usual in DEA research to emphasize counterintuitive results that challenge current paradigms.

Nevertheless, the term "model" represents our assumptions (or hypotheses) about how a particular part of the world works. The value in modeling rests in how *useful* the model is in terms of shedding light on an issue or problem. As Deming among others has stated that all models are wrong and some are useful. What has been missing from the DEA literature is an assessment of how useful specific DEA models have been.Nevertheless, system thinkers focus on choosing numbers that are simple and easy to understand and that make sense relative to one another. They justify doing this because they believe that insight arises from understanding the *relationships* between numbers rather than the absolute numbers themselves.   This is a fundamental notion in systems thinking, which is consistent with the concept of relative efficiency performance in

DEA.Furthermore, seasoned system thinkers continually resist the pressure to "validate" their models (i.e., prove truth) by tracking history. Instead, they work hard to become aware of where their models cease to be useful for guiding decision-making and to communicate their models to the decision-makers. This means that they are mostly concerned with face-validity (model structure and robustness). Face-validity tests assess how well the structure of a system model matches the structure of the reality of the model is intended to represent (Richmond (2000)). For example, do the input/output variables truly represent the underlying production technology?

Systems thinkers also pay a lot of attention to model robustness, i.e., they torture-test their models. They want to know under what circumstances their models "break down". They also want to know, "Does it break down in a realistic fashion. There is a whole set of variables in a production environment whose behavior patters and interrelationships show an internal consistency. To judge robustness of the model, one asks whether its variables show the same kinds of internal consistency (Richmond (2000)). These issues of model structure and robustness need to be addressed further by DEA researchers in the context of assessing the scientific contributions of their DEA research.

5.3   Relationship of the DEA Model with the Real World

During the DEA research process, the relationship between the real and virtual world represented by the DEA model should be constantly kept in mind. A DEA modeling effort should have the primary goal of establishing policies or procedures that can be implemented in the real world. However, many modelers and performance measurement teams often lose sight of the real world implementation.

In the virtual world the system structure is known and controllable. Information is complete, accurate and the feedback is almost instantaneous from actions within the system. Decisions and policy implementations are usually perfect. The goal of the virtual world is learning (Sterman (2000)). One interesting question to investigate within the DEA research community is the extent of learning that is achieved by the DEA modeling process.

The goal of real world systems is performance. In the real world, systems often have unknown structures and experiments are uncontrollable. Given these factors, policies that are implemented in the real world system first often do not realize their full impact until long after the policy is implemented. Information from real world systems is often incomplete due to long delays, and data is often biased or contains errors (Sterman (2000)).

The erroneous or incomplete data often lead to inaccurate mental models. Mental models serve as the basis for one's beliefs of how a system works. Receiving inaccurate or incomplete data about the system often leads to erroneous beliefs of how the system performs. This condition leads to the establishments of rules, system structure, or strategies that govern the system in a sub-optimal manner. When strategies, policies, structure, or decision rules are inaccurate, decisions made within the system are also incorrect. This creates negative consequences for the system processes and leads to a vicious circle for declining system performance.

However, DEA modeling should offer the decision-maker an opportunity to learn about system behavior, ask pertinent questions, collect accurate information and define more effective policies. The constant and effective interface between the virtual and real world allows for better decision-making over the long run.

REFERENCES

1. Adolphson, D, Cornia, G., and Walters, L., (1990), "A Unified Framework for Classifying DEA Models, in H. Bradley (ed.), *Operational Research 90*, Oxford, United Kingdom, Pergamon Press, 647-657
2. Aigner, D. J., C. A. K. Lovell, and P. Schmidt, 1977, Formulation and Estimation of Stochastic Frontier Production Functions, *Journal of Econometrics*, 6(1): 21-37.
3. Akiyama, T. and, Shao, C-F, 1993, Fuzzy Mathematical Programming for Traffic Safety Planning on an Urban Expressway, *Transportation Planning and Technology*, 17, 179-190.
4. Akosa, G., Franceys, R., Barker P., and T. Weyman-Jones, T., 1995, Efficiency of Water Supply and Sanitation Projects in Ghana, *Journal of Infrastructure Systems*, 1(1), 56-65.
5. Al-Majed, M., 1998, Priority-Rating of Public Maintenance Work in Saudi Arabia, M.S. Thesis, King Fahd University of Petroleum and Minerals, Saudi Arabia.
6. Almond, R.G., 1995, Discussion: Fuzzy Logic: Better Science or Better Engineering? *Technometrics*, 37(3), 267-270.
7. Amos, J., 1996, Transformation to Agility (Manufacturing, Aerospace Industry, Ph.D. Dissertation, The University of Texas at Austin.
8. Anandalingam, G., 1988, A Mathematical Programming Model of Decentralized Multi-Level Systems, *Journal of the Operational Research Society*, 39, 1021-1033.
9. Anderson, T. and K. Hollingsworth, 1997, An Introduction to Data Envelopment Analysis in Technology Management, *Portland Conference on Management of Engineering and Technology*, D. Kacaoglu and T. Anderson, eds. 773-778, IEEE.
10. Athanassopoulos, A., 1995, Goal Programming and Data Envelopment Analysis (GoDEA) for Target-Based multi-level Planning: Allocating Central Grants to the Greek Local Authorities, *European Journal of Operational Research*, 87, 535-550.
11. Athanassopoulos, A., Lambroukos, N. and L. Seiford, 1999, Data Envelopment Scenario Analysis for Setting Targets to Electricity Generating Plants, *European Journal of Operational Research*, 115(3), 413-428.
12. Bagdadioglu, N., Price, C., and T. Weymanjones, 1996, Efficiency and Ownership in Electricity Distribution-A Nonparametric Model of the Turkish Experience, *Energy Economics*, 18(1/2), 1-23.
13. Baker, R., and S. Talluri, 1997, A Closer Look at the Use of Data Envelopment Analysis for Technology Selection, *Computers & Industrial Engineering*, 32(1), 101-108.
14. Banker, R. D., Datar, S. M., and C. F. Kermerer, 1987, Factors Affecting Software Maintenance Productivity: An Exploratory Study, *Proceedings of the 8th International Conference on Information Systems*, 160-175, Pittsburgh, PA.
15. Banker, R. D., Datar, S. M., and C. F. Kermerer, 1991, A Model to Evaluate Variables Impacting the Productivity of Software Maintenance Projects, *Management Science*, 37(1), 1-18.
16. Banker, R. D. and C. F. Kermerer, 1989, Scale Economies in New Software Development, *IEEE Transactions on Software Engineering*, 15(10), 1199-1205.

17. Banker, R. D., Charnes, A., and W.W. Cooper, 1984, Some Models for Estimating Technical and Scale Efficiencies in Data Envelopment Analysis, *Management Science,* 30(9), 1078-1092.

18. Bannister G. and C. Stolp, Regional Concentration and Efficiency in Mexican Manufacturing, 1995, *European Journal of Operational Research,* 80(3), 672-690.

19. Battese, G. S. and G. S. Corra, 1977, Estimation of a Production Frontier Model: With Application to the Pastoral Zone of Eastern Australia, *Australian Journal of Agricultural Economics,* 21, 169-179.

20. Bellman, R. E. and Zadeh L. A., 1970, Decision-Making in a Fuzzy Environment, *Management Science,* 17(4), 141-164.

21. Boggs, R.L., Hazardous Waste Treatment Facilities: Modeling Production with Pollution as Both an Input and Output, 1997, Ph.D. Dissertation, The University of North Carolina at Chapel Hill.

22. Boile, M.P., 2001, Estimating Technical and Scale Inefficiencies of Public Transit Systems, *Journal of Transportation Engineering,* May/June 2001, 127(3), ASCE, 187-194.

23. Bookbinder, J.H. and W.W. Qu, 1993, Comparing the Performance of Major American Railroads, *Journal of the Transportation Research Forum,* 33(1), 70-83.

24. Borja, A., 2002, Outcome Based Measurement of Social Service Organizations: A DEA Approach," Ph.D. Dissertation, *Virginia Tech, Department of Industrial and Systems Engineering,* Falls Church, VA.

25. Bowen, W.M., 1990, The Nuclear Waste Site Selection Decision-A Comparison of Two Decision-Aiding Models, Ph.D. Dissertation, Indiana University.

26. Bowlin, W.F., 1987, Evaluating the Efficiency of US Air Force Real-Property Maintenance Activities, *Journal of the Operational Research Society,* 38(2), 127-135.

27. Bowlin, W.F., Charnes, A., and W.W. Cooper, 1988, Efficiency and Effectiveness in DEA: An Illustrative Application to Base Maintenance Activities in the U.S. Air Force, In: Davis, O.A., ed. *Papers in Cost Benefit Analysis,* Carnegie-Mellon University.

28. Braglia, M. and A. Petroni, 1999, Data Envelopment Analysis for Dispatching Rule Selection, *Production Planning & Control,* 10(5), 454-461.

29. Braglia, M. and A. Petroni, 1999, Evaluating and Selecting Investments in Industrial Robots, *International Journal of Production Research,* 37(18), 4157-4178.

30. Bulla, S., Cooper, W.W., Wilson, D., and K.S. Park, 2000, Evaluating Efficiencies of Turbofan Jet Engines: A Data Envelopment Analysis Approach, *Journal of Propulsion and Power,* 16(3), 431-439.

31. Busby, J.S., Williams, G.M., and A. Williamson, 1997, The Use of Frontier Analysis for Goal Setting in Managing Engineering Design, *Journal of Engineering Design,* 8(1), 53-74.

32. Byrnes, P. Färe, R., and S. Grosskopf, 1984, Measuring Productive Efficiency: An Application to Illinois Strip Mines, *Management Science,* 30(6), 671-681.

33. Campbell, D.G., Production Frontiers, 1993, Technical Efficiency and Productivity Measurement in a Panel of United States Manufacturing Plants, Ph.D. Dissertation, University of Maryland.

34. Caporaletti, L. and E. Gillenwater, 1995, The Use of Data Envelopment Analysis for the Evaluation of a Multiple Quality Characteristic Manufacturing Process, *37ᵗʰ Annual Meeting-Southwest Academy of Management,* C. Boyd, ed., 214-218, Southwest Academy of Management.

35. Carbone, T.A., 2000, Measuring efficiency of semiconductor manufacturing operations using Data Envelopment Analysis (DEA), *Proceedings of IEEE International Symposium on Semiconductor Manufacturing Conference,* IEEE, Piscataway, NJ, USA, 56-62.

36. Cardillo, D. and F. Tiziana, 2000, DEA Model for the Efficiency Evaluation of Non-dominated Paths on a Road Network, *European Journal of Operational Research,* 121(3), 549-558.

37. Carotenuto, P., Coffari A., Gastaldi, 1997, M., and N. Levialdi, Analyzing Transportation Public Agencies Performance Using Data Envelopment Analysis, *Transportation Systems IFAC IFIP IFORS Symposium,* Papageorgiou, M. and A. Poulieszos, editors, 655-660, Elsevier.

38. Carotenuto, P. Mancuso, P. and L. Tagliente, 2001, Public Transportation Agencies Performance: An Evaluation Approach Based on Data Envelopment Analysis, NECTAR Conference No 6 European Strategies in the Globalizing Markets; Transport Innovations, Competitiveness and Sustainability in the Information Age, 16-18 May, Espoo Finland.

39. Chai, D.K. and D.C. Ho, 1998, Multiple Criteria Decision Model for Resource Allocation: A Case Study in an Electric Utility, *INFOR*, 36(3), 151-160.

40. Chang, K.-P. and P-H Kao, 1992, The Relative Efficiency of Public-versus Private Municipal Bus Firms: An Application of Data Envelopment Analysis, *Journal of Productivity Analysis*, 3, 67-84.

41. Chang, Y.L., Sueyoshi, T. and R.S. Sullivan, 1996, Ranking Dispatching Rules by Data Envelopment Analysis in a Job Shop Environment, *IIE Transactions*, 28(8), 631-642.

42. Charnes, A., Cooper W.W. and Rhodes E. (1978), "Measuring the Efficiency of Decision-Making Units," *European Journal of Operational Research* 2, 429-444.

43. Charnes, A., Cooper, W.W., Lewin, A. and Seiford, L. editors, 1994, *Data Envelopment Analysis: Theory, Methodology and Applications*, Norwell, MA, Kluwer Academic Publishers.

44. Chen, T -Y., 2002, An Assessment of Technical Efficiency and Cross-Efficiency in Taiwan's Electricity Distribution Sector, *European Journal of Operational Research*, 137(2), 421-433.

45. Chen, W., 1999, The Productive Efficiency Analysis of Chinese Steel Firms: An Application of Data Envelopment Analysis, Ph.D. Dissertation, West Virginia University.

46. Chen, T.Y., 1999, Interpreting Technical Efficiency and Cross-Efficiency Ratings in Power Distribution Districts, *Pacific & Asian Journal of Energy*, 9(1), 31-43.

47. Chen, T.Y. and O.S. Yu, 1997, Performance Evaluation of Selected U.S. Utility Commercial Lighting Demand-Side Management Programs, *Journal of the Association of Energy Engineers*. 94(4), 50-66.

48. Chismar, W.G., 1986, Assessing the Economic Impact of Information Systems Technology on Organizations, Ph.D. Dissertation, Carnegie-Mellon University.

49. Chitkara, P., 1999, A Data Envelopment Analysis Approach to Evaluation of Operational Inefficiencies in Power Generating Units: A Case Study of Indian Power Plants, *IEEE Transactions on Power Systems*, 14(2), 419-425.

50. Chu, X. Fielding, G.J., and B. Lamar, 1992, Measuring Transit Performance using Data Envelopment Analysis, *Transportation Research Part A: Policy and Practice*, 26(3), 223-230.

51. Clarke, R.L., 1992, Evaluating USAF Vehicle Maintenance Productivity Over Time, *Decision Sciences*, 23(2), 376-384.

52. Clarke, R.L., 1988, Effects of Repeated Applications of Data Envelopment Analysis on Efficiency of Air Force Vehicle Maintenance Units in the Tactical Air Command and a Test for the Presence of Organizational Slack Using Rajiv Banker's Game Theory Formulations, Ph.D. Dissertation, Graduate School of Business, University of Texas.

53. Clarke, R.L. and K.N. Gourdin, 1991, Measuring the Efficiency of the Logistics Process, *Journal of Business Logistics*, 12(2), 17-33.

54. Co, H.C. and K.S. Chew, 1997, Performance and R&D Expenditures in American and Japanese Manufacturing Firms, *International Journal of Production Research*, 35(12), 3333-3348.

55. Collier, D. and J. Storbeck, 1993, Monitoring of Continuous Improvement Performance Using Data Envelopment Analysis, *Proceedings of the Decision Sciences Institute*, 3, 1925-1927.

56. Cook, W. D. and D.A. Johnston, 1991, Evaluating Alternative Suppliers for the Development of Complex Systems: A Multiple Criteria Approach, *Journal of the Operations Research Society*, 43(11), 1055-1061.

57. Cook, W. D., Johnston, D.A. and D. McCutcheon, 1992, Implementation of Robotics: Identifying Efficient Implementors, *Omega: International Journal of Management Science*, 20(2), 227-239.

58. Cook, W. D., Roll, Y. and A. Kazakov, 1990, A DEA Model for Measuring the Relative Efficiency of Highway Maintenance Patrols, *INFOR*, 28(2), 113-124.

59. Cook, W. D., Kazakov, A., and Y. Roll, 1994, On the Measuring and Monitoring of Relative Efficiency of Highway Maintenance Patrols, In: Charnes, A., Cooper, W.W., Lewin, A. and Seiford, L., editors, 1994, *Data Envelopment Analysis: Theory, Methodology and Applications*, Norwell, MA, Kluwer Academic Publishers.

60. Cook, W. D., Kazakov, A., Roll, Y. and L.M. Seiford, 1991, A Data Envelopment Approach to Measuring Efficiency: Case Analysis of Highway Maintenance Patrols, *Journal of Socio-Economics*, 20(1), 83-103.

61. Cook, W D., Kazakov, A. and B.N. Persaud, 2001, Prioritizing Highway Accident Sites: A Data Envelopment Analysis Model, *Journal of the Operational Research Society*, 52(3), 303-309.
62. Cooper, W.W., 1999, OR/MS: Where It's been. Where It Should Be Going? *Journal of the Operational Research Society*, 50, 3-11.
63. Cooper, W. W., Park, K. S.G. Yu, 2001, An Illustrative Application of IDEA (Imprecise Data Envelopment Analysis) to a Korean Mobile Telecommunication Company, *Operations Research*, 49(6), 807-820.
64. Cooper, W. W., Seiford, L. and K. Tone, 2000, *Data Envelopment Analysis: A Comprehensive Text with Models, Applications, References and DEA-Solver Software*, Kluwer Academic Publishers, Boston.
65. Cooper, W. W., Sinha, K. K., and R. S. Sullivan, 1992, Measuring Complexity in High-Technology Manufacturing: Indexes for Evaluation, *Interfaces*, 4(22), 38-48.
66. Coyle, R.G., 1996, Systems Dynamics Modeling: A Practical Approach, (1st edition) London, Great Britain: Chapman & Hall.
67. Cowie, J., and D., Asenova, Organization Form, Scale Effects and Efficiency in the British Bus Industry, *Transportation*, 26(3), 231-248.
68. Criswell, D R. and R.G. Thompson, 1996, Data Envelopment Analysis of Space and Terrestrially Based Large Commercial Power Systems for Earth: A Prototype Analysis of their Relative Economic Advantages, *Solar Energy*, 56(1), 119-131.
69. Debreu, G., 1951, The Coefficient of Resource Utilization, *Econometrica*, 19(3), 273-292.
70. Deprins, D., Simar, L. and H. Tulkens, 1984, Measuring Labor-Efficiency in Post Offices, in Marchand, M, Pestieau, P. and H. Tulkens, editors, The Performance of Public Enterprises: Concepts and Measurement, Elsevier Science Publishers B.V. (North Holland).
71. Dervaux, B., Kerstens, K. and P. Vanden Eeckaut, 1998, Radial and Non-radial Static Efficiency Decompositions: A Focus on Congestion Management, *Transportation Research-B*, 32(5), 299-312.
72. Doyle, J.R. and R.H. Green, 1991, Comparing Products Using Data Envelopment Analysis, *Omega: International Journal of Management Science*, 19(6), 631-638.
73. Dubois, D. and H. Prade, H., 1986, Fuzzy Sets and Statistical Data, *European Journal of Operations Research*, 25, 345-356.
74. Dyson, R.G., Allen, R., Camanho, A.S., Podinovski, V.V., Sarrico, C.S. and E.A. Shale, 2001, Pitfalls and Protocols in DEA, *European Journal of Operational Research*, 132, 245-259.
75. Ewing, R., 1995, Measuring Transportation Performance, *Transportation Quarterly*, 49(1), 91-104.
76. Färe, R. and S. Grosskopf, S., 1996, Intertemporal Production Frontiers: With Dynamic DEA, Kluwer Academic Publishers Boston.
77. Färe, R., Grosskopf, S., and C.A.K. Lovell, 1994, Production Frontiers, Cambridge University Press, Cambridge, MA.
78. Färe, R. and C.A.K. Lovell, 1978, Measuring the Technical Efficiency of Production, *Journal of Economic Theory*, 19(1), 150-162.
79. Färe, R. and D. Primont, D., 1995, Multi-Output Production and Duality: Theory and Applications, Kluwer Academic Publishers, Boston, MA.
80. Färe, R. and S. Grosskopf, 2000, Network DEA, *Socio-Economic Planning Sciences*, 34. 35-49.
81. Färe, R., Grosskopf, S. and J. Logan, 1987, The Comparative Efficiency of Western Coal-Fired Steam Electric Generating Plants; 1977-1979, *Engineering Costs and Production Economics*, 11, 21-30.
82. Färe, R., Grosskopf, S. and C. Pasurka, 1986, Effects on Relative Efficiency in Electric Power Generation Due to Environmental Controls, *Resources and Energy*, 8, 167-184.
83. Färe, R., Grosskopf, S., Lovell, C.A.K., and C. Pasurka, 1989, Multilateral Productivity Comparisons When Some Outputs are Undesirable: A Nonparametric Approach, *The Review of Economics and Statistics*, 71(1), 90-98.
84. Farrell, M.J., 1957, The Measurement of Productive Efficiency, *Journal of the Royal Statistical Society*, Series A (General), 120(3), 253-281.
85. Ferrier, G.D. and J.G. Hirschberg, 1992, Climate Control Efficiency, Energy Journal, 13(1), 37-54.
86. Fielding, G.J., 1987, *Managing Public Transit Strategically*, Jossey-Bass Publishers.

87. Fisher, 1997, An Integrated Methodology for Assessing Medical Waste Treatment Technologies (Decision Modeling), D. ENG., Southern Methodist University.
88. Forsund, F. R. and E. Hernaes, E., 1994, A Comparative Analysis of Ferry Transport in Norway, in Charnes, A., Cooper, W.W., Lewin, A. and Seiford, L., editors, *Data Envelopment Analysis: Theory, Methodology and Applications*, Norwell, MA, Kluwer Academic Publishers.
89. Forsund, F. and S. Kittelsen, 1998, Productivity Development of Norwegian Electricity Distribution Utilities, *Resource & Energy Economics*. 20(3), 207-224.
90. Forrester, J. W., 1961, Industrial Dynamics, MIT Press, Cambridge, MA.
91. Forrester, J. W., 1968, Principles of Systems, MIT Press, Cambridge, MA.
92. Fried, H., Lovell, C.A.K. and S. Schmidt, S., editors, 1993, *The Measurement of Productive Efficiency*, Oxford University Press.
93. Frisch, R., 1935-36, On the Notion of Equilibrium and Disequilibrium, *Review of Economic Studies*, 100-106.
94. Gathon, H-J., 1989, Indicators of Partial Productivity and Technical Efficiency in European Transit Sector, *Annals of Public and Co-operative Economics*, 60(1), 43-59.
95. Gillen, D. and A. Lall, 1997, Developing Measures of Airport Productivity and Performance: An Application of Data Envelopment Analysis, *Transportation Research Part E-Logistics and Transportation Review*, 33(4), 261-273.
96. Giokas, D.I. and G.C. Pentzaropoulos, 1995, Evaluating the Relative Efficiency of Large-Scale Computer Networks-An Approach via Data Envelopment Analysis, *Applied Mathematical Modeling*, 19(6), 363-370.
97. Girod, O., 1996, Measuring Technical Efficiency in a Fuzzy Environment, Ph.D. Dissertation, Department of Industrial and Systems Engineering, Virginia Polytechnic Institute and State University.
98. Girod, O. and K. Triantis, 1999, The Evaluation of Productive Efficiency Using a Fuzzy Mathematical Programming Approach: The Case of the Newspaper Preprint Insertion Process, *IEEE Transactions on Engineering Management*, 46(4), 1-15.
99. Golany, B., Roll, Y. and D. Rybak, 1994, Measuring Efficiency of Power-Plants in Israel by Data Envelopment Analysis, *IEEE Transactions on Engineering Management*, 41(3), 291-301.
100. Golany, B. and Y. Roll, 1994, Incorporating Standards Via Data Envelopment Analysis, in Charnes, A., Cooper, W.W., Lewin, A. and Seiford, L., editors, *Data Envelopment Analysis: Theory, Methodology and Applications*, Norwell, MA, Kluwer Academic Publishers.
101. Haas, D.A., 1998, Evaluating the Efficiency of Municipal Reverse Logistics Channels: An Application of Data Envelopment Analysis (Solid Waste Disposal), Ph.D. Dissertation. Temple University.
102. Hayes, K. E., Ratick, S., Bowen, W.M., and J. Cummings-Saxton, 1993, Environmental Decision Models: U.S. Experience and a New Approach to Pollution Management, *Environment International*, 19, 261-275.
103. Hjalmarsson, L. and J. Odeck, 1996, Efficiency of Trucks in Road Construction and Maintenance: An Evaluation with Data Envelopment Analysis, *Computers & Operations Research*, 23(4), 393-404.
104. Hollingsworth, K.B., 1995, A Warehouse Benchmarking Model Utilizing Frontier Production Functions (Data Envelopment Analysis), Ph.D. Dissertation, Georgia Institute of Technology.
105. Hoopes, B. and K. Triantis, 2001, Efficiency Performance, Control Charts and Process Improvement: Complementary Measurement and Evaluation, *IEEE Transactions on Engineering Management*, 48(2), 239-253.
106. Hoopes, B., Triantis, K., and N. Partangel, 2000, The Relationship Between Process and Manufacturing Plant Performance: A Goal Programming Approach, *International Journal of Operations and Quantitative Management*, 6(4), 287-310.
107. Hougaard, J., 1999, Fuzzy Scores of Technical Efficiency, *European Journal of Operational Research*, 115, 529-541.
108. Husain, N., Abdullah, M. and S. Kuman, 2000, Evaluating Public Sector Efficiency with Data Envelopment Analysis (DEA): A Case Study in Road Transport Department, Selangor, Malaysia, *Total Quality Management*, 11(4/5), S830-S836.
109. IEEE Transactions on Fuzzy Systems (1994), February 1994, volume 2, number 1, pp. 16-45.

110. Inuiguchi, M. and T. Tanino, 2000, Data Envelopment Analysis with Fuzzy Input and Output Data, *Lecture Notes on Economic Mathematics*, 487, 296-307.
111. Kabnurkar, A., 2001, Math Modeling for Data Envelopment Analysis with Fuzzy Restrictions on Weights, M.S. Thesis, *Virginia Tech, Department of Industrial and Systems Engineering*, Falls Church, VA.
112. Kao, C. and S. T. Liu, 2000, Fuzzy Efficiency Measures in Data Envelopment Analysis, *Fuzzy Sets and Systems*, 113(3), 427-437.
113. Karsak, E.E., 1999, DEA-based Robot Selection Procedure Incorporating Fuzzy Criteria Values, *Proceedings of the IEEE International Conference on Systems, Man and Cybernetics*, 1, I-1073-I-1078, IEEE.
114. Kazakov, A., Cook, W D. and Y. Roll, 1989, Measurement of Highway Maintenance Patrol Efficiency: Model and Factors, *Transportation Research Record*, 1216, 39-45.
115. Kemerer, C.F., 1987, Measurement of Software Development, Ph.D. Dissertation, Graduate School of Industrial Administration, Carnegie-Mellon, University.
116. Kemerer, C.F., 1988, Production Process Modeling of Software Maintenance Productivity, *Proceedings of the IEEE Conference on Software Maintenance*, p. 282, IEEE Computer Society Press, Washington, DC, USA.
117. Kerstens, K., 1996, Technical Efficiency Measurement and Explanation of French Urban Transit Companies, *Transportation Research-A*, 30(6), 431-452.
118. Khouja, M., 1995, The Use of Data Envelopment Analysis for Technology Selection, *Computers and Industrial Engineering*, 28(1), 123-132.
119. Kim, S.H., Park, C.-G. and K.-S. Park, 1999, Application of Data Envelopment Analysis in Telephone Offices Evaluation with Partial Data, *Computers & Operations Research*, 26(1), 59-72.
120. Kleinsorge, I.K., Schary, P.B. and R.D. Tanner, 1989, Evaluating Logistics Decisions, *International Journal of Physical Distribution and Materials Management*, 19(12),
121. Koopmans, T., 1951, Analysis of Production as an Efficient Combination of Activities, *Activity Analysis of Production and Allocation*, New Haven, Yale University Press, 3-97.
122. Kumar. M., 2002, A Preliminary Examination of the use of DEA (Data Envelopment Analysis) for Measuring Production Efficiency of a Set of Independent Four Way Signalized Intersections in a Region, MS Thesis, Virginia Polytechnic Institute and State University, Department of Civil Engineering, Advanced Transportation Systems.
123. Kumar, C. and B.K. Sinha, 1998, Efficiency Based Decision Rules for Production Planning and Control, *International Journal of Systems Science*, 29(11), 1265-1280.
124. Land, Lovell, and Thore, 1993, Chance-Constrained Efficiency Analysis, *Managerial and Decision Economics*, 14, pp. 541-553.
125. Laviolette, M., J. W. Seaman, J.D. Barrett, and W.H. Woodall, 1995, A Probabilistic and Statistical View of Fuzzy Methods, *Technometrics*, 37, 249-261.
126. Laviolette, M. and J.W. Seaman, 1994, Unity and Diversity of Fuzziness-From a Probability Viewpoint, *IEEE Transactions on Fuzzy Systems*, vol. 2, No.1, 1994, pp. 38-42.
127. Lebel, L.G., 1996, Performance and Efficiency Evaluation of Logging Contractors Using Data Envelopment Analysis, Ph.D. Dissertation, Virginia Polytechnic Institute and State University.
128. Lelas, V., 1998, Chance Constrained Models for Air Pollution Monitoring and Control (Risk Management), Ph.D. Dissertation, The University of Texas at Austin.
129. Liangrokapart, J., 2001, Measuring and Enhancing the Performance of Closely Linked Decision Making Units in Supply Chains Using Customer Satisfaction Data, Ph. Dissertation, Clemson University.
130. Linton, J.D. and W.D. Cook, 1998, Technology Implementation: A Comparative Study of Canadian and US Factories, *INFOR*, 36(3), 142-150.
131. Löthgren, M. and M. Tambour, 1999, Productivity and Customer Satisfaction in Swedish Pharmacies: A DEA Network Model, *European Journal of Operational Research*, 115, 449-458.
132. Lovell, C. A. K (1997), "What a Long Strange Trip It's Been," *Fifth European Workshop on Efficiency and Productivity Analysis*, October 9-11, 1997, Copenhagen, Denmark.
133. Mahmood, M.A., Pettingell, K.J., and A.I. Shaskevich, 1996, Measuring Productivity of Software Projects-A Data Envelopment Analysis Approach, *Decision Sciences*, 27(1), 57-80.

134. Martinez, M., 2001, Transit Productivity Analysis in Heterogeneous Conditions Using Data Envelopment Analysis with an Application to Rail Transit, Ph.D. Dissertation, New Jersey Institute of Technology.
135. Majumdar, S.K., 1995, Does Technology Adoption Pay-Electronic Switching Patterns and Firm-Level Performance in US Telecommunications, *Research Policy*, 24(5), 803-822.
136. Majumdar, S.K., 1997, Incentive Regulation and Productive Efficiency in the US Telecommunication Industry, *Journal of Business*, 70(4), 547-576.
137. McMullen, P.R. and G.V. Frazier, 1999, Using Simulation and Data Envelopment Analysis to Compare Assembly Line Balancing Solutions, *Journal of Productivity Analysis*, 11(2), 149-168.
138. McMullen, P.R. and G.V. Frazier, 1996, Assembly Line Balancing Using Simulation and Data Envelopment Analysis, *Proceedings of the Annual Meeting-Decision Sciences Institute*, volume 3.
139. Meeusen, W. and J. Vanden Broeck, 1977, Efficiency Estimation from Cobb-Douglas Production Functions with Composed Error, *International Economic Review*, 18(2): 435-444.
140. Miyashita, T. and H. Yamakawa, 2002, A study on the collaborative design using supervisor system, *JSME International Journal Series C-Mechanical Systems Machine Elements & Manufacturing*, 45(1), 333-341.
141. Morita, H., Kawasakim T., and S. Fujii, 1996, Two-objective Set Division Problem and Its Application to Production Cell Assignment,
142. Nakanishi, Y.J. and J.R. Norsworthy, J.R., 2000, Assessing Efficiency of Transit Service, *IEEE International Engineering Management Conference*, IEEE, Piscataway, NJ, USA, 133-140.
143. Nijkamp, P., and Rietveld, P., "Multi-Objective Multi-Level Policy Models: An Application to Regional and Environmental Planning", *European Economic Review*, No. 15 (1981), pp. 63-89.
144. Nolan, J.F. (1996), "Determinants of Productive Efficiency in Urban Transit," *Logistics and Transportation Review*, 32(3), 319-342.
145. Nozick, L.K., Borderas, H., and Meyburg, A.H. (1998), "Evaluation of Travel Demand Measures and Programs: A Data Envelopment Analysis Approach," *Transportation Research-A*, 32(5), 331-343.
146. Obeng, K., Benjamin, J. and A. Addus, 1986, Initial Analysis of Total Factor Productivity for Public Transit, *Transportation Research Record*, 1078, 48-55.
147. O'Connor, J. and I. McDermott, 1997, *The Art of Systems Thinking: Essential Skills for Creativity and Problem Solving*, Thorsons
148. Odeck, J., 1993, Measuring Productivity Growth and Efficiency with Data Envelopment Analysis: An Application on the Norwegian Road Sector, Ph.D. Dissertation, Department of Economics, University of Goteborg, Goteborg, Sweden.
149. Odeck, J. 1996, Evaluating Efficiency of Rock Blasting Using Data Envelopment Analysis, *Journal of Transportation Engineering-ASCE*, 122(1), 41-49.
150. Odeck, J., 2000, Assessing the Relative Efficiency and Productivity Growth of Vehicle Inspection Services: An Application of DEA and Malmquist Indices, *European Journal of Operational Research*, 126(3), 501-514.
151. Odeck, J. and L. Hjalmarsson, 1996, The Performance of Trucks-An Evaluation Using Data Envelopment Analysis, *Transportation Planning and Technology*, 20(1), 49-66.
152. Olesen, O.B. and N.C. Petersen, 1995, Chance Constrained Efficiency Evaluation, *Management Science*, 41, 3, pp. 442-457.
153. Otis, P.T., 1999, Dominance Based Measurement of Environmental Performance and Productive Efficiency of Manufacturing, Ph.D. Dissertation, Department of Industrial and Systems Engineering, Virginia Polytechnic Institute and State University.
154. Papahristodoulou, C., 1997, A DEA Model to Evaluate Car Efficiency, *Applied Economics*, 29(11), 14913-1508.
155. Paradi, J.C., Reese, D.N. and D. Rosen, 1997, Applications of DEA to Measure of Software Production at Two Large Canadian Banks, *Annals of Operations Research*, 73, 91-115.
156. Paradi, J.C., Smith, S. and C. Schaffnit-Chatterjee, 2002, Knowledge Worker Performance Analysis using DEA: An Application to Engineering Design Teams at Bell Canada, *IEEE Transactions on Engineering Management*, 49(2), 161-172.

157. Peck, M.W., Scheraga, C.A. and R.P. Boisjoly, 1998, Assessing the Relative Efficiency of Aircraft Maintenance Technologies: An Application of Data Envelopment Analysis, *Transportation Research Part A-Policy and Practice*, 32(4), 261-269.
158. Peck, M.W., Scheraga, C.A. and R.P. Boisjoly, 1996, The Utilization of Data Envelopment Analysis in Benchmarking Aircraft Maintenance Technologies, *Proceedings of the 38th Annual Meeting-Transportation Research Forum*, 1, 294-303.
159. Pedraja-Chaparro, F. Salinas-Jiménez, and P. Smith, 1999, On the Quality of the Data Envelopment Analysis, *Journal of the Operational Research Society*, 50, 636-644.
160. Polus, A. and A.B. Tomecki, 1986, Level-of-Service Framework for Evaluating Transportation System Management Alternatives, *Transportation Research Record*, 1081, 47-53.
161. Pratt, R.H. and T.J. Lomax, 1996, Performance Measures for Multi modal Transportation Systems, *Transportation Research Record*, 1518, 85-93.
162. Ramanathan, R., 2002, Combining Indicators of Energy Consumption and CO_2 Emissions: A Cross-Country Comparison, *International Journal of Global Energy Issues*, 17(3), 214-227.
163. Ray, S.C. and X.W. Hu, 1997, On the Technically Efficient Organization of an Industry: A study of US Airlines, *Journal of Productivity Analysis*, 8(1), 5-18.
164. Resti, A. 2000, Efficiency Measurement for Multi-Product Industries: A Comparison of Classic and Recent Techniques Based on Simulated Data, *European Journal of Operational Research*, 121(3), 559-578.
165. Richmond, B., 2000, *The "Thinking" in Systems Thinking: Seven Essential Skills*, Pegasus Communications, Inc., Waltham, MA
166. Ross, A. and M.A. Venkataramanan, 1998, Multi Commodity-Multi Echelon Distribution Planning: A DSS Approach with Application, *Proceedings of the Annual Meeting-Decision Sciences*,
167. Rouse, P., Putterill, M., and D. Ryan, 1997, Towards a General Managerial Framework for Performance Measurement: A Comprehensive Highway Maintenance Application, *Journal of Productivity Analysis*, 8(2), 127-149.
168. Ryus, P., Ausman, J., Teaf, D., Cooper, M. and M. Knoblauch M., 2000, Development of Florida's Transit Level-of-Service Indicator, *Transportation Research Record*, 1731, 123-129.
169. Samuelson, P.A., 1947, Foundations of Economic Analysis, Harvard University Press, Cambridge, MA.
170. Sarkis, J., 1997, An Empirical Analysis of Productivity and Complexity for Flexible Manufacturing Systems, *International Journal of Production Economics*, 48(1), 39-48.
171. Sarkis, J., 1997, Evaluating Flexible Manufacturing Systems Alternatives Using Data Envelopment Analysis, *The Engineering Economist*, 43(1) 25-47.
172. Sarkis, J., 1999, Methodological Framework for Evaluating Environmentally Conscious Manufacturing Programs, *Computers & Industrial Engineering*, 36(4), 793-810.
173. Sarkis, J. and J. Cordeiro, 1998, Empirical Evaluation of Environmental Efficiencies and Firm Performance: Pollution Prevention Versus End-of-Pipe Practice, *Proceedings of the Annual Meeting-Decision Sciences Institute.*
174. Sarkis, J. and S. Talluri, 1996, Efficiency Evaluation and Business Process Improvement through Internal Benchmarking, *Engineering Evaluation and Cost Analysis*, 1, 43-54.
175. Sarkis, J. and S. Talluri, 1999, A Decision Model for Evaluation of Flexible Manufacturing Systems in the Presence of Both Cardinal and Ordinal Factors, *International Journal of Production Research*, 37(13), 2927-2938.
176. Sarkis, J. and J. Weinrach, 2001, Using Data Envelopment Analysis to Evaluate Environmentally Conscious Waste Treatment Technology, *Journal of Cleaner Production*, 9(5), 417-427.
177. Scheraga, C.A. and P.M. Poli, 1998, Assessing the Relative Efficiency and Quality of Motor Carrier Maintenance Strategies: An Application of Data Envelopment Analysis, *Proceedings of the 40th Annual Meeting-Transportation Research Forum*, Transportation Research Forum, 1, 163-185.
178. Seaver B. and K. Triantis, 1989, The Implications of Using Messy Data to Estimate Production Frontier Based Technical Efficiency Measures, *Journal of Business and Economic Statistics*, Vol. 7, No. 1, 51-59, 1989

179. Seaver, B., and K. Triantis, K., 1995, The Impact of Outliers and Leverage Points for Technical Efficiency Measurement Using High Breakdown Procedures, *Management Science*, 41(6), 937-956.
180. Seaver, B. and K. Triantis, 1992, A Fuzzy Clustering Approach Used in Evaluating Technical Efficiency Measures in Manufacturing, *Journal of Productivity Analysis*, Volume 3, 337-363.
181. Seaver, B., Triantis, K. and C. Reeves, 1999, Fuzzy Selection of Influential Subsets in Regression, *Technometrics*, 41(4), 340-351.
182. Seiford, L., 1999, An Introduction to DEA and a Review of Applications in Engineering, *NSF Workshop on Engineering Applications of DEA*, Union College, NY, December, 1999.
183. Sengupta, J.K., 1987, Data Envelopment Analysis for Efficiency Measurement in the Stochastic Case, *Computers in Operational Research*, Vol. 14, No. 2, 1987, pp. 117-129.
184. Sengupta, J.K., 1992, A Fuzzy Systems Approach in Data Envelopment Analysis, *Computers and Mathematical Applications*, 24(8/9), 259-266.
185. Shafer, S.M., and J.W. Bradford, 1995, Efficiency Measurement of Alternative Machine Components Grouping Solutions via Data Envelopment Analysis, *IEEE Transactions on Engineering Management*, 42(2), 159-165.
186. Shao, B., 2000, Investigating the Value of Information Technology in Productive Efficiency: An Analytic and Empirical Study, Ph.D. Dissertation, State University of New York in Buffalo.
187. Shash, A.A.H., 1988, A Probabilistic Model for U.S. Nuclear Power Construction Times, Ph.D. Dissertation, Department of Civil Engineering, University of Texas.
188. Shephard, R.W., 1953, Cost and Production Functions, Princeton University Press, Princeton, New Jersey.
189. Shephard, R.W., 1970, Theory of Cost and Production Functions, Princeton University Press, Princeton, New Jersey.
190. Sheth, N., 1999, Measuring and Evaluating Efficiency and Effectiveness Using Goal Programming and Data Envelopment Analysis in a Fuzzy Environment, M.S. Thesis, *Virginia Tech, Department of Industrial and Systems Engineering*, Falls Church, VA
191. Shewhart, W.A., 1980, *Economic Control of Quality in Manufacturing*, D. Van Nostrand, New York (Republished by the American Society for Quality Control, Milwaukee, WI, 1980).
192. Sinha, K.K., 1991, Models for Evaluation of Complex Technological Systems: Strategic Applications in High Technology Manufacturing, Ph.D. Dissertation, Graduate School of Business, University of Texas.
193. Sjvgren, S., 1996, Efficient Combined Transport Terminals-A DEA Approach, Department of Business Administration, University of Götenberg.
194. Soloveitchik, D., Ben-Aderet, N. Grinman, M., and A. Lotov, 2002, Multiobjective Optimization and Marginal Pollution Abatement Cost in the Electricity Sector - An Israeli Case Study, *European Journal of Operational Research*, 140(3), 571-583.
195. Smith, J.K., 1996, The Measurement of the Environmental Performance of Industrial Processes: A Framework for the Incorporation of Environmental Considerations into Process Selection and Design, Ph.D. Dissertation, Duke University.
196. Sterman, J. D., 2000, Business Dynamics: Systems Thinking and Modeling for a Complex World, Irwin McGraw-Hill, Boston, MA.
197. Storto C.L., 1997, Technological Benchmarking of Products Using Data Envelopment Analysis: An Application to Segments A' and B' of the Italian Car Market, *Portland International Conference on Management of Engineering and Technology*, D.F. Kocaoglu and T.R. Anderson, editors, 783-788.
198. Sueyoshi, T., 1999, Tariff Structure of Japanese Electric Power Companies: An Empirical Analysis using DEA, *European Journal of Operational Research*, 118(2), 350-374.
199. Sueyoshi, T., Machida, H., Sugiyama, M., Arai, T., and Y., Yamada, 1997, "Privatization of Japan National Railways: DEA Time Series Approaches," *Journal of the Operations Research Society of Japan*, 40 (2), 186-205.
200. Sun, S., 2002, Assessing Computer Numerical Control Machines using Data Envelopment Analysis, *International Journal of Production Research*. 40(9), 2011-2039.
201. Talluri, S., Baker, R.C. and J. Sarkis, 1999, A Framework for Designing Efficient Value Chain Networks, *International Journal of Production Economics*, 62(1-2), 133-144.

202. Talluri, S., Huq, F. and W.E. Pinney, 1997, Application of Data Envelopment Analysis for Cell Performance Evaluation and Process Improvement in Cellular Manufacturing, *International Journal of Production Research*, 35(8), 2157-2170.

203. Talluri, S. and J. Sarkis, 1997, Extensions in Efficiency Measurement of Alternate Machine Component Grouping Solutions via Data Envelopment Analysis, *IEEE Transactions on Engineering Management*, 44(3), 299-304.

204. Talluri, S. and K.P. Yoon, 2000, A Cone-Ratio DEA Approach for AMT Justification, *International Journal of Production Economics*, 66, 119-129.

205. Talluri, S., 1996, A Methodology for Designing Effective Value Chains: An Integration of Efficient Supplier, Design, Manufacturing, and Distribution Processes (Benchmarks), Ph.D. Dissertation, The University of Texas at Arlington.

206. Talluri, S., 1996, Use of Cone-Ratio DEA for Manufacturing Technology Selection, *Proceedings of the Annual Meeting-Decision Sciences Institute*.

207. Technometrics, 1995, volume 37, no.3, August 1995, pp.249-292.

208. Teodorovic, D., 1994, Invited Review: Fuzzy Sets Theory Applications in Traffic and Transportation, *European Journal of Operational Research*, 74, 379-390.

209. Teodorovic, D., 1999, Fuzzy Logic Systems for Transportation Engineering: The State of the Art, *Transportation Research*, 33A, 337-364.

210. Teodorovic, D. and K. Vukadinovic, K., 1998, *Traffic Control and Transport Planning: A Fuzzy Sets and Neural Networks Approach*, Kluwer Academic Publishers, Boston/Dordrecht/London.

211. Thanassoulis, E. and R.G. Dyson, 1992, Estimating Preferred Target Input-Output Levels using Data Envelopment Analysis, *European Journal of Operational Research*, 56, 80-97.

212. Thompson, R.G., Singleton, F.D. Jr., Thrall, R.M., and B.A. Smith, 1986, Comparative Site Evaluation for Locating a High-Energy Physics Lab in Texas, *Interfaces*, 16(6), 35-49.

213. Tone, K. and T. Sawada, 1991, An Efficiency Analysis of Public Vs. Private Bus Transportation Enterprises, *Twelfth IFORS International Conference on Operational Research*, 357-365.

214. Tofallis, C., 1997, Input Efficiency Profiling: An Application to Airlines, *Computers and Operations Research*, 24 (3), 253-258.

215. Tran, A. and K. Womer, 1993, Data Envelopment Analysis and System Selection, *The Telecommunications Review*, vol. ?, 107-115.

216. Triantis, K., 1984, Measurement of Efficiency of Production: The Case of Pulp and Linerboard Manufacturing, Ph.D. Dissertation, Columbia University.

217. Triantis, K., 1987, Total and Partial Productivity Measurement at the Plant Level: Empirical Evidence for Linerboard Manufacturing, *Productivity Management Frontiers - I*, edited by D. Sumanth, Elsevier Science Publishers, Amsterdam, 113-123.

218. Triantis, K., 1990, An Assessment of Technical Efficiency Measures for Manufacturing Plants, *People and Product Management in Manufacturing, Advances in Industrial Engineering*, No. 9, edited by J. A. Edosomwan, Elsevier Science Publishers, Amsterdam, 149-166.

219. Triantis, K., 2003, Fuzzy Non-Radial DEA Measures of Technical Efficiency, forthcoming, *International Journal of Automotive Technology and Management*.

220. Triantis, K. and O. Girod O., 1998, A Mathematical Programming Approach for Measuring Technical Efficiency in a Fuzzy Environment, *Journal of Productivity Analysis*, 10, 85-102.

221. Triantis, K. and P. Otis, 2003, A Dominance Based Definition of Productive Efficiency for Manufacturing Taking into Account Pollution Prevention and Recycling, forthcoming, *European Journal of Operational Research*

222. Triantis, K. and A. Medina-Borja, 1996, Performance Measurement: The Development of Outcome Objectives: Armed Forces Emergency Services," *American Red Cross*, Chapter Management Workbook, Armed Forces Emergency Services, System Performance Laboratory.

223. Triantis, K. and R. NcNelis, 1995, The Measurement and Empirical Evaluation of Quality and Productivity for a Manufacturing Process: A Data Envelopment Analysis (DEA) Approach, *Flexible Automation and Intelligent Manufacturing-5th International Conference*, Schraft, R.D., editor, 1134-1146, Begell House Publishers.

224. Triantis, K. and P. Vanden Eeckaut, 2000, Fuzzy Pairwise Dominance and Implications for Technical Efficiency Performance Assessment, *Journal of Productivity Analysis*, 13(3), 203-226.

225. Triantis, K., Coleman, G., Kibler, G., and Sheth, N., 1998, Productivity Measurement and Evaluation in the United States Postal Service at the Processing and Distribution Center Level, System Performance Laboratory, distributed to the *United States Postal Service*.

226. Triantis, K., Sarangi, S. and D. Kuchta, 2003, Fuzzy Pair-Wise Dominance and Fuzzy Indices: An Evaluation of Productive Performance, *European Journal of Operational Research*, 144, 412-428.

227. Triantis, K., Seaver, B., and B. Hoopes, 2003, "Efficiency Performance and Dominance in Influential Subsets: An Evaluation using Fuzzy Clustering and Pair-wise Dominance," forthcoming, *Journal of Productivity Analysis*.

228. Tyteca, D., 1995, Linear Programming Models for the Measurement of Environmental Performance of Firms - Concepts and Empirical Results, *Intitut d'Administration et de Gestion Université Catholique de Louvain*, Place des Doyens, 1, B-1348, Louvain-la-Neuve, Belgium, September.

229. Tyteca, D., 1996, On the Measurement of the Environmental Performance of Firms - A Literature Review and a Productive Efficiency Perspective" *Journal of Environmental Management*, 46, 281-308.

230. Uri, N D., 2001, Changing Productive Efficiency in Telecommunications in the United States, *International Journal of Production Economics*, 72(2), 121-137.

231. Vaneman, W., 2002, Evaluating Performance in a Complex and Dynamic Environment, Ph.D. Dissertation, Department of Industrial and Systems Engineering, Virginia Polytechnic Institute and State University.

232. Vaneman, W. and K. Triantis, 2003, The Dynamic Production Axioms and System Dynamics Behaviors: The Foundation for Future Integration, *Journal of Productivity Analysis*, 19 (1), 93-113.

233. Vargas, V.A. and R. Metters, 1996, Adapting Lot-Sizing Techniques to Stochastic Demand through Production Scheduling Policy, *IIE Transactions*, 28(2), 141-148.

234. Wang, B., Zhang, Q. and F. Wang, 2002, Using DEA to Evaluate Firm Productive Efficiency with Environmental Performance, *Control & Decision*, 17(1), 24-28.

235. Wang, C.H., Gopal, R.D. and S. Zionts, 1997, Use of Data Envelopment Analysis in Assessing Information Technology Impact on Firm Performance, *Annals of Operations Research*, 73, 191-213.

236. Wang, C.H., 1993, The Impact of Manufacturing Performance on Firm Performance, the Determinants of Manufacturing Performance and the Shift of the Manufacturing Efficiency Frontier, Ph.D. Dissertation, State University of New York in Buffalo.

237. Ward, P., Storbeck J.E., Magnum S.L. and P.E. Byrnes, 1997, An Analysis of Staffing Efficiency in US Manufacturing: 1983 and 1989, *Annals of Operations Research*, 73, 67-90.

238. Wilson, P.W., 1993, Detecting Outliers in Deterministic Nonparametric Frontier Models with Multiple Outputs, *Journal of Business and Economic Statistics*, 11, 319-323.

239. Wilson, P.W., 1995, Detecting Influential Observations in Data Envelopment Analysis, *Journal of Productivity Analysis*, 6, 27-45.

240. Wolstenholme, E.F., 1990, System Enquiry: A System Dynamics Approach, New York: John Wiley & Sons.

241. Wu, L. and C. Xiao, 1989, Comparative Sampling Research on Operations Management in Machine Tools Industry between China and the Countries in Western Europe (in Chinese), *Journal of Shanghai Institute of Mechanical Engineering*, 11(1), 61-67.

242. Ylvinger, S., 2000, Industry Performance and Structural Efficiency Measures: Solutions to Problems in Firm Models, *European Journal of Operational Research*, 121(1), 164-174.

243. Zadeh, L. A., 1965, Fuzzy Sets, *Information and Control*, 8, 338-353.

244. Zeng, G., 1996, Evaluating the Efficiency of Vehicle Manufacturing with Different Products, *Annals of Operations Research*, 66, 299-310.

245. Zhu, J. and Y. Chen, 1993, Assessing Textile Factory Performance, *Journal of Systems Science and Systems Engineering*, 2(2), 119-133.

Chapter 15

BENCHMARKING IN SPORTS
Bonds or Ruth: Determining the Most Dominant Baseball Batter Using DEA

Timothy R. Anderson
Department of Engineering and Technology Management, Portland State University, Portland, OR 97207 USA   email: tima@etm.pdx.edu

Abstract: Operations research, management science, and economics analyses have frequently been conducted in the area of sports. Not surprisingly, DEA has also been used frequently in sports. These applications provide convenient and accurate data that a broad audience can understand. As such it provides a useful pedagogical tool for explaining DEA as well as demonstrating methodological extensions and implications. This chapter uses DEA to provide insights into the debate of who had the most dominant baseball batting season: Babe Ruth or Barry Bonds. The analysis highlights important characteristics of DEA such as the ability to adjust to changing operating conditions.

Key words: Data envelopment analysis (DEA); Super-Efficiency; Sports; Baseball

1.   INTRODUCTION

1.1   Motivation

Regardless of the sport, fans enjoy debating over who was the best player ever. Baseball lends itself particularly well to these debates because the rules of the game have not changed much in the past century compared to other sports. While people may debate questions such as who was the best base stealer, the best pitcher, or the best fielding third baseman, there has been little disagreement over who had the best batting season ever. Babe Ruth had two seasons in 1920 and 1921 which essentially rescued the sport of baseball from the 1919 scandal of the Chicago "Black Sox" that tainted the sport. In 1920 Babe Ruth hit 54 home runs, shattering the major league baseball record of 29 that he had set the previous year. He followed this up with 59 home runs in 1921. Babe Ruth was not just a power hitter though, in both years he led the league in walks (also known as bases on balls) and he was among the league leaders in batting average. Thorn and Palmer (1984, pg. 226) go so far as to say "Ruth's 1920 season is the best any mortal has ever had... Anyone who would disagree with calling this season the best ever would have to choose Ruth's 1921 or 1927 [seasons.]"

In 2001, another batter had a season that received frequent comparison to Babe Ruth. Barry Bonds set a new major league record by hitting 73 home runs, getting the most walks in a season, and having the highest slugging percentage of all time. In other words, he set new records in all three categories just as Babe Ruth had done 81 years earlier.

The goal of this study is to examine whether DEA could provide a fresh perspective in comparing these two batters and demonstrate additional insights by examining the numerical results in detail.

1.2   A brief history of baseball and management science

Baseball has been examined by management science and operations research professionals for a long time. Some of the earliest work can be traced back to George Lindsey's computer simulation models in the early 1960s (1963). The first DEA study of baseball was presented by Mazur in 1989 and later published (1994). Howard and Miller (1993) evaluated the linkage between salary and performance using DEA. More recently, Sueyoshi (1999) built a sophisticated model incorporating the batter's equivalent of a common pitcher's statistic into DEA to provide greater discriminatory power between efficient baseball batters. This paper builds upon the baseball batting model originally developed by Anderson and Sharp (1997) but extends the interpretation by using super-efficiency rather than ordinary DEA efficiency to measure a batter's dominance. We are emphasizing comparisons with batters who are DEA efficient.

1.3   Super-efficiency

We will use an extension of DEA introduced by Andersen and Peterson (1993) later called super-efficiency. The basic idea behind super-efficiency is to exclude the decision making unit, DMU, being evaluated from the comparison set. Normally in DEA an efficient DMU can be compared against itself and therefore receives an efficiency score of one. In the case of an input-oriented super-efficiency evaluation, efficient DMUs can receive scores better than one. The amount by which the efficiency score exceeds one indicates by how much the efficient DMU is able to outperform the best possible combination of other DMUs.

Conversely, in the output-oriented DEA model, the efficient DMUs will usually receive scores that are less than 1.0 indicating the amount by which they can exceed the performance of their peers.

For this application, we will use an output-oriented DEA model with constant returns to scale (CCR). The formulation is based upon (1.9) from Chapter 1 with the additional constraint that a DMU cannot use itself as a peer for evaluation,

$$
\begin{aligned}
&\max \phi + \varepsilon(\sum_{i=1}^{m} s_i^- + \sum_{r=1}^{s} s_r^+) \\
&\text{subject to} \\
&\sum_{j=1}^{n} x_{ij}\lambda_j + s_i^- = x_{io} \quad i = 1,2,...,m; \\
&\sum_{j=1}^{n} y_{rj}\lambda_j - s_r^+ = \phi y_{ro} \quad r = 1,2,...,s; \\
&\lambda_o = 0, \\
&\lambda_j \geq 0 \qquad\qquad j = 1,2,...,n,\ j \neq o.
\end{aligned}
\tag{15.1}
$$

For the purpose of this study, we are primarily concerned with the radial measure of efficiency, ϕ. The slack variables, s^+ and s^-, while important in many DEA applications will not greatly affect our results.

Let us next examine how this super-efficiency output-oriented DEA model affects the first numerical example described in chapter 1 with a single input, (X), and a single output, (Y). From table 1-2, it can be seen that each DMU's optimal solution had a non-negative envelopment variable for DMU P2 (i.e. $\lambda_2 > 0$ and $\lambda_2 = 0$). The imposition of the constraint that a DMU cannot be compared against itself clearly indicates that only P2 needs to be evaluated in terms of super-efficiency. (The other four DMUs cannot be super-efficient because they were not efficient to start with.)

Figure 15-1 shows there are now two efficient frontiers both consisting of lines drawn out from the origin. The higher frontier is based upon P3's performance and given that we are focusing on the output-oriented CCR model, the four other DMUs would all be compared against this frontier in the same way, as shown in Figure 1-3.

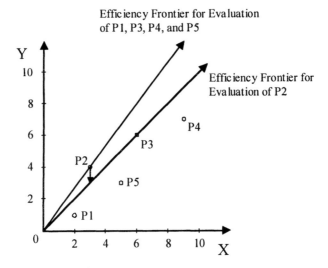

Figure 15-1. Super-efficiency frontiers and projection to the frontier for DMU P2

Notice that while all of the inefficient DMUs are projected up to the efficiency frontier, the efficient DMU, P2, is now projected down to the lower efficient frontier as determined by P3. This highlights that the performance of P2 will exceed that of its target output based upon the best available alternative technology (P3).

These target levels of performance, (or projections), for all five DMUs are shown in Table 15-1 below. As would be expected, all of the inefficient DMUs had target output levels that exceeded their actual output, Y. On the other hand, the best comparison for P2 consists of half of P3 (λ_3=1/2) and produced only three fourths as much output as P2 (ϕ=3/4) while using no more input than P2.

Table 15-1. Optimal solution values for the CCR model

DMU	X	Y	ϕ^*	λ^*	Target X	Target Y
P1	2	1	8/3	λ_2=2/3	2	8/3
P2	3	4	3/4	λ_3=1/2	3	3
P3	6	6	4/3	λ_2=2	6	8
P4	9	7	12/7	λ_2=3	9	12
P5	5	3	20/9	λ_2=5/3	5	20/3

Note that while it is possible for the constant returns to scale super-efficiency model to be infeasible for certain DMUs, this will only occur in very rare situations as demonstrated by Thrall (1996) and further elaborated upon by Zhu (1996). First, where a DMU has a value of zero for an input for which every other DMU has a positive value or second, where the DMU is the only DMU with a positive value for an output. Neither of these situations occur for the model of baseball batting used in this study. The situation is more complex under other returns to scale assumptions as demonstrated by Seiford and Zhu (1999).

2.   DEA MODEL FOR BASEBALL BATTING

In this application, the *n* DMUs correspond to the number of batters in each league. Major League Baseball is composed of the American League (AL) and National League (NL). Each league has slight rule variations, different umpires, and more importantly, plays against a very different mix of teams. Playing against different teams is important because this affects the quality of pitching a batter faces and the stadiums that they play in to such an extent that player statistics may vary significantly from one league to the next. Historically, teams in each league did not play against any teams in the other league during the regular season. In the last few years, teams have started playing a small number of games against several teams from the other league but, for all practical matters, the leagues still play under very different conditions and should therefore remain separate. For the sake of this analysis, we limited the dataset to compare players only against the players in the same league.

The only input used was plate appearances, (PA), which is the number of official at-bats plus the number of walks. Plate appearances represent the number of opportunities that the batter had to attempt to produce a walk or a hit.

The outputs were the number of walks (BB), singles (1B), doubles (2B), triples (3B), and home runs (HR) that the batter produced in those plate appearances. In baseball, the goal is to produce runs for a team and the value of these outputs are in increasing order of value, achievement, and production. This relationship is critical to incorporate into the application, and as discussed in section 1.4.3, there are a number of ways to incorporate this judgmental information into a DEA model.

Given that these are ordinal weight restrictions, we chose to use a special technique for incorporating ordinal weight restrictions by Ali, *et al.* (1991) that is based upon aggregating outputs. The same approach was used by Anderson and Sharp in their earlier study (1997) of baseball batters. Each output is replaced by the sum of all outputs at least as valuable as itself. In other words, the output of triples was replaced by the sum of triples and home runs. Similarly, the output of doubles was replaced by the sum of doubles, triples, and home runs. The output of singles was equal to the sum of singles, doubles, triples, and home runs. Lastly, the output of walks was equal to all the other outputs added together. The outputs used for the analysis were then labeled as (HR, 3B+, 2B+, 1B+, BB+).

3. BASEBALL DATASET

The source of the data for this study was a 45 megabyte Microsoft Access database containing a collection of baseball statistics from 1871 through 2002 developed by Sean Lahman (2002).

Many batters played less than full-time for a variety of reasons such as being pitchers or platoon batters who only get to bat against certain types of pitchers. In any case, baseball batting is an inherently probabilistic event, and therefore batters (or DMUs) with fewer opportunities may have substantially greater variation in apparent production. While techniques have been developed to extend DEA to deal with stochastic production technologies, this would not provide greater insight to the question of finding the most dominant baseball batting season. Therefore, we limited our dataset to only the batters with over 350 plate appearances in that league. Batters who played for two or more different teams in the same season had season totals used for their statistics in that league.

4. RESULTS

Since an output-oriented DEA model was used, efficiency scores would normally range from one and greater, with scores greater than one indicating relative inefficiency and scores of one indicating at least weak technical efficiency. The super efficiency output-oriented DEA model extends this to allow scores of less than one to indicate super-efficiency.

4.1 1920 American League

Sixty players met the minimum plate appearance criteria and were included in the analysis. The worst score was Ivy Griffin with an efficiency score of 1.655. This indicates that if Ivy Griffin had been efficient, as demonstrated by the efficient batters, he would have produced at least 65.5% more of each output using the same number of plate appearances.

Only two batters were found to be efficient: George Sisler and Babe Ruth. These two and the next three most efficient batters, Tris Speaker, Joe Jackson, and Eddie Collins, are listed in Table 15-2. It should be noted that all five players are considered among the best players of all time, accounting for four of the first twenty-six players elected to the Baseball Hall of Fame. The only player among the five not in the Hall of Fame is Joe Jackson. While Joe Jackson compiled statistics worthy of Hall of Fame consideration, he was permanently banned from baseball for his involvement (whether direct or indirect) in the 1919 Chicago White Sox scandal. Coincidentally, Bill James (2002, pg. 132) pointed out that both George Sisler and Babe Ruth started their careers as successful major league pitchers before their teams realized that they had more value as full-time batters.

Table 15-2. The five most efficient baseball batters in the 1920 American League. (Percentage leaders among the top five batters are marked in bold.)

Name	PA	BB	%	1B	%	2B	%	3B	%	HR	%
Ruth	608	*150*	***24.7%***	*73*	*12.0%*	*36*	*5.9%*	*9*	*1.5%*	*54*	***8.9%***
Sisler	677	*46*	*6.8%*	*171*	***25.3%***	*49*	*7.2%*	*18*	*2.7%*	*19*	*2.8%*
Speaker	649	*97*	*14.9%*	*145*	*22.2%*	*50*	***7.7%***	*11*	*1.7%*	*8*	*1.2%*
Jackson	626	*56*	*8.9%*	*144*	*23.0%*	*42*	*6.7%*	*20*	***3.2%***	*12*	*1.9%*
Collins	671	*69*	*10.3%*	*170*	***25.3%***	*38*	*5.7%*	*13*	*1.9%*	*3*	*0.4%*

Next we will how and why the ordinal weight restrictions were incorporated into the model. Notice in Table 15-2 that Tris Speaker was the best "producer" of doubles (2B) among the top five batters, producing 50 doubles in only 649 plate appearances (or 7.7% of the time compared to 7.2% for Sisler) but was not deemed efficient. Normally in DEA, any DMU that has the best ratio of one output to one input would be deemed efficient. An advantage in doubles alone should not be sufficient for Tris Speaker to be efficient though and can be corrected by aggregating the outputs in BB+, 1B+, 2B+, 3B+, and HR as described earlier.

The resulting data are shown in Table 15-3. This demonstrates that Tris Speaker's comparative advantage in doubles was more than compensated for by Babe Ruth's and George Sisler's much better performances in home runs (HR) and triples (3B) respectively. Now Babe Ruth is the leader in four of the five outputs (BB+, 2B+, 3B+, HR) per plate appearance as shown by the percentage columns of Table 15-3.

Table 15-3. Transformed 1920 output data to incorporate weight restrictions.

Name	BB+	%	1B+	%	2B+	%	3B+	%	HR	%
Ruth	322	***53.0%***	172	28.3%	99	***16.3%***	63	***10.4%***	54	***8.9%***
Sisler	303	44.8%	257	***38.0%***	86	12.7%	37	5.5%	19	2.8%
Speaker	311	47.9%	214	33.0%	69	10.6%	19	2.9%	8	1.2%
Jackson	274	43.8%	218	34.8%	74	11.8%	32	5.1%	12	1.9%
Collins	293	43.7%	224	33.4%	54	8.0%	16	2.4%	3	0.4%

As shown in table 15-4, Babe Ruth and George Sisler were the only two efficient batters in the 1920 American League. George Sisler was efficient on the basis of his ability to get a single, double, triple, or home run per plate appearance. George Sisler's score of 0.902 indicates that he would have produced approximately 10% more of one or more outputs than the best possible comparison that could be developed.

Table 15-4. 1920 Results from the CCR Output-Oriented Super-Efficiency

Name	Efficiency	Nonzero Envelopment Multipliers
Ruth	0.31599	$\lambda_{Sisler}=0.898$
Sisler	0.90216	$\lambda_{Jackson}=0.986$, $\lambda_{Ruth}=0.098$
Speaker	1.01406	$\lambda_{Ruth}=0.499$, $\lambda_{Sisler}=0.510$
Jackson	1.04975	$\lambda_{Ruth}=0.150$, $\lambda_{Sisler}=0.790$
Collins	1.06910	$\lambda_{Ruth}=0.259$, $\lambda_{Sisler}=0.758$

Table 15-4 tells us that the best possible comparison for Babe Ruth is composed of 0.898 of George Sisler. While this comparator may exceed the production of Babe Ruth in one or more outputs, it is still only able to reach at most 31.599% of Ruth's production for at least one output (not surprisingly, this was based on home runs.) To illustrate these relationships in more detail, the output constraint for home runs from formulation (15.1) used in the analysis of Babe Ruth is illustrated by (15.2). Not that the summation includes zero λ values for every term other than George Sisler. For notational convenience we have replaced the summation with just the term corresponding to Sisler in (15.2).

$$y_{HR,\,Sisler}\,\lambda_{Sisler} - s_{HR}^{+} = \phi y_{HR,o} \qquad (15.2)$$

Substituting the home run totals from table 15-3 and the results for Ruth from table 15-4 results in (15-3). Note that slacks such as s_{HR}^{+} were not reported in table 15-4 due to the emphasis on radial efficiency in this study but in this case, $s_{HR}^{+} = 0$ was found.

$$19 \cdot 0.898 - 0 = 0.31599 \cdot 54 \qquad (15.3)$$

$$17.06 = 17.06 \qquad (15.4)$$

The left hand side of (15.3) corresponds to the number of home runs hit by the best comparator for Ruth. The right hand side indicates this relates to the actual home run production of Babe Ruth. In other words, the comparator only produces 31.599 % of the output of home runs as did Ruth. In fact, (15.4) shows that Babe Ruth would be efficient by hitting just over 17 home runs. (Since fractional home runs are not allowable, 18 home runs would have sufficient for Babe Ruth to be efficient.)

A natural question is why is the comparator for Ruth 0.898 of Sisler ($\lambda_{Sisler}=0.898$) rather than being 1.000. The reason for this is that Sisler used many more plate appearances to produce his outputs than did Ruth. The value of λ is essentially equalizing the input of plate appearances between Ruth and the comparator based upon Sisler (608/677=0.898).

To summarize the 1920 results, the performance of Babe Ruth was truly exceptional.

4.2   1921 American League

There were sixty-four American League baseball batters with 350 or more plate appearances in 1921. Table 15-5 summarizes the results for the top five players and another player, Ken Williams for reasons discussed later. For the sake of brevity, the untransformed data and the results of the analysis with the ordinal weight restrictions applied are given.

Table 15-5. The five most efficient American League batters in 1921 and Ken Williams.

Name	PA	BB	1B	2B	3B	HR	Eff.	Nonzero Envelopment Multipliers
Ruth	685	145	85	44	16	59	0.44870	$\lambda_{Williams}=1.103$
Heilman	655	53	161	43	14	19	0.95299	$\lambda_{Cobb}=1.050$, $\lambda_{Ruth}=0.093$
Cobb	563	56	132	37	16	12	1.00712	$\lambda_{Heilman}=0.733$, $\lambda_{Ruth}=0.121$
Sisler	616	34	148	38	18	12	1.03189	$\lambda_{Heilman}=0.940$
Speaker	574	68	114	52	14	3	1.06537	$\lambda_{Heilman}=0.573$, $\lambda_{Ruth}=0.290$
Williams	*621*	*74*	*128*	*31*	*7*	*24*	*1.10197*	$\lambda_{Heilman}=0.583$, $\lambda_{Ruth}=0.349$

Just as in 1920, the top five most efficient batters, Babe Ruth, Harry Heilman, Ty Cobb, George Sisler, and Tris Speaker, have all been elected to the Hall of Fame. It is interesting to note that while in 1920 Babe Ruth was compared to a mix of an efficient player and an inefficient player (Sisler and Jackson respectively), in 1921, he was compared only against an inefficient player (Ken Williams). While Ken Williams was not even among the top five players, instead ranking eighth in the league, he was tied for second best in home runs. The implication is that while Ken Williams provided the best comparison to Babe Ruth, he was a very poor imitation.

Looking at the numbers in more detail indicates that giving Ken Williams 10.3% more plate appearances still would have under-produced Babe Ruth by 44.87% in one or more outputs – once again, home runs. The interested reader can conduct the same calculations as (15.2) through (15.4) to demonstrate this.

4.3   2001 National League

There were 118 National League batters with at least 350 plate appearances and the top ten most efficient batters are listed in Table 15-6. The number of players increased from 60 and 64 to 118 because there are more teams in each league as well as more games being played by each team in the season but this should not have a major impact on this study for two reasons. First, all of the top ten batters greatly exceeded the minimum requirement of plate appearances so a shorter season would not have dropped them from the evaluation. Secondly, these players in 2001 were excellent batters and they would have undoubtedly been major league players if there were fewer teams as well.

Of the top ten listed in table 15-6, only the first five have efficiency scores less than 1.0. Hence the last five are not super-efficient. It is somewhat surprising that the number of super-efficient players increased from two in both the 1920 and 1921 analyses to five in 2001 but this is consistent with the trends noted in Anderson and Sharp (1997) which observed that the number of radially efficient players each year tended to increase from the 1900s through the 1990s.

The efficient batters included Barry Bonds, Todd Helton, Larry Walker, Sammy Sosa, and Juan Pierre. Three of these five efficient players play for the Colorado Rockies. This is consistent with the reputation of the Colorado Rockies playing in a field that inflates batting statistics. While adjustments could be made to adjust for home field advantage, the "Colorado effect" does not influence the score for Barry Bonds since Bonds' only peer in the super-efficiency analysis is Sammy Sosa of the Chicago Cubs.

Table 15-6. The ten most efficient National League batters in 2001.

Name	PA	BB	1B	2B	3B	HR	Eff.	Nonzero Envelopment Multipliers
Bonds	653	177	49	32	2	73	0.82611	$\lambda_{Sosa}=0.942$
Helton	685	98	92	54	2	49	0.95813	$\lambda_{Sosa}=0.888, \lambda_{Walker}=0.120$
Walker	579	82	98	35	3	38	0.96254	$\lambda_{Alou}=0.159, \lambda_{Helton}=0.713$
Sosa	693	116	86	34	5	64	0.97792	$\lambda_{Bonds}=0.377, \lambda_{Gonazlez}=0.547, \lambda_{Walker}=0.101$
Pierre	658	41	163	26	11	2	0.98247	$\lambda_{Aurilia}=0.963$
Aurilia	683	47	127	37	5	37	1.00029	$\lambda_{Pierre}=0.189, \lambda_{Walker}=0.965$
Gonzalez	709	100	98	36	7	57	1.00668	$\lambda_{Sosa}=0.714, \lambda_{Walker}=0.370$
Vidro	517	31	105	34	1	15	1.01238	$\lambda_{Pierre}=0.365, \lambda_{Walker}=0.479$
Pujols	659	69	106	47	4	37	1.01287	$\lambda_{Helton}=0.174, \lambda_{Walker}=0.932$
Alou	570	57	111	31	1	27	1.01373	$\lambda_{Pierre}=0.244, \lambda_{Walker}=0.707$

Table 15-6 shows that Barry Bonds was the most super-efficient National League batter in 2001. Barry Bonds' superior performance in home runs (HR and HR%), resulted in his being the best producer in all of the aggregated outputs (2B+, 3B+, and BB+) on a percentage basis except for singles and longer (1B+).

The most surprising result is the inclusion of Juan Pierre among the efficient batters and the reason for his success. Juan Pierre was the leader in the rate of producing singles and longer per plate appearance. Pierre's success in this measure was particularly surprising given that he was only ninth in the National League in batting average. The difference is attributable to the fact that walks are included in the denominator for the former but left out of the latter. This benefits Juan Pierre because he had a particularly low number of walks (BB) for a hitter with a high batting average.

Barry Bonds' had the best super-efficiency in the National League with 0.8264 meaning that he was able to produce at least 17% of one or more outputs than could his best possible peer, in this case, Sammy Sosa.

5.   DISCUSSION

With efficiency scores of 0.31599 and 0.44872 in 1920 and 1921 compared to 0.82611, the results have clearly demonstrated that Babe Ruth in both of these years was more dominant than Barry Bonds in 2001 on the basis of super-efficiency. The results would have been different had the question been, who dominated the largest percentage of batters. Neither Ruth nor Bonds was the most frequently used peer for the evaluation of the batters in their leagues. Examining the results showed the following reference count results:

- 98% of the batters had a non-zero envelopment variable for Sisler, (every batter except himself) compared to 85% for Ruth in 1920,
- 97% of the batters had a non-zero envelopment variable for Heilman (every batter but Ruth and himself) compared to 75% for Ruth in 1921, and
- 97% of the batters had a non-zero envelopment variable for Walker, (all but four batters) compared to 21% for Bonds in 2001.

It may be surprising that the most super-efficient batters were not used as frequently in developing targets of performance as were other efficient, but less super-efficient, batters. Care must be taken in employing the concept of super-efficiency since it tends to reward extreme performance that is unlike that of other DMUs. Also, unlike traditional DEA models, in the case of super-efficiency, efficient DMUs may be compared against one or more inefficient peers, as demonstrated by Ruth in 1921 being compared to Ken Williams. In this case, the most super-efficient DMUs were all extreme in what was known to be the most valuable output, home runs.

The analysis also demonstrated some of the strengths of DEA. For example, DEA was found to adjust to the changing circumstances of the game.

REFERENCES

1. Ali, A.I., W.D. Cook, and L.M. Seiford, 1991, Strict vs. weak ordinal relations for multipliers in data envelopment analysis. *Management Science*. **37**(6): p. 733-8.
2. Andersen, P. and N.C. Petersen, 1993, A procedure for ranking efficient units in data envelopment analysis. *Management Science*. **39**(10): p. 1261-4.
3. Anderson, T.R. and G.P. Sharp, 1997, A new measure of baseball batters using DEA. *Annals of Operations Research*. **73**: p. 141-155.
4. Howard, L.H. and J.L. Miller, 1993, Fair pay for fair play: estimating pay equity in professional baseball with data envelopment analysis. *Academy of Management Journal*. **36**(4): p. 882-94.
5. James, B., 2002, *The New Bill James Historical Baseball Abstract*: Free Press. 998.
6. Lahman, S., 2002, The Lahman Baseball Database, Version 5.0, http://www.baseball1.com/.
7. Lindsey, G.R., 1963, An investigation of strategies in baseball. *Operations Research*. **11**(4): p. 477-501.
8. Mazur, M.J., 1994, Evaluating the relative efficiency of baseball players?, in *Data Envelopment Analysis: Theory, Methodology and Applications*, A. Charnes, W.W. Cooper, A. Lewin, L.M. Seiford, Editors. Kluwer Academic Publishers: Boston. p. 369-91.
9. Seiford, L.M. and J. Zhu, 1999, Infeasibility of super-efficiency data envelopment analysis models. *INFOR*. **37**(2): p. 174-187.
10. Sueyoshi, T., K. Ohnishi, and Y. Kinase, 1999, A benchmark approach for baseball evaluation. *European Journal of Operational Research*. **115**(3): p. 429-448.
11. Thorn, J. and P. Palmer, 1984, *Hidden Game of Baseball*. New York: Doubleday.
12. Thrall, R.M., 1996, Duality, classification, and slacks in DEA. *Annals of Operations Research*. **66**: p. 109-138.
13. Zhu, J., 1996, Robustness of the efficient DMUs in data envelopment analysis. *European Journal of Operational Research*. **90**(3): p. 451-460.

Chapter 16

ASSESSING THE SELLING FUNCTION IN RETAILING
Insights from Banking, Sales forces, Restaurants & Betting shops

Antreas Athanassopoulos*
Athens Laboratory of Business Administration (ALBA),
Athinas Avenue & Areos Str. 2A
166 71 Vouliagmeni, Greece
aathana@alba.edu.gr

Abstract: In this chapter we focus on the performance of for-profit retail service industries. The assessment of the performance in retail service industries has gained considerable attention is recent years. In this paper we report results regarding efficiency from a panel of selected retail service industries. Particular emphasis is given to the development of a unified methodological framework for assessing the operating efficiency of retail networks. Data envelopment analysis, which is used for their performance assessment, is a linear programming methodology for estimating empirically based efficiency frontiers and its results will be contrasted with more traditional measures of performance such as profitability. A sequence of 512 retail outlets, from four different retail chains, was examined for their performance and only 164 were found to operate on the best practice frontier.

Key words: Service industries, Retailing, data envelopment analysis, profitability.

1.   INTRODUCTION

The assessment of the effectiveness of the marketing function in both manufacturing and service organizations has gained popularity as a direct result of the increasing role of the marketing function. This role of marketing has brought forward issues of marketing control and, in particular,

* I would like to acknowledge the comments and support provided by Dr. W.W. Cooper during the preparation of the chapter. Omissions and mistakes remain the author's responsibility.

its relationship with traditional control yardsticks emanating mainly from the marketing literature, Achabal et al. (1985). In retailing organizations there is a strong influence of cost and labour productivity measures which reflect particular strategic decisions on behalf of the corresponding retail firms. That is, a retail firm that emphasizes its control function on cost containment measures has, implicitly, accepted a cost focus strategy which may not be in line with the perceived strategy of the firm.

This gap between perceived and realized strategies as a result of how performance is assessed detaches the empirical findings from the strategic marketing literature in which marketing orientation is associated with better performing firms. There is a need, therefore, to bring the marketing orientation at the operations level of individual firms. In this chapter we propose measures of marketing performance applied to different retail service organisations promoting their practical managerial usefulness. For this purpose we draw upon previous research attempts by Athanassopoulos (1995a) and (1998b) in which marketing performance measures were suggested at different management levels of a retail environment spaning the whole marketing spectrum i.e. from strategic retail marketing to line management levels.

Marketing performance has been distinguished at two levels by a series of research scholars. Ingene (1984b) argues that marketing efficiency concerns minimum input use for given levels of output, while marketing effectiveness concerns the maximum output for given levels of input. In the same vein, Gronroos (1990) makes the distinction between the internal and external efficiency of retail stores. In the management science literature, Mahajan (1991) and Athanassopoulos (1995b, 1998a) propose a distinction between market efficiency and cost efficiency (profitability) in an attempt to address extrinsic and intrinsic oriented performance measures in retailing. Alternative conceptualisations of marketing performance have strong resemblances to the economic notion of a production function, Charnes et al. (1985). The production function (marketing efficiency) approach proposed by Berne et al. (1995) contrasts retail output with various endogenous input factors of retail outlets (e.g. wages, size, capital and maintenance). The demand function (marketing effectiveness) is the complement to retail productivity, as it seeks to contrast the generation of sales with various endogenous (e.g. size, advertisement) and exogenous (e.g. market size, competition) input factors. The search for the most appropriate measures of retail output is another area of research interest (Ingene (1982) and Berne et al. (1995) and Athanassopoulos (1998b)). On the retail input side there is research advocating the use of market potential of individual outlets, Lusch and Moon (1984), Ingene (1984a) and Doyle et al. (1979).

The remainder of this chapter is organized as follows. In section 2 we give an overview of methodologies proposed for assessing performance in retail organizations. We focus on the distinction between the concepts of market and cost efficiency as being the external and internal image of retail outlets. The operational assessment of the market and cost efficiency of individual branches is pursued using frontier analysis methods. In section 3 we include the empirical results of a study for each of five retail firms included in the study. In section 4 we present results regarding the association between the marketing efficiency and the profitability of four service chains. Section 5 concludes the paper.

2. ASSESSING THE RETAIL SALES FUNCTION

The assessment of retail efficiency in a for-profit environment is inextricably linked with the ability of individual branches to contribute to the corporate objectives of their headquarters. Profitability is an important criterion of branch performance which, however, does not necessarily contain the full picture regarding the effort made by each branch. To better demonstrate the operating characteristics of retail branches we use the diagrammatic representation in Figure *16-1*.

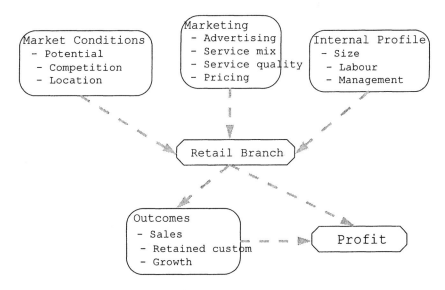

Figure 16-1. Productive process in retail branches

The performance of retail branches needs to be assessed by formulating the multi-input and multi-output operations of each branch. In a service setting it is not always feasible to aggregate outputs into a single profit figure (e.g. the outcomes of a bank branch or sales representative). Furthermore, service operations are not characterized by a single objective and therefore one needs to take into account the coexistence of objectives such as cost minimization, service adequacy and market penetration on the assessment of retail branch performance.

The information in Figure *16-1* describes the forces characterizing the operations of retail firms. The operating environment of individual retail firms always needs to be considered and in any case to elements of market size, competition and location attractiveness need to be taken into account. The exact definition of the size of the market surrounding each branch is an area of continuous research effort in the disciplines of marketing and regional economics, (Ghosh and McLafferty, 1987). The marketing effort made by individual restaurant branches, for instance, depends to a large extent on issues of marketing-mix and also the particular features of the services/products (i.e. in a restaurant the mix is between sales of foods and drinks). Finally, the internal characteristics of each restaurant include issues of branch size and staff resources, as well as the quality of management that is in charge of each branch.

The objectives of individual branches include *inter alia*
- Location in commercially viable places and selection of effective service and size mix (Strategic marketing)
- Generation of sales via the appropriate penetration at the local level (Line-marketing)
- The retention of customers and exploitation of referrals, loyalty and cross-selling benefits
- The cost-control in the operation of individual branches such that the generation of sales will be converted into profits.

In this research we have sought to represent the objectives of service operations at an operational level and thus provide a framework for assessing the strategic and line marketing efficiency of multi-branch service organizations.

The *Strategic Marketing efficiency* of the branches of service operations seeks to assess the overall performance profile of the branches which include the entire selection of the marketing mix for the corresponding branches. Control over the marketing mix of individual branches is primarily the responsibility of the central marketing function. For example, the selection of location and size of a branch, the staffing levels, the product range and corresponding pricing policies, and the promotion and advertising.

The *Line-marketing efficiency*, seeks to create an intermediate level of marketing accountability that concerns the elements of the marketing mix that are under sole or partial control of line management. For example, in the service sector the elements of marketing mix concerning customer service, entrepreneurial climate in the interior of the branch, smooth service processes and operations, and product/service differentiation to customer segments are prime responsibilities of the line-management.

Finally, a complementary dimension of branch performance, which draws both upon strategic and line-marketing, concerns cost control and profitability at the branch level. The assessment of cost control slices across the whole spectrum of strategic and line management and thus cost efficiency results can be contrasted with the corresponding marketing efficiency assessment. The complex function operated by service organizations needs to be appreciated and thus the need for comprehensive and integrative systems of marketing control are next proposed.

2.1   Benchmarketing frontiers for assessing the selling function

On the measurement side the assessment of retail performance has been primarily addressed via central tendency methods, namely regression analysis, (Jones and Mock, 1984). Regression has been widely used for assessing service productivity of bank branches Olsen and Lord (1979), Doyle et al. (1979), clothing stores, Davies (1973) and grocery stores, Berne et al. (1995). To circumvent some of the limitations of ordinary regression analysis an alternative optimization methodology has been proposed in the literature. The method was initially developed by Charnes et al. (1978) and was suggested later by Golany and Roll (1993) as a means of assessing service productivity. These methods were used to assess the market efficiency of consumer products by Kamakura et al. (1988), the market efficiency of sales forces by Mahajan (1991) and Horsky and Nelson (1996), the market efficiency and profitability of public houses by Athanassopoulos and Thanassoulis (1995a), alternative market efficiency indices of restaurants by Athanassopoulos (1995), and the market and cost efficiency of bank branches by Athanassopoulos (1998a). The successful use of optimisation methodologies to assess branch efficiency has also been extended to address planning problems of retail management by Banker and Morey (1993) and Athanassopoulos (1998a) who developed market efficiency models to guide the selection of alternative sites to open new outlets while Athanassopoulos and Thanassoulis (1995b), developed optimization models to assess the marginal impacts of investments on the market efficiency of retail outlets.

Aside from the strict measurement of relative efficiency the intuition behind the use of Data Envelopment Analysis tools can yield the following outcomes with marketing support implications:

o Accommodates *controllable* and *uncontrollable* input factors: Controllable factors include those inputs that are under the strategic and/or line marketing discretion (e.g. pricing, promotion, product mix, personnel). Uncontrollable factors are mainly market related factors that are outside the direct control of management (e.g. the degree of competition). The frontier assessment methodology used in DEA enables the user to pre-specify controllable and uncontrollable inputs and thus takes into account the extent to which management affects the levels of certain categories of resources used. Marketing efficiency studies are particularly concerned with using market related factors that are beyond the control of local retail branches.

o *Target projections:* The benchmarking mechanism. In addition to the relative efficiency values, DEA also yields target positions for inefficient firms according to a projection formula, which shows the extent to which inputs / outputs need improvement compared to their benchmarks.

o *Economic assumptions*: Frontier analysis models encapsulate alternative returns to scale assumptions between inputs and outputs. That is, the technical efficiency of individual firms can be assessed under the assumptions of constant, non-increasing, non-decreasing and variable returns to scale. (Details on the treatment of returns to scale in DEA can be found in Chapter 2 of this handbook.)

o *Strategic Marketing Perspective*: Overall evaluation of branch operations is best described by the constant returns to scale assessment of efficiency. In that perspective, one gets measurable outcomes related to the overall positioning of a retail outlet encapsulating factors that are in and beyond the control of local retail management.

o *Line-marketing Perspective.* This perspective seeks to disentangle the controllable from the uncontrollable elements of the operation of a branch. As the assessment of marketing efficiency is primarily focused on maximisation of sales, the degree of controllability is addressed at the level of "size-control" over the operation of the branch. After, having controlled for the size of operations, one can focus on the performance of line-marketing within the branch. In practical terms, a branch manager is assessed for efficiency after having controlled for the size, location, market and other factors that directly affect branch performance.

3.   CASE RESULTS FROM RETAIL-SERVICE INDUSTRIES

The empirical illustrations we use for the assessment of retail marketing performance are based on a unique data base that comprises micro-analytic data from four retail networks: commercial banking, sales forces of brewing products, restaurants, and betting shops. All retail networks operate in the UK and the information refers to the financial year 1995/96. Four case studies are presented separately since they are each related to a different body of literature.

3.1 Commercial banking

Chapter 13 by Paradi, Vela and Yang in this handbook provides a detailed discussion of uses of DEA in evaluating bank branch performances. Based primarily on experience with Canadian banks this chapter develops DEA models for use in evaluating their performance. Here, however, we are concerned with using results from DEA studies of banking, sales forces, restaurants chains and betting shops in order to evaluate the degree of uniformity in these results. For this purpose we start with DEA evaluations on the performances of individual branch banks.

✓ Staff size
✓ # of current accounts
✓ Affluence measure
✓ Competition
✓ Branch attractiveness
✓ Marketing effort

Sales of credit cards
Sales of mortgages
Sales of insurances
Sales of financial products
Sales of current accounts
Sales of loans

Figure 16-2. Inputs – outputs used for assessing the marketing efficiency of commercial bank branches

In our case study we report results from the market efficiency of a sample of 120 commercial bank branches in the UK. That is, we seek to assess the ability of individual branches to penetrate their market and generate sales. The input-output framework for assessing the production efficiency of individual bank branches is listed in Figure 16-2.

The proposed input framework is tailored for the assessment of the marketing efficiency of bank branches in line with Athanassopoulos (1998a) models. That is, we have represented the market size surrounding individual branches (we have taken the size of the market on a 0.5 miles radius from each branch) using sociodemographic measures of affluence supplied by the marketing department of the bank. Furthermore, the number of current accounts was used as an indication of the customer base of individual branches that seek to promote new products. Competition is represented by the number of bank branch establishments surrounding each branch. Staff size is used to capture the capacity of each branch to promote products (it was found to be correlated with the size of each branch). The input side also included two qualitative measures regarding the attractiveness and marketing effort of each branch. Both measures are aggregates of multi-item questions completed by area managers of each branch and seek to measure the location attractiveness of each branch and the marketing effort made in terms of promotion, customer service and local advertising. The input side of the model consisted of controllable (staff size, number of controllable accounts, branch attractiveness and marketing effort) and uncontrollable (market affluence and competition) subsets.

On the output side we used new sales measures of groups of products which were decided jointly with the marketing department of the bank. The product categories used as outputs were based on the similarity of effort required for their management at individual branches. We used six output categories in an attempt to account for the different effort that is necessary to "produce" each type of output.

3.2   Sales forces

The management of sales forces is a well recognized part of the marketing management function of many organizations. In both product and service organizations the performance of sales teams remains an open question despite the wide coverage given to the problem even in elementary marketing textbooks (Kotler 2002). One common problem in managing sales forces concerns the development of appropriate mechanisms for assessing the sales effort function and, subsequently, for setting performance targets applied to individual sales representatives (or teams).

The existing literature on sales force management has had inputs from alternative perspectives to deal with issues of measuring the response of sales to the levels of sales effort (for a review see Ryans and Weinberg (1987). Specific issues in this area concern the allocation of salesperson time to existing and prospective customers and/or across different products, Lodish et al. (1988). The management of sales territories by means of optimization models are also of interest, Rangaswamy et al. (1990). On the control side, Mahajan (1991) used frontier methods, as in DEA, in order to assess the effectiveness of regional sales forces in insurance services. In a recent research contribution, Horsky and Nelson (1997) combined optimization with frontier benchmarking methods in order to support the evaluation of the salesforce size and productivity.

In our case study we use data from 80 sales representatives that covered sales territories in central England. The range of products in the portfolio of the sales representatives consists of brewing products. The nature of their work entails sales visits to their current accounts with a dual role of replacement of stocks and the promotion of new product lines. At the same time the sales representatives spend time visiting prospects in which they seek to sell their products. The two objectives of the sales people were represented by the following output-definition of the proposed input-output model. Both the input and output framework that describes the operation process of sales force representatives is given in Figure 16-3.

✓ # of current accounts
✓ Size of category
✓ # of competing brands
✓ Promotions
✓ Years of experience

New customers gained
Sales of promoted goods
Sales of non-promoted goods

Figure 16-3. Input-output set for evaluating sales forces efficiency

The input framework in the assessment of the sales representatives conveys information about market conditions in the trade market of each sales person and also the input resources that were committed for the same purpose. The number of accounts represents the customer base while size of the category is used to control for the consumption patterns of the local consumers for the bundle of products that are marketed by the particular sales manager. The size of the category was calculated from market research figures and entails the total consumption for each geographical area.

The number of competing brands takes into account the presence of national and local competing brands while the intensity of competition was assessed via a qualitative scale from [1, 11] with 1 indicating low and 11 high competition. This qualitative scale was obtained from the national co-ordinator of the sales force and reflected the perception of the firm's sales planning unit about the nature of competition in the 80 trade areas of the sales forces. The input side also included attributes related to the promotion expenditure at the level of individual product lines (i.e. different types of beer and food related products). Finally, years of experience of the sales people was considered an important factor related both to ability to perform and relationship to marketing elements. The input side of the model used controllable (Promotions and Years of experience) and uncontrollable (number of accounts, size of category and competition) subsets.

The output side of the particular sales force was modeled to represent the objective of sales growth and account penetration. Thus we used measures reflecting the growth in sales after controlling for products with and without being promoted by the marketing department of the firm. Furthermore, we used a measure reflecting the number of new accounts opened. This was a composite measure that took into account not only new accounts opened but also the extent to which this penetration could be regarded as feasible. This was accomplished by creating a composite measure that accounted for the number of new accounts, the number of accounts lost to competing brands and the total number of accounts available for penetration. The brewing market in the UK has the peculiarity that not all accounts are penetrable since there are pubs owned and managed by breweries, pubs linked via financial and other ties to breweries and free pubs. In essence new accounts are obtained from the free pubs while the overall consumption is related to the total number of pubs in the area.

3.3   Restaurant chains

The operation of restaurants is a very active part of retail service offering encapsulating many alternative forms of business structures, Doutt (1984). It includes, chains of restaurants with centrally employed managements operating from headquarters offering franchising operations with many independent operators. There are also mixed cases with large networks controlling independent outlets via pricing and loan agreements. Irrespective of the structure of the operation, restaurant operations are in a low margin and high volume industry. We concentrated on a chain of restaurant outlets in the UK which is controlled by one large brewery and offers mixed (food and drink) services. National chains of restaurants and pubs operate under uniform pricing and product mix policies and thus no attempt was made to incorporate these factors on the assessment of restaurant efficiencies. The input-output set for assessing the marketing efficiency of this chain of restaurants is given in Figure 16-4.

✓ Bar area
✓ State of repair
✓ Car park facilities

Food sales
Drink sales

✓ Consumption of alcohol
    in the surrounding area
✓ Propensitiy to visit
    restaurants
✓ Competition

Figure 16-4. Input and output for evaluating restaurant marketing efficiency

The proposed framework of marketing efficiency entails a set of market indicators that were used to encapsulate the potential of the market surrounding each restaurant. The trade area was estimated from annual survey statistics of the firm and thus the market figures were derived for a 1.5-mile radius from each restaurant. The consumption of alcohol was supplied by independent market research consultants while the propensity to visit restaurants corresponds to a matching between the targeted socio-economic profile and the corresponding volume of this type of potential customers in the surrounding area. Competition simply reflects the number of operating establishments within the restaurant's trade area. The input side of the model also distinguished between controllable (Bar area and State of repair) and uncontrollable (consumption of alcohol, propensity to visit restaurants and competition)) subsets. On the output side we used turnover generated for two core products sold, i.e. food and drinks. The distinction between two outputs was made since some restaurants seemed to emphasize to different output categories in response to the different demand profile in their market environment.

3.4 Betting shops

✓ Competition
✓ Potential customers
✓ Adults betting

✓ Revenue / Turnover      Revenue

No. of bets

✓ Age of shop
✓ Size
✓ Shop profile

Figure 16-5. Input and output set for assessing betting shop marketing efficiency

The provision of retail services draws on a series of activities which are little understood or at least researched. The operation of betting shops is in a heavily regulated industry in the UK with at least 4 major chains controlling over 3,000 betting shops along with a number of independents. Betting shop operations, despite this heavy regulation, are businesses facing intense competition. The profitability of a betting shop depends primarily on its ability to attract customers, as the majority of its costs are inelastic (labour and rent). We have adopted a prototype for assessing the market efficiency of the betting shops included in this study which is based on the input-output framework shown in Figure 16-5.

The output of the DEA model proposed included the realized revenue per betting shop and the number of bets serviced. These two outcomes encapsulated an element of monetary returns (revenue) and an element of sales transactions representing the number of bets taking place at the shop. The number of bets conveys information about propensity to bet but does not give any indication of the monetary value of each bet. Both of these output variables give a fair representation of the ability of the shop to attract players and also its ability to yield sufficient volume of bets.

The input side of the proposed model was distinguished between three groups of variables: (i) market conditions, (ii) betting shop attractiveness and (iii) microeconomic risk.

i.   Market conditions

Information concerning the market environment in the close neighborhood (1 mile radius) of each betting shop was obtained from commercial databases. The number of adults betting was obtained from dedicated market research information while the potential population to bet was obtained as the proportion of the local population with socio-demographic profiles that match betting shop customers. Competition was represented as the inverse number of competitors in the trade area of each branch. All these three input variables were considered to be uncontrollable factors since their size is not under the control of the management of each betting shop.

ii.   Betting shop attractiveness

We have used a direct measure of size, the age of the shop and a qualitative assessment of the shop profile given by the company's internal auditors regarding the state of repair of the facilities of each branch. These three input factors were considered as controllable from the management of the betting shop chain in the long term. That is, the size of the betting shop can be increased by means of relocation, the state of repair can be altered by means of investment, while the age of the branch which is a factor reflecting local establishment and acquaintance with the local shop can have increasing positive effect on output generated with diminishing returns after a certain point in time. Thus, such variables have an association with the generation of output and statistical investigation was deemed appropriate prior to using them in DEA assessments.

iii.   Microeconomic risk

The rate of winnings at individual shops is taken into account as an extra input factor (revenue/turnover). The objective of the assessment was to maximise Revenue and No. of bets. The revenue, however, may vary due to the varying levels of winning or losing across different shops over time (In the long run these variations reach equilibrium but in the short run variations can be considerable). For the purpose of the benchmarking exercise the risk variable was reversed in the actual assessment so that the higher the risk of a shop the lower should be its revenue.

3.5 The marketing-efficiency results

The assessment of the cost and market efficiency of the five chains of retail stores opens the field for attempting to draw some overall conclusions regarding the state-of-the art on identifying regularities in evaluating the performance of retail operations. Despite the limited sample of retail operations we have calculated some statistics about the extent to which there seem to be statistical similarities between the efficiencies of the retail stores of different operations. We also seek to investigate a hypothesis which states that the operation of retail networks entails some slack that accounts for excessive costs or under-utilised market potentials. Results on the marketing efficiency of each service chain are exhibited in Table 16-1.

Table 16-1. Comparative DEA efficiency results of service operations: Strategic & Line marketing perspectives

	Bank branches		Sales forces		Restaurants		Betting shops	
	SMP	LMP	SMP	LMP	SMP	LMP	SMP	LMP
Mean (%)	87	90	83	88	74	83	82	86
Median (%)	89	97	82	90	74	90	83	92
St.Dev	12	11	9	11	19	18	21	17
Minimum (%)	62	64	55	68	32	38	45	47
# efficient	37	53	32	38	39	71	36	61
# branches	120		80		160		152	

SMP: Strategic marketing perspective, **LMP**: Line-marketing perspective

The efficiency profile of the four multi-branch service operations exhibits an information pattern regarding the state of competitiveness of each service operation in terms of strategic (SMP) and line marketing (LMP). The results show some consistency regarding the existence of a notable difference between the best practice branches (within each operation) and the remaining branches. Thus for the strategic marketing (SMP) side we have an average efficiency gap that varies from 74% to 87% while in the line-marketing assessment (LMP) the average efficiency gap varies from 83% to

90%. These results suggest that notable opportunities exist for improving the sales performance of each chain of branches. Furthermore, the results show the presence of a consistent pattern within each of the four service operations in which strong and weak operating practices have been exhibited.

The minimum efficiency levels for all branch networks are indeed alarming for the management of these organisations since the levels of below 60% marketing efficiency cast serious doubts about the viability of individual outlets. The retailing market in the UK, however, has moved many retail giants in the direction of market share competition. Furthermore, the licensing war that exists between competing firms within similar industries makes it imperative for some chains to sustain their presence in local markets with low viability.

Since the empirical results in Table 16-1 were drawn from a single period of time they may not give a fair representation of the dynamic aspect of performance in the service operations. This kind of dynamic analysis was feasible only for the chain of restaurants. Thus we have estimated the marketing efficiency of the restaurant network for a period of four years as given in Table 16-2.

Table 16-2. A dynamic view on marketing efficiency: the case of restaurant chains

	Year 1		Year 2		Year 3		Year 4	
	SMP	LMP	SMP	LMP	SMP	LMP	SMP	LMP
Mean (%)	75.1	84.8	71.2	80.8	74.4	81.6	74.9	83.8
Median (%)	74.2	100.0	70.4	87.6	73.3	88.6	74.4	90.8
St. Dev	19.2	19.0	20.2	20.7	19.1	19.6	19.4	18.2
Minimum (%)	31.3	31.6	31.0	32.4	32.7	33.7	32.6	35.0
# efficient	22	61	25	62	30	62	39	71
# branches	108		127		152		160	

SMP: Strategic marketing perspective; LMP: Line marketing perspective

The panel of data in Table 16-2 refers to the operation of one network of restaurants over a period of four years. Although the number of these restaurants has grown over the four year period the strategic and line marketing efficiency nevertheless give a consistent pattern of indicators. That is, there is a consistent pattern of under-performance which accounts for about 25% of opportunity losses in sales. This pattern of efficiency has even more significant variations in the median efficiency pattern. In any event the evidence from this panel of marketing efficiency results is that the increased number of branches (from 108 in year 1 to 160 in year 4) did not alter the performance profile of the restaurant network. Upon closer inspection it was found that new restaurants did not reach efficiencies at the level of 100% (and thus were not used as benchmarks). It seems that the newer branches did not enjoy first-mover advantages since they were opened in areas with established competition. However, they managed to yield sufficient revenue to avoid undermining the average marketing efficiency of the existing network. Finally, a more general observation that emanates from this dynamic assessment of efficiency is the limited ability of the marketing function to promote decisions that alter the marketing efficiency of this particular organization.

The assessment of the marketing efficiency of the four service networks also gives insights about the extent to which economies/diseconomies of scale prevailed in the market. Results regarding the returns to scale exhibited by the four branch networks are given in Table 16-3.

Table 16-3. Returns to scale of service chains

	Increasing	Constant	Decreasing	# of outlets
Bank branches	23	37	60	120
Sales forces	10	32	38	80
Restaurants	81	39	40	160
Betting Shops	66	36	51	152
Total	180	164	189	512

The overall assessment from the results in Table 16-3 concerns the presence of increasing returns to scale in networks of restaurants and betting shops. Increasing returns to scale from the marketing efficiency perspective indicates the presence of lost sales opportunities as a result of the inadequate scale size of operation of the corresponding branches. By virtue of the results listed in Table 16-3 the networks of restaurants and betting shops can ameliorate their performance by increasing the scale size of operation of the corresponding branches.

The case of the bank branch returns to scale is different because the majority of branches has exhibited decreasing returns to scale, which indicates that a further increase in the scale size of operation will lead in a disproportionate way to diminishing output increases from scaling up the individual branches. The case of bank branches and their size is a notable case in the UK as most commercial banks are currently undertaking large scale rationalisation projects seeking to reduce the size of their branch networks and also to reduce the physical size of their individual branches in an attempt to master their operating costs.

4.   MARKETING EFFICIENCY VS. PROFITABILITY

An important issue addressed in our studies involved the relationship between efficiency and the profitability of individual shops. For efficiency scores we used those obtained from DEA assessments. Profitability was obtained as the ratio of net operating profit divided by the shop revenue. The contrast between efficiency and profitability sought to facilitate the assessment of the viability of individual shops. Profitability versus efficiency defines long run viability while local efficiency concentrates on the short run viability of individual shops. The pictorial representation (see Figure 16-6) of the efficiency-profitability matrix shows the status of individual shops in two vital performance dimensions. As can be seen there is no direct linkage between shop efficiency and profitability. This demands some detailed discussion.

As the size of outlets in service operations can be excessive a taxonomy for classifying the network of outlets into clusters with similar viability prospects was deemed necessary. In the context of the present study, market efficiency and profitability are used as criteria for assessing the viability of each unit by using a two dimensional portfolio matrix. This is demonstrated in Figure 16-6 where operating outlets are classified into four clusters based on their market efficiency and profitability. Each cluster represents outlets with different operating profiles and thus different viability prospects.

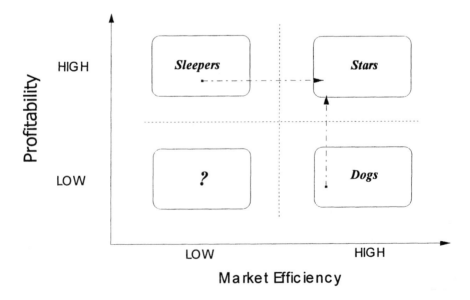

Figure 16-6. Market efficiency Vs. Profitability

Market efficiency is used in Figure 16-6 to reflect the extent to which individual outlets realise their market potential by generating turnover. Profitability as a percent of revenue is a relative measure (as opposed to gross profit which is an absolute measure) which focuses on an outlet's relative ability to convert generated turnover into profit (cost efficiency).

The top right quadrant in Figure 16-6 corresponds to "STAR" shops since they are efficient in terms of sales and also profitability. Shops in this quadrant will typically be used as the benchmarks of the organisation. The bottom right quadrant contains shops with low profitability and high efficiency. These shops are in a bad condition since they seem to generate sales adequate to their market conditions without however converting these sales into profits. One needs to look at issues of cost control but also to investigate whether the size of the market is not sufficient for covering the set-up costs of the particular shops. The bottom left quadrant contains shops with all around poor performance. Shops in this quadrant should focus on increasing their customer base in order to allow them to improve their sales so they will have a better chance for higher profits.

The top left quadrant contains shops with high profitability and low sales efficiency. These shops seem to enjoy high profitability due either to their low physical size or due to very favourable market conditions. The low market utilisation, however, indicates that these shops have considerable capacity improving their sales and thereby increase net profit. The high profitability of these shops may also be a bad indicator in case their low costs are related to low quality of services. In such a case the shops should incur additional costs in order to attract more customers and increase their net profit with an even lower profitability.

A methodological problem arises from the selection of the appropriate model to represent market efficiency. Profitability reflects an ability to control costs and, therefore, it is a measure with a primary appeal to local management (given the scale size of the unit). Therefore, assessment of profitability against line-marketing efficiency focuses on outlet viability from a local management's point of view. The likelihood of improving performance relies on ability to improve local management and control costs. We shall refer to this as the short run viability of individual units.

On the other hand profitability against a measure of strategic marketing efficiency brings scale size factors into the assessment of viability. This kind of assessment allows for the importance of central management decisions on an outlet's long run viability. This type of assessment also concentrates on long run viability of individual outlets since potential performance improvements would require capital investment decisions which have a long time horizon.

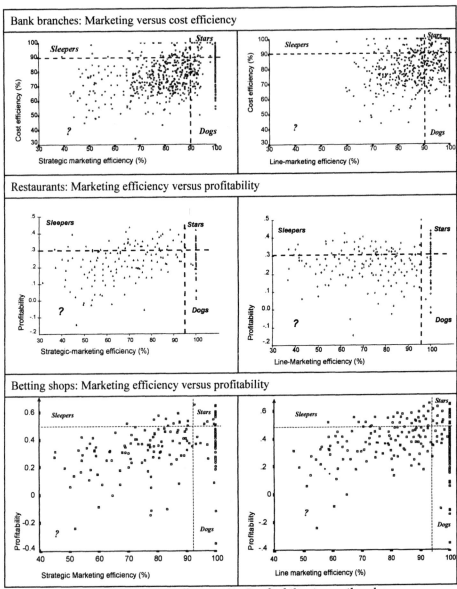

Figure 16-7. Market efficiency Vs. Profitability in retail outlets

The series of scatter diagrams listed in Figure 16-7 provide a panorama of the efficiency - profitability relationship in these two of the four branch networks because the sales forces network we did not have information about profitability or operating costs. Also for the bank branch networks we used a DEA based measure of cost efficiency as a profitability surrogate.

The representation of the profitability of bank branches is a complex problem because it is related to the internal transfer pricing mechanisms of particular banks. In our empirical application it was decided to derive a measure of the cost efficiency of each bank branch as a surrogate of its profitability. The assessment of cost efficiency was done using the same methodology as for the assessment of marketing efficiency. A different input-output set was put in place, however, drawing on the Berger et al. (1997) and Athanassopoulos (1998a) models. That is, we used operating costs and number of computer facilities as inputs while for outputs we used the number of credit and debit transactions, the number of loan transactions and the number of non-traditional banking transactions. Cost efficiency was assessed under the assumption of variable returns to scale with an average of 84% which indicates the scope for reducing the operating costs in the branch network.

The efficiency-profitability matrix was derived using as efficiency indicators the strategic and line marketing efficiency. The matrices derived from the strategic marketing framework yielded to information related to a "causal-mapping" between profitability of the branch and the effectiveness of long term decisions as is depicted in the strategic marketing framework. The extension of this framework to the use of the line-marketing efficiency indicators was motivated by the need to get better insights regarding the controllable side of efficiency and its relationship to the profitability of the branches.

Simple comparisons of the strategic and line marketing maps in Figure 16-7 for each type of branch shows that a shift in the cluster-membership is due to the change in marketing efficiency and is attributable to the distinction between the strategic and line marketing efficiency of each branch. This difference in marketing efficiency corresponds to the different layer of controls that corresponds to each level of assessment. A more specific discussion for the three branch networks is as follows:

- **Bank branch efficiency-profitability matrix**

A weak but significant association seems to hold between marketing efficiency and the profitability of individual branches with a Mann-Whitney rank correlation coefficient of 0.38 in strategic marketing and 0.31 in line marketing efficiency. A vast proportion of the bank branches lie within the quadrant termed "?" in Figure 16-6 since they have exhibited performance below the median efficiency in terms of marketing and costs. The picture is significantly improved when we consider the case of the line-marketing efficiency where we can observe a reduction of the branches in the "?" quadrant shown in the second column of Figure 16-7. Overall, there seems to be sufficient scope for improving both the cost and marketing efficiency of this branch network.

- **Restaurant efficiency-profitability matrix**

The strategic marketing efficiency is associated with the profitability indices in a lesser extend compared to the corresponding results obtained from the line-marketing efficiency. Undoubtedly, low profitability is associated with low marketing efficiency and the use of the strategic and line marketing efficiency maps can aid in the investigation of individual branches and the assessment of the controllability of causes that lead to underperformance. For example, a restaurant with low profitability, low strategic marketing efficiency and high marketing efficiency indicates an area whereby divestment should occur since the market of this restaurant cannot support its long term viability despite its management's effort.

- **Betting shops efficiency-profitability matrix**

The marketing efficiency of betting shops seems to determine a large part of their profitability. To a large extent betting shops incur fixed and labour costs that are standardized since most of them are in firm-owned properties and labour costs follow a firm-wide pattern. Thus we witness a pattern similar to that of the restaurants with each branch having very low marketing efficiency and profitability. Clearly, there are lessons for the multi-branch operators regarding the minimum sales revenue that is necessary for running individual branches at a profit. Of course, in the context of the branching policy there maybe more strategic issues entering into the formula. For instance some badly performing branches could be retained due to company wide reasons for supporting the performance of nearby branches that belong to the same firm.

Although the evidence provided in Figure 16-7 is only an illustration for a particular set of data, the general message is that the assessment of outlet viability needs to take into account alternative definitions of efficiency. The product portfolio matrix alluded to above gives preliminary signals concerning the relative positioning of individual outlets, and therefore the general status of each network. A set of alternative scenarios needs to be developed for groups of outlets on the product portfolio matrix. These scenarios should seek to address performance issues of the network taking into account corporate and strategic issues.

5. SUPPORTING THE OPENING OF NEW OUTLETS

The management of retail networks is required to adjust its policies in line with the strategy of the organisation to open or close branches at various places. Any decision to open new branches is followed by uncertainty regarding expected returns from the operation of these outlets. Regression-based models are customarily used in order to estimate the expected returns based on extrapolations from the performance of the existing network (see Clawson (1974), Olsen and Lord (1979) and Mahajan et al. (1985)). An optimisation based framework is thereby proposed for assessing marketing efficiency returns from future retail outlets. For this instance let us consider $j=1,...,n$ retail outlets that use quantities from the input set $I=1,...,m$ (x_{ij} is the input i of outlet j) in order to produce quantities from the output set $O=1,...,s$ (y_{rj} is the output r of outlet j). Let us also consider an hypothesized outlet j_o with known inputs but unknown outputs since it has not started operating. The expected revenue $y_{rj_o}^*$ of the hypothetical outlet j_o with input profile ($x_{ij_o}^c$, $x_{ij_o}^f$) can be obtained by solving (16.1).

$$\underset{q,k_j}{M\ \alpha\chi}\quad q$$

$$S.t. \sum\nolimits_{j=1}^{n} x_{ij}^c \lambda_j + w_i^c = x_{ij_o}^c \qquad \forall i \in I_c$$

$$\sum\nolimits_{j=1}^{n} \left(x_{ij}^f - x_{ij_o}^f \right) \lambda_j = 0 \quad \forall i \in I_f$$

$$\sum\nolimits_{j=1}^{n} y_{rj}^c \lambda_j - w_r^c = q y_{rj_o}^c \quad \forall r \in O_c \qquad (16.1)$$

$$\sum\nolimits_{j=1}^{n} \left(y_{rj}^f - y_{rj_o}^f \right) \lambda_j = 0 \quad \forall r \in O_f$$

$$\lambda_j, w_i^c, w_r^c \geq 0 \ \text{ and } \ q \ \text{ free}$$

The proposed model (16.1) treats some outputs $r \in O_c$ of each assessed outlet as an unknown variable to be estimated by the solution of the optimization model. The mechanism of the revenue prediction model in (16.1) is next illustrated for the restaurant chain that has been discussed earlier. Selecting a candidate retail with known input levels one can pursue the assessment of the expected revenue as listed in Table 16-4. That is our illustration is limited to the case of multiple inputs and a single output.

A demonstration of the information derived from the solution in model (1), is shown in Table 16-4. Given the input profile of the particular outlet an estimate of the efficient level of turnover of the outlet concerned can be obtained. The information provided in Table 16-4 can be read as follows. The column headed "basic profile" includes the input variables as measured for the candidate site that its output is due to be assessed. Solving the optimization problem in (1) we obtain an assessment of the expected turnover for the outlet and we also obtain optimal solutions for the input side. Since the problem in (1) is an output maximization one, the optimal solution on the input side can take values equal or less than the observed ones. For instance, the outlet can generate the estimated turnover while the propensity to visit restaurants can be reduced from 317 to 281 without affecting the optimal solution and the level of generated turnover by the restaurant. In our example, such inputs are the bar area and the propensity to visit restaurants. For those inputs that are fully utilized within the current optimal framework (car park, state of repair, competition) the model gives upper and lower tolerance levels upon which, the current optimal solution (not objective function) remains stable.

Table 16-4. Evaluating the input-output profile of future outlets

| | Basic profile | Estimated profile | Level of variation[§] | |
			Minimum	Maximum
Bar area (ft²)	835	737	737	none
Car park facilities	10	10	7	16
State of repair	3	3	2	6
No. of competitors	8	8	6	20
Propensity to visit restaurants	317	281	281	none
Consumption of alcohol	27813	27813	15467	31459
Turnover (£)	---	148128	---	---

[§] Allowable level of variation to one input at a time before a new solution in (16.1) is obtained.

Retail managers gain competitive insights by knowing the extent to which future market conditions may affect the performance of individual outlets. Note here that, more advanced optimization aids can be employed, in order to explore simultaneous variation of market conditions of individual outlets.

6. CONCLUSIONS

In this chapter we sought to provide a better understanding regarding the progress that has been made in the management science literature regarding the monitoring and assessment of the performance of large retail networks. On the methodology side the assessment has given emphasis to a relatively new concept of market efficiency which, in conjunction with the internal performance of individual branches, characterizes their overall performance. The means for assessing performance of these networks is primarily Data Envelopment Analysis because the method can accommodate without any difficulties operating processes with multiple inputs and outputs.

We have chosen five case studies from different retail service organizations in order to explore the applicability of the proposed framework. The paper has also explored similarities in the results obtained for the five cases seeking to investigate the hypothesis of connection between the performance of branch networks irrespective of the industry in which they operate. The marketing performance of chains of service retail outlets is not a widely researched subject partly because there are problems of information accessibility. On the other hand, the increasing acceptance of marketing in the line management of retail outlets poses strong demands for accountability of the marketing function.

The empirical results of the present study draw from four different cases of multi-branch operations and it can be argued that some notable evidence was found regarding the consistent underperformance of the four chains in terms of marketing efficiency at both the strategic and also line-marketing levels. The use of tools can monitor marketing and cost efficiency of service chains enhances both the accountability of the marketing function and the decision making ability of individual firms to improve the performance of individual branches. The limitations exhibited by these types of assessments concerns the quality of data that is supplied in the performance monitoring services and should be subject to constant control and periodic upgrades. Future research should link the assessment of the marketing efficiency of branch management with the decision making elements of the marketing mix that entail elements of pricing and promotion, as well as local advertisement. In other words, future research should seek to give a more accurate representation of the marketing mix of individual branches such that the marketing efficiency will provide better managerial insights.

REFERENCES

1. Achabal, D., J. Heineke, S. McIntyre, (1985), 'Productivity measurement and the output of retailing', *Journal of Retailing,* 61, 83-88.
2. Athanassopoulos A.D., (1998a), Multivariate and frontier analysis for assessing the market and cost efficiency of large scale bank branch networks. *Journal of Money and Credit Banking*, Volume 30, Number 2, May 1998, pp. 30-51
3. Athanassopoulos, A., (1995), "Developing Performance Improvement Decision Aid Systems in retailing organisations using Data Envelopment Analysis", *Journal of Productivity Analysis*, 6 (2), 153-170.
4. Athanassopoulos, A., and E. Thanassoulis, (1995a), "Assessing the marginal impacts of investments in the performance of organisational units", *International Journal of Production Economics*, 39, 149-164.

5. Athanassopoulos, A., and E. Thanassoulis, (1995b), "Separating market efficiency from profitability and its implications for planning", *Journal of Operational Research Society*, 46(1), 20-35.
6. Athanassopoulos, Antreas D., (1998b), "Optimization models for assessing marketing efficiency in multi-branch organizations", *The International Review of Retail, Distribution and Consumer Research* 8(4), pp 415-443.
7. Banker, R. and R. Morey, (1993), 'Integrated system design and operational decisions for service sector outlets', *Journal of Operations Management*, 11, 81-98.
8. Berger, A., and D., Hamprey, (1997), "Efficiency of Financial Institutions: International Survey and Directions for Future Research", *European Journal of Operational Research*, pp. 175-212.
9. Berne, C., J. Mugica and J. Yague, (1995), "Productivity analysis in Retailing", *The Second International Research Workshop on Service Productivity*, Stockholm University, Sweden.
10. Charnes A., W. Cooper, and E. Rhodes (1978), "Measuring the efficiency of decision making units", *European Journal of Operations Research*, 2, 429-444.
11. Charnes, A., W. Cooper, B. Golany, D. Learner, F. Philips and J. Rousseau, (1994), 'A multiperiod analysis of market segments and brand efficiency in the competitive carbonated beverage industry', in A. Charnes, W. Cooper, and A. Lewin and L. Seiford, eds., *New management application of data envelopment analysis*, Kluwer Publishers.
12. Clawson, J. (1974), "Fitting branch locations, performance standards and marketing strategies to local conditions", *Journal of Marketing*, 38, 8-14.
13. Davies, R., (1973), "Evaluation of retail store attributes and sales performance", *European Journal of Marketing*, 7, 89-102.
14. Doutt, J., (1984), 'Comparative productivity performance in fast-food retail distribution', *Journal of Retailing*, 60, 98-106.
15. Doyle, P., I. Fenwick and G. Savage, (1979), 'Management planning and control in multi-branch banking', *Journal of the Operational Research Society*, 30, 105-111.
16. Ghosh, A. and S. McLafferty, (1987), *Location strategies for retail and service firms*, Lexington, MA: Lexington Books.
17. Golany B., and Roll (1993), "Some extensions of techniques to handle non-discretionary factors in data envelopment analysis", *Journal of Productivity Analysis*, 4, 419-432.
18. Gronroos, C., (1990), *Service management and Marketing. Managing the moments of truth in service competition*, Lexington Books.

19. Horsky, D. and P. Nelson, (1996), "Evaluation of salesforce size and productivity through efficient frontier benchmarking", *Marketing Science,* Vol. 15, No. 4, pp. 301-320.
20. Ingene, C., (1982), 'Productivity in retailing', *Journal of Marketing,* 46, 75-90.
21. Ingene, C., (1984a), 'Structural determinants of market potential', *Journal of Retailing,* 60(1), 37-64.
22. Ingene, C., (1984b), 'Productivity and functional shifting in spatial retailing: private and social perspectives', *Journal of Retailing,* 60(3), 15-35.
23. Jones, K. and D. Mock, (1984), "Evaluating retail trade performance", in R. Davies and D. Rogers (eds.), Store Location and Store Assessment Research, New York, John Wiley and Sons.
24. Kamakura, Wagner A. and Brian T. Ratchford, (1988), "Measuring market efficiency and welfare loss", *Journal of Consumer Research* 15(3), Dec.
25. Kotler, P., (2002), *Marketing Management,* Prentice Hall.
26. Lodish, L., E. Curtis, N. Ness and M. Simpson, (1988), "Sales force sizing and deployment using a decision calculus model at syntex laboratories", *Interfaces,* Vol. 18, pp. 5-20.
27. Lusch, R. and S. Moon, (1984), 'An exploratory analysis of the correlates of labour productivity in retailing', *Journal of Retailing,* 60, 37-61.
28. Mahajan, J., (1991), "A data envelopment analysis model for assessing the relative efficiency of the selling function", *European Journal of Operational Research,* 53, 189-205.
29. Mahajan, V., S. Sharma, D. Srinivas, (1985), 'An application of portfolio analysis for identifying attractive retail locations', *Journal of Retailing,* 61, pp. 19-34.
30. Olsen, L. and J. Lord, (1979), "Market area characteristics and branch bank performance", *Journal of Bank Research,* 10, 102-110.
31. Rangaswamy, A., S. Prabhakant and Z. Andris, (1990), "An Integrated Model-Based Approach for Sales Force Structuring", Marketing Science, Vol. 9, No. 4, pp. 279-298.
32. Ryans, A. and C. Weinberg, (1987), "Territory sales response models: stability over time", *Journal of Marketing Research,* Vol. 24, pp. 229-233.

Chapter 17

HEALTH CARE APPLICATIONS
From Hospitals to Physicians, From Productive Efficiency to Quality Frontiers

Jon A. Chilingerian[1] and H. David Sherman [2]
[1] *Heller School for Social Policy and Management, MS-35, Brandeis University Waltham, Massachusetts, 02254-9110 USA   Chilingerian@Brandeis.edu*

[2] *College of Business Administration, Northeastern University, Boston, Massachusetts, 02115 USA   H.Sherman@NEU.edu*

Abstract: This chapter focuses on health care applications of DEA. The paper begins with a brief history of health applications and discusses some of the models and the motivation behind the applications. Using DEA to develop quality frontiers in health services is offered as a new and promising direction. The paper concludes with an eight-step application procedure and list of Do's and Don'ts when applying DEA to health services.

Key words: Health Services Research, Physicians, Hospitals, HMOs, Frontier Analysis, Health Care Management, DEA, Performance, Efficiency, and Quality

1.   INTRODUCTION

Throughout the world, health care delivery systems have been under increasing pressure to improve performance: that is, to control health care costs while guaranteeing high quality services and better access to care. Improvements in health care performance are important because they can boost the well being, as well as the standard of living and the economic growth of any nation. The quest for high performance in health care has been a difficult and intractable problem historically. Efforts to reduce costs and improve service quality and access have been only marginally successful (Georgopoulos 1986; Newhouse 1994; Shortell et al. 2000).

Although no theoretically correct or precise measures of "health" exist, there is a great deal of interest in studying and understanding health care costs, outcomes, and utilization. This interest in understanding health outcomes is associated with our continual desire to improve health care. While we may have no precise measures of health care performance, the desire to improve health care can be seen in the unrelenting increase in the quantity of health products and services available to patients. Yet, even with these increases, health care seems to offer fewer perceived benefits to patients in relation to their perceived sacrifices (see Anderson et al. 2003, Bristol Royal Infirmary Inquiry Final Report 2002, Newhouse 2002).

As one economist has proclaimed, "Despite the lack of a summary measure of its efficiency, many seem convinced that that the industry's performance falls short" (Newhouse 2002: 14). For example, one medical center in England received special funding to become a Supra Regional Service center for pediatric cardiac surgical care (Bristol Royal Infirmary Inquiry Final Report 2002). This program received funding over a fourteen-year period, despite significantly higher mortality and morbidity rates and poor physician performance. Should a clinical manager have detected these poor practices sooner? In addition, what internal control systems are available to measure and evaluate individual physician performance?

In 2000, although the United States spent far more on health care per capita than any other country in the world, the number of physicians per 1000 population, primary care visits per capita, acute beds per capita, hospital admissions per capita, and hospital days per capita were below the median of most other developed countries (Anderson et al. 2003). Are quality and productive efficiency of some national health care systems really lower than that of many other industrialized nations? What accounts for performance differences?

Twenty years ago, research studies that questioned the patterns or cost of care rarely attempted to estimate the amount and sources of inefficiency and poor performance. Today that has changed. Several hundred productive efficiency studies have been conducted in countries such as Austria, Finland, the Netherlands, Norway, Spain, Sweden, United Kingdom, and the United States. These studies have found evidence that technical inefficiency in these systems is significant. For example, in the United States such inefficiencies result in billions of "wasted" United States dollars (USD) each year. While other statistical techniques have also been used, Data Envelopment Analysis (DEA)) has become the researchers method of choice for finding best practices and evaluating productive inefficiency.

While there are several techniques for estimating best practices such as stochastic frontier analysis, fixed effects regression models, and simple ratios, DEA has become a preferred methodology when evaluating health care providers. DEA is a methodology that estimates the degree to which observed performance reaches its potential and/or indicates how well resources have been utilized (see Chapter 1 in this handbook). Benchmarked against actual behavior of decision making units, DEA finds a best practices frontier—i.e., the best attainable results observed in practice—rather than central tendencies. The distance of the DMUs (decision-making units) to the frontier provides a measure of overall performance. Although criticized by some health economists (see Newhouse 1994), the late Harvey Leibenstein praised DEA as a primary method for measuring and partitioning X-inefficiency (Lebenstein and Maital, 1992).

DEA offers many advantages when applied to the problem of evaluating the performance of health care organizations. First, the models are non-parametric and do not require a functional form to be prescribed explicitly i.e., linear, non-linear, log-linear, and so on (Charnes et al, 1994). Second, unlike statistical regressions that average performance across many service providers, DEA estimates best practice by evaluating the performance behavior of each individual provider, comparing each provider with every other provider in the sample. The analysis identifies the amount of the performance deficiency and the source. Third, unlike regression and other statistical methods, DEA can handle multiple variables, so the analysis produces a single, overall measure of best results observed.

Finally, in order to identify those providers who achieved the best results, DEA groups providers into homogenous sub-groups. Providers that lie on the frontier achieved the best possible results and are rated 100% efficient. Providers that do not lie on the surface under-performed, and their performance is measured by their distance from the frontier. The analysis not only provides a measure of their relative performance, but also uncovers sub-groups of providers similar in their behavior or similar in the focus of their attention to performance.

This chapter focuses on health care applications of DEA. We begin with a brief history of health care applications. Next, several health care models and approaches are discussed, followed by new directions for future studies. We conclude with a summary of do's and don'ts when applying DEA to evaluate the performance of health care organizations.

2.   BRIEF BACKGROUND AND HISTORY

For health care organizations, finding a single overall measure of performance has been difficult. Goals of most health care services are multiple, conflicting, intangible, vague and complex. Virtually every study of efficiency could be criticized for failing to look at quality, clinical innovation, or the changing nature of the services (Newhouse 1994). The worldwide demand to provide health care services at lower cost made this an ideal focus for DEA research and health care (like banking) has had innumerable DEA applications.

The first application of DEA in health care began with H. David Sherman's Doctoral dissertation in 1981. W.W. "Bill" Cooper had been a professor at Harvard Business School in 1979, David Sherman was a doctoral candidate and he and Rajiv Banker happened to take Bill Cooper's seminar. Bill Cooper mentioned DEA as a new technique, and both Banker and Sherman used that technique in their dissertations. Working with the Massachusetts Rate Setting Commission to ensure relevance in the complex task of evaluating hospital performance, David Sherman applied DEA to evaluate the performance of medical and surgical departments in 15 hospitals. When Dr. Sherman became a professor at MIT in 1980, his doctoral student and research assistant, Jon Chilingerian helped him to compare DEA results with other statistical models and to look for interesting and novel health care applications.

In 1983, Nunamaker published the first health application using DEA to study nursing services (Nunamaker 1983). In 1984, the second DEA paper (Sherman 1984) evaluated the medical and surgical departments of seven hospitals. By 1997, Hollingsworth et al. (1999) counted 91 DEA studies in health care. The health applications include: health districts, HMOs, mental health programs, hospitals, nursing homes, acute physicians, and primary care physicians.

DEA applications in health have been evolving over the last two decades. As access to bases and information technology have improved, so have the quality of the studies. In the next section, we review some of the DEA literature in health care starting with acute hospitals, nursing homes, other health organizations and physicians.

2.1   Acute General Hospitals and Academic Medical Centers

Acute hospitals have received the most research attention using DEA. These hospital studies use DEA alone. They measure overall technical efficiency using the CCR model. Finally, they define outputs as patient days or discharges and do not measure clinical outcomes. A comprehensive review of efficiency studies in health care by Hollingsworth et al. (1999) found systematic differences among the average efficiency and range of DEA scores by ownership type and national hospital systems. For example, public hospitals had the highest mean efficiency scores (0.96) and not-for-profit hospitals had lower mean DEA scores (.80). When comparing U.S.

studies with studies from other European countries Hollingsworth *et al.* found a greater potential for improvement in the U. S. with an average efficiency score of 0.85, and a range of 0.60 – 0.98, in contrast to Europe with an average efficiency score of 0.91, and a range of 0.88-0.93.

Although comparing DEA scores among various hospital studies is useful for hypothesis generation, there are many limitations. First, these studies used different input and output measures during different time periods. Second, the distribution of DEA scores is so skewed, (given the huge spike of efficient units), that reliance on the usual measures of central tendency will be misleading. By excluding the efficient units, the average inefficiency score may be a more reasonable comparison. Third, the output measures in these studies are vastly different. Many did not use the same type of case mix adjusted data, and some studies used crude case mix adjusted data such as age-adjusted discharges, so the results are likely affected by unaccounted case mix differences.

One recent study of 22 hospitals in the National Health Service in the United Kingdom used a four output, five input model (Kerr et al. 1999). The outputs were defined as: (1) Surgical inpatients and visits, (2) Medical inpatients and visits, Obstetrics/Gaenaecology patients and visits, Accidents and Emergency visits. Without knowing the complexity and severity of patients, raw measures of output will lead to distorted results. If Hospital A receives a lower DEA score because Hospital A admits more "fevers of unknown origin," and performs more combined liver-kidney transplants, hip replacements, and coronary by-pass grafts and Hospital B has more tooth extractions, vaginal deliveries without complications, and circumcisions, it is an unfair comparison.

Acute hospitals are among the most complex organizations to manage. Periodically, there are random fluctuations and chaos is everywhere—the emergency room, the operating rooms, the intensive care units, and the like. Hospital studies are complex because of the amount of input and output information needed to describe the clinical activities and services and the patients' trajectories. Most of care programs are not really under the control of the hospital manager (see Chilingerian and Glavin 1994; Chilingerian and Sherman 1990). Therefore, traditional DEA hospital studies may not have been useful to practicing managers.

Most of the hospital studies have merely illustrated DEA as a methodology and demonstrated its potential. Unfortunately, they use very different hospital production models. Some combine patient days with discharges as outputs, and others separate the manager-controlled production process from the clinical-controlled process. Few studies have tested any clinical or organizational theory.

Evaluating acute hospitals requires a large and complex DEA model. To identify under-performing hospitals DEA makes non-testable assumptions such as no random fluctuations, no measurement errors, no omitted outputs and output homogeneity (see Newhouse 1994). If additional inputs would improve quality, omitting an output variable such as the quantity of case-mix adjusted mortalities can distort DEA results. Perhaps bringing DEA inside the hospital to compare departments, care programs, care teams, diseases, specific procedures and physicians' practice patterns would be more useful for practice. One might conclude from this that hospital-level comparisons are not the best application for DEA.

Nevertheless, there have been innovative hospital-level studies potentially useful for policy makers. For example, dozens of DEA papers have focused on the association between hospital ownership and technical inefficiency studying several thousand hospitals as DMUs (see for example, Burgess and Wilson 1996). DEA studies have also focused on critical health policy issues such as: regional variations (Perez 1992), rural hospital closures (Ozcan and Lynch 1992), urban hospital closures (Lynch and Ozcan 1994), hospital consolidations (Luke et al., 1995), and rural hospital performance (Ferrier and Valdmanis 1996). O'Neill (1998) recently made an important methodological contribution by developing an interesting DEA performance measure that allows policy makers to make fair comparisons of teaching and non-teaching hospitals. The DEA work on hospital performance continues to be important and needs more development.

2.2   Nursing Homes

Sexton et al. and Nyman and Bricker published the first two DEA studies of nursing homes in 1989. Sexton et al. ran a model that relied on two output measures (Medicaid and Other) and six inputs that only included labor; consequently, the study had some limitations. Nyman and Bricker (1989) used DEA to study 195 for profit (FP) and not-for-profit (NFP) U.S. nursing homes. Employing four categories of labor hours as inputs (e.g. nursing hours, social service hours, therapist hours, and other hours), and five outputs (skilled nursing patients (SNF), intermediate care patients (ICF), personal care patients, residential care patients, and limited care patients), they regressed the DEA scores in an ordinary least squares regression analysis and reported that NFP nursing homes were more efficient.

Another study in the United States by Nyman *et al.* (1990) investigated the technical efficiency of 296 nursing homes producing only intermediate care, no skilled nursing care, and relying on 11 labor inputs. They defined only one output, the quantity of intermediate care patients produced. Fizel and Nunnikhoven (1993) investigated the efficiency of U.S. nursing home chains ignoring non-labor inputs and focusing on two outputs: the quantity of intermediate and skilled nursing patients. They study found that chains were more efficient. Kooreman's 1994 study of 292 nursing homes in the Netherlands utilized six labor inputs and four case-mix adjusted outputs: the quantity of physically disabled patients, quantity of psychogeriatrically disabled, quantity of full care, and quantity of day care patients. He found that fifty percent were operating efficiently, and the inefficient homes used 13% more labor inputs per unit of output and quality and efficiency seemed to be going in the opposite direction.

A study of 461 nursing homes by Rosko et al., in 1995 employed five labor inputs, and two outputs, ICF patients and SNF patients. The study found that the variables associated with nursing home efficiency were managerial and environmental. Differences in efficiency were not associated with quality measures.

The nursing home studies are among the better applications of DEA. Although they encounter the same case mix problems when modeling outputs, most of these studies regress the DEA scores to identify the variables associated with inefficiency. Since the outputs are often adjusted by payment types, or are crude patient types, many of these studies controlled for quality, patient characteristics, while exploring effects of ownership, operating environment and strategic choice on performance. These studies have found that managerial and environmental variables (the location of the home, nurse training, size of the homes, and wage rates) are strongly associated with the DEA scores, rather than quality of care, or patient mix.

2.3 Department Level, Team-level, and General Health Care Studies

In addition to acute care hospitals and nursing homes, DEA has been applied in a wide variety of health services and activities. For example, previous work has studied the productive efficiency of: Obstetrics units (Finkler and Wirtscafter 1993), pharmacies (Fare et al 1994), intensive care units (Puig-Junoy 1998), organ procurement programs (Ozcan et al., 1999), and dialysis centers (Ozgen and Ozcan 2002). All of these studies have used DEA to measure technical efficiency using basic models.

Ozcan et al.'s exploratory study conducted in 1999 on 64 organ procurement organizations employed four inputs (a capital proxy, FTE development labor, FTE Other Labor, and operating expenses), and 2 outputs (kidneys recovered, extra-renal organs recovered). The study found some evidence of scale efficiency not only did the larger programs produce

2.5 times more outputs, the average DEA efficiency scores were 95% for the large programs versus 79% for the smaller programs.

In 2000, a paper comparing the efficiency of 585 HMOs from 1985-1994 computed the estimates using data envelopment analysis, stochastic frontier analysis, and fixed effects regression modeling (Bryce et al 2000). Unfortunately, they relied on a single output (total member-years of coverage) four input (hospital days, ambulatory visits, administrative expenses, and other expenses) model. They concluded that the three techniques identify different firms as more efficient.

In 2002, Ozgen and Ozcan reported a study of 791 dialysis centers in the United States. Constructing a multiple output (quantity of dialysis treatments, dialysis training, and home visits) and multiple input model (including clinical providers by type, other staff, operating expenses, and the number of dialysis machines--a proxy for capital), the study evaluated pure technical efficiency assuming variable returns to scale. The study found that the wide variations in efficiency were associated with ownership status—for-profit dialysis centers were less inefficient than not for profit.

2.4   Physician-Level Studies

A new and interesting development in health care has been taking DEA down to the workshop-level, focusing on individual physician performance. The first application of DEA at the individual physician level was in 1989, the analysis identified the non-technical factors associated with technical efficiency of surgeons and internists (Chilingerian 1989).

Physician studies have developed new conceptual models of clinical efficiency (Chilingerian and Sherman 1990), as well as using DEA to explore new areas such as: most productive scale size of physician panels (Chilingerian 1995); benchmarking primary care gatekeepers (Chilingerian and Sherman 1996); and preferred practice styles (Chilingerian and Sherman 1997, Ozcan 1998). Getting physicians to see how their practice behavior ranks in relation to their peers is a step toward changing the culture of medicine and offering insights for a theory of clinical production management (Chilingerian and Glavin 1994).

In 1989, Chilingerian used DEA to investigate the non-technical factors associated with clinical efficiency of 12 acute hospital surgeons and 24 acute hospital internists. Since clinical efficiency assumes a constant quality outcome, the study classified each physician's patients into two outcomes: (1) satisfactory cases—i.e., patients discharged alive without morbidity; and (2) unsatisfactory cases—i.e., patients who experienced either morbidity, or who died in the hospital. Relying on a two output (low and high severity discharges with satisfactory outcomes), two input (ancillary service, and total length of stay) model each of the 36 physicians were evaluated (Chilingerian 1989).

The study reported that 13 physicians were on the best practice frontier (i.e., 13 physicians produced good outcomes with fewer resources). After controlling for case mix complexity and the severity mix of the patients, the DEA scores could not be explained by the type of patients treated. Moreover, younger physicians (age 40 and under) who belonged to a group practice HMO were more likely to practice efficiently.

In 1995, Chilingerian reported a six-month follow-up study on the 36 acute care physicians (Chilingerian 1994). A tobit analysis revealed that physicians affiliated with a group practice HMO were likely to be more efficient as well as physicians whose medical practices focused on a narrow range of diagnoses. Using DEA to investigate the most productive scale size of hospital-based physicians, the study reported that on average one high-severity patient utilized five times the resources of a low-severity patient.

In 1997, Chilingerian and Sherman evaluated the practice patterns of 326 primary care physicians in an HMO. They employed a seven output (gender and age-adjusted patients) 8 input (acute hospital days, ambulatory surgery, office visits, sub-specialty referrals, mental health visits, therapy units, tests, and emergency room visits) model. They demonstrated that the HMO's belief that generalists are more efficient than sub-specialists was not supported. Using clinical directives to establish a cone ratio model, practice styles were identified that should have increased their office visits and reduced hospitals days. Thus, a DEA-based definition of efficiency identified relatively "efficient" physicians who could change their styles of practice by seeing more of their patients in their offices rather than in the hospital.

In 1998, Ozcan studied 160 primary care physicians' practice styles using DEA with weight restrictions to define preferred practice styles for a single medical condition. The study identified three outputs (low, medium and high severity patients with otitis media, a disease that affects hearing) and five inputs (primary care visits, specialist visits, inpatient days, drugs, and lab tests). The study found 46 efficient and 114 less efficient practices. After defining preferred practice styles, even the efficient physicians could have a reduction in patient care costs of up to 24%.

2.5 Data Envelopment Analysis versus Stochastic Frontier Analysis

Several health care papers have compared DEA results with other techniques such as stochastic frontier analysis (SFA). These techniques use two vastly different optimization principles. SFA has one overall optimization across all the observations in order to arrive at the estimates of inefficiency. DEA runs a separate optimization for each hospital, thus allowing a better fit to each observation and a better basis for identifying sources of inefficiency for each hospital.

The differences in optimizing principles used in DEA and regression estimates suggest that one might be preferred over the other to help solve different problems. For example, SFR may be more helpful to understand the future behavior of the entire population of hospitals. DEA might be used when the policy problem centers on individual hospitals how specific inefficiency can be eliminated. One problem with comparing DEA to SFA is that SFA requires a specification such as a translog function with a single dependent variable. Forcing DEA to utilize the same dependent variable as a single output will lead to the conclusion that the results are similar but not exactly the same (see Bryce et al 2000).

2.6   Reviewer Comments on the Usefulness of DEA

As the life cycle of DEA matures, we have seen an increase in the health applications published. However the skeptics remain. The most skeptical view on frontier estimation has come from Joseph P. Newhouse at the Harvard School of Public Health. In 1994, in the Journal of Health Economics, Newhouse, raises several fundamental questions about frontier studies (Newhouse 1994). For example, he asks if one could define a frontier with certainty what purpose would it serve? Can frontier studies be used to set reimbursement rates? Can frontier analysis create benchmarks?

Unlike kilowatt-hours, Newhouse also suggests that because the product in health care is neither homogeneous, nor uni-dimensional the output problem is serious. He also identifies three other problems with modeling medical care delivery as a production process. First, frontier studies often omit critical inputs such as physicians, contract nurses, capital inputs, students and researchers. Employing case mix measures such as Diagnostic Related Groups, or Ambulatory Cost Groups, can also hide severity within a diagnosis or illness. He points out that DEA assumes no random error or random fluctuations, which is often not the case in health care.

The problem of measurement errors, or unclear, ambiguous, and omitted inputs and outputs will haunt every DEA health care application. The threshold question is--how seriously do these issues affect the results? While Newhouse claims a serious distortion, the question is largely empirical. Researchers must put that issue to rest each time they conduct a DEA study. Do the results make sense?

To advance the field, every DEA health care study should test the following hypothesis:

Ho: Variations in DEA scores (i.e., excess utilization of clinical resources) can be explained by complexity, severity, and type of illness.

Though researchers cannot prove that DEA scores are measuring efficiency, they can disconfirm the case mix hypothesis. If DEA scores are not associated with severity, patient characteristics or poor quality the DEA results become very interesting. One analytic strategy for disconfirming the casemix/quality hypothesis is to regress the DEA scores against the best available measures of case mix and patient characteristics. We will discuss how to use tobit models to validate DEA scores in another section of this chapter.

2.7   Summary

One conclusion to be reached from all of the health care studies is that even when it appears that a substantial amount or resources could be "saved" if every hospital, nursing, or physician were as good as those who use medical techniques the least, DEA scores must always be interpreted with care. Researchers should assume that there is unaccounted case mix until that issue is put to rest.

In summary, there are two obstacles to advancing applications of DEA to health services research. The first impediment is the confusion around modeling hospitals, physicians, nursing homes, and other health care providers. The findings from DEA studies lose credibility when inputs and outputs are defined differently from study to study. A second impediment that interferes with the development of generalizations from DEA applications is the lack of stability in the results from different studies. A variable like quality of care, associated with productive efficiency in one study, disappears in another study (see Kooreman 1994; Rosko et al 1995). Rarely have two researchers studied the same problem and when they have, rarely have they employed the same categories of inputs and outputs.

One way to deal with these difficulties is to try to ascertain how much of the efficiency scores is explained by case mix. For example, studies that blend DEA with tobit or other statistical models can sharpen the analysis of best practices. If adequate controls are included in the model, the challenge of output heterogeneity in health care can be laid to rest (see Rosko et al 1995, Kooreman 1994, Chilingerian 1995).

This review suggests that DEA is <u>capable</u> of producing new knowledge advancing the science of health care management. As chapter one has argued:

"...DEA proves particularly adept at uncovering relationships that remain hidden from other methodologies. For instance, consider what one wants to mean by "efficiency", or more generally, what one wants to mean by saying that one DMU is more efficient than another DMU. This is accomplished in a straightforward manner by DEA without requiring expectations and variations with various types of models such as are required in linear and nonlinear regression models...." (See Chapter 1 of his handbook).

Before DEA becomes a primary tool to help policy makers and practicing clinical managers, researchers must shift from health care "illustrations of DEA" to advancing the field of performance improvement in the delivery of health services. If DEA is to reach its potential for offering new insights and making a difference to clinicians and patients, DEA research has to overcome the many pitfalls in health care. The remainder of this chapter identifies the issues, and offers some practical suggestions for dealing with them. We conclude with a new approach that investigates quality frontiers in health care.

3.   HEALTH CARE MODELS

Medical care production is different from manufacturing. A traditional factory physically transforms raw materials into finished products. The customer is absent, so there is no participation or co-production. If demand is higher than capacity, inventory can be stored. In manufacturing, quality can be built into design of a product. Since goods can be inspected and fixed at every workstation, quality can improve productive efficiency.

In health care after patients are admitted to a care facility (or visit a clinic) there are three major clinical processes: (1) investigation/diagnosis, (2) treatment/therapy, and recovery. Health care services are intangible "performances," that can only be experienced or used (Teboul 2002). They must be consumed immediately following production and they cannot be stockpiled. Unused capacity—i.e., nursing care, empty beds, idle therapists and the like--are sources of inefficiency. However, overutilization—i.e., unnecessary tests, x-rays, and surgeries, or unnecessary days in the intensive care unit, or nursing home--are also sources of inefficiency. Once delivered, clinical services cannot be taken back and corrected patient perceptions are immediate. In the next section, efficiency in the health care context will be defined and discussed.

3.1   CLINICAL EFFICIENCY DEFINITIONS

Clinical inefficiency in the provision of health care services occurs when a provider uses a relatively excessive quantity of clinical inputs when compared with providers treating a similar case load and mix of patients. There are many categories of efficiency--technical, scale, allocative, and overall efficiency. Most studies in health care have measured the overall technical and scale components of clinical efficiency, which is represented by the "average productivity attainable at the most productive scale size...." (Banker, Charnes, and Cooper 1984, p. 1088). Most simply, technical inefficiency refers to the extent to which a decision-making unit (DMU) fails to produce maximum output from its chosen combination of factor inputs, and scale inefficiency refers to sub-optimal activity levels.

Evaluating a health care provider's clinical efficiency requires an ability to find "best practices"--i.e., the minimum set of inputs to produce a successfully treated patient. Technical inefficiency occurs when a provider uses a relatively excessive quantity of clinical resources (inputs) when compared with providers practicing with a similar size and mix of patients. Scale inefficiency occurs when a provider is operating at a sub-optimal activity level--i.e., the unit is not diagnosing and/or treating the most productive quantity of patients of a given case mix. Hence, hospital providers will be considered 100% efficient if they cared for patients with fewer days of stay and ancillary services and at an efficient scale size. Primary care providers will be considered efficient if they cared for their panels of patients with fewer visits, ancillary tests, therapies, hospital days, drugs, and sub-specialty consults.

Researchers can use a variety of DEA models to measure and explain overall technical and scale efficiency. The CCR model, initially proposed by Charnes, Cooper and Rhodes (1978) is considered a sensitive model for finding inefficiencies (Golany and Roll 1989). In 1984, Banker et al. added another very useful model (BCC model) for health care studies. The BCC model can be used to separate technical from scale efficiency. Both models (if formulated as input-minimizing) can be used to explore some of the underlying reasons for inefficiency--e.g., to estimate divergence from most productive scale size and returns to scale (Banker et al. 1984). Consequently, DEA can yield theoretical insights about the managerial problems or decision choices that underlie efficient relationships: e.g., magnitude of slack, scale effects of certain outputs on the productivity of inputs, marginal rates of substitution and marginal rates of transformation and so on (Cooper et al., 2000).

Once DEA rates a group of providers efficient and inefficient, researchers, managers and/or policy makers can use this information to benchmark best practice by constructing a theoretical production possibility set. Analysts or researchers could use the DEA linear programming formulations to estimate potential input savings (based on a proportional reduction of inputs). Analysts or researchers can use the ratios of the weights u_r and v_i to provide estimates of marginal rates of substitution and marginal rates of transformation of outputs, measured on a segment of the efficient frontier (Zhu 2000). Again, analysts or researchers could use the BCC model to evaluate returns to scale--i.e., in the case of physicians, the effects of a small versus large proportion of high severity cases. Furthermore, the production possibility set is used to estimate alternative outputs of low severity and high severity patients that can be offered clinical services having utilized a mix of clinical services (for a primary care example see Chilingerian 1995).

3.2   How to Model Health Care Providers: Hospitals, Nursing Homes, Physicians

The threshold question for DEA is what type of production model to choose. Depending on the type of health care organization, there are many ways of conceptualizing the inputs and outputs of production. Since the selection of inputs and outputs often drives the DEA results, it is important to develop a justification for selecting inputs and outputs. In the next section, we review DEA models used in various health applications. First, we will differentiate clinical from managerial efficiency.

3.3   Managerial and Clinical Efficiency Models

In health care, technical efficiency is not always synonymous with managerial efficiency. On the one hand, technical efficiency in nursing homes, rehabilitation hospitals, and mental health facilities can be equated with managerial efficiency. On the other hand, medical care services especially in acute and primary care settings are fundamentally different in that there are two medical care production processes, and consequently types of technical and scale efficiency: managerial and clinical. Managerial efficiency requires practice management—i.e., achieving a maximum output from the resources allocated to each service department, given clinical technologies. Clinical efficiency requires patient management—i.e., physician decision making that utilizes a minimal quantity of clinical resources to achieve a constant quality outcome, when caring for patients with similar diagnostic complexity and severity.

In the case of acute hospitals, the role of the manager is to set up and manage a decision-making organization whose basic function is to have clinical services ready for physicians and other clinical providers. For example, hospital managers must have admitting departments, dietary and diagnostic departments, operating rooms, and ICU and regular beds staffed and ready to go. Physicians make decisions to use these intermediate products and services, and managers make it all available. The major challenge facing hospital managers is to decide how much reserve capacity is reasonable given fluctuating patient admissions (daily and seasonal), and stochastic emergency events. Managerial efficiency can be equated with producing nursing care, diagnostic and therapeutic services, and treatment programs of satisfactory quality, using the least resources. Good practice management achieves managerial efficiency.

Physicians are fundamentally different from other caregivers. They are not only providers of medical care, they also enjoy the primary "decision rights" for patient care, with little interference from management. They are the decision-making unit that steers a patient through various phases of patient care—such as, office visits, primary care and diagnostic services, hospital admissions, consultations with other physicians, surgeries, drugs, discharges, and follow-up visits. Physicians organize and direct the entire production process, drawing on the talents of a hidden network of providers—nurses, therapists, dietary specialists, and the like. The reason that physician practice patterns are of interest is that 80% to 90% of the health care expenditures in every system are the result of physician decision making. These are dominant and highly influential DMUs. Clinical inefficiency then, as it is used here, refers to physicians who utilize a relatively excessive quantity of clinical resources, or inputs, when compared with physicians practicing with a similar size and mix of patients.

Figure 17-1 describes the medical service production system as a two-part service process: (1) a manager-controlled DMU, and (2) a physician-controlled DMU. In the diagram below, the intermediate outputs of the manager-controlled production process become the clinical inputs for the physician-controlled production process. A discharged patient is the final product, and the clinical inputs are the bundle of intermediate services that the patient received.

Hospital managers set up and manage the assets of the hospital. They control the labor, the medical supplies, and all expenditures related to nursing care, intensive care, emergency care, and ancillary services (such as lab tests, radiology, and other diagnostic services), pharmacy, dietary, as well as laundry, central supplies, billing, and other back office functions. However, these departments (or functions) merely produce intermediate services that are available for utilization by physicians (see Chilingerian and Glavin, 1994; Chilingerian and Sherman, 1990; Fetter and Freeman, 1986). Physician decision making determines how efficiently these assets are utilized. Once a patient is admitted to hospital, physicians decide on the care program--i.e., the mix of diagnostic services and treatments, as well as the location and intensity of nursing care, and the trajectory of the patient. Physicians decide how and when to utilize nursing care, intensive care, emergency care, ancillary services, and other clinical inputs.

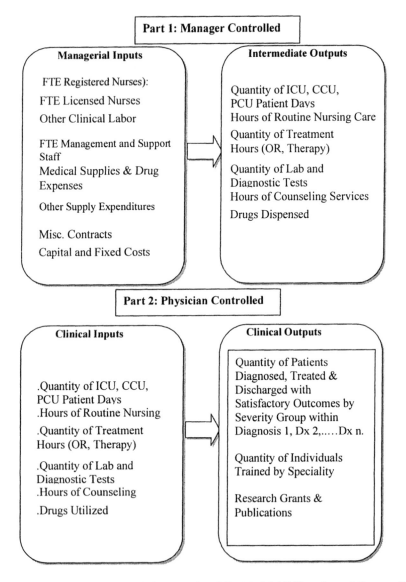

Figure 17-1. Acute Hospital As Two-Part DEA Model (Chilingerian and Sherman 1990)

The productive efficiency of the hospital is complicated. A hospital can be clinically efficient, but not managerially efficient. A hospital can be managerially efficient, but not clinically efficient. More often, both parts of the production process are inefficient. If physicians over utilize hospital services the cost-per-patient day, cost-per-nursing hour, cost-per-test is reduced and the hospital appears to making the best use of its inputs.

To be efficient, clinical and non-clinical managers must perform two tasks very well. Clinical managers must manage physicians' decision making (i.e., patient management) and non-clinical managers make the best use of all hospital assets by managing operations (i.e., practice management). Therefore, patient and practice management require an extraordinary amount of coordination and commitment to performance improvement.

3.3.1 Medical Center and Acute Hospital Models: Examples of Managerial Efficiency

Three examples of acute production models appear below. The first model (Burgess and Wilson 1998) includes five types of labor inputs and weighted beds as a proxy for capital, but excludes drugs, medical supplies and other operating expenses. The second model (Sexton et al 1989) collapses nurses into one category, but adds physicians and residents and excludes beds. The third model collapses labor into one variable, includes other operating expenses and beds, but also adds a proxy measure of capital based on a count of the number of specialty and diagnostic services.

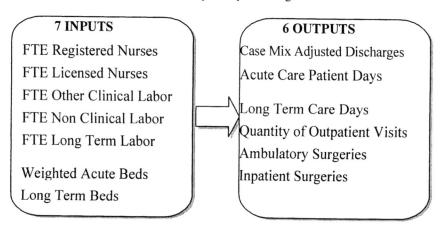

7 INPUTS	6 OUTPUTS
FTE Registered Nurses	Case Mix Adjusted Discharges
FTE Licensed Nurses	Acute Care Patient Days
FTE Other Clinical Labor	
FTE Non Clinical Labor	Long Term Care Days
FTE Long Term Labor	Quantity of Outpatient Visits
	Ambulatory Surgeries
Weighted Acute Beds	Inpatient Surgeries
Long Term Beds	

Figure 17-2. Variables in General Acute Hospital Model (Burgess and Wilson 1998)

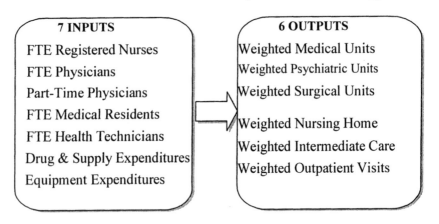

Figure 17-3. Variables in A Medical Center Study (Sexton et al., 1989)

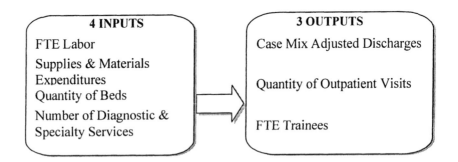

Figure 17-4. Variables in Urban Hospital Model (Ozcan and Luke 1993)

The outputs are different in all three models. Conceptually if the inputs are costs, then the input/output ratios are cost per case, cost per procedure, cost per visit, or cost per nursing day. If the inputs are beds or FTE labor, then the input-output ratio is represented by labor utilized per admission, labor utilized per patient day, labor per surgery, and the like. Although mixing managerial inputs with clinical outputs is acceptable, managerial and clinical inefficiencies become indistinguishable.

3.3.2   Nursing Homes: Another Example of Managerial Efficiency

Nursing home studies in the United States typically segment the outputs by sources of payments: quantity of residents supported by the state, or people without insurance. Figure 17-5 displays the inputs and outputs often used in DEA nursing home studies. The nine resource inputs are full time equivalent (FTE) registered nurses, FTE licensed practical nurses, and FTE nurse aides, FTE other labor, and medical supplies and drugs, clinical and other supplies, and claimed fixed costs (a proxy for capital). Since DEA can handle incommensurable data, the FTEs are in quantities, and the supplies, drugs and fixed costs are measured by the amount of dollars spent. The outputs are the quantity of nursing home days produced during a given time period. In the Figure 17-5 below, the outputs are the quantity of resident days broken into three payment classification groups: Medicare patients (a national program to pay for elderly care), Medicaid patients (a state program to pay for impoverished residents), and Private patients (residents without financial assistance).

Problems arise when the outputs are not homogeneous due to unaccounted case mix. For example, if the nursing home has skilled nursing beds, an Alzheimer unit, or a large proportion of patients: older than 85, confused, requiring feeding, bathing and toilet assistance, then the model is not measuring differences in pure productive efficiency. One solution is to collect information on patients' characteristics and regress the DEA scores against the patient, environmental, and managerial factors to be sure that the DEA scores are not due to case mix. For an example of this type of study, see Rosko et al. 1995.

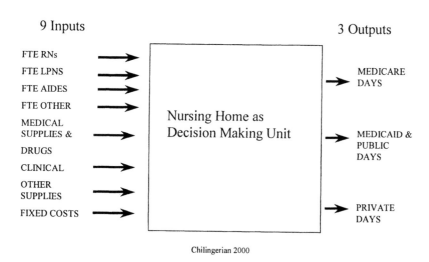

Chilingerian 2000

Figure 17-5. Nursing Home Inputs and Outputs (Chilingerian 2000)

3.3.3   Primary Care Physician Models: An Example of Clinical Efficiency

With growth of managed care, the primary care physician has emerged as an important force in the struggle for efficient and effective medical care. Since lab and radiology tests, prescription drugs, surgeries, and referrals to hospitals all require a physician's approval, physicians report cards or profiles have become a way to benchmark physician practice patterns. Managers could use DEA as a tool to profile and evaluate physicians.

Previous research has found that three patient variables drive managed care costs. They are: patients' age, gender, and geographic location. Consequently managed care organizations set their budgets and prices based on these variables. The final product produced by physicians in managed care organizations is one year of comprehensive care for their patients. To care for patients, primary care physicians utilize office visits, hospital days, lab tests, and therapy units. Figure 17-6 is an example of how one large HMO conceptualized a physician DEA application.

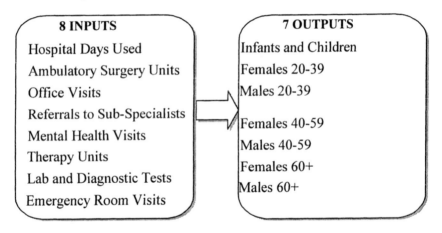

8 INPUTS	7 OUTPUTS
Hospital Days Used	Infants and Children
Ambulatory Surgery Units	Females 20-39
Office Visits	Males 20-39
Referrals to Sub-Specialists	Females 40-59
Mental Health Visits	Males 40-59
Therapy Units	Females 60+
Lab and Diagnostic Tests	Males 60+
Emergency Room Visits	

Figure 17-6. Variables in A Primary Care Physician Study (Chilingerian and Sherman 1997)

3.4   Hospital Physician Models: Another Example of Clinical Efficiency

To study the practice behavior at the individual physician-level, analysts and researchers can use a variety of DEA models. For example, one recent study a study of 120 cardiac surgeons evaluated how efficiently they performed 30,000 coronary artery by-pass grafts (CABG) on patients over a two-year period (see Chilingerian et al., 2002). Figure 17-7 illustrates a two-input, four-output clinical production model used in that study. The two inputs are defined as (1) the total length of stay (days) for the CABG cases handled, and (2) the total ancillary and other charges (dollars) for the CABG cases handled. The ancillary and other charges input category includes ancillary, drug, equipment, and miscellaneous charges. The first input, length of stay, represents a measure of the duration of CABG admissions and the utilization of clinical inputs such as nursing care and support services. The second input, ancillary and other charges, represents a measure of the intensity of CABG admissions and the utilization of operating rooms, laboratory and radiological testing, drugs, and so on.

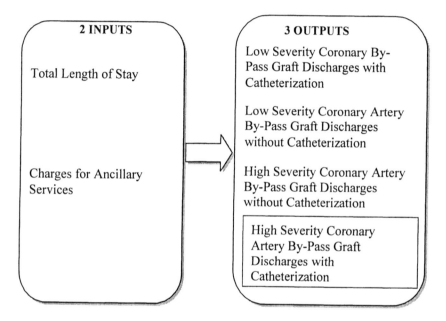

Figure 17-7. Variables in A Cardiac Surgeon Model of DRG 106 and 107 (Chilingerian et al. 2002)

The four classes of clinical outputs represent completed CABG surgery cases. Since patients with more severe clinical conditions will likely require the use of more clinical inputs, the efficiency analysis must account for variations in case mix in order to be fair to surgeons or hospitals treating relatively sicker CABG patient populations. Accordingly, the outputs are defined by diagnostic category and severity level within diagnosis. In this example, a system of case mix classification called Diagnostic Related Groups (DRG) are used to segment outputs by complexity; moreover, a severity system called MEDSGRPS was used to further segment each DRG into low and high severity categories. The researchers treated DRG 106 and

DRG 107 as separate clinical outputs because a CABG procedure with catheterization is more complicated and requires more clinical resources. As explained above, each DRG was further divided into low-severity and high-severity cases.

3.5   Profitability Models: A Nursing Home Example

Although "profit" is still a dirty word in health care, there is a need to do more performance studies that look at revenue and expenses, and investigate the factors affecting profitability. In these studies, the maximum profit includes actual profit, plus maximum overall inefficiency (see Cooper, Park and Pastor 2000). Figure 17-8 illustrates an example of a profitability model. for a nursing home.

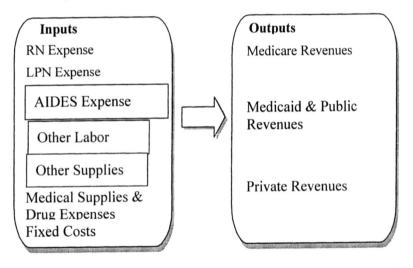

Figure 17-8. Hypothetical Example of Nursing Home Inputs and Outputs in the Profitability Model

Since the performance measure, takes the form of Profit = Revenues – Expenses, an additive mode could be used (see early chapters in this handbook). The additive model shown below has several advantages.

$$\max z = \sum_{r=1}^{s} \mu_r y_{ro} - \sum_{i=1}^{m} v_i x_{io} + u_o$$

subject to

$$\sum_{r=1}^{s} \mu_r y_{rj} - \sum_{i=1}^{m} v_i x_{ij} + u_o \leq 0$$

$$\mu_r, v_i \geq 1$$

One advantage of the additive model is that the objective function, max z, can be interpreted as maximizing profit, or maximizing an excess of revenue over expenses. The model contains an unconstrained variable, u_o, which forces the model to assume variable return to scale (see early chapters in this handbook). Given the advantages and disadvantages of having a great deal of capacity, this assumption works for a profitability model. The optimal value of v_r^* can provide insights with respect to the trade-offs associated with adding skilled nursing, or reducing other labor or fixed expenses.

Employing profitability models to evaluate health care organizations is a promising area for research for two reasons. First, profit models make the evaluation of the complex delivery systems such as acute care hospitals more manageable. Second, this approach opens up the opportunity to use a multi-stage evaluation approach. With a three-stage approach, DEA could be used to investigate the underlying reasons for the profitability. During stage one, an average DEA score of the clinical efficiency of the each hospital's busiest physicians could be obtained by studying the individual physician level. For the second stage, a profitability model would be run, obtaining DEA scores for each hospital in the comparison set. Finally, the third stage would regress the hospital's profitability scores against the explanatory variables, including the average clinical efficiency of physicians.

The variables that influence health care profit margins include: cost drivers, prices (relative to competitors), relative scale and capacity, the market share, quality of care, and technical delivery mode (Neuman, Suver, and Zelman 1988). According to the literature, the main cost drivers are: the volume of patients, within diagnostic categories, adjusted by severity; the practice style and clinical behavior of the medical staff; and the degree of managerial and clinical efficiency (Chilingerian and Glavin 1994).

4.   SPECIAL ISSUES FOR HEALTH APPLICATIONS

There are conceptual, methodological, and practical problems associated with evaluating health care performance with DEA. First, conceptualizing clinical performance involves identifying appropriate inputs and outputs. Selecting inputs and outputs raises several questions--Which inputs and outputs should the unit be held accountable? What is the product of a health care provider? Can outputs be defined while holding quality constant? Should intermediate and final products be evaluated separately?

Another conceptual challenge involves specifying the technical relationship among inputs. In analyzing clinical processes, it is possible to substitute inputs, ancillary services for routine care days, more primary care prevention for acute hospital services, and the like. Within the boundaries of current professional knowledge, there are varieties of best practices. Consequently, an evaluation model should distinguish best practices from alternative practice styles.

Methodological problems also exist around valid comparisons. Teaching hospitals, medical centers, and community hospitals are the same. Nursing homes and skilled nursing facilities may not be the same. These institutions have a different mix of patients and have different missions. Finally, the information needed to perform a good study may not be available. Much of the interesting DEA research has not relied on an easily accessible data set, but required merging several data bases (see Burgess and Wilson 1998; Rosko et al 1995).

There are problems about choice of inputs and outputs, and especially finding an "acceptable" concept of product/service. Which inputs and outputs should physicians be held accountable? In addition, there are other issues about measures and concepts. For example:

- Defining Models from Stakeholder Views
- Selection of Inputs and outputs
- Should inputs include environmental and organizational factors?
- Problems on the Best Practice Frontier
  - o Are the input factors in medical services substitutable?
  - o Are returns to scale constant or variable?
  - o Do economies of scale and scope exist?

We discuss each of these issues in the next section.

4.1   Defining Models from Stakeholder Views

The models used in DEA depend on the type of payment system and the stakeholder perspective. Every health care system faces different payment and financing schemes that determine the location of financial risk and, subsequently, the interests of the stakeholders. Examples of payment and finance systems include fee-for-service reimbursement, prospective payment, fixed budgets and capitation. All of these payment systems have one thing in common--they are all seriously flawed. They can imperil quality and often have contributed to increased health care costs.

There are also many stakeholders, each with different and sometimes conflicting interests. For example, there are institutional providers (hospitals, nursing homes, clinics, etc) as well as individual providers (physicians, nurses, therapists, etc), governments (local, regional, and national health authorities), third party payers (insurance companies, sick funds, etc), and patients and their families. All stakeholders want better quality and better access for patients. Although stakeholders talk about more efficient care, they focus most of their attention on how to stem their rising costs.

In many countries, ambulatory physician services are reimbursed on a fee-for-service basis. Since there is no incentive to reduce patient care, from a provider, self-interest standpoint under an unrestricted fee-for-service the care philosophy becomes *more care = better care*, which can lead to over utilization of clinical services and rising health costs. Unless physician decision making is managed by strong professional norms and benchmarking best practices, systems that create incentives to maximize patient visits, hospital days and procedures leads to rising costs and less efficient care.

Pre-payment is the flip side of fee-for-service. With a pre-payment system, there is a contractually determined, fixed price/payment for a defined set of services. For example, countries like Germany, Italy and the United States have prospective payment systems that reimburse the hospital for each patient admission based on diagnosis irrespective of the actual costs. Taking the hospital's perspective, to enhance revenue the hospital must keep the beds full (increase admissions), and minimize the hospital's cost-per-admission, which of course can lead to over utilization of clinical services and rising health costs.

Capitation and gate keeping is another payment scheme that places physicians' interests at risk. The idea behind this design is that by having the primary care physician (PCP) assume responsibility for all aspects of care for a panel of patients, the patients will enjoy a greater continuity of care while the HMO achieves a more efficient care process at a more consistent level of quality. The 'gatekeeper' receives a fixed monthly payment for each patient (adjusted actuarially by age and gender) as an advance payment to provide all the primary care the patient needs. The incentives can be designed so that primary care physicians prosper if they keep their patients out of the hospital and away from specialists. However, the less care the PCPs provide the greater the financial gain. Failure to provide needed care in a timely manner to HMO patients may make these patients' treatments more expensive when their illnesses become more advanced. This type of payment scheme can lead to under-serving patients and rising health costs.

The payment systems described above can create weak or even distorted motivating environments. DEA models should be defined from a particular stakeholder point-of-view; moreover, models should be selected to ensure that patients are neither under-served nor over-served.

4.2   Selecting Appropriate Health Care Outputs and Inputs: The Greatest Challenge for DEA

Acute care hospitals and medical centers are complex medical care production processes, often bundling hundreds of intermediate products and services to care for each patient (Harris 1979). Major surgery might also require major anesthesia, blood bank services, hematology tests, pathology & cytology specimen, drugs, physical therapy, an intensive care bed, and time in routine care beds. Given the complexity, the greatest challenge for DEA health care applications lies in conceptualizing and measuring the inputs and outputs (see Chilingerian and Sherman 1990; Newhouse 1994).

Dazzling new technologies, desires to improve efficiency, and new consumer attitudes have changed service capacities, practice behaviors and outcomes. The explosive growth in outpatient surgeries is a good example. In the United States in 1984 only 400,000 outpatient surgeries were performed; by 2000 that number grew to 8.3 million (Lapetina and Armstrong 2002). The intensity and mix of outpatient and inpatient surgeries are continually changing. When outpatient surgeries shift from tooth extractions to heterogeneous procedures such as: hernia repairs, cataract and knee surgeries, and non-invasive interventions, simple counts of the quantity of surgical outputs are misleading. The mix and type of surgeries must be taken into account as well as the resources and technological capabilities of the decision-making units.

The first DEA paper published by Nunamaker (1983) used a one input, three-output model. The outputs used crude case mix adjustments (pediatric days, routine days, and maternity days). The single input was an aggregate measure of inpatient routine costs, which will lead to unstable results unless the inputs among the comparison set homogeneous. We know that hospital salaries, education, experience and mix of the nurses and other staff, the vintage of the capital and equipment, the number of intensive beds, medical supplies and materials, are not the same (see Lewin 1983).

To make the DEA results useful for practice, it makes sense to segment inputs into a few familiar managerial categories that make up a large percent of the expenditures. Since acute hospitals have high labor costs, it makes sense to distinguish the types of personnel: i.e., the number of full time equivalent (FTE) registered nurses, FTE licensed practical nurses, FTE nurses aids, FTE therapists, other FTE clinical, and general administrative staff. Selecting managerially relevant categories is necessary if analysts are to use DEA to identify the sources waste and inefficiency.

4.2.1   Take Two Aspirin and Call Me in the Morning

With respect to outputs, there are two requirements. First, health care outputs cannot be adequately defined without measures of case mix complexity and severity.   Second, it makes no sense to evaluate the efficiency of a medical service that results in an adverse event: such as morbidity, mortality, readmissions, and the like. To construct the "best practices" production frontier, observed behavior is evaluated as clinically inefficient if it is possible to decrease any input without increasing any other input and without decreasing output; and if it is possible to increase any output without decreasing any other output and without increasing any input (see Chapter One in this handbook).   A decision-making unit will be characterized as perfectly efficient only when both criteria are satisfied.

If hospital discharges include both satisfactory and unsatisfactory outcomes, a hospital could be considered efficient if they produced more output per unit of input.   If a patient die in the operating room and few clinical resources are utilized, the outcome falls short of the clinical objective.   Clinical efficiency requires that outcomes be considered in the performance standard.   Therefore, as a concept, clinical efficiency makes sense if and only if the clinical outputs achieve constant quality outcomes. Therefore, the best attainable position for a health care organization is when a unit achieves maximum the outputs should guarantee constant quality outcomes. Unsatisfactory outcomes should be taken out of the DEA analysis and evaluated separately (see Chilingerian and Sherman 1990). Likewise, if a primary care physician uses fewer office visits, fewer hospitals days, and fewer tests because she or he are postponing care, they should not be on the best practice frontier.

4.2.2   Using DEA to Adjust Outputs for Patient Characteristics and Case mix

If we are to go beyond simple illustrations, a DEA study should provide some guarantee that the outputs are similar. For example, it has long been established that acute patient care cannot be measured by routine patient days, maternity days, and pediatric days alone because patient days are an intermediate output and a poor indicator for other services such as blood work, x-rays, drugs, intensive care days, physical therapy and the like.

Therefore, it is important that both clinical and ancillary services be included. In contrast to acute hospitals, nursing home studies that define nursing days segmented by payer-mix and/or age-adjusted may be an acceptable proxy of final outputs.

To help guide policy makers or practitioners, researchers might consider the following four-part DEA analytic strategy. The first part would begin by running the some DEA models and the second part by regressing the DEA scores against the case mix and patient characteristic variables using a censored regression model such as Tobit. If the goodness of fit test is significant, adjust each health care provider outputs by multiplying them by the ratio of the original DEA score to the Tobit's predicted DEA score. Some providers operating with a less complex case mix will have their outputs lowered and other providers operating with a more complex mix will have their outputs increased. A second DEA model would be run during the third part and the last step would be regressing the "new" efficiency scores again to validate that they are unrelated to any control variables other than the critical managerial or policy variables of interest. Using DEA with this analytic strategy might provide some additional insights to policy makers or managers searching for answers to questions like what is the impact of ownership structure or leadership on the productivity of the industry.

4.3   SHOULD ENVIRONMENTAL AND ORGANIZATIONAL FACTORS BE USED AS INPUTS?

Random variations in the economic environment (such as labor markets, accidents, epidemics, equipment failures, and weather) and organizational factors (such as leader behavior, employee know-how, and coordination techniques) can influence organizational performance. However, there are so critical inputs and outputs that must be included in health applications, that non-discretionary variables should be omitted from the DEA model. For a strong health application, the clinical production model should only include resources that clinical decision makers utilize and manage. Environmental (and other organizational) factors omitted from the DEA model should be included in a second stage model that investigates variables associated with the DEA scores.

Every researcher must decide on a reasonable conceptual model. To guide research, collaborating with practicing managers or policy makers to identify relevant environmental and explanatory factors can only strengthen a health application.

4.4 PROBLEMS ON THE BEST PRACTICE FRONTIER: A PHYSICIAN EXAMPLE[1]

As described in Chapter one, of this book, DEA relies on the Pareto-Koopmans definition of efficiency. For example, to construct the "best practice" production frontier for primary care physicians, observed behavior is evaluated by using the following input-output criteria:

a) A physician is clinically inefficient if it is possible to utilize fewer clinical resources without increasing any other resources and without decreasing the number of patients cared for;

b) A physician is clinically inefficient if it is possible to care for more of a given segment of patients without decreasing the care given to any other segment of patients and without utilizing more clinical resources.

A physician can be characterized as clinically efficient only when both criteria are satisfied and when constant quality outcomes can be assumed.

A brief example with a small group of pediatricians illustrates how DEA can be used to define a best practice production frontier for primary care physicians. Consider fourteen primary care pediatricians who cared for 1000 female children 1-4 years old from a homogeneous socio-economic background. Figure 17-9 plots the amount of medical-surgical hospital days used and primary care visits for one full year, for the 1000 children,

The Pareto-Koopmans efficiency criteria suggests that any pediatrician in Figure 17-9 who was lower and to the left of another pediatrician was more successful because he or she used fewer clinical inputs to care for the "same" type of children. By floating a piecewise linear surface on top of the actual observations, Figure 17-9 also plots a best practices production frontier (BPPF) for each of the fourteen pediatricians. As shown in Figure 17-9, pediatricians Pa, Pb, Pc, and Pd dominate Pe, Pf, Pg, Ph, Pi, Pj, Pk, Pl, and Pm. Accordingly, Pa, Pb, Pc, and Pd, physicians not dominated by any other physicians lie on a best practices production frontier and all designated as frontier points.

DEA defines a best practice frontier by constructing a set of piecewise linear curves, such that any point on the BPPF is a weighted sum of observed DMUs. DEA measures the relative efficiency of each physician by the relative distance of that physician from the frontier. Physicians on the frontier (points Pa, Pb, Pc, and Pd) have a DEA score of 1 and physicians off the frontier have a value between 0.0 and 1.0. For example, in Figure 17.9, the efficiency score of Pe, a pediatrician less efficient than Pb and Pc,

---

[1]. The next section draws heavily on Chilingerian and Sherman (1995), Rosko (1990) and Charnes, Cooper and Rhodes (1978).

can be measured by constructing a line segment from the origin to point Pe. This line segment crosses the BPPF at point e'. The efficiency rating of Pe is equal to the ratio of the length of line segment Oe* to the length of OPe. Since the numerator Oe* is less than the denominator OPe, the efficiency rating of physician Pe will be less than 1.0.

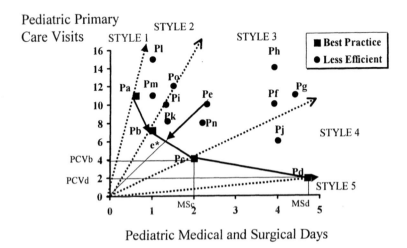

Figure 17-9. Two-Dimensional Picture of Pediatrician Practice Behavior

The practice of pediatric primary care medicine is a very complicated production problem where there are many slightly different ways to mix inputs to care for a population of patients. To solve the problem, each physician adopts a "style of practice"[2] that can be represented by the mix of clinical inputs used. The geometry of convex cones provides a pictorial language for interpreting the practice style of physicians. In mathematics, a cone is a collection of points such that if a physician P is at any point in the collection, then μP is also in the collection for all real scalars $\mu \geq 0$ (Charnes, Cooper and Henderson 1953). However, cones can also be interpreted as a

---

[2]. Practice style, as it used here, refers to the treatment patterns defined by the specific mix of resource inputs used by a physician (or group of physicians) to care for a given mix of patients.

linear partition of physician practice styles based on a set of linear constraints such as a range of substitution ratios.

For example, in Figure 17-9 the ray that begins at the origin (0) and intersects the frontier at Pb and the ray that begins at the origin and intersects the frontier at Pc define the upper and lower boundaries of a cone or *practice style 3*. All points contained on the line that begins at the origin (0) and intersects the frontier at Pb have the same ratio of medicine and surgery days to primary care office visits. Therefore, a physician practicing *style 3* will have a ratio of visits to hospital days that falls between the line that begins at the origin (0) and intersects the frontier at Pb and the line that begins at the origin (0) and intersects the frontier at Pc.[3]

So, each cone can also be thought of as a different practice style. For example, practice style 2 represents physicians who use relatively more visits than hospital days. In contrast, practice style 5 represents physicians who use relatively more medical-surgical days than primary care office visits. By modifying the standard DEA model, it can be used to locate cones or natural clustering of DMUs based on the optimal weights assigned by the linear program. The next section will illustrate how the new DEA concept of assurance regions can be used to identify and investigate practice styles.

4.4.1   The Concept of A Preferred Practice Cone or Quality Assurance Region

The largest single expense item for an HMO is hospital days. Lower utilization rates for hospital days represents more desirable practice style. Physicians who succeed in substituting primary care visits and ambulatory surgeries for medical-surgical days will contribute to a more efficient delivery system. Understandably, most HMO managers would prefer their physicians to reduce hospital days and increase their primary care visits.

Microeconomics offers a way to quantify the clinical guideline that primary care physicians (PCPs) should trade-off medical surgical days (input X_2) for more primary care visits (input X_1) while maintaining a constant output rate. Economists refer to the change in medical-surgical days (ΔX_2) per unit change in primary care visits (ΔX_1) as the marginal rate of technical substitution--i.e., $-dx_2/dx_1$., which equals the marginal product of medical surgical days (ω_h) divided by the marginal product of primary care visits (ω_v).[4]

---

[3]. Although the two-input single output problem can be solved graphically, multi-input multi-output problems require a mathematical formulation that can only be solved by using a linear programming model.

[4]. In economics, holding constant all other inputs, the marginal product of an input is the addition to total output resulting from using the last unit of input. The ratio of

In Figure 17-9 every physician on the BPPF practices an input minimizing style of care -- i.e., their DEA efficiency score = 1. However, not all of these physicians on the BPPF are practicing minimum-cost care. In other words, the mix of inputs used by some of these physicians is not the most cost effective. Consider the physicians Pd in practice style 5 and Pc in practice style 4. Since style 4 cares for the same patients with fewer hospital days, cost conscious HMO managers would probably want their primary care physicians practicing to the left of style 5.

Now, consider the BPPF in Figure 17.9. If the quantities of medical-surgical days are increased from MSc to MSd, the same panel of patients will be cared for with slightly fewer of the less expensive primary care visits and much more of the expensive medical and surgical days. This points out a weakness in using DEA to estimate the best practice production frontier. From an HMO's perspective, although the practice style of physician Pd represents an undesirable practice region, the standard DEA model would allow Pd to be on the BPPF.

Modified versions of additive DEA models allows managers to specify bounds on the ratio of inputs such as the ratio of hospital days to primary care office visits (see Cooper, Park and Pastor 2000). These bounds can be called a preferred practice cone (Chilingerian and Sherman 1997). This new development in DEA bounds the optimal weights (or marginal productivities) and narrows the set of technically efficient behaviors. By incorporating available managerial knowledge into a standard model, an assurance region truncates the BPPF frontier -- i.e., tightens the production possibility set -- and opens up the potential to find more inefficiency.

4.4.2 Constant versus Variable Returns to Scale

One difficult issue with every health application is whether to use DEA models that assume variable or constant returns to scale. Researchers should address this question based on prior knowledge and logical inferences about the production context. While imaginative guesses are tolerable, it is unacceptable to pick a model in order in order to get "better looking" DEA results.

Although a health care organization's production function may exhibit variable returns to scale, there are intuitive reasons for expecting that a physician's clinical production function exhibit constant returns to scale (Pauly 1980). Physicians are taught that similar patients with common conditions should be taken through the same clinical process. Thus if a surgeon operates on twice as many patients with simple inguinal hernias or

---

the marginal products is the marginal rate of technical substitution, defined as - dx_2/dx_1.

performs twice as many coronary by-pass grafts, they would expect to use twice as many clinical resources. Consequently, scaling up the quantity of patients should result in a doubling of the inputs.

Hospitals and physician practices both vary in size. In North America, the largest stand-alone hospitals have less than 1,500 beds. In Europe, there are medical centers with as many as 2,000 and 3,000 beds. What is the minimal efficient bed size or the most productive bed size? Some primary care physician practices vary from 1200 to 4000 patients. From a quality standpoint, what is the most optimal patient size? Questions about returns to scale should be addressed in future DEA research and would benefit from cross-cultural comparisons.

4.4.3 Scale and Scope Issues

Health care providers and organizations are multi-service firms, offering many clinical services to provide convenient, one-stop shopping, to connect diagnostic services with treatments, etc. Is it more efficient to offer many services under one roof? If offering many services together is more efficient, then economies of scope exist.

On the other hand, the rise in "focused factories" in health care such as hernia clinics, heart clinics, hip replacement centers, and the like, is stirring a great deal of interest throughout the world. Are patients better off going to a super-focused clinic? For example, is a cataract hospital more efficient than a general hospital performing fewer cataract procedures in its operating rooms? Are the efficiency gains of a focus strategy large or small? Questions about scale versus scope should also be addressed in future DEA research.

4.5 ANALYZING DEA SCORES WITH CENSORED REGRESSION MODELS

DEA's greatest potential contribution to health care is in helping managers, researchers, and policy makers understand why some providers perform better or worse than others do. The question can be framed as follows--How much of the variations in performance are due to: (1) the characteristics of the patients, (2) the practice styles of physicians, (3) the micro-processes of care, (4) the managerial practices of the delivery systems, or (4) other factors in the environment? The following general model has been used in this type of health care study:

DEA Score = f (ownership, competitive pressure, regulatory pressure, demand patterns, wage rates, patient characteristics, physician or provider practice characteristics, organizational setting, managerial practices, patient illness characteristics, and other control variables).

The DEA score depends on the selection of inputs and outputs. Hence every health application is obliged to disconfirm the hypothesis that DEA is not measuring efficiency, but is actually picking up differences in case mix or other non-discretionary variables. The best way to validate or confirm variations in DEA scores is to regress the DEA scores against explanatory and control variables. But what type of regression models should be used?

If DEA scores are used in a two-stage regression analysis to explain efficiency, a model other than ordinary least squares (OLS) is required. Standard multiple regression assumes a normal and homoscedastic distribution of the disturbance and the dependent variable; however, in the case of a limited dependent variable the expected errors will not equal zero. Hence, standard regression will lead to a biased estimate (Maddala 1983). Logit models can be used if the DEA scores are converted to a binary variable—such as efficient/inefficient. However, converting the scores < 1 to a categorical variable results in the loss of valuable information; consequently logit is not recommended as a technique for exploring health care problems with DEA.

Tobit models can also be used whenever there is a mass of observations at a limiting value. This works very well with DEA scores which contain both a limiting value (health care providers whose DEA scores are clustered at 1) and some continuous parts (health care providers whose DEA scores fall into a wide variation of strictly positive values < 1). No information is lost, and tobit fits nicely with distribution of DEA scores as long as there are enough best practice providers. If, for example, in a sample of 200 providers less than 5 were on the frontier, a tobit model would not be suitable.

In the econometrics literature, it is customary to refer to a distribution such as DEA as either a truncated or a censored normal distribution. There is, however, a basic distinction to be made between truncated and censored regressions. According to one source:

> "*The main difference arises from the fact that in the censored regression model the exogenous variables x_i are observed even for the observations for which $y_i > L_i$. In the truncated regression model, such observations are eliminated from the sample.*" (Maddala 1983:166)

Truncation occurs when there are no observations for either the dependent variable, y, or the explanatory variables, x. In contrast, a censored regression model has data on the explanatory variables, x, for all observations; however the values of the dependent variable are above (or below) a threshold are measured by a concentration of observations at a single value (Maddala 1983). The concentration of threshold values is often based on an actual measure of the dependent variable-- i.e., zero arrests, zero expenditures-- rather than an arbitrary value based on a lack of information.

DEA analysis does not exclude observations greater than 1, rather the analysis simply does not allow a DMU to be assigned a value greater than 1. Hence, Chilingerian (1995) has argued that DEA scores are best conceptualized as a censored, rather than a truncated distribution. The censored model would take the following form:

Efficiency score = actual score if score < 1
Efficiency score = 1 otherwise

There is a substantial literature on modeling data with dependent variables whose distributions are similar to DEA scores. For example, empirical studies with the number of arrests after release from prison as the dependent variable (Witte 1980) or the number of extra marital affairs as the dependent variables (Fair 1978) are among the best known examples in the published literature on the Tobit censored model. Each of these studies analyzes a dependent variable censored at a single value (zero arrests, zero marital affairs) for a significant fraction of observations. Just as the women in Fair's study can do no better than have zero extra-marital affairs, neither could a relatively efficient health care provider be more efficient than 1. Thus, one could equate zero extra-marital affairs or zero arrests to a "best practicing" hospital or provider.

A censored Tobit model fits a line which allows for the possibility of hypothetical scores > 1. The output can be interpreted as "adjusted" efficiency scores based on a set of explanatory variables strongly associated with efficiency. To understand why censored regression models make sense here, one must consider how DEA evaluates relative efficiency.

DEA scores reflect relative efficiency within similar peer groups (i.e., within a "cone of similar decision making units") without reference to relative efficiency among peer groups (i.e. cones). For example, an efficient provider scoring 1 in a peer group using a different mix of inputs (i.e., rates of substitution) may produce more costly care than a provider scoring 1 in a peer group using another mix of inputs (Chilingerian and Sherman 1990). Superior efficiency may not be reflected in the DEA scores because the constraints in the model do not allow a decision making unit to be assigned a value greater than 1. If DEA scores could be re-adjusted to compare efficiencies among peer groups, some physicians could have a score that is likely to be greater than 1. Despite the advantages to blending nonparametric

DEA with censored regression models in practice, some conceptual problems do arise.

The main difficulty of using Tobit to regress efficiency scores is that DEA does not exactly fit the theory of a censored distribution. The theory of a censored distribution argues that due to an underlying stochastic choice mechanism or due to a defect in the sample data there are values above (or below) a threshold that are not observed for some observations (Maddala 1983). As mentioned above, DEA does not produce a concentration of ones due to a defect in the sample data, rather it is embedded in the mathematical formulation of the model.

A second difficulty of using Tobit is that it opens up the possibility of rank ordering superior efficiency among physicians on the frontier--in other words "hypothetical" scores > 1. In production economics, the idea that some DMUs with DEA scores of 1 may possibly have scores > 1 makes no sense. It suggests that some candidates for technical efficiency (perhaps due to random shifts such as luck, or measurement error) are actually less efficient.

Despite these drawbacks, blending DEA with Tobit model's estimates can be informative. Although DEA does not fit the theory of a censored regression, it easily fits the tobit model and makes use of the properties of a censored regression in practice. For example, the output can be used to adjust efficiency scores based on factors strongly associated with efficiency.

Tobit may have the potential to sharpen a DEA analysis when expert information on input prices or exemplary DMUs are not available. Thus in a complex area like physician utilization behavior, tobit could help researchers to understand the need to introduce boundary conditions for the DEA model's virtual multipliers.

The distribution of DEA scores is never normally distributed, and often skewed. Taking the reciprocal of the efficiency scalar, (1/DEA score), helps to normalize the DEA distribution (Chilingerian 1995).

Greene (1993) points out that for computational reasons, a convenient normalization in Tobit studies is to assume a censoring point at zero. To put a health care application into this form, the DEA scores can be transformed with the formula:

Inefficiency score = (1/DEA score) - 1

Thus, the DEA score can become a dependent variable that takes the following form:

DEA Inefficiency score = $x B + u$ if efficiency score > 0
DEA Inefficiency Score = 0 otherwise

Once health care providers' DEA scores have been transformed, tobit becomes a very convenient and easy method to use for estimating efficiency. The slope coefficients of tobit are interpreted as if they were an ordinary least squares regression. They represent the change in the dependent variable with respect to a one unit change in the independent variable, holding all else constant.

When using tobit models they can tested with a log-likelihood ratio test. This statistic is calculated by $-2 \log(\lambda)$, where $\log \lambda$ is the difference between the log of the maximized value of the likelihood function with all independent variables equal to zero, and the log of the maximized values of the likelihood function with the independent variables as observed in the regression. The log-likelihood ratio test has a chi square distribution, where the degrees of freedom are the number of explanatory variables in the regression.

5. NEW DIRECTIONS: FROM PRODUCTIVE EFFICIENCY FRONTIERS TO QUALITY-OUTCOME FRONTIERS

Although most applications of DEA have been applied to estimations of technical efficiency and production frontiers, the methodology offers an empirical way to estimate quality frontiers. For example, any dimension of quality can be assessed by employing multiple indicators in a DEA model and comparing a provider against a composite unit projected onto a frontier.

Table 17-1. Nursing Home Outcomes: 100 Patients Traced for Six Months

Nursing home	Functional Status Q1	Functional Status Q3
N1	.15	.40
N2	.15	.14
N3	.35	.35
N4	.30	.15
N5	.40	.50
N6	.20	.30

Although DEA could be applied to any type of quality evaluation, the following example is from a nursing home, demonstrating how to measure improvements in functional status. Drawing heavily on a paper by Chilingerian (2000), the following illustration explains how DEA works in estimating outcome frontiers. Consider a group of six nursing homes each with 100 patients whose bed mobility, eating, and toilet use have been traced for a period of six months.

Table 17-1 displays the overall functional status of the 100 patients in each of the six homes from quarter 1 to quarter three. The functional status represents a vector of improvements or deteriorations of functional status in terms of the proportion of residents in quarter 1 completely independent and the changes to that group in quarter 3. That situation can be depicted graphically in Figure 17-10 as a piecewise linear envelopment surface.

Identifying a Quality Frontier

**Tracing Changes in Patient's
Functional Status Q3**

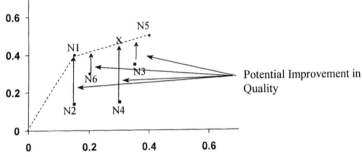

Chilingerian 2000

Figure 17-10. Two-Dimensional Picture of a Quality Frontier

In Figure 17-10, DEA identifies nursing homes that had the greatest improvement in functional quality over the six-month period by allowing a line from the point of origin to connect the extreme observations in a 'northwesterly' direction. Nursing homes N1 and N5 had the greatest improvement in the functional independence of their patient populations and the dotted lines connecting N1 and N5 represent the best practice frontier.

The vertical lines above N2, N6, N4, and N3 represents the potential improvements in quality outcomes if N2, N4, N6, and N3 had performed as well as a point on the line between N1 and N5. Note that since the initial functional status is a non-discretionary variable, the potential improvement can only occur along the vertical line. Associated with each under-performing nursing home is an optimal comparison point on the frontier that is a convex combination of the nursing homes. For example, N4 could be projected onto the best practice frontier at point X. The performance measure is the linear distance from the frontier expressed as a percent such as: 90%, 84% and so on. To be rated 100%, the nursing homes must be on the best practice surface.

In this simple two-dimensional example, the nursing homes with the greatest improvements in functional status were identified. DEA is capable of assessing quality with dozens of indicators in n-dimensional space. For a complete mathematical explanation, see Chapter One in this book.

A model that could be used to introduce an outcome-quality frontier is called the additive model (described in Chapter One). The additive to be used is based on the idea of subtracting the functional status of the patient at the outset from the status attained after the care process. The resulting measure is expressed as a change in functional status based on the (functional status achieved during quarter 3) minus (the functional status at quarter 1). The numerical differences from Q_1 to Q_3 can be interpreted as improvements, deteriorations, or no change in outcomes.

The optimal value $w*_o$ is a rating that measures the distance that a particular nursing home being rated lies from the frontier. A separate linear programming model is run for each nursing home (or unit) whose outcomes are to be assessed. The additive model is shown below:

(ADDITIVE MODEL)

$$\max_{u, v, uo} \quad w_o = u^T Y_o - v^T X_o + u_o$$

$$\text{s.t.} \quad u^T Y - v^T X + u_o \vec{1} \leq 0$$

$$-u^T \leq \vec{-1}$$

$$-v^T \leq \vec{-1}$$

Here the vector Y represents the observed functional status variables at Q_3 and X represents the initial functional status variables observed at Q_1. The additive model subtracts the functional status variable at Q_3 from those at Q_1. The variables, u^T *and* v^T are the weights assigned by the linear program so $u^T Y$ is the weighted functional status at Q_3 and $v^T X$ is the weighted functional status at Q_1. The model is constrained so that every nursing home is included in the optimization such that the range of scores will be between 0 and 100%. The u^T and v^T are constrained to be non-negative.

For the problem of developing a frontier measure of improvements or deterioration in functional status, the additive has number of advantages (see Cooper et al., 2000). The model does not focus on a proportional reduction of inputs or an augmentation of outputs. It offers a global measure of a distance from a frontier, by giving an 'equal focus on functional status before and after by maximizing the functional improvement between time periods (characterized as the difference between Q_1 and $Q_{3)}$. Another advantage of the additive model is that it is translation invariant which means we can add a vector to the inputs and outputs and though we get a new data set, the estimates of best practice and the outcome measures will be the same. This model will be applied to a nursing home data set to find a DEA outcome frontier and a DEA decision-making efficiency frontier.

5.1 A FIELD TEST: COMBINING OUTCOME FRONTIERS AND EFFICIENCY FRONTIERS

To illustrate the ideas discussed above, the following DEA model will be used to find a quality-outcome frontier for 476 nursing homes in Massachusetts (United States). We developed outcome measures for the nursing homes from the Management Minutes Questionnaire (MMQ). This is the case-mix reimbursement tool used in several states in the United States and in particular is used in the state of Massachusetts to pay nursing homes for the services they provide Medicaid residents. The MMQ collects information on the level of assistance that nursing home residents need from staff members to carry out activities of daily living such as dressing, eating, and moving about. Fries (1990) explains that the MMQ index is constructed for each resident based on a spectrum of resident characteristics, each with a specified weight. Values are supposed to correspond to actual nursing times so the total should correspond to total staffing needs for the resident. Weights are derived from expert opinion rather than statistical analysis, and total weights are adjusted with time values added for each of the items measured.

Changes in overall resident functioning (determined by measuring the change in MMQ scores over two quarters) were used as a proxy for quality of care. These variables depict the direction of functional status change (improvement, maintenance or decline) experienced by the residents during the last six months. Changes in a positive or static direction (improvement or maintenance) will be used as proxies for high quality care (controlling for health status) and changes in a negative direction (decline) will be used as a proxy for a decrease in the quality of care.

In each of the nursing homes, the residents' functional status was evaluated. The proportion of residents who were independent in mobility and eating, and continence were traced for 6 months from quarter 1 to quarter 3. If we consider how a patient's functional status changes over time, whether a nursing home is improving, maintaining or declining, these changes become an outcome measurement tool. In particular, resident functional improvement was monitored by three activities of daily living:

bed mobility, eating, and toilet use. The model used and the variables are displayed below in Figure 17-11 and 17-12.

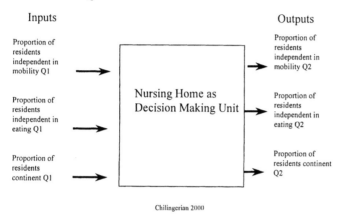

Figure 17-11. Three- Input Three- Output Model Used to Estimate Quality

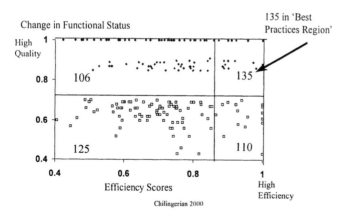

Figure 17-12. Plot of Change in Functional Status and Productive Efficiency (source: Chilingerian 2000)

The DEA model identified 41 or 9% of the nursing homes on the best practice quality frontier. The average quality score was 74%, which means that the functional status outcomes could potentially be improved by 26%.

Figure 17-12 plots the DEA measure of the changes in functional status against a DEA measure of the decision-making efficiency for the 476 nursing homes in the data set. By partitioning this two-dimensional summary of nursing home performance at the means (81% and 74%), four categories of performers emerge: 135 with high quality outcome and high efficiency; 106 with high quality outcomes but lower efficiency; 110 with low quality outcomes but high efficiency; and 125 with lower quality outcomes and low efficiency. Each of these unique and exhaustive categories can be further analyzed to explore the factors associated with each category of performance. By studying the interaction of these two dimensions of performance, it may be possible to gain new insights about quality and efficiency.

6.   A HEALTH DEA APPLICATION PROCEDURE: EIGHT STEPS

Several studies have suggested analytic procedures for DEA studies (Golany and Roll 1989; Lewin and Minton 1986). Drawing heavily on these papers, this last section highlights how the researcher connects the empirical and conceptual domains with real world problems in health policy and management. The structure of DEA research can be seen as a sequence of the following eight applied research steps:

6.1   Step 1: Identification of Interesting Health Care Problem and Research Objectives

Applied work always begins with a real-life policy and/or management problem. Interesting health policy and management research questions are needed to guide the data to be collected. If we take a close look at all of the DEA studies, many have been illustrations of uses of DEA based on available data. Therefore, the first step is to find questions of practical importance, and never let the available data drive the research. Identify what is new and interesting about the study.

Set challenging research goals--an interesting question is always preferable to a researchable question from an available data set. The goal should always be to inform both theory and practice. The purpose of a good research question is to get to some answers.

6.2   Step 2: Conceptual Model of the Medical Care Production Process

There is no final word on how to frame a health care production model. Each DEA study claims to capture some part of reality. Hence, the critical question in this step is--Does the production model make sense? Most likely, the problem has been studied before. So an obvious first step is to ask--How has it been modeled in previous work? What aspects were missing? There should always be a justification of the inputs and outputs selected. This can be based on the literature, prior knowledge, and/or expert knowledge (see O'Neill 1998; Chilingerian and Sherman 1997). The use of clinical experts is critical if the application is to become useful for practice.

A related question is choice of the DEA models. Because of strong intuitive appeal, the CCR (clinical and scale efficiency) and BCC (pure clinical efficiency) models have been used in most health care studies. The multiplicative model (for cobb-douglas production functions) and the additive model hardly appear in the health care literature. Researchers should consider the underlying production technology and take a fresh look at the choice of DEA models. If the justification for the final choice of a DEA model is sketchy, consider running more than one.

6.3   Step 3: Conceptual Map of Factors Influencing Care Production

Step three identifies the set of variables and some empirical measures based on several important questions. What is the theory of clinical production or successful performance? Do medical practices vary because some patients are sicker, poorer, or socially excluded? What explains best practices? For this step, researcher should identify the environmental and other factors out of the control of the managers, organizational design and managerial factors, provider and patient characteristics and case-mix variables (see Chilingerian and Glavin 1994). The goal is to build a conceptual map that identifies some of the obstacles, or explanatory variables associated with best practices from the literature and expert knowledge. If the theory is weak, try some simple maps or frames that raise theoretical issues.

6.4   Step 4: Selection of Factors

Now the researcher is ready to search for databases or collect the data. There is always some difficulty obtaining the variables for the study files: inputs, outputs, controls, explanatory variables, and the like. Sometimes physician, hospital, medical association, and insurance databases must be merged into one study file to link DEA with other variables.

6.5   Step 5: Analyze Factors using Statistical Methods

DEA assumes that a model is assessing the efficiency of "comparable units," not product differences. Before running an efficiency analysis, if there is reason to believe that outputs are heterogeneous, it is recommended that peer groups be developed (Golany and Roll 1989). In health care, a variety of peer groups could be developed based on medical sub-specialty (orthopedic surgeons versus cardiac surgeons), diagnostic complexity, and other product differences.

The mathematics of DEA assumes that there is an isotonic (order preserving) relation between inputs and outputs. An increase in an input should not result in a decrease in an output. Inputs should be correlated. Golany and Roll (1989) have argued that running a series of regressions (variable by variable) can help to reduce these problems. If there is multicollinearity among inputs (or among outputs), one remedy is to eliminate one or more inputs or outputs.

Finally, there has rarely been a DEA study that has not had to deal with the problem of zeroes. In any case, there are two crude ways of dealing with problem. One is to throw out all of the DMUs with missing values and reduce the number of DMUs. The other way is to substitute the zero with a very small number such as .001. In addition to this handbook, Charnes et al. (1991) offer better ways of dealing with the problem of zeros in the data set.

6.6   Step 6: Run Several DEA Models

Are the results reasonable? If the data set finds a three-fold difference in costs among the units, something other than efficiency is being measured. If you are running different DEA models check for the stability of results.

Sometimes there are many self-referring units on the best practice frontier. One solution to this problem is to impose cone ratio conditions that reflect preferred practice styles (see Chilingerian and Sherman 1997; Ozcan 1998). As a rule of thumb, if the majority of the DMUs are showing up as 100% efficient and they are mostly self-referencing, there are two possible explanations. Perhaps the production technology is so complex, that there are many slightly different ways of practicing. On the other hand, the free choice of weights is giving the providers the 'benefit of the doubt' by hiding unacceptable practice styles that care for very few patients, and/or utilize a very high quantity of clinical resources.

6.7 Step 7: Analyze DEA Scores with Statistical Methods

The next step is to use the DEA results to test hypotheses about inefficiency. Blending DEA with various statistical methods has been all the rage in health care studies. This has been a health trend. To convince the reader that the DEA scores are valid, every health care study must test the hypothesis that there is unaccounted case mix. If the explanatory and control variables have no significant association with the DEA scores, something may be wrong with the production model.

6.8 Step 8: Share Results with Practitioners and Write It Up

Traditionally the DEA work in health care focused more on methodology than the issue of usefulness. To advance the field more quickly, research needs to spend time with clinicians and practitioners (who are not our students). Before writing up the results show the models and findings to real clinical managers and listen to their advice about performance effectiveness. This is a suggestion for everyone.

7. DEA HEALTH APPLICATIONS: DO'S AND DON'TS

Given the twenty-year history we have with DEA health applications, there are several do's and don'ts. Based on our experience reviewing and reading DEA papers--three will be discussed.

7.1 ALMOST NEVER INCLUDE PHYSICIANS AS A LABOR INPUT

One of the concerns in DEA hospital studies is including physicians as an input in health care (Burgess and Wilson 1998; Chilingerian and Sherman 1990). In many health care systems such as the United States, Belgium, Switzerland, there are academic medical centers that have salaried physicians reporting to a medical director in the hospital; there are also community general hospitals that have established cooperative, as opposed to hierarchical relations with physicians. Though physicians are granted "admitting privileges" and can use the hospital as a workshop to care for patients, some physicians may not be very active in admitting patients. Physicians can enjoy these privileges at several hospitals (Burns et al. 1994). Including a variable for the quantity of FTE physicians is legitimate input if they are a salaried such as in academic medical centers. However, in many countries community hospitals do not employ physicians so they should not be included as an input.

7.2 USE CAUTION WHEN MODELLING INTERMEDIATE AND FINAL HOSPITAL OUTPUTS

Although modeling hospitals with DEA can be complex, the DEA literature offers some exemplars for research purposes. While none of these papers is perfect, the models are reasonable given the available data. For example:

- Nunamaker (1983) used three outputs: age adjusted days, routine days, maternity days.
- Sherman (1984) used three outputs: age-adjusted patient days, and nurses and interns trained as outputs.
- Banker, Conrad and Strauss (1986) used three outputs: aged adjusted patient days
- Sexton et al (1989) used six categories of workload-weighted units (for example, medical work load weighted units (WWU), psychiatric WWU, surgical WWU, OPD WWU, etc.)
- Chilingerian and Sherman (1990) used two outputs: high severity cases of DRG 127 with satisfactory outcomes and low severity cases of DRG 127 with satisfactory outcomes.

Some of these papers focus on 'inputs per patient day' others on 'inputs per discharged case'. Note that none of these papers mixed final products i.e., discharges with intermediate products i.e., patient days. If both discharges and patient days are the outputs, DEA will obtain results based on a composite of optimal--cost per day/cost per case/ cost per visit. DEA will give misleading results.

There have been several papers published in the health care literature that combine intermediate and final outputs in a single model. While there are arguments on both sides, there are conceptual problems with how DEA would evaluate efficiency using these models. Let us consider the model used by several researchers (see Burgess and Wilson 1996, Ferrier and Valdmanis 1996) combining case-mix adjusted discharges with total patient days. Assuming constant inputs, if a hospital maintains its case-mix adjusted admissions, while increasing its patient days, its average length of stay (ALOS) is increasing. The DEA model, however, would rate the hospital with higher length of stay (LOS) as more efficient, though these hospitals would, by most managerial definitions, be less efficient because they could have discharged some patients sooner, and <u>utilized their capacity better</u> by admitting <u>more</u> patients. In a high fixed cost service, longer lengths of stay might suggest poorer quality (more morbidity, nosocomial infections, etc.) Since quality measures are not available, one cannot assume maximum output from inputs with constant quality outcomes.

Now consider the alternative model with case mix adjusted days and teaching outputs, excluding discharges (see Sherman 1994). Assuming constant inputs, if a hospital maintains the number of interns being taught and increases its case mix adjusted patient days, the hospital is producing more outputs with fewer inputs--i.e., becoming more efficient. That is, DEA would rate that hospital more efficient. Although the ALOS is not known, the Sherman model is not rating hospitals with longer lengths of stay as better. This type of model works better conceptually and in practice.

DEA Output Frontier

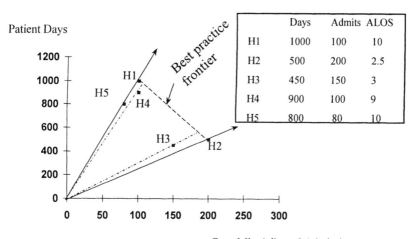

Figure 17-13. Two-Dimensional Picture of An Output Frontier That Combines Intermediate and Final Hospital Outputs

To illustrate the point, consider the following data from a policy maker's perspective. In Figure 17-13 five teaching hospitals, each with <u>similar inputs</u> and the same case mix, and the <u>same total ancillary charges, but with</u> <u>variable patient days and case-mix adjusted admissions.</u> In the plot of the four hospitals, the DEA output frontier identifies the best practice frontier to be the segment connecting H1 and H2. H4 for example, is less efficient than H1 because it had fewer patient days, even though it discharged the patients sooner and presumably used its available capacity better.

H5 is inefficient and would become more efficient if H5 could move along the ray (represented by the solid lines) and toward H1. The efficiency rating is based on a radial measure of efficiency that maintains the rate of transformation. This measure forces H5 to increase days and admissions, and ignores the need to improve clinical efficiency—e.g., by reducing length of stay and increasing admissions. Do the definitions of efficiency underlying this model make sense?

H4 has the same admissions but fewer patient days; it has the same cost per case, but a higher cost per patient day than H1 and a lower ALOS. From a policy perspective, should not H4 be ranked at least as efficient as H1? DEA suggests that H4 becomes efficient by maintaining its ALOS and increasing days and admissions accordingly.

Now, when H2 is thought of as an extreme point, policy makers might consider H1 as the only frontier point. H2, 3, 4, & 5 all could improve their capacity utilizations (by treating more cases) and improve their clinical efficiency (by reducing lengths of stay). There are DEA models that define preferred practice regions (or cone ratios) to handle these situations.

Therefore, if the model includes **both** patient days and cases as DEA outputs (as the author's have done), DEA defines two type of efficiency—one based on cost per case, and another based on cost per admission. H1 is rated 100% because it produced more of the variable output—patient days. H2 is rated 100% efficient because it produced more admissions.

Hospitals with the same case mix and admissions, but higher lengths of stay would be rated as 100% efficient along side hospitals with the same case mix, patient days, and lower lengths of stay. This runs counter to what policy makers would want and would confuse everyone. There are two solutions. One is to use the managerial inputs with the clinical outputs (see Figure 17-4). Another approach is to use a two-part production model: one DEA for practice management with managerial inputs and intermediate outputs; and another DEA model for patient management with clinical inputs and clinical outputs (see Figure 17-1).

7.3   Do Check the Distribution of DEA Scores and Influence of Best Practice Providers on Reference Sets

DEA will yield some approximation of an efficiency score with most data sets. Consequently, every health care application requires a careful check of the distribution of DEA scores. For example, whenever the range includes efficiency scores below .50, or whenever there are more than 50% or 60% of the DMUs on the frontier, there is a strong likelihood that something is wrong. The following common sense test appears to be helpful—Given the health care context, do these results make sense? Though one can hardly imagine finding a hospital, nursing home, or physician operating at 17% efficiency and surviving another day, it is possible. Low scores always require a plausible explanation, which might be a non-competitive environment, or the existence of subsidies. Newhouse points out that Zuckerman et al. (1994) found productive inefficiency to account for 14% of total hospital costs (1994). Should policy-makers reduce hospital budgets by this amount? Would this force many hospitals out of business? Before anyone takes this findings too seriously, the inefficiencies have to be valid.

In health care, a high proportion of very low DEA scores are likely to be due to unaccounted case mix, heterogeneity of output measures, or returns to scale (Chilingerian 1995). Alternatively, a very high proportion of DMUs on the best practice frontier (40-50%+) could be the result of caring for many different types of patients, using slightly different styles of practice. It is important to examine the number of physicians appearing in only one reference set and to identify the most influential physicians—those physicians who appear in the most reference sets. When a priori information is available on best practices, it is possible to reduce the number of efficiency candidates in any given analysis. For example, a cone ratio DEA model allows for a meaningful upper and lower-bound restriction to be placed on each input virtual multiplier. The restriction reduces the number of efficient candidates, bringing the DEA measure of technical efficiency closer to a measure of overall efficiency. Imposing cone ratio models on the results will usually reduce the number of self-referencing DMUs on the frontier (see Chilingerian and Sherman 1997; Charnes, Cooper, Huang, and Sun 1990). Running a second DEA model without the most influential observation will reveal how robust the DEA results.

For example, one DEA study of 326 primary care physicians conducted by Chilingerian and Sherman (1997) found 138 on the best practice frontier. In Figure 17-14, 45 physicians appear in one reference set and are, therefore, self-referencing physicians. Expert opinion from the clinical director helped to identify a preferred practice region. The medical director's criterion for 'best practice' was a primary care physician who: (1) performed under budget; (2) utilized less than 369 medical/surgical days per 1000 members; (3) had referral rates of less than 1.2 per member; and (4) provided at least 1760 primary care office visits per 1000 members. A second DEA model would be run setting the minimum and maximum marginal rates of substitution for referral rates, medical surgical hospital days, and primary care visits. The second DEA model reduced the number of efficient physicians from 138 to 85 and found most of the 45 self-referring physicians less efficient.

How Many Reference Sets Do the 'Efficient' Physicians Influence?

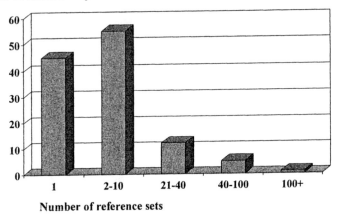

Figure 17-14. Number of Efficient Physicians By the Number of Reference Sets They Influence

8.   A FINAL WORD

This chapter has looked at some of the conceptual and methodological challenges associated with measuring and evaluating health care performance with DEA. While DEA offers many advantages, the critics have raised fundamental questions such as: How useful is DEA? What purpose does a DEA study serve? (Newhouse 1994).

There are several purposes for conducting DEA studies. One is to develop better descriptions and analyses about practice patterns and styles. Insights about most optimal caseloads, effects of severity on scale inefficiency, or other sources of inefficiency are helpful to health care managers.

A second purpose for undertaking a health application with DEA is to identify best practices to create insights and new ideas that explain the successes and failures of clinical providers for policy makers. Examining why some provider firms succeed while others fail and identifying the sources of performance improvement remains an important societal problem, especially if the sources of failure are rooted in the payment and financing systems. Finally, once best practices are identified, health care mangers need help in finding ways to reduce waste in the utilization of clinical resources to achieve health goals. This strategic objective ranks high on many health managers' agenda.

While the DEA toolbox has a potential to serve these goals, the work has been more illustration than practicable and theory developing. The body of DEA research that has accumulated is substantial. Is there anyway to order and account for the panorama of information? Is there an acceptable general DEA model or subset of models for health and hospital studies? Have DEA applications improved the delivery of care for patients? At this time, the answer to all of these questions is "No." There appears to be neither a taxonomy nor a general single model capable of handling all of the issues that critics can raise. Most distressing is the fact that very few DEA applications in health have documented any significant improvements to quality, efficiency, or access.

Since the early 1990's, the pace of DEA research has been rapid. Has the information been mined? The problems inherent in deepening understanding of health care operations and performance today are more intricate than those encountered when DEA research began "exploring and testing" DEA applications to understand health care performance problems.

The main problem has been the lack of rigor in developing models. Simple measurement errors are less of a problem than selecting patently different input and output variables. To prevent the critics from challenging the models and to help advance policy and management, models should be based on good theoretical ideas. Otherwise, there is a stalemate in the theoretical development of health applications.

This chapter has uncovered research and managerial opportunities and challenges and noted that there is a great deal of work that remains to be done. Health care has a goldmine of clinical information and medical records that document cost, quality, and access. On the one hand, every single diagnosis and illness could be studied with DEA at the physician/patient level. On the other hand, in a single general hospital, there are the more than 5000 distinct products and services--no single DEA model will ever be able to analyze all of those activities.

In 1996, Seiford asked the experts to nominate "novel" DEA application. Four health care papers were included among the handful of innovative studies. Every one of these DEA studies focused on individual physicians as decision-making units. This remains a very promising area, since physicians are not only an important provider of care, they are the principle decision makers and entrepreneurs for the care programs and clinical DMUs.

A few years ago Scott (1979) proposed a multi-output-multi-input measure of clinical efficiency that divided the amount of improvement for each type of patient by the cost of each hospitalization (input). Measuring the change in severity of illness from admission to discharge would make a DEA efficiency measure clinically relevant by capturing a more complete picture of a physician's clinical performance relative to the resources used.

Finally, Farrell (1957) has argued that activity analysis can be used to evaluate the productive efficiency of various economic levels ranging from small workshops to an entire industry. Is DEA the right tool to motivate a general model of performance (quality, access, and productive efficiency) in health care? Clearly DEA is proving to be an effective approach to evaluate individual efficiency as well as organizational efficiency; consequently, DEA could be used to evaluate quality and efficiency at all four levels of the health are industry:

(1) The individual patient's experience,
(2) Individual physician level
(3) The department or organizational level
(4) The entire industry level

As information systems become more integrated and more available, and as DEA evolves from mere illustration to real health services research, future studies could attempt to connect individual patient changes in health status, to physician efficiency, to department efficiency, to overall hospital efficiency, and then connect hospital efficiency to the efficiency of the entire hospital industry.

REFERENCES

1. Anderson, G. F., U. E. Reinhardt, P. S. Hussey, and V. Petrosyan., 2003, It's the prices, stupid: why the United States is so different from other countries, Health Affairs, 22, No. 3, 89-105.
2. Banker, R. D., 1984, Estimating most productive scale size using data envelopment analysis, European Journal of Operational Research 17, 35-44.
3. Banker, R. D., R. Conrad, and R. Strauss, 1986, A comparative application of data envelopment analysis and translog methods: an iolustrative study of hospital production, Management Science, 32, No.1, 30-44.

4. Banker, R., A. Charnes and W.W. Cooper, 1984, Some models for estimating technical and scale inefficiencies in data envelopment analysis, Management Science 30, 1078-1092.
5. Banker, R.D. and R.C. Morey, 1986a, Efficiency analysis for exogenously fixed inputs and outputs, Operations Research 34, No. 4, 513-521.
6. Banker, R.D. and R.C. Morey, 1986b, The use of categorical variables in data envelopment analysis, Management Science 32, No. 12, 1613-1627.
7. Bristol Royal Infirmary Inquiry Final Report 2002, The Report of the public inquiry into the children's heart surgery at the Bristol Royal Infirmary, Learning from Bristol, 2002, Presented to Parliament by the Secretary of State for Health by command of her Majesty, Crown, London, England.
8. Bryce, C. L., J. B. Engberg, and D. R. Wholey, 2000, Comparing the agreement among alternative models in evaluating HMO efficiency, Health Services Research 35, 509-528.
9. Burgess, J. F., P. W. Wilson, 1996, Hospital ownership and technical efficiency, 42, No. 1, 110-123.
10. Burgess, Jr., James F., and Paul W. Wilson, 1998, Variation in inefficiency among U.S. hospitals, INFOR 36, 84-102.
11. Burns L. R., J. A. Chilingerian, and D. R. Wholey, 1994, The Effect of Physician Practice Organization on Efficient Utilization of Hospital Resources, Health Services Research, 29, No. 5, 583-603.
12. Caves, R. E., Industrial Efficiency in Six Nations, Cambridge: MIT Press, 1992.
13. Charnes A., W. W. Cooper, and A. Henderson, 1953, An introduction to linear programming, John Wiley & sons, New York.
14. Charnes, A., W.W. Cooper, and E. Rhodes, 1978, Measuring the efficiency of decision-making units, European Journal of Operations Research 3, 429-444.
15. Charnes, A. W.W. Cooper, and R. Thrall, 1991, A structure for classifying and characterizing efficiencies in data envelopment analysis, Journal of Productivity Analysis, 2, 197-237.
16. Charnes, A., W.W. Cooper, A.Y. Lewin, and L. M. Seiford, 1994, Basic DEA models, Data Envelopment Analysis: Theory, Methodology, and Application, Kluwer Academic Publishers, Boston, Massachusetts.
17. Charnes, A., W.W. Cooper, D.B. Sun, and Z.M. Huang, 1990, Polyhedral cone-ratio DEA models with an illustrative application to large commercial banks, Journal of econometrics 46, 73-91.
18. Chilingerian, J.A. 1989, Investigating non-medical factors associated with the technical efficiency of physicians in the provision of hospital services: a pilot study, Best Paper Proceedings, Annual Meeting of the Academy of Management, Washington, D.C

19. Chilingerian, J. A., 1995, Evaluating physician efficiency in hospitals: A multivariate analysis of best practices, European Journal of Operational Research 80, 548-574.
20. Chilingerian, J. A and Glavin, M. 1994, Temporary Firms in Community Hospitals: Elements of a Managerial Theory of Clinical Efficiency. Medical Care Research and Review, 51 No.3.
21. Chilingerian, J.A. and H.D. Sherman, 1990, Managing Physician Efficiency and Effectiveness in Providing Hospital Services, Health Services Management Research, Vol. 3, No. 1, 3-15.
22. Chilingerian, J. A., and H. David Sherman, 1996, Benchmarking physician practice patterns with DEA: A multi-stage approach for cost containment, Annals of Operations Research 67, 83-116.
23. Chilingerian, J. A., and H. David Sherman, 1997, DEA and primary care physician report cards: Deriving preferred practice cones from managed care service concepts and operating strategies, Annals of Operations Research 73, 35-66.
24. Chilingerian, J.A.. 2000, Evaluating quality outcomes against best practice: a new frontier. The Quality Imperative: Measurement and Management of Quality. Imperial College Press, London, England.
25. Chilingerian, J., A., M. Glavin, and S. Bhalotra, 2002, Using DEA to Profile Cardiac Surgeon Efficiency, Draft of Technical Report to AHRQ, Heller School, Brandeis University.
26. Cooper, W.W., K.S. Park, and J.T. Pastor, 2000, Marginal Rates and elasticities of substitution with additive models in DEA, Journal of Productivity Analysis, 13, 105-123.
27. Cooper, W.W., Seiford, L.M. and Tone, K., 2000, Data Envelopment Analysis: A Comprehensive Text with Models, Applications, References and DEA-Solver Software, Kluwer Academic Publishers, Boston.
28. Fair, R., 1978, A Theory of Extramarital Affairs, Journal of Political Economy, 86, 45-61.
29. Fare, R., S. Grosskopf, B. Lindgren, and Pontius Roos, 1994, Productivity developments in Swedish pharmacies: a malmquist output index approach, Data Envelopment Analysis: Theory, Methodology, and Applications, Kluwer Academic Publishers, Boston, Massachusetts.
30. Farrell, M.J., 1957, The measurement of productive efficiency, Journal of Royal Statistical Society A 120, 253-281.
31. Ferrier, G. D., and V. Valdanis, 1996, Rural hospital performance and its correlates, The Journal of Productivity Analysis, 7, 63-80.
32. Fetter, R.B., and J. Freeman, 1986, Diagnostic related groups: product line management with in hospitals, Academy of Management Review, XI, 1, 41-54.
33. Finkler, Merton D., and David D. Wirtschafter, 1993, Cost-effectiveness and data envelopment analysis, Health Care Management Review 18, no.3, 81-89

34. Fizel, J. L., and T.S. Nunnikhoven, Technical efficiency of nursing home chains, Applied Economics, 25, 49-55.
35. Fries, B., E., 1990, Comparing case-mix systems for nursing home payment, Health Care Financing Review. 11, 103-114.
36. Georgopoulos, B. S., 1986, Organizational Problem Solving Effectiveness: A Comparative Study of Hospital Emergency Services. Jossey-Bass Publishers, San Francisco.
37. Golany, B. and Roll, Y., 1989, An application procedure for DEA, Omega, 17, No. 3: 237-250.
38. Greene, W.H., 1983, Econometric Analysis, Second Edition, Macmillan Publishing Company, New York.
39. Harris, J.E. 1977, The internal organization of hospitals: some economic implications, The Bell Journal of Economics, VIII, 467-482.
40. Harris, J.E. 1979, Regulation and Internal Control In Hospitals, Bulletin. of the New York Academy of Medicine Vol. 55, No. 1, 88-103.
41. Hao, Steve (Horng-Shu), and C. Carl Pegels, 1994 Evaluating relative efficiencies of veteran's affairs medical centers using data envelopment, ratio, and multiple regression analysis, Journal of Medical Systems 18, 55-67.
42. Hollingsworth, B., P.J. Dawson, and N. Maniadakis, 1999, Efficiency measurement of health care: a review of non-parametric methods and applications, Health Care Management Science, 2, No. 3, 161-172.
43. Kerr, C., J. C. Glass, G. M. McCallion, and D. G. McKillop, 1999, Best-practice measures of resource utilization for hospitals: a useful complement in performance assessment, Public Administration, 77, No.3, 639-650.
44. Kooreman, P., 1994, Nursing home care in The Netherlands: a non[parametric efficiency analysis, Journal of Health Economics, 13, 301-316.
45. Lapetina, E. M., and E. M. Armstrong, Preventing errors in the outpatient setting, 2002, Health Affairs, 21, no. 4, 26-39.
46. Leibenstein, H. and S. Maital, 1992, Empirical estimation and partitioning of x-inefficiency: a data envelopment approach, American Journal of Economics, Vol. 82, No.2, 428-433.
47. Lewin, A.Y. and J. W. Minton. 1986, Determining organizational effectiveness: another look and an agenda for research, Management Science. 32, No. 5: 514-538.
48. Lynch, J., Y. Ozcan, 1994, Hospital Closure: AN efficiency analysis, Hospital and Health Services Administration, 39, No. 2, 205-212.
49. Luke, R. D., Y. A. Ozcan, 1995, Local markets and systems: hospital consolidations in metropolitan areas, Health Services Research, 30, No. 4, 555-576.
50. Maddala, G.S. 1983. Limited dependent and qualitative variables in econometrics, Cambridge University Press, New York.

51. Morey, D., and Y. Ozcan, 1995, Medical Care, 3, no.5, 531-552.
52. Newhouse, J. P., 1994, Frontier estimation: how useful a tool for health economics?, Journal of Health Economics, 13, 317-322.
53. Newhouse J., 2002, Why is there a quality chasm?, Health Affairs, 21, No. 4, 13-25.
54. Neuman, B.R., Suver, J.D., Zelman, W.N. 1988, Financial Management: Concepts and Applications for Health Care Providers. National Health Publishing, Maryland.
55. Nunamaker, T., 1983, Measuring routine nursing service efficiency: A comparison of cost per day and data envelopment analysis models, Health Services Research, XVIII (2) Part 1, 183-205.
56. Nyman, J.A., and D.L. Bricker, 1989, Profit incentives and technical efficiency in the production of nursing home care, The Review of Economics, and Statistics, 56, 586-594.
57. Nyman J.A., D. L. Bricker, D. Link, 1990, Technical efficiency in nursing homes, Medical Care, 28, 541-551.
58. Ozcan, Y., 1998, Physician benchmarking: measuring variation in practice behavior in treatment of otitis media, Health Care Management Science, 1, 5-18.
59. Ozcan, Y., J. Lynch, 1992, Rural hospital closures: an inquiry into efficiency, Advances in Health Economics and Health Services Research, 13, 205-224.
60. Ozcan, Yasar A., James W. Begun, and Martha M. McKinney, 1999, Benchmarking organ procurement organizations: A national study, Health Services Research 34, 855.
61. Ozcan, Yasar A., and Roice D. Luke, 1992, A national study of the efficiency of hospitals in urban markets, Health Services Research 27, 719-739.
62. Ozgen, H., and Y. A. Ozcan, 2002, A national study of efficiency for dialysis centers: an examination of market competition and facility characteristics for production of multiple dialysis outputs, Health Services Research 37, 711-722.
63. Perez, E.D., 1992, Regional variations in VAMC's operating efficiency, Journal of Medical Systems, 16, No. 5, 207-213.
64. Puig-Junoy, J.., 1998, Technical efficiency in the clinical management of critically ill patients, Health Economics, 7, 263-277.
65. Rosko, Michael D., 1990, Measuring technical efficiency in health care organizations, Journal of Medical Systems 14, no. 5, 307-322.
66. Rosko, Michael D., Jon A. Chilingerian, Jacqueline S. Zinn, and William E. Aaronson, 1995, The effects of ownership, operating environment, and strategic choices on nursing home efficiency, Medical Care 33, no. 10, 1001-1021.
67. Scott, R. W., 1979, Measuring outputs in hospitals, Measuring and Interpreting Productivity, National Research Council, 255-275.

68. Seiford, L., 1996, Data envelopment analysis: the satt6e of the art, Journal of Productivity analysis, 7, NO.2/3: 99-138.

69. Sexton, T., A. Leiken, A. Nolan, S. Liss, A. Hogan, R. Silkman, 1989, Evaluating managerial efficiency of veterans administration medical centers using data envelopment analysis, Medical Care, 27, No.12, 1175-1188.

70. Sherman, H.D., Hospital efficiency measurement and evaluation, Medical Care, 22, No. 10, 922-928

71. Shortell, S., A. Kaluzny., 2000, Health Care Management, 4th edition, Delmar Publishers Inc, Albany, New York.

72. Teboul, J. 2002, Le temps des services: une nouvelle approche de management, Editions d'Organisation, Paris, France.

73. Witte, A., 1980, Estimating an Economic Model of Crime with Individual Data, Quarterly Journal of Economics, 94, 57-84.

74. Zhu, Joe, 2000, Further discussion on linear production functions and DEA, European Journal of Operational Research 127, 611-618.

75. Zuckerman, S., J. Hadley, L. Iezzoni, 1994, Measuring hospital efficiency with frontier cost functions, Journal of Health Economics, 13, 255-280.

Acknowledgements

We are very grateful to Dianne Chilingerian and W.W. Cooper for their encouragement and thoughtful comments. Part of the material for this chapter was adapted from: Chilingerian, J. A., 1995, Evaluating physician efficiency in hospitals: A multivariate analysis of best practices, European Journal of Operational Research 80, 548-574; Chilingerian, J.A. 2000, Evaluating quality outcomes against best practice: a new frontier. The Quality Imperative: Measurement and Management of Quality. Imperial College Press, London, England; and Chilingerian, J. A., and H. David Sherman, 1997, DEA and primary care physician report cards: Deriving preferred practice cones from managed care service concepts and operating strategies, Annals of Operations Research 73, 35-66.

Chapter 18

DEA SOFTWARE TOOLS AND TECHNOLOGY
A State-of-the-Art Survey

Richard S. Barr

Department of Engineering Management, Information, and Systems; Southern Methodist University, Dallas, TX 75275 USA   barr@engr.smu.edu

Abstract: Today's DEA practitioners and researchers have a wide range of solution technology choices. Clearly, software is no longer an impediment to the incorporation of DEA into decision-support systems and benchmarking processes. This is evidenced by the availability of interoperable tools with a variety of user interfaces, advanced modeling options, and the power to evaluate large-scale data sets on inexpensive computing platforms. This survey describes and critiques prominent DEA software packages, both commercial and non-commercial, and summarizes the current state of the art.

Key words: Data envelopment analysis (DEA); Software; Optimization

1.   INTRODUCTION

As the field of Data Envelopment Analysis has grown and blossomed, so have the varieties of models, data, and types of analyses. Similarly, as DEA software technology has emerged from its academic roots into production usage, it has been accompanied by expectations of advanced modeling options and professional implementations, including graphical user interfaces, interoperability with other applications, and the ability to quickly evaluate large populations.

This chapter surveys the best of the commercial and non-commercial DEA tools available today. Provided are descriptions of individual packages, comparisons of their features and capabilities, and links to further information on each.

While close to 20 software options were explored, eight packages emerged as viable alternatives for general DEA application.[1] Half of the eight are commercial packages and half are distributed at no charge (some with use restrictions). Most were developed for Windows platforms. Each excelled on one or more dimensions, and trailed in at least one other, i.e. were technically efficient without assurance region constraints.

The packages provided a wide range of available models, features and capabilities, user interfaces, reporting options, model solution speeds, and acquisition costs. The categories below describe the commercial and non-commercial packages individually, and evaluate each one in terms of model selection, DEA features, user interface, interoperability, reporting capabilities, documentation, test performance, and affordability.

A summary table of key elements invites the reader's own comparisons of the various software options for his or her own needs. Further information is available from an accompanying website, including links to software sources.

2.   EVALUATION CRITERIA

To compare the various software options, over 70 criteria were identified, and organized into eight categories, A-H. These criteria and categories are listed in Figures 18-1 and 18-2, along with values for each DEA package. The Categories below describe these evaluation criteria in more detail.

2.1   Available Models (Category A)

The software packages vary widely in their selection of available DEA models. The classics—CCR/CRS and BCC/VRS—are universally included, but others vary by package. Evaluation Criteria Category A contains a non-exhaustive list of DEA models and indicates those that are included with each code. (See other chapter 1 for model formulations and descriptions.)

---

[1] The others, listed in the Appendix (Section 6), were either no longer available, not designed specifically for DEA models, proprietary applications, suitable only for research, or required other specialized software to operate.

	COMMERCIAL				NON-COMMERCIAL			
	DEA Solver Pro	Frontier Analyst	OnFront	Warwick DEA	DEA Excel Solver	DEAP	EMS	Pioneer
Version	4.0	3.1.5	2.02	1.0	1.0	2.1	1.3.0	2.0
From	SAITECH	Banxia Software	EMQC	Warwick Univ.	Zhu	Colletti	Scheel	Barr, McLoud
A. Models								
1 CCR/CRS	●	●	●	●	●	●	●	●
2 BCC/VRS	●	●	●	●	●	●	●	●
3 NIRS,NDRS,GRS	●	·	●	·	●	·	●	●
4 Additive/slack-based method	●	·	·	●	●	·	●	·
5 Malmquist	●	●	●	·	●	●	●	·
6 Non-convex	●	·	·	·	●	·	●	·
7 Non-radial	●	·	·	·	●	·	·	·
8 Preference-structure	·	·	·	·	●	·	·	·
9 Undesirable-measure	·	·	·	·	●	·	·	·
10 Context-dependent	·	·	·	·	●	·	·	·
11 Free-disposal hull (FDH)	●	·	·	·	●	·	●	●
12 Cost efficiency	●	·	●	·	●	●	·	·
13 Revenue efficiency	●	·	●	·	●	·	·	·
14 Profit, revenue/cost efficiency	●	·	·	·	●	·	·	·
15 Target, mixed improvement	●	·	·	●	·	·	·	·
16 Capacity utilization	·	·	●	·	●	·	·	·
17 Variable-benchmark	·	·	·	·	●	·	·	·
18 Fixed-benchmark	·	·	·	·	●	·	·	·
19 Minimum-efficiency	·	·	·	·	●	·	·	·
20 Value chain	·	·	·	·	●	·	·	·
21 Weak disposability	·	·	●	·	●	·	·	·
22 New cost, revenue, profit	●	·	·	·	·	·	·	·
23 Congestion	●	·	·	·	●	·	·	·
24 Scale elasticity	●	·	·	·	·	·	·	·
B. Features								
1 Orientation (i/o) control	●	●	●	●	●	●	●	●
2 Window/multi-period analysis	●	●	●	·	·	·	●	·
3 Weight constraints	Conical, UB, LB	UB,LB	·	Conical CRS	·	·	Conical	Conical
4 Super-efficiency scores	●	·	·	●	●	·	●	·
5 Non-discretionary/fixed factors	●	●	●	●	●	·	●	·
6 Categorical variables	●	·	·	·	·	·	·	·
7 Variable priorities	·	·	·	●	·	·	·	·
8 Sensitivity analysis	·	·	·	·	●	·	·	·
9 Multi-phase/multi-step	·	·	·	●	·	·	·	·
10 Nested frontiers (tiers, layers)	·	·	·	·	●	·	·	●
11 Disposability controls	·	·	●	·	●	·	·	·
12 Scenario comparison	·	·	●	·	·	·	·	·
13 Efficiency components analysis	·	·	●	·	·	●	·	·
14 Zero substitution	·	●	·	·	·	·	·	·
15 Benchmarking comparisons	●	·	●	·	●	·	·	●

● Included/available   · Not included/not available   ○ Limited capability

Figure 18-1. Evaluation Criteria, Categories A and B

2.2   Key DEA Features and Capabilities (Category B)

There are variants of the models listed in Category A—such as input- or output-orientation—listed in Category B as DEA features.[2] All codes allow selection of input or output orientation (and possibly non-oriented). Some provide a window analysis or other multi-period evaluation capability (B2), super-efficiency scores[3] (B3), non-discretionary or categorical factors (B5, B6), priorities on variables (B7), sensitivity analysis (B8), and multi-phase or multi-step analysis (B9).

Criteria B3 indicates the types of user-specified constraints—if any—that can be imposed on the input and output factor weights. Such assurance-region or cone-ratio-envelopment restrictions can take the form of upper and lower bounds on individual factor weights (UB, LB) or the more general conical form (Conical): $\sum_r a_r u_r + \sum_i b_i v_i \leq 0$, where u_r and v_i are variables in an input-oriented DEA multiplier model, and a_r and b_i are arbitrary real numbers.

Tiered DEA[4] involves recursively uncovering a nested hierarchy of efficient frontiers, or layers (B10). Controls on disposability (e.g., weak, strong) may be included (B11), feature B12 allows accumulation of results from multiple scenarios, and B13 reports on each DMU's technical efficiency components. The zero substitution option (B14) permits users to specify a constant to replace any zero data values, and B15 lets users define baseline/reference DMUs for benchmarking analyses

2.3   Platform and Interoperability (Category C)

Evaluation criteria C1 indicates the operating system(s) supported: Microsoft Windows (▦), PC DOS (▮),[5] Unix (▱), and Microsoft Excel for Windows (☒). The acceptable forms of data input and results output (C2, C3) are indicated for each code as: a Microsoft Excel spreadsheet (SS), a text file (TXT), the clipboard (Clip), the current selection in an Excel worksheet (SS select), an SPSS-formatted file (SPSS), direct manual entry (manual), Adobe's Portable Document Format (PDF), and the HyperText Markup Language (HTML).

---

[2] Some authors refer to such features as separate DEA models.
[3] As documented in Andersen and Petersen (1993)
[4] Per Barr, Durchholz, and Seiford (1994)
[5] Note that DOS programs can be run in a Command Prompt window under MS Windows.

	COMMERCIAL				NON-COMMERCIAL			
	DEA Solver Pro	Frontier Analyst	OnFront	Warwick DEA	DEA Excel Solver	DEAP	EMS	Pioneer
Version	4.0	3.1.5	2.0	1.0	1.0	2.1	1.3.0	2.0
From	SAITECH	Banxia Software	EMQC	Warwick Univ.	Zhu	Colletti	Scheel	Barr, McLoud
C. Platform and Interoperability								
1 Platform(s)	▦⊠	▦	▦	▦	▦⊠	▣	▦	▣▦
2 Input file types: ⊠=Excel, TXT=text file, Clip=clipboard, select=current SS selection, SPSS=SPSS data file, manual=direct manual entry	SS, manual	TXT,Clip, SS, SS select, SPSS, manual	TXT, Clip, manual	TXT	SS, manual	TXT	TXT, SS	TXT
3 Output file types	SS	SS, Clip, TXT, PDF, HTML, ⊠	TXT	TXT	SS	TXT	TXT, Clip	TXT
D. User Interface								
1 GUI	⊠	●	●	●	⊠	·	●	·
2 Spreadsheet format	⊠	●	●	●	⊠	·	·	·
3 Interactive data manip	⊠	●	●	●	⊠	·	·	·
4 Individual observation editing	⊠	●	●	●	⊠	·	·	·
5 DMU, i/o factor subset control	⊠	●	●	●	⊠	·	O	·
6 Data filters	·	●	·	·	·	·	·	·
7 Data and results sorting	⊠	●	●	O	⊠	·	·	·
E. Reporting								
1 # standard reports	12	5	9	5	3	8	2	3
2 Custom reports	⊠	⊙	·	O	⊠	·	·	·
3 # std. graphs and charts	4	12	·	·	·	·	·	·
4 Custom charts	⊠	O	●	·	⊠	·	·	·
5 Efficiency scores report	●	●	●	●	●	●	●	●
6 Projected/target factors report	●	●	·	●	●	●	·	●
7 Optimal factor weights report	●	●	●	●	●	●	·	●
8 Efficient reference sets report	●	●	●	●	●	●	●	●
9 Slacks report	●	·	●	●	●	●	●	●
10 RTS analysis	●	·	●	●	●	·	·	·
11 Cross-efficiencies	·	⊙	·	●	●	·	·	·
F. Documentation and Support								
1 Tutorial (pages)	·	26	·	·	·	·	·	·
2 Users guide (pages)	·	99	38	19	43	49	12	2
3 Reference manual (pages)	43	·	52	⊙	▢	·	·	·
4 Built-in help	·	●	●	·	·	·	·	·
5 Technical support	●	●	·	●	·	·	·	·
6 Web site tech info	·	●	·	·	●	·	·	·
G. Testing								
1 Time 431 dmu, CCR̄i (mm:ss)	0:17	0:05	0:06	0:53	2:30	<0:01	0:11	<0:01
2 Time 431 dmu, BCC̄i, superE	0:16	·	·	18:45	4:40	·	0:12	<0:01
H. Availability								
1 Free demo available	▢	●	●	●	·	●	●	●
2 Comm'l license cost	$1,600	£395 – £2395	$1,750	£200 – £500	▢	$0	·	·
3 Academic cost	$800	£195 – £595	$750	£800+ (site)	▢	$0	$0	$0
4 Maintenance available	·	●	●	●	·	·	·	·

▦ Microsoft Windows ▣ DOS ● Included/available ⊙ Optional
⊠ Uses Microsoft Excel as a GUI ▦ Unix · Not included/not available
▢ Included with book purchase O Limited capability

Figure 18-2. Evaluation Criteria, Categories C-H

2.4   User Interface (Category D)

User interfaces can range from elaborate graphical user interfaces (GUIs) to simple command-line controls. Two of these DEA packages use Microsoft Excel as a GUI for data entry, reporting, and graphics and its Solver tool as a linear program optimizer. The advantages of Excel-based products are: input and output are simplified, reports and graphs can be easily added by the developer and user, interoperability with other Microsoft products is straightforward, and connection to enterprise data sources via ODBC is possible. The disadvantages of the Excel environment are: the default Excel Solver limits application to small populations, larger instances require purchase of a replacement optimizer, and the interpreted execution speed is typically slower than compiled, stand-alone applications.

Category D criteria include: availability of a GUI (D1), ability to display data in spreadsheet or matrix format (D2) and ability to edit datasets and individual observations within the application (D3, D4). D5 indicates whether a DEA can be executed for user-selected subsets of the DMUs and the input/output factors and whether such subsets can be based on a user-defined rule or data filter (D6). D7 is flagged when a user can interactively sort the population data or the results of an analysis

2.5   Reporting (Category E)

The reporting criteria in Category E include the number of standard reports included (which may be tabs in a worksheet or separate tables within a longer report (E1), whether the user can create ad hoc customized reports (E2), and similar information about graphs and charts (E3, E4). Criteria E5-E11 indicate whether specific reports are available from the application.

2.6   Documentation and Support (Category F)

Category F reports on the length, in pages, of a hands-on software tutorial for new users (F1), a users guide to the operation of the software (F2), and a technical reference manual with modeling details (F3). Also indicated is the availability of a built-in help facility (F4), technical support by phone or email (F5) and technical support information on the provider's web site (F6).

2.7 Test Performance (Category G)

Each of the packages was installed and tested on a Dell Precision Workstation 340, 1.8 GHz Pentium 4, with 512MB RAM running under Windows 2000. The standard Excel solver was replaced with the Premium Solver Platform V5.0 from Frontline Systems.[6]

The Unix version of one code was also tested on a DEC Personal Workstation 600 AU with a 600 MHz Alpha processor and 576 MB RAM, running OSF1 (Unix) and using the cxx compiler.

To evaluate the software's operating speed, the following problems were solved and the run times reported in minutes and seconds in Category G:

1. 1991 Quarter 4, U.S. Bank data set, with six inputs, three outputs, and 431 DMUs; model: CCR-I (G1)[7] and
2. The same dataset processed with BCC-I and super-efficiency scores (G2).

This dataset is available from the web site listed at the end of this chapter. Since processing effort is affected by report generation, run times were collected with minimal reporting options selected.

2.8 Availability (Category H)

Category H reports on the availability of a demonstration version (H1), the cost range for a single-user commercial license as of this writing (H2), and similar cost information for academic users (H3). Also indicated is whether annual software maintenance contracts are available (H4).

3. COMMERCIAL DEA SOFTWARE

The commercial packages explored are: DEA-Solver-Pro 4.0 from SAITECH, Frontier Analyst[®8] 3.1.5 from Banxia Software, OnFront 2.02 from Economic Measurement and Quality Corporation, and Warwick DEA from Warwick University. The evaluation criteria values for each are given in the first four columns of Figures 18-1 and 18-2. The individual product sections below detail the system requirements, known developers, distributor, current (2003) single-user licensing costs, and points of contact

---

[6] Cost: $1295, August 2003, details at: http://solver.com.
[7] See Barr et al (2001) for model details.
[8] A registered trademark of Banxia Software

(✉ mailing address, ☎ phone numbers, 🏯 web site URL, and ✉ email address.)

3.1   DEA-Solver-Pro

- *Software title:* DEA-Solver-Pro,Version 4.0
- *System requirements*: Microsoft Windows, Microsoft Excel 97 or newer
- *Available from:* SAITECH, Inc; demonstration version included with Cooper, Seiford, and Tone (2000)
- *Licensing:* $1600 (190,000¥) plus shipping and taxes; $800 (100,000¥) academic
- *Contact:*
    ✉ SAITECH, Inc., 1 Bethany Road, Hazlet, NJ 07330
    ☎ (732) 264-4700, fax (732) 264-1538
    🏯 www.saitech-inc.com
    ✉ dea@saitech-inc.com

DEA-Solver-Pro from SAITECH, is an application that runs within Microsoft Excel. Population data is input via a worksheet, arranged in a specific format, and results are generated in tab sheets of a separate workbook. The process is guided by an onscreen wizard (initiated when the application spreadsheet is loaded), which prompts for file names and processing options. Hence the operation is straightforward.

The input data for most models must be in a single Excel worksheet, with variable labels in the first row and individual DMU values in subsequent rows, each with a label in the first column. Input variables' labels must begin with "(I)" and output variables are indicated by a label beginning with "(O)." Variants of this format are used to specify assurance region constraints, multi-period values, and new-cost/revenue/profit data.

DEA-Solver-Pro provides an enormous compendium of models, including some not yet available in the literature. The documentation lists "118 models in 31 clusters" (e.g., models CCR-I and CCR-O make up the CCR cluster). The feature set is strong, including window analysis, conical weight constraints, non-discretionary variables, and three types of categorical variables.

A dozen different reports can be generated, along with four standard graphs and charts (efficiency score histogram, plus time-windows-statistics, and Malmquist-index plots). See Figure 18-3 for sample reports. The Excel environment permits the sorting of result tables and the creation of ad hoc, custom reports and graphs.

The indexed 43-page manual, SAITECH (2003), is an updated and expanded version of the appendix to Cooper, Seiford, and Tone (1999). It focuses on documenting the various mathematical models and output values, rather than program operation.

m91-q4dspOut.xls

No.	DMU	Score	(I)FTEmp Projection	Change(%)	(I)SalaryE Projection	Change(%)
1	DMU44	0.852277	4.39672	-26.72%	132.1029	-14.77%
2	DMU62	0.622433	1643.223	-37.76%	46216.04	-58.23%
3	DMU98	0.753498	18.08395	-24.65%	534.0933	-27.33%
4	DMU126	0.635068	170.1982	-36.49%	5374.147	-45.33%
5	DMU181	1	3	0.00%	108	0.00%
6	DMU226	0.764121	26.74423	-23.59%	717.3329	-55.36%
7	DMU239	1	20	0.00%	485	0.00%
8	DMU245	0.800998	36.84591	-19.90%	970.0086	-19.90%
9	DMU275	0.848541	4.242705	-15.15%	183.0172	-69.95%
10	DMU366	0.705981	7.059808	-29.40%	168.5027	-30.66%
11	DMU388	0.952115	15.4429	-22.79%	501.1567	-8.88%
12	DMU408	0.825807	97.44517	-17.42%	1987.045	-56.70%
13	DMU416	0.680907	10.89451	-31.91%	203.3032	-41.41%
14	DMU423	0.841117	24.64847	-35.14%	591.3051	-15.89%
15	DMU462	1	70	0.00%	2412	0.00%
16	DMU475	0.62688	8.498668	-39.30%	206.8704	-37.31%
17	DMU562	0.712984	6.824597	-47.50%	187.5148	-28.70%
18	DMU623	0.843865	6.148671	-23.14%	173.8362	-51.61%
19	DMU794	0.720073	42.48428	-27.99%	792.0799	-27.99%
20	DMU813	0.72739	1125.959	-27.26%	33455.52	-46.87%
21	DMU821	0.781417	5.46992	-21.86%	158.9243	-45.20%
22	DMU864	0.842377	46.33075	-15.76%	1399.541	-27.67%
23	DMU876	0.839911	484.6287	-16.01%	13007.7	-16.01%
24	DMU898	0.834047	70.89396	-16.60%	2310.309	-16.60%
25	DMU1237	0.9479	12.32271	-5.21%	363.6987	-6.98%
26	DMU1244	0.99609	42.83188	-0.39%	1200.549	-10.14%
27	DMU1248	0.788675	60.72797	-21.13%	1622.843	-28.26%
28	DMU1276	0.944286	24.61677	-23.07%	802.642	-5.57%
29	DMU1301	0.878824	3.555703	-28.89%	114.6771	-23.04%
30	DMU1337	0.869699	29.46999	-20.35%	719.2415	-13.03%
31	DMU1445	0.866579	35.91965	-38.07%	981.8343	-13.34%
32	DMU1449	0.948363	19.91562	-5.16%	602.2106	-5.16%
33	DMU1478	0.948802	5.784095	-27.70%	189.3345	-27.46%

Model Name = DEA-Solver Pro4 0/ CCR(CCR-I) Returns to Scale = Constant
Workbook Name = C \deasurvey\DEAsolver\m91-q4dspOut.xls

m91-q4dat1bcciOut.xls

Returns to Scale = Variable (Sum of Lambda = 1)

Statistics on Input/Output Data

	FTE	SalyExp	PrFxAsts	NoIntExp	IntExp	PurchFund Cc
Max	12031	433236	380856	513160	1694392	6247887 1
Min	0	0	0	0	0	0
Average	119.7216	3946.94	4160.246	5804.833	13203.97	51418.98
SD	745.4171	25047.82	25859.72	33738.86	93210.39	432386.8

Correlation

	FTE	SalyExp	PrFxAsts	NoIntExp	IntExp	PurchFund Cc
FTE	1	0.927227	0.959947	0.958553	0.899637	0.957599 0
SalyExp	0.927227	1	0.973461	0.979932	0.988924	0.918896 0
PrFxAsts	0.959947	0.973461	1	0.981587	0.944721	0.917883 0
NoIntExp	0.958553	0.979932	0.981587	1	0.965323	0.957134 0
IntExp	0.899637	0.988924	0.944721	0.965323	1	0.925865 0
PurchFund	0.957599	0.918896	0.917883	0.957134	0.925865	1
CoreDep	0.914769	0.988015	0.946619	0.956427	0.993031	0.92037
EarnAssts	0.976586	0.964961	0.957434	0.973186	0.963266	0.976911 0
IntIncome	0.946052	0.981212	0.960003	0.987704	0.986208	0.970509 0

DMUs with inappropriate Data with respect to the chosen Model

No.	DMU
None	

No. of DMU	431
Average	0.656381
SD	0.203507
Maximum	1
Minimum	0.231067

Frequency in Reference Set

Reference	Frequency to other DMUs
DMU62	0
DMU181	5
DMU239	2
DMU275	0

Figure 18-3. DEA-Solver Pro

As expected, DEA-Solver-Pro's solution time for the 431-DMU model was slower than most of the compiled programs (17 seconds vs. one to five seconds), but certainly fast enough for all but the largest of instances. Note that the standard Excel solver is limited to small problems, and the Premium Solver Platform V5.0 from Frontline Systems was used in this testing.

The pricing is mid-to-upper range for the commercial packages, but its internal optimizer eliminates the need to purchase an Excel Solver replacement. A size-limited demonstration version accompanies Cooper, Seiford, and Tone (2000), so a test drive is readily available.

With its strong technical capabilities and full compliment of features, DEA-Solver-Pro should appeal to DEA researchers, sophisticated program managers, and decision-support-system developers. Documentation geared to non-technical users would enhance its appeal to an even wider audience.

3.2   Frontier Analyst®

- *Software title:* Frontier Analyst, Version 3.1.5
- *System requirements*: Microsoft Windows 95/98/2000/NT/XP
- *Available from:* Banxia Software Ltd.
- *Licensing:* £395-£2395 (£195-£595 academic) plus shipping and taxes; "Plus Packs" for publish-to-web and cross-efficiencies are an additional £95-£195 each; annual maintenance £229.
- *Contact:*
    ⊠ Banxia Software Ltd., P.O. Box 134, Kendal, LA9 4TP, UK
    ☎ +44 (0) 870 787 2994, fax +44 870 787 2995
    🏠 www.banxia.com
    ✉ info@banxia.com

Frontier Analyst from Banxia Software is a stand-alone Windows application, with the most professional user interface and documentation of those packages evaluated. Analyses are organized as projects, with population data accessible from a wide range of sources, including text, Excel, and SPSS files, the Windows clipboard, the current Excel selection, and direct entry. Once input, DMU data is displayed as a matrix where individual DMUs and factors can easily be included, excluded, deleted, added, or edited. Each factor is interactively categorized as output, input, or uncontrolled input (non-discretionary); population DMUs can be screened by filtering rules to form subsets for analysis. Bounds on individual factor weights can be specified.

The interface is intuitive, provides strong user control over the data and reports, and has exceptional drill-down analytical tools at the DMU level. For example, a given inefficient unit can have its factors charted against those of its efficient-reference-set counterparts, and a series of other charts and graphs displayed to help identify the sources of inefficiency and promising areas of improvement. Language localization enables customization of the terminology displayed in the user interface.

A large number of useful and attractive graphs and tables provide a wide variety of analytical viewpoints for the results and make analysis easy and even fun. Graphs of the efficiency-scores distribution, reference frequencies for the efficient units, efficiency and x-y plots for all factors, four DMU-level exploratory graphs, and a frontier plot for small instances quickly provide new insights into the population's dynamics. All graphics can be transferred via the clipboard to other applications.

Frontier Analyst generates a set of standard reports: DMU efficiency scores, actual versus target factors by DMU (with and without charts), and complete details by unit; each report can be sorted four ways and can include efficient, inefficient, or all units (see Figure 18-4 for an example report). Extra-cost report options are: a custom report designer, publish-to-web HTML generator, and a cross-efficiency analysis feature, which recalculates each DMU's efficiency score using other DMUs' optimal weights, for additional peer and efficiency analysis.

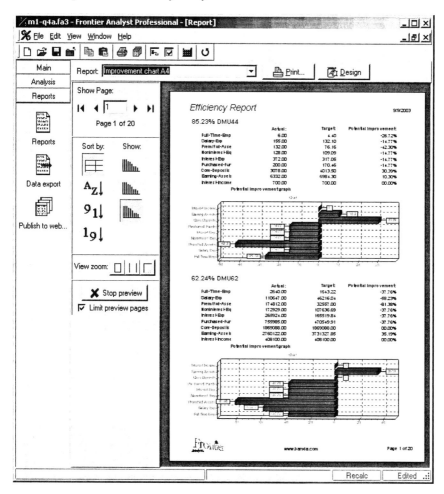

Figure 18-4. Frontier Analyst

The indexed users guide, Banxia (2003), is excellent: well-written and user-oriented, with clear explanations and examples of each program feature, accompanied by screen shots. A separate tutorial guide, Hussain and Jones (2001), is designed to help new users become productive quickly. The two optional modules are accompanied by their own documentation [Banxia (2001a, 2001b)].

The program's built-in optimizer is fast, requiring only five seconds to evaluate the 431-DMU test problem. Only the DOS applications were faster.

The only disappointing aspect of Frontier Analyst is its limited selection of DEA models and a few missing features. It has the classic CCR (CRS) and BCC (VRS) models, performs window analysis, computes Malmquist indices, and supports non-discretionary variables and a limited form of weight constraints. These by themselves are sufficient to handle a majority of business and not-for-profit applications. The developers have created such an attractive analytical environment that it seems unfortunate that some of the newer techniques (see Categories A and B) are not included. From Pareto and marketing aspects, this may be the right product for them (new model types do not help sales as much as flashy analysis graphics). It just could be so much better with a modest effort on the modeling side.

The cost of Frontier Analyst varies with the maximum number of DMUs to be considered. The lowest price is for 75 DMUs, the highest posted price is for 2500 units. An unlimited version is available at an unspecified price. While it is one of the more expensive tools available, it has many features that the others do not.

With its exceptional user interface, wide-ranging interoperability options, handsome and insightful graphs and charts, and excellent documentation, this package should appeal to managers and other practitioners for putting classic DEA to work for program analysis, benchmarking, process improvement, etc. While the limited repertoire of models and lack of some important supporting features—such as categorical variables, super-efficiency scores, and conical weight constraints—limits its range of application and audience appeal, Banxia has created an impressive DEA application.[9]

3.3   OnFront

- *Software title:* OnFront, Version 2.02
- *System requirements:* Microsoft Windows 95/98/2000/NT, 16 MB memory, 8MB disk space
- *Developers:* Rolf Färe and Shawna Grosskopf
- *Available from:* Economic Measurement and Quality Corporation
- *Licensing:* $1750 ($750 academic), plus shipping and taxes

---

[9] For another evaluation of Frontier Analyst, see Nyhan (1998), available at www.informs.org

- *Contact:*
  ✉ EMQ AB, Box 2134, SE-220 02 Lund, Sweden;
  ☎ +46 40 369 565, fax +46(0)40 164 511
  🏠 www.emq.com
  ✉ emq@emq.com

OnFront 2.02, from the Economic Measurement and Quality Corporation (EMQC), is a fast and elegant Windows DEA package. Its operation is straightforward: load population data from a text file or the clipboard into a data table, click the table's rows and columns to select factors and DMUs for evaluation, choose the desired model and output options, request optimization, and view the resulting tables. All results can be saved in Excel-readable text files. All tables can be sorted, based on any column; any table's columns can be selected and graphed within the program. See Figure 18-5 for an example.

While nine different tables can be generated, the "reporting" capability consists of printing out these spreadsheets as is, with no special formatting. Users will likely prefer to transfer these values to another application with report-generation features.

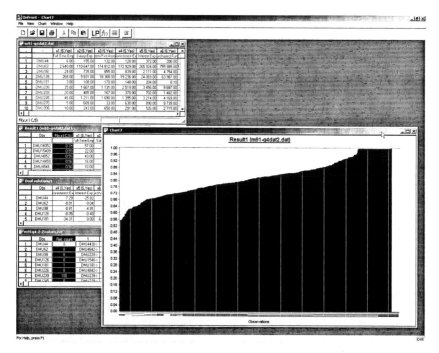

Figure 18-5. OnFront Example Windows

OnFront provides a good selection of models and features, including window analysis and congestion models, although it is missing any form of weight constraints for managerial applications. Users will need to learn the authors' terminology and abbreviations to choose the desired model. For example, for an input-oriented CCR analysis, one selects the *F(y,x)* efficiency and *CRS* technology-scale options. To designate an input variable as non-discretionary, uncheck the "Subvector Include" box.

Fortunately, the models and features are nicely documented in the accompanying reference guide, Färe and Grosskopf (2000). The attractively produced users guide, EMQC (2000), is clearly and succinctly written. The built-in help consists of having these two PDF documents accessible from the Help menu.

Solution times with the built-in optimizer were quick, taking only six seconds for our 431-DMU test-problem, as well as subsequent runs for different models and orientations. A nice feature is OnFront's ability to accumulate different efficiency scores for the same population in the same table, allowing comparisons of alternative models, features, or DMU scenario values. The rapid solution times permit extensive use of such interactive features.

While other packages seem oriented to managerial application, OnFront's emphasis appears to be economists, researchers, and DEA-savvy users. Its clean, minimalist user interface gets the job done quickly, once the terminology and notation are mastered. While some users may prefer a more full-featured interface, others will find OnFront's elegant approach refreshing.

3.4   Warwick DEA

- *Software title:* Warwick DEA
- *System requirements*: Microsoft Windows, DOS
- *Developers:* Emmanuel Thanassoulis, Keith Halstead, Mike Stelliaros, Robert Dyson, A. Athanassopoulos, A. Emrouznejad
- *Available from:* University of Warwick; demonstration version included with Thanassoulis (2001)
- *Licensing:* £200-£500, site license £800-£1600 (£400-£800 academic) plus shipping and taxes
- *Contact:*
  ✉ Emmanuel Thanassoulis, Aston Business School, University of Aston, Birmingham B4 7ET, UK
  ☎ +44 (0) 121 359 3611 Ext 5033, fax +44 (0) 121 359 5271

www.deazone.com
dea.et@btinternet.com

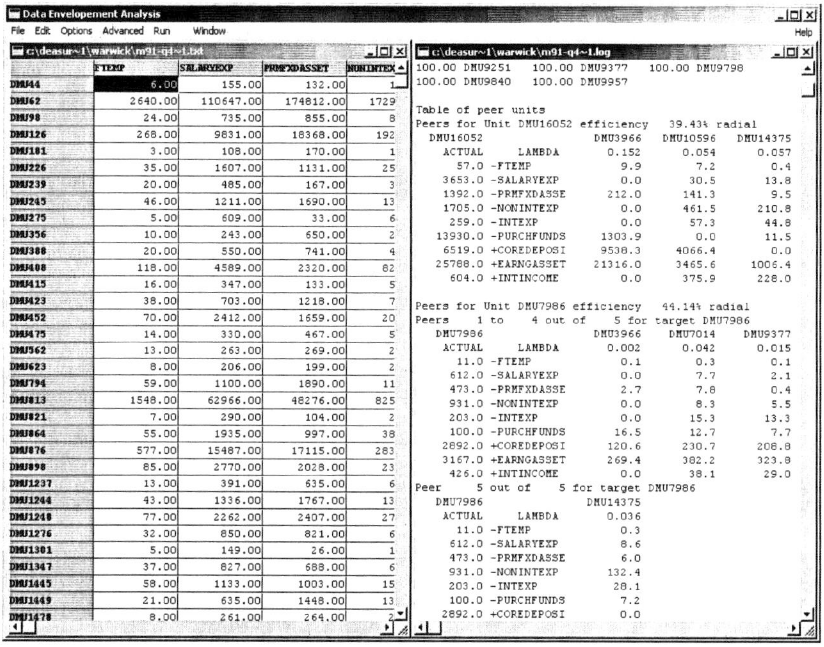

Figure 18-6. Warwick DEA

The Warwick DEA package, developed at the University of Warwick, has a straightforward graphical user interface consisting of two windows: a data table and output ("log file") text display. Population data is read from text files (prepared in a specific format) and displayed in the table. Two dialog windows enable selection of DMU subsets for analysis and designation of input and output variables; the latter can be avoided through variable naming conventions. Model types, orientations, variable priorities, and other options are selected via the menu bar and on the execution dialog box prior to performing the analysis.

The model selection is limited, but covers the most important cases. The feature set includes priorities for variables, super-efficiency scores, and conical weight constraints (for CCR models only).

The selected text reports are generated, displayed on-screen, and saved on disk. The reporting options include: (1) a list of DMUs, sorted by efficiency score; (2) a peer report giving—for each inefficient unit—the optimal λ-weights, inputs, and outputs for all efficient-reference-set members; (3) a report of each unit's optimal μ, ν multipliers; and (4) each unit's projected inputs and outputs, with associated percentage deviations from current values. See examples in Figure 18-6. Report customization consists of DMU sorting options, partial reports based on efficiency ranges and DMU name, and peer-scaling choices. Results can be exported in text files formatted for spreadsheet or database importing.

The software documentation consists of a reprint of Chapter 10 from Thanassoulis (2001). While this is sufficient to operate the software, the source book (sold separately) forms a comprehensive reference manual and user guide, with detailed model developments, illustrative examples, and advanced operating instructions for the software, including how to derive Malmquist indices from other calculated values. The book includes a demonstration version of the software.

Warwick DEA's execution speed for the test problem was slower than the other commercial packages, requiring 53 seconds for our base test problem. Surprisingly, the same problem required over 18 minutes when super-efficiency scores were requested. Other codes did not exhibit such behavior.

The pricing of Warwick's package was the lowest of the commercial offerings, and varied with the maximum number of DMUs accommodated. Academic discounts are only available for site licenses. Technical support responded promptly to questions.

4.   NON-COMMERCIAL SOFTWARE

The following DEA solution systems are distributed at little or no cost: DEA Excel Solver, DEAP, EMS, and PIONEER 2. Their evaluation criteria are given in the rightmost four columns of Figures 18-1 and 18-2.

4.1   DEA Excel Solver

- *Software title:* DEA Excel Solver
- *System requirements*: Microsoft Windows, Microsoft Excel
- *Developer:* Joe Zhu
- *Available from:* Included Zhu (2002)
- *Licensing:* included with Cooper, Seiford, and Zhu (2003)
- *Contact:*
  ⊠ Joe Zhu, Department of Management, Worcester Polytechnic Institute, 100 Institute Road, Worcester, MA 01609
  ☎ (508) 831-5467, fax (508) 831-5720
  ⚲ www.deafrontier.com
  ✉ jzhu@wpi.edu

The DEA Excel Solver software is included with, and documented in, Zhu (2002). When loaded, this Microsoft Excel add-in appends a "DEA" drop-down menu to the standard menu bar, from which specific models and actions can be invoked.

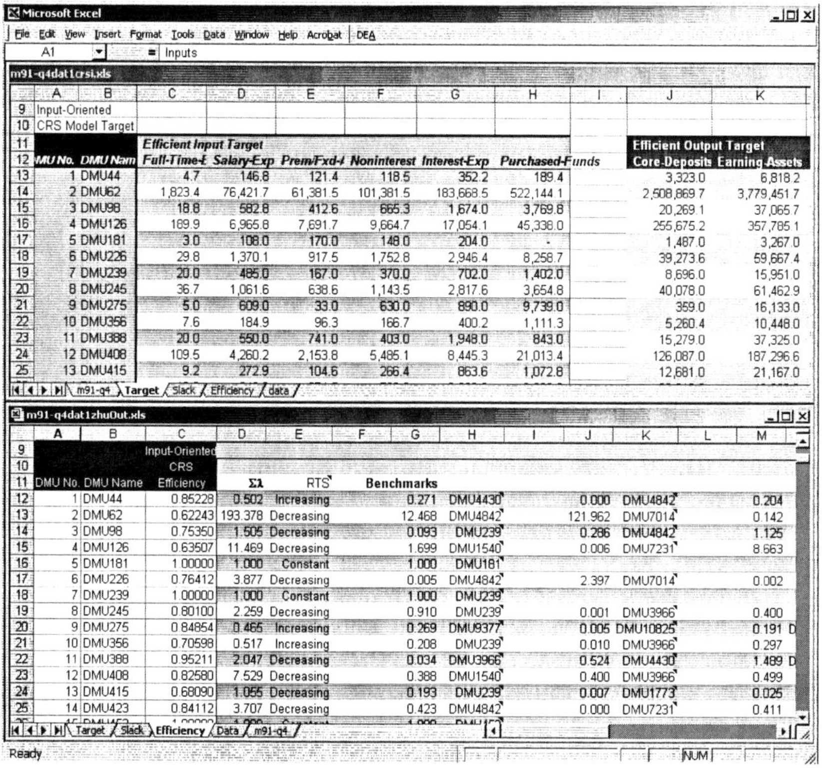

	A	B	C	D	E	F	G	H	I	J	K
9	Input-Oriented										
10	CRS Model Target										
11			Efficient Input Target							Efficient Output Target	
12	MU No.	DMU Nam	Full-Time-E	Salary-Exp	Prem/Fxd-A	Noninterest	Interest-Exp	Purchased-Funds		Core-Deposits	Earning-Assets
13	1	DMU44	4.7	146.8	121.4	118.5	352.2	189.4		3,323.0	6,818.2
14	2	DMU62	1,823.4	76,421.7	61,381.5	101,381.5	183,668.5	522,144.1		2,508,869.7	3,779,451.7
15	3	DMU98	18.8	582.8	412.6	665.3	1,674.0	3,769.8		20,269.1	37,065.7
16	4	DMU126	189.9	6,965.8	7,691.7	9,664.7	17,054.1	45,338.0		255,675.2	357,785.1
17	5	DMU181	3.0	108.0	170.0	148.0	204.0	-		1,487.0	3,267.0
18	6	DMU226	29.8	1,370.1	917.5	1,752.8	2,946.4	8,258.7		39,273.6	59,667.4
19	7	DMU239	20.0	485.0	167.0	370.0	702.0	1,402.0		8,696.0	15,951.0
20	8	DMU245	36.7	1,061.6	638.6	1,143.5	2,817.6	3,654.8		40,078.0	61,462.9
21	9	DMU275	5.0	609.0	33.0	630.0	890.0	9,739.0		359.0	16,133.0
22	10	DMU356	7.6	184.9	96.3	166.7	400.2	1,111.3		5,260.4	10,448.0
23	11	DMU388	20.0	550.0	741.0	403.0	1,948.0	843.0		15,279.0	37,325.0
24	12	DMU408	109.5	4,260.2	2,153.8	5,485.1	8,445.3	21,013.4		126,087.0	187,296.6
25	13	DMU415	9.2	272.9	104.6	266.4	863.6	1,072.8		12,681.0	21,167.0

	A	B	C	D	E	F	G	H	I	J	K	L	M
9			Input-Oriented										
10			CRS										
11	DMU No.	DMU Name	Efficiency	Σλ	RTS	Benchmarks							
12	1	DMU44	0.85228	0.502	Increasing		0.271	DMU4430		0.000	DMU4842		0.204
13	2	DMU62	0.62243	193.378	Decreasing		12.468	DMU4842		121.962	DMU7014		0.142
14	3	DMU98	0.75350	1.505	Decreasing		0.093	DMU239		0.286	DMU4842		1.125
15	4	DMU126	0.63507	11.469	Decreasing		1.699	DMU1540		0.006	DMU7231		8.663
16	5	DMU181	1.00000	1.000	Constant		1.000	DMU181					
17	6	DMU226	0.76412	3.877	Decreasing		0.005	DMU4842		2.397	DMU7014		0.002
18	7	DMU239	1.00000	1.000	Constant		1.000	DMU239					
19	8	DMU245	0.80100	2.259	Decreasing		0.910	DMU239		0.001	DMU3966		0.400
20	9	DMU275	0.84854	0.465	Increasing		0.269	DMU9377		0.005	DMU10825		0.191 D
21	10	DMU356	0.70598	0.517	Increasing		0.208	DMU239		0.010	DMU3966		0.297
22	11	DMU388	0.95211	2.047	Decreasing		0.034	DMU3966		0.524	DMU4430		1.489 D
23	12	DMU408	0.82580	7.529	Decreasing		0.388	DMU1540		0.400	DMU3966		0.499
24	13	DMU415	0.68090	1.055	Decreasing		0.193	DMU239		0.007	DMU1773		0.025
25	14	DMU423	0.84112	3.707	Decreasing		0.423	DMU4842		0.000	DMU7231		0.411

Figure 18-7. DEA Excel Solver

User-provided data and processing results are stored as different worksheets in a common file. The population data is supplied as columns containing: DMU names, input values, a blank column, and output values. The generated report sheets follow prescribed naming conventions.

The DEA Excel Solver provides an abundance of model choices, many of which are unique to this package (see Figure 16-1). Its rich feature set provides sensitivity analyses, benchmarking evaluations, superefficiencies, and nested frontiers; missing are categorical variables, multi-period analyses, and weight constraints.

The three generated results tables provide as much information as other programs do in five to eight separate reports. Color-coding designates different sections within a table. In some reports, DMU text-name cells have comments with the corresponding DMU index numbers, to expedite cross-references—a nice touch. Figure 18-7 shows two example reports.

The solution times in tests were slowest of all of the codes on problem G1, and second slowest on problem G2. As with Warwick DEA, but unlike other codes, super-efficiency scores increased runtimes significantly. While the Excel framework imposes a processing overhead and DEA Excel Solver does extensive cell formatting, some improvements are needed in this area.

Zhu (2002) provides a user guide, comprehensive reference manual, and the software itself. The only other expense might be the cost of a replacement optimizer if problem instances exceed the capacity of the standard Excel Solver.

In summary, DEA Excel Solver provides researchers and practitioners an extensive collection of models (some unique) and an appealing user interface. It is an impressive contribution to DEA research and industry application.

4.2   DEAP

- *Software title:* DEAP, Version 2.1
- *System requirements*: DOS
- *Developer:* Tim Coelli
- *Available from:* Centre for Efficiency and Productivity Analysis
- *Licensing:* Free download
- *Contact:*
    ⊠ Tim Coelli, CEPA, School of Economics, University of Queensland, Brisbane Australia
    ☎ (+61 7) 336 56470
    🕸 www.uq.edu.au/economics/cepa/software.htm
    ✉ t.coelli@economics.uq.edu.au

The Data Envelopment Analysis Program (DEAP) is a fast, batch-oriented DOS program, distributed by its author, Tim Coelli, and the Center for Efficiency and Productivity Analysis. DEAP is guided by two text files: one with input, output, and price values (where appropriate) and another with operating instructions: numbers of DMUs, inputs, outputs, and time periods; model selection; and file names. Execution of the program creates a text file of reports.

Analysis options include the classic CCR and BCC models, plus cost-efficiency and Malmquist formulations. The reports are simple and complete. DMUs are referenced by their sequence numbers, rather than by alphanumeric data labels.

DEAP's operation was too fast to allow timing of the 431-DMU test problem. The DOS window was on-screen so briefly that the program appeared to have malfunctioned; it was just finished.

This freeware is distributed with Coelli (1996), a straightforward user guide with examples of each model type. The user can find further information on the software and models in Coelli, Rao, and Battese (1997).

DEAP's rudimentary user interface disguises a program that is quite capable, providing all of Frontier Analyst's models and more. Its remarkable speed enables efficient operation on old and new PCs alike and its cost is, well, hard to beat.

4.3   EMS: Efficiency Measurement System

- *Software title:* EMS: Efficiency Measurement System, Version 1.3
- *System requirements:* Microsoft Windows
- *Developer:* Holger Scheel
- *Available from:* Holger Scheel
- *Licensing:* Free download for academic users
- *Contact:*
  ⊠ Holger Scheel, Operations Research und Wirtschaftsinformatik, Universität Dortmund, D-44221 Dortmund, Germany
  ☎ fax +49-(0)231-755-5408
  🏠 www.wiso.uni-dortmund.de/lsfg/or/scheel/ems
  ✉ h.scheel@wiso.uni-dortmund.de

The Efficiency Measurement System (EMS) is a standalone Windows application, developed by Holger Scheel and distributed at no cost to academic users. The user interface is clean and efficient, simplifying the tasks of importing data values from a tab-delimited-text or Excel file, selecting processing options, and running the solver to create a data table.

All data editing must be handled outside of EMS and the inability to display the input data is disconcerting at first. Inputs, outputs, and non-discretionary values are indicated by data labeling conventions and DMU subsets may be selected for analysis. Results are displayed in a single data table (in one of two formats, per Figure 18-8), which can be saved to a text file or copied to the clipboard.

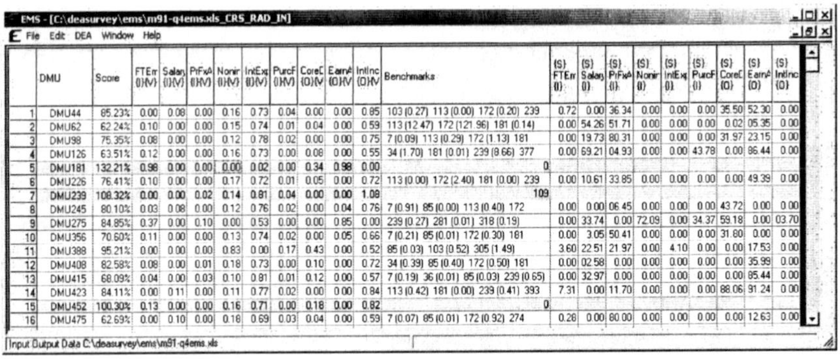

EMS - [C:\deasurvey\ems\m91-q4ems.xls_CRS_RAD_IN]

File Edit DEA Window Help

	DMU	Score	FTErr (I/M)	Salan (I/M)	PrFvA (I/M)	Nonir (I/M)	IntEx (I/M)	PurcF (I/M)	CoreL (I/M)	Earn (I/M)	Intinc (I/M)	Benchmarks	{S} FTErr (I)	{S} Salan (I)	{S} PrFvA (I)	{S} Nonir (I)	{S} IntEx (I)	{S} PurcF (O)	{S} CoreL (O)	{S} Earn (O)	{S} Intinc (O)
1	DMU44	85.23%	0.00	0.08	0.00	0.16	0.73	0.04	0.00	0.00	0.85	103 (0.27) 113 (0.00) 172 (0.20) 239	0.72	0.00	36.34	0.00	0.00	0.00	35.50	52.30	0.00
2	DMU62	62.24%	0.10	0.00	0.00	0.15	0.74	0.01	0.04	0.00	0.59	113 (12 47) 172 (121.96) 181 (0 14)	0.00	54.26	51.71	0.00	0.00	0.00	0.02	05.35	0.00
3	DMU98	75.35%	0.08	0.00	0.00	0.12	0.78	0.02	0.00	0.00	0.75	7 (0.09) 113 (0.29) 172 (1.13) 181	0.00	19.73	80.31	0.00	0.00	0.00	31.97	23.15	0.00
4	DMU126	63.51%	0.12	0.00	0.00	0.16	0.73	0.00	0.08	0.00	0.55	34 (1.70) 181 (0.01) 239 (8.66) 377	0.00	69.21	04.93	0.00	0.00	43.78	0.00	86.44	0.00
5	DMU181	132.21%	0.98	0.00	0.00	0.00	0.02	0.00	0.34	0.98	0.00	0									
6	DMU226	76.41%	0.10	0.00	0.00	0.17	0.72	0.01	0.05	0.00	0.72	113 (0 00) 172 (2 40) 181 (0.00) 239	0.00	10.61	33.85	0.00	0.00	0.00	0.00	49.39	0.00
7	DMU239	108.32%	0.00	0.00	0.02	0.14	0.81	0.04	0.00	0.00	1.08	109									
8	DMU245	80.10%	0.03	0.08	0.00	0.12	0.76	0.02	0.00	0.04	0.76	7 (0.91) 85 (0.00) 113 (0.40) 172	0.00	0.00	06.45	0.00	0.00	0.00	43.72	0.00	0.00
9	DMU275	84.85%	0.37	0.00	0.10	0.00	0.53	0.00	0.00	0.85	0.00	239 (0 27) 85 (0.01) 318 (0.19)	0.00	33.74	0.00	72.09	0.00	34.37	59.18	0.00	03.70
10	DMU356	70.60%	0.11	0.00	0.00	0.13	0.74	0.02	0.00	0.05	0.66	7 (0.21) 85 (0.01) 172 (0.30) 181	0.00	3.05	50.41	0.00	0.00	0.00	31.80	0.00	0.00
11	DMU388	95.21%	0.00	0.00	0.00	0.83	0.00	0.17	0.43	0.00	0.52	85 (0 03) 103 (0 52) 305 (1 49)	3.60	22.51	21.97	0.00	4.10	0.00	0.00	17.53	0.00
12	DMU408	82.58%	0.08	0.00	0.01	0.18	0.73	0.00	0.10	0.00	0.72	34 (0 39) 85 (0 40) 172 (0 50) 181	0.00	02.58	0.00	0.00	0.00	0.00	0.00	35.99	0.00
13	DMU415	68.09%	0.04	0.00	0.03	0.10	0.81	0.01	0.12	0.00	0.57	7 (0.19) 85 (0.01) 85 (0.03) 239 (0 65)	0.00	32.97	0.00	0.00	0.00	0.00	0.00	85.44	0.00
14	DMU423	84.11%	0.00	0.11	0.00	0.11	0.77	0.02	0.00	0.00	0.84	113 (0 42) 181 (0.00) 239 (0.41) 393	7.31	0.00	11.70	0.00	0.00	0.00	88.06	91.24	0.00
15	DMU452	100.30%	0.13	0.00	0.00	0.16	0.71	0.00	0.18	0.00	0.82	0									
16	DMU475	62.69%	0.00	0.10	0.00	0.00	0.69	0.03	0.04	0.00	0.59	7 (0.07) 85 (0.01) 172 (0.92) 274	0.28	0.00	80.00	0.00	0.00	0.00	0.00	12.63	0.00

Input Output Data C:\deasurvey\ems\m91-q4ems.xls

Figure 18-8. Efficiency Measurement System (EMS)

EMS has a good selection of available models and strong feature-set., including conical weight constraints, window analysis, Malmquist indices, super-efficiency scores, and program efficiencies from selecting evaluation and technology DMU subsets. This is all documented clearly and succinctly in the user's manual, Scheele (2000).

Problems are optimized with Csaba Mészáros' BPMPD interior-point solver (available from www.netlib.org), and problem sizes are limited only by available memory. While solution times are slower than most of the other native optimizers, EMS is significantly faster than the Excel-based packages and quick enough to be acceptable even for large instances.

EMS is a solid, straightforward DEA solution tool with many models and features to recommend it. While the niceties of a more professional user interface are missing, this software is a valuable contribution to the field of DEA.

4.4   PIONEER 2

- *Software title:* PIONEER, Version 2.0
- *System requirements:* Microsoft Windows, DOS, or Unix
- *Developers:* Thomas McLoud and Richard Barr, under sponsorship by Eclectic Computing Concepts, McKinney, TX
- *Available from:* Richard Barr
- *Licensing:* Free for non-commercial use
- *Contact:*
  ⊠ Richard Barr, Engineering Management, Information, and Systems Department, Southern Methodist University, P.O. Box 750123, Dallas, TX 75275
  ☎ (214) 768-2605, fax (214) 768-1112

🏛 faculty.smu.edu/barr/pioneer
✉ barr@engr.smu.edu

PIONEER 2 is a new C++ implementation of the research software by Barr and Durchholz (1997) for large-scale DEA problems.[10] It implements the Hierarchical Decomposition (HDEA) algorithm, a divide-and-conquer method for expediting the solution of large problem instances, and automates the Tiered DEA (TDEA) procedure for identifying nested frontiers and rank-ordering all DMUs based on tier number and super-efficiency scores, per Barr, Durchholz, and Seiford (1994). Convex constraints on the multiplier weights can be included, with several options on how they are to be applied.

The text-based user interface accepts problems from text files, while user options (model, orientation, super-efficiency, and constraint choices) are selected interactively. Program execution generates text files with three test-file reporting options: a summary report, efficiency scores by tier, and DMU detail report (with multipliers, slacks, efficient reference set, λ values, and projection for each unit, tier-by-tier). See Figure 18-9 for example reports, as displayed from Unix Xterm windows. Being a research code, documentation is minimal and assumes familiarity with DEA models and terminology.

The code is fast and designed for large data sets. This speed—due to the HDEA method and built-in optimizer—is needed for TDEA processing, which requires separate optimization of the dataset for each tier.

PIONEER 2 is distributed at no cost to non-commercial users. Both Windows and Unix versions are available from its web site.

5. SUMMARY AND FURTHER INFORMATION

Today's DEA practitioners and researchers have a wide range of solution technology choices. Clearly, software is no longer an impediment to the incorporation of DEA into decision-support systems and benchmarking processes. This is evidenced by the availability of interoperable tools with a variety of user interfaces, advanced modeling options, and the power to evaluate large-scale data sets on inexpensive computing platforms.

Besides the eight prominent DEA software packages highlighted in this survey, other solution options are documented in the Appendix below. Updated results, resources, and links to additional information are maintained at the accompanying DEA Software Portal:

http://faculty.smu.edu/barr/deahandbook/

---

[10] Pioneer 1 was written in Fortran for parallel computers and tested on problem instances with 25,000 DMUs.

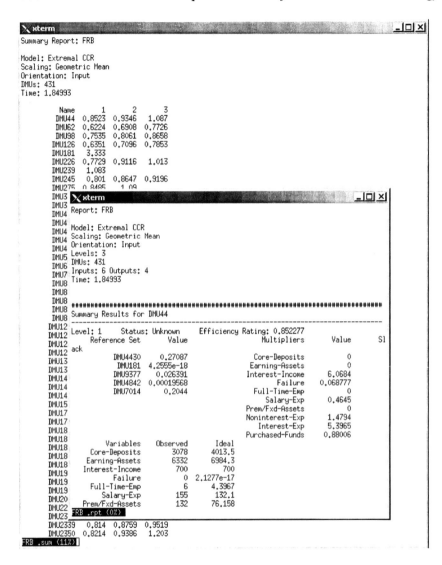

Figure 18-9. PIONEER 2 Example Reports

6. APPENDIX. OTHER DEA-RELATED SOFTWARE

The following software packages and libraries related to the solution of DEA models were not included in this survey, for the reasons noted below. Many are research codes, some address more general problems.

6.1   BYU-DEA

This early PC Fortran code for DEA problems was developed by Donald L. Adolphson and Lawrence C. Walters at Brigham Young University in Provo, UT. It solves CCR, BCC, and multiplicative models and handles non-discretionary and categorical variables. It is not being distributed at this time.

6.2   cdd/cdd+/cddlib

- *Software title:* cdd/cdd+/cddlib
- *System requirements:* C++ compiler
- *Developer:* Komei Fukuda of McGill University, Montreal, Canada
- *Licensing:* Free source code available from website
- *Contact:*
  - http://www.cs.mcgill.ca/~fukuda/soft/cdd_home/cdd.html
  - fukuda@cs.mcgill.ca

The program cdd+ is a C++ implementation of the Double Description Method of Motzkin et al. (1953) for generating all vertices, extreme rays, and the convex hull of a general convex. Hence, cdd/cdd+/cddlib is general software that can be used for DEA, but it is not tailored for this type of analysis.

6.3   Frame DEA

- *Software title:* Frame DEA
- *System requirements*: Fortran compiler and CPLEX callable library
- *Developer:* José Dulá
- *Contact:*
  - José Dulá, School of Business Administration, University of Mississippi, University, MS
  - jdula@olemiss.edu

These research codes were written to test the author's "frame algorithms" for large-scale variable-returns-to-scale (BCC) models "using additive LP formulations." Dulá (1997) documents the methodology and computational testing on problem instances with up to 10,000 DMUs.

6.4   GAMS Implementations

- *Software title:* Generalized Algebraic Modeling System (GAMS)
- *System requirements:* Microsoft Windows, Linux, Solaris, HP-UX 10, Digital Unix 4.0, AIX 4.3, or IRIX
- *Developers:* Anthony Brooke, David Kendrick, and Alex Meeraus
- *Available from:* GAMS Development Corporation
- *Licensing:* $3200 ($640 academic) + optional optimizers
- *Contact:*
  ✉ GAMS Development Corporation, 1217 Potomac Street, NW, Washington, DC 20007 USA
  ☎ (202) 342-0180, fax: (202) 342-0181
  🕸 www.gams.com
  ✉ sales@gams.com

The Generalized Algebraic Modeling System (GAMS) was one of the first modeling languages for optimization problems. Algebraic model formulations can be solved by a variety of commercial and noncommercial packages for linear, mixed integer, and nonlinear optimization, with user reports generated from the results.

Several DEA implementations have been developed in GAMS. These include:

- *GAMS/DEA module*, providing language extensions specifically for DEA. The Cplex solver is required, and the "slice" modeling technique is documented in Ferris and Voelker (2001). Example applications, including standard DEA models, cross-validations, and quadratic programming models for support vector machines are available at www.gams.com/contrib/gamsdea/dea.htm
- dea.gms, free contributed software for classic DEA formulations, available from www.gams.com and documented in Kalvelagen (2002)
- *GAMS DEA code*, published in Olesen and Petersen (1995)

6.5   IDEAS

This strong commercial DEA package was developed by Agha Iqbal Ali of the University of Massachusetts at Amherst and distributed by 1 Consulting of Amherst, MA. The company is closed for business and the software is no longer distributed.

6.6   Qhull

- *Software title:* qhull
- *System requirements:* DOS
- *Licensing:* Free source code available from website
- *Contact:*
    🕸 http://www.thesa.com/software/qhull/

This software computes convex hull based on the Quickhull algorithm of Barber, Dobkin and Huhdanpaa (1996). Website description: "Qhull computes convex hulls, Delaunay triangulations, halfspace intersections about a point, Voronoi diagrams, furthest-site Delaunay triangulations, and furthest-site Voronoi diagrams." Hence it can be adapted to determine an envelopment surface.

6.7   SAS/OR Implementations

- *Software title:* SAS/OR
- *Available from:* SAS Institute Inc.
- *Contact:*
    ✉ SAS Institute Inc., 100 SAS Campus Drive, Cary, NC 27513-2414 USA
    ☎ (919) 677-8000, fax: (919) 677-4444
    🕸 www.sas.com

Two DEA implementations for the SAS/OR software system have been published: SAS/DEA, a CCR-I macro, and SAS/MALM, with macros for CCR-O and Malmquist index calculations. The papers—Emrouznejad (2000, 2002)—contain source-code listings and are available at www.deazone.com.

6.8   Web Implementation

- *Software title:* Internet-Based Data Envelopment Analysis System (iDEAs)
- *System requirements:* Internet access
- *Developer:* Leon McGinnis
- *Availability:* For registered warehousing companies only
- *Contact:*
    ✉ Leon McGinnis, School of Industrial and Systems Engineering, Georgia Institute of Technology, Atlanta, GA
    🕸 www.isye.gatech.edu/ideas
    ✉ leon.mcginnis@isye.gatech.edu

Georgia Institute of Technology's website for warehouse benchmarking, iDEAs, "makes the DEA methodology accessible via the Internet, and enables firms to perform a self-assessment and benchmark themselves against other firms in their industry" (from website). The current database of 200 warehouse operations supports benchmarking of all participating organizations against each other.

REFERENCES

1. Andersen, P., and N. Petersen, 1993, A Procedure for Ranking Efficient Units in Data Envelopment Analysis, Management Science 39, No. 10, 1261-1264.
2. Banxia Software Limited, 2001a, Banxia Frontier Analyst® Plus Pack 1 User's Guide, Banxia Software Ltd., Kendal, UK.
3. Banxia Software Limited, 2001b, Banxia Frontier Analyst Plus Pack 2 User's Guide, Banxia Software Ltd., Kendal, UK.
4. Banxia Software Limited, 2003, Banxia Frontier Analyst User's Guide: Professional Edition, Banxia Software Ltd., Kendal, UK.
5. Barr, R., and M. Durchholz, 1997, Parallel and Hierarchical Decomposition Approaches for Solving Large-Scale Data Envelopment Analysis Models, Annals of Operations Research 73, 339-372.
6. Barr, R., M. Durchholz, and L.M. Seiford, 1994, Peeling the DEA Onion: Layering and Rank-Ordering DMUs Using Tiered DEA, Southern Methodist University technical report, Dallas, Texas.
7. Barr, R.S., K.A. Killgo, T.F. Siems, and S. Zimmel, 2002, Evaluating the Productive Efficiency and Performance of U.S. Commercial Banks, Managerial Finance 28, No. 8, 3-25.
8. Barber, C.B., D.P. Dobkin, and H. Huhdanpaa, 1996, The Quickhull Algorithm for Convex Hulls, ACM Transactions on Mathematical Software 22, No. 4, 469—483.
9. Brooke, A., D. Kendrick, A. Meeraus, and R. Raman, 1998, GAMS: A Users Guide, GAMS Development Corporation, Washington, DC.
10. Coelli, T., 1996, A Guide to DEAP Version 2.1: A Data Envelopment Analysis (Computer) Program, CEPA Working Paper 96/08, Centre for Efficiency and Productivity Analysis, University of New England, Armidale, NSW, Australia.
11. Coelli, T., D.S.P. Rao, and G.E. Battese, 1997, An Introduction to Efficiency and Productivity Analysis, Kluwer Academic Publishers, Boston.

12. Cooper, W.W., L.M. Seiford, and K. Tone, K., 2000, Data Envelopment Analysis: A Comprehensive Text with Models, Applications, References and DEA-Solver Software, Kluwer Academic Publishers, Boston.

13. Cooper, W.W., L.M. Seiford, and J. Zhu, eds., 2003, Handbook of Data Envelopment Analysis, Kluwer Academic Publishers, Boston.

14. Dulá, J.H, 1997, An Algorithm for Data Envelopment Analysis (DEA), Working Paper, School of Business Administration, University of Mississippi, University, MS.

15. Economic Measurement and Quality Corporation, 2000, OnFront 2 Users Guide, Economic Measurement and Quality Corporation, Lund, Sweden.

16. Emrouznejad, A., 2000, An Extension to SAS/OR for Decision System Support, in Proceeding of SAS Users Group International, 25th Annual Conference, Indianapolis, Indiana, USA

17. Emrouznejad A., 2002, A SAS Application for Measuring Efficiency and Productivity of Decision Making Units, in Proceedings of the 27th SAS International Conference, pp 259-27, USA.

18. Färe, R. and S. Grosskopf, 2000, Reference Guide to OnFront, Economic Measurement and Quality Corporation, Lund, Sweden.

19. Ferris, M.C. and M.M. Voelker, 2001, Cross-Validation, Support Vector Machines and Slice Models, Data Mining Institute Technical Report 01-07, Department of Computer Science, University of Wisconsin, Madison, WI.

20. Herrero, I. and S. Pascoe, 2002, Estimation of Technical Efficiency: A Review of Some of the Stochastic Frontier and DEA Software, Computers in Higher Education Economics Review 15, No. 1.

21. Hussain, A. and M. Jones, 2001, An Introduction to Frontier Analyst: Frontier Analyst Workbook 1, Banxia Software Ltd., Kendal, UK.

22. Kalvelagen, E., 2002, Efficiently Solving DEA Models with GAMS, technical report, GAMS Development Corporation, Washington, DC (available at: http://www.gams.com/~erwin/dea/dea.pdf)

23. Motzkin, T.S., H. Raiffa, G. L. Thompson, and R. M. Thrall, 1953, The Double Description Method, in H. W. Kuhn and A. W. Tucker, editors, Contributions to the Theory of Games -- Volume II, number 28 in Annals of Mathematics Studies, Princeton University Press, Princeton, New Jersey, 51-73.

24. Nyhan, R., 1998, Software Review: Frontier Analyst, ORMS Today 25, No. 2.

25. Olesen, O.B., and N.C. Petersen, 1995, A Presentation of GAMS for DEA, Computers & Operations Research 23, No. 4, 323-339.

26. SAITECH, Inc., 2003, Introduction to DEA-Solver-Pro Professional Version 4.0, SAITECH, Inc., Hazlet, New Jersey.
27. Scheel, H., 2000, EMS: Efficiency Measurement System Users Manual, Version 1.3, Universität Dortmund, Dortmund, Germany.
28. Thanassoulis, E., 2001, Introduction to the Theory and Application of Data Envelopment Analysis, Kluwer Academic Publishers, Boston.
29. Zhu, Joe, Quantitative Models for Performance Evaluation and Benchmarking: Data Envelopment Analysis with Spreadsheets, Kluwer Academic Publishers, Boston. 2002

Notes about Authors

Rachel Allen is a business consultant within the manufacturing industry. She holds a PhD from Warwick Business School in the area of Performance measurement and has co-authored a number of papers on the theory and applications of DEA.

Timothy R. Anderson is an Associate Professor of Engineering and Technology Management at Portland State University. He received electrical and industrial engineering degrees from the University of Minnesota and the Georgia Institute of Technology. He has been the Program Chair for PICMET, the Portland International Conference on the Management of Engineering in 1999, 2001, and 2003. PICMET draws about 500 attendees and typically has a track dedicated to engineering and technology management applications of DEA. Dr. Anderson's current research focuses on applications of productivity analysis, technology forecasting, quantitative benchmarking, and new product development in diverse areas including construction projects, telecommunications, enterprise database systems, microprocessors, and the electric utility industry.

Antreas D. Athanassopoulos is the Konstantine Kapsaskis Chair in Financial Services in Athens Laboratory of Business Administration (ALBA), Greece. He holds a BSc in Applied Mathematics (Patras, Greece), MSc in Statistics and Operational Research (Essex, UK) and PhD in Industrial and Business Studies (Warwick, UK). He lecturered at Warwick Business School (1992-1996), and Athens University of Economics and Business 1997-1998. He has also consulted various private and public sector firms including Allied Breweries Ltd., Coral Ltd., Home Office in the UK,

Siemens, Pireaus Bank, HBID Bank, Hellenic Telecommunications Organisation, and Piraeus Bank. His primary research, teaching and consulting activities are in the areas of Strategic management, benchmarking and Performance measurement. His academic research has appeared in various academic journals including *Management Science, Journal of Regional Science, Journal of Money and Credit Banking, European Journal of Operational Research, Journal of Operational Research Society, International Journal of Production Economics, Journal of Business Research, International Journal of Bank Marketing, Annals of Operations Research, Journal of Productivity Analysis and European Journal of Marketing.* Dr. Athanassopoulos is a member of the Institute for Management Science and Operations Research (INFORMS), and the Royal Operational Research Society.

Rajiv D. Banker is the Ashbel Smith Chair in Accounting and Information Management at the University of Texas at Dallas. He received a Doctorate in Business Administration from Harvard University and many academic awards, including those for standing first at the B.S. degree examination of the University of Bombay and professional accountancy examinations in India. His previous appointments include professorship at Carnegie Mellon University and the Arthur Andersen/Andersen Consulting Chair at the University of Minnesota. Dr. Banker is one of the most prolific and influential interdisciplinary researchers in management. He has received numerous awards for his research and published more than 100 articles in prestigious research journals. His research articles are cited over 100 times each year by other researchers in a wide range of disciplines. Dr. Banker is recognized as a leader in analytical modeling and statistical analysis of data collected from different companies to address complex or emerging problems of importance to managers. His research addresses questions pertaining to performance measurement, incentive compensation, e-business competitive strategy, valuation of information technology and software productivity and quality management. His research has been supported by the National Science Foundation, the Financial Executives Research Foundation, the Institute of Management Accountants, and several leading corporations. He has supervised seven award winning doctoral dissertations and lectured to executives and academics at institutions around the world.

Dick Barr (B.S.E.E., M.B.A, Ph.D. from U. Texas at Austin) is Associate Professor and Chair of the Engineering Management, Information, and Systems department at Southern Methodist University, Dallas, Texas. He is past chair of the INFORMS Computing Society and his research has been funded by NSF, DARPA, THECB/ATP, and various defense contractors. Dr.

Barr's research interests include large-scale optimization, DEA benchmarking models, network algorithms for telecommunications and logistics, parallel algorithms, and the empirical analysis of algorithms. He is the author of over 40 journal articles, which have appeared in *Operations Research, Mathematical Programming, Parallel Computing*, and the *ORSA Journal on Computing*. Past and current editorial work includes *INFORMS Journal on Computing, Interfaces, Journal of Heuristics*, and *Interactive Transactions on OR/MS*. His business background includes the establishment of several start-up businesses and a consulting practice whose client list includes CONRAIL, Coopers & Lybrand, Cray Research, E-Systems, General Motors, Heritage Foundation, IBM, Peat Marwick Main & Co., Raytheon, and the U.S. Departments of Commerce, Health and Human Services, Interior, and Treasury.

Jon A. Chilingerian is a tenured professor at Brandeis University. He is director of the MD-MBA Program, and Co-Director of the AHRQ-funded Doctoral Training Program in Health Services Research at Brandeis University. He is a visiting Professor of Health Management and member of the Health Management Initiative at INSEAD. He received his Ph.D. in Management from MIT's Sloan School of Management. He teaches graduate courses and executive education sessions in Organizational Theory and Behavior, Management of Health Care Organizations, and Health Services Research. Dr. Chilingerian is the author of *The Lessons and the Legacy of the Pew Health Policy Program*, with Corinne Kay, published in 1997 by the Institute of Medicine National Academy Press. He has scholarly papers and review essays published in journals such as: *Annals of Operational Research, Medical Care, European Journal of Operational Research, Health Services Research, Health Care Management Review, Medical Care Research and Review, Inquiry, Health Services Management Research*, and *The Journal of Health Politics, Policy and Law*. Dr. Chilingerian was former chair of the Health Care Management Division of the Academy of Management, and currently sits on the Academy Council. He has been a consultant for a number of health care organizations. His research focuses on managing health care organizations, ranging from studies of executive leadership and management of professionals to the measurement of performance (i.e., productive efficiency, quality, etc.), identification of physician best practices and the analysis of effective operating strategies. His most recent research is on understanding capacity problems in terms of technical, political and cultural systems inside health care organizations. In addition to authoring many articles, and a book for the IOM, he is on the editorial board of several leading health care and management journals.

Wade Cook is a Professor and Head of Management Science in the Schulich School of Business, York University, Toronto, Canada. He also currently serves as Associate Dean of Research in the Schulich School of Business. Dr. Cook holds a doctorate in Mathematics and Operations Research, and has published more than 100 articles in a wide range of academic and professional journals and books. (Journals include Management Science, Operations Research, JORS, EJOR, IIE Transactions, etc.). His areas of specialty include Data Envelopment Analysis, and Multi Criteria Decision Modeling. He is the author of the book Ordinal Information and Preference Structures. He is also an Editor of the Journal of Productivity Analysis, and former Editor of INFOR. Professor Cook has consulted widely with various companies and government agencies.

William W. Cooper is the Foster Parker Professor of Finance and Management (Emeritus) in the Red McCombs School of Business, University of Texas at Austin. Author or coauthor of 23 books and 490 scientific professional articles he has written extensively on DEA and consulted on its uses with numerous business firms and government agencies. He holds honorary D.Sc. degrees at Ohio State and Carnegie Mellon Universities in the U.S. and the degree of Doctor Honoris Causa from the University of Alicante in Spain. A fellow of the Econometric Society and of INFORMS (Institute of Operations Research and Management Science), he is also a member of the Accounting Hall of Fame.

Honghui Deng received the degree of Bachelor of Engineering from Chongqing University in July 1990 and the Bachelor degree of Business Administration in July 1994. Then he worked in the Ministry of Education of China in Beijing as a Project Director. After three years with Ministry of Education of China, he joined the Marketing department of the Red McCombs School of Business of the University of Texas at Austin as a Visiting Scholar supported by China Government. In 1999, he accepted by the department of Management Science & Information Systems of the Red McCombs School of Business at University of Texas at Austin as a Ph.D. student and got his degree of Doctor of Philosophy at May, 2003. He is now author or coauthor of 8 articles in the scientific-professional literature. He is currently a tenure-track Assistant Professor in the department of Management Information Systems of the College of Business at the University of Nevada at Las Vegas.

Rolf Färe is Professor of Economics and Agricultural Economics at Oregon State University. After receiving his Filosofie Licentiat from Lund

University under Professor Bjorn Thalberg he continued his studies at U.C. Berkeley under Professor Ronald W. Shephard. There his work with Shephard laid the foundations for an axiomatic approach to production theory. Over the years this has led to the publication of 10 books and more than 200 articles. See also Who's Who in Economics.

Shawna Grosskopf is a Professor of Economics at Oregon State University. She received her M.S. and Ph.D. at Syracuse University where her areas specialization were public economics and the economics of education. Over the last 15 years she has worked in the area of performance measurement and its applications to her areas of interest. See also Who's Who in Economics.

Zhimin Huang is Professor of Operations Management in the School of Business at Adelphi University. He received his BS in Industrial Engineering from The Beijing University of Aeronautics and Astronautics, MS in Economics from The Renmin University of China, and Ph. D. in Management Science from The University of Texas at Austin. He is currently a Research Fellow at the Center for Management of Operations and Logistics, The University of Texas at Austin. He was a Research Fellow at the Center for Cybernetic Studies, The University of Texas at Austin. His research interests are mainly in supply chain management, data envelopment analysis, distribution channels, game theory, chance constrained programming theory, and multi-criteria decision making analysis. He has published articles in *Naval Research Logistics, Decision Sciences, Journal of Operational Research Society, European Journal of Operational Research, Journal of Economic Behavior and Organization, Optimization, Omega, Research in Marketing, Annals of Operations Research, International Journal of Systems Science, Journal of Productivity Analysis, Journal of Economics, Journal of Mathematical Analysis and Applications, Computers and Operations Research, International Journal of Production Economics*, and other journals.

Shanling Li is an associate professor of Operations Management and Management Science at the Faculty of Management of McGill University. She is also the director of Management Science Research Center. Prof. Li received her M.S. from Georgia Institute of Technology and her Ph.D. from University of Texas at Austin. Her research interests include capacity expansion, technology investment, Data Envelopment Analysis, supply chain management, and production control. She is currently an associate editor of IEEE Engineering Management and International Journal of Information Technology and Decision Making.

Susan X. Li is Professor of Management Information Systems in the School of Business at Adelphi University. She received her BS in Economics from The Renmin University of China and her Ph. D. in Management Science and Information Systems from the Graduate School of Business at The University of Texas at Austin. She was an Assistant Manager of the Chemical Machinery and Plant Division at MITSUI & CO., LTD. (Japanese General Trading Company) for a year before she went to The University of Texas at Austin for her Ph. D studies. Her research interests are mainly in supply chain related information systems, distribution channels, data envelopment analysis, chance constrained programming theory in investment and insurance portfolio analysis, and multi-criteria decision making analysis. She has published articles in *Decision Sciences, Journal of Operational Research Society, European Journal of Operational Research, Annals of Operations Research, Omega, Information Systems and Operational Research, Computers and Industrial Engineering Journal, Journal of Optimization Theory and Applications, Journal of Productivity Analysis, Computers and Operations Research, International Journal of Production Economics*, and other journals.

Ram Natarajan is an Assistant Professor of Accounting and Information Management at the University of Texas at Dallas. Dr. Natarajan received a Post Graduate Diploma in Management from the Indian Institute of Management, Calcutta, and a Ph.D. in Accounting from the Wharton School of University of Pennsylvania. Dr. Natarajan's research addresses questions pertaining to analysis of productivity and efficiency in organizations, the determinants of CEO compensation, use of accounting performance measures for performance evaluation and link between corporate disclosure policies and managerial incentives. His academic honors include a Dean's Fellowship of Distinguished Merit from the Wharton School of University of Pennsylvania, A dissertation Fellowship from Ernst and Young, a runner-up award for the outstanding doctoral dissertation from the Management Accounting Section of the American Accounting Association and a Junior Faculty Teaching and Research Fellowship from Arthur Andersen. He has presented his research in many U.S universities and academic conferences.

Joseph C. Paradi was educated at the University of Toronto, Canada where he received his B.A.Sc., M.A.Sc. and Ph.D. degrees in Chemical Engineering. He spent 20 years in the computer services industry where he was a founder and president of a large Canadian firm. He has served as the President and the Chairman of the Board of CADAPSO, the Canadian Association for the Data Processing Services Organisations, he is the director of several high-tech firms and is involved in venture capital activities. Since 1989 he has been at the University of Toronto where he is a Professor and Executive Director of the Centre for Management of Technology and Entrepreneurship at the University of Toronto, Canada. His research focus is on the measurement and improvement of productivity in the services elements of telecommunications and financial services firms using Data Envelopment Analysis. He has published papers in EJOR, IEEE Engineering Management, OMEGA and JPA. He is a member of the IEEE, IAMOT, INFORMS, IIE and several other academic oriented organisations.

Maria Conceição A. Silva Portela teaches at the Catholic University, Porto, Portugal. She has a BSc. degree in *Management* from the Portuguese Catholic University, in Portugal and an MSc degree in *Management Science and Operational Research* from Warwick University in the United Kingdom. She has co-authored a number of papers on the theory and application of DEA and is currently completing her research for a PhD in the field of DEA.

John Ruggiero is an Associate Professor of Economics at the University of Dayton. He received his B.A. in economics and mathematics with a concentration in computer science from the State University of New York at Cortland and his M.S. and Ph.D. in economics from Syracuse University. Dr. Ruggiero's current research interests are in Data Envelopment Analysis theory and applications, applied microeconomics, public economics and applied econometrics. He has recently served on a panel of experts analyzing performance of Ohio school districts and has presented at academic and policy-oriented conferences on performance evaluation in the public sector. He has published articles in Review of Economics and Statistics, Public Choice, European Journal of Operational Research, Journal of the OR Society, Economics of Education Review, Contemporary Economic Policy, Managerial and Decision Economics, Journal of Public Administration Review and Theory, and other journals.

Lawrence M. Seiford is Professor and Chair of Industrial and Operations Engineering at the University of Michigan. Prior to joining the University of Michigan he was Program Director of the *Operations Research* and *Production Systems* programs at the National Science Foundation (1997-2000) and was a member of the faculty at the University of Massachusetts, the University of Texas at Austin, the University of Kansas, and York University. Professor Seiford's teaching and research interests are primarily in the areas of quality engineering, productivity analysis, process improvement, distributed-systems design issues, and performance measurement. He has written and co-authored three books and approximately one hundred articles in the areas of quality, productivity, operations management, process improvement, decision analysis, and decision support systems. Professor Seiford holds editorial positions for a number of journals and has received teaching and research excellence awards. In November, 2000, he was awarded the degree *Docteur Honoris Causa* from the National Ministry of Education of France in a special recognition ceremony at the Universite de la Mediterranee, Aix-Marseille II.

David Sherman is an Associate Professor at Northeastern University, College of Business Administration (Boston, Massachusetts) and Adjunct Associate Professor at the Tufts University School of Medicine. He began working with DEA as a Doctoral Student. His dissertation explored applications of DEA to hospitals. Professor W.W. Cooper served as his dissertation chair at Harvard Business School. In addition to work on health care productivity, he has managed several DEA applications in banks, brokerage firms and government organizations. Among these were three of the largest 50 U.S. banks, the largest U.S. discount brokerage, and the Government of Canada. Four of these DEA applications have generated measurable benefits that have been documented in the press and other publications by the clients using DEA. He teaches executive and MBA courses in accounting, control and global financial statement analysis with a focus on high technology, medical technology, financial services, health care and nonprofit organizations. He has served as board member and manager of several new ventures from startup to IPO and sale to acquiring private and public companies. His current research focuses on financial literacy issues for management and boards in global businesses, financial reporting issues in the modern economy, improving shareholder returns in bank mergers, and managing healthcare service cost and quality. He is an expert on developing and using financial and non-financial measures to evaluate and manage performance of businesses and non-profit organizations. He served as director, manager, and consultant to financial, healthcare and high technology businesses. Clients and executive teaching include: BankBoston, EMC, KLA, A.T.Kearney (BankAmerica), U.S. Bancorp; Fidelity Brokerage Services (FMR Corp.); CRESAP, US Department of Defense – managed health care programs, Department Of Supply and Services - Government of Canada and KPMG. Hereceived his DBA and MBA degrees from Harvard Business School and an AB in economics from Brandeis University. He previously was on the faculty of the MIT Sloan School of Management. He has made presentations at business and academic meetings in the U.S., Canada, Europe, Asia and India. In 1999, he was a visiting faculty at INSEAD (Fontainebleau, France). He is a Certified Public Accountant

(CPA), practiced with Coopers & Lybrand, and served on an advisory task force of the Financial Accounting Standards Board (FASB). He authored one book with (Young and Collingwood): *Profits You Can Trust: Spotting and Surviving Accounting Landmines* (2003-Prentice Hall-Financial Times). He also authored two monographs: *Accountability – A Key Obligation of Management* and *Service Organization Productivity Management*, which documents adoption of DEA in the Government of Canada (both published by the Society of Management Accountants of Canada). His research has been published in academic and management journals including: *Harvard Business Review* (three articles), *Sloan Management Review, Accounting Review, Bankers Magazine, Interfaces, Management Accounting, Journal of Banking and Finance, American Banker, Medical Care, and Auditing*. In 2002, he served as the chair of the Health Applications Section of INFORMS.

Léopold Simar is Professor of Statistics at The Univeristé Catholique de Louvain, Louvain-la-Neuve, Belgium where he is the Founder-Chairman of the Institute of Statistics. He is also Professor at The Facultés Universitaires Saint-Louis, Brussels, Belgium. He is teaching mathematical statistics, multivariate analysis, bootstrap methods in statistics and econometrics. His research focuses on non-parametric and semi-parametric methods (boundary estimation, density estimation, single index models, bootstrap), and multivariate statistics. He is an elected member of the ISI (International Statistical Institute) and the past President of the Belgian Statistical Society. He is currently associate editor of the Journal of Productivity Analysis.

Emmanuel Thanassoulis is Professor in Management Sciences at Aston Business School of Aston University in Birmingham, England. He has published over 70 refereed and working papers in the theory and application of DEA and is author of *Introduction to the Theory and Application of Data Envelopment Analysis: A foundation text with integrated software* (Kluwer academic publishers, 2001). He has extensive research, consultancy and teaching experience in DEA. He has taught DEA at undergraduate and postgraduate level and runs specialist training courses on DEA for practitioners. He initiated and co-developed the *Warwick DEA Software*. He has had extensive involvement within the UK on DEA applications for regulated utilities, regulators of such utilities, the Home Office, the Treasury, the Higher Education Funding Council for England and Wales and the Department for Education and Skills.

Kaoru Tone Graduated from Department of Mathematics, the University of Tokyo, in 1953. After serving as an Associate Professor at Keio University, he joined Graduate School of Policy Science at Saitama University in 1977. Currently he is a Professor at National Graduate Institute for Policy Studies. He is an Honorary Member of the Operations Research Society of Japan and served as President of the Operations Research Society of Japan, 1996-1998. His research interests include theory and applications of optimizations, evaluation methods for managerial and technical efficiency of performances, and decision making methods. He is conducting cooperative studies on the above subjects with Institute of Posts and Telecommunications Policy and National Land Agency, Japan. Also, he is on the Editorial Boards of Omega, Socio-Economic Planning Sciences (International Journal of Public Sector Decision Making), and Encyclopedia of Operations Research and Management Science among others. His works on the above subjects have appeared in *European Journal of Operational Research, Omega, Journal of the Operational Research Society, Journal of Productivity Analysis, Journal of the Operational Research Society of Japan, Mathematical Programming* and *Policy and Information.*

Konstantinos (Kostas) Triantis has been educated at Columbia University, where he received his BS and MS in Industrial and Management Engineering, and M.Phil. and Ph.D. in Industrial Engineering and Operations Research. He is a Professor in the Grado Department of Industrial and Systems Engineering, Virginia Polytechnic Institute and State University, where he also serves as the Research Director of the System Performance Laboratory (SPL), the Director of the Systems Engineering Program, and the Academic Director of the of the Engineering Administration Program. His teaching and research interests cover the areas of performance measurement and evaluation, the design of performance management systems, data envelopment analysis, systems thinking and modeling, and fuzzy methods. His research has been published in a number of journals including *Management Science, European Journal of Operational Research, Technometrics, IEEE Transactions on Engineering Management, Annals of Operations Research, Journal of Productivity Analysis,* among others. He has received research funding from a number of organizations that include the *Office of Naval Research, Department of Energy, American Red Cross, United States Postal Service,* among others. He is a member of INFORMS, ASQ, and IIE.

Sandra Vela was educated at the University of Toronto where she received her B.A.Sc., M.A.Sc. and Ph.D. in Chemical Engineering. Her research focus was on productivity and efficiency in the Canadian Insurance Industry and the evaluation of a bank merger in Canada. She is currently working in the Marketing Department at TD Bank Financial Group in Toronto.

Gerald Whittaker earned a J.D. from Northwestern School of Law, Lewis and Clark College, Portland OR. He was a computer programmer and economist for Data Resources Incorporated in Washington, DC, then worked as an economist at the Economic Research Service, USDA. He is now a research hydrologist at the National Forage Seed Production Research Center, Agricultural Research Service, USDA in Corvallis, OR.

Paul W. Wilson is Professor of Economics at The University of Texas at Austin, where he specializes in theoretical an applied econometrics. His theoretical research interests include bootstrap methods, outlier detection, and computational methods. His applied research includes studies of mergers, failures, and efficiency among commercial banks, returns to scale in the banking and hospital industries, demand for health-care services, and urban transportation issues. He is currently editor of *Journal of Productivity Analysis*.

Zijiang Yang received her Bachelor's degree in Computer Science from the Beijing Institute of Technology in China. Upon arrival in Canada, enrolled in the University of Toronto where she received her M.A.Sc and Ph.D. in Industrial Engineering. Her focus of research was on information technology and banking productivity respectively. She is currently an Assistant Professor at York University, Toronto, Canada.

Joe Zhu is Associated Professor of Operations, Department of Management at Worcester Polytechnic Institute, Worcester, MA. His research interests include issues of performance evaluation and benchmarking, information technology and productivity, supply chain design and efficiency, and Data Envelopment Analysis. His research has appeared in such journals as *Management Science, Operations Research, IIE Transactions, Annals of Operations Research, Journal of Operational Research Society, European Journal of Operational Research, Information Technology and Management Journal, Computer and Operations Research, OMEGA, Socio-Economic Planning Sciences, Journal of Productivity Analysis, INFOR, Journal of Alternative Investment* and others. He is the author of *Quantitative Models for Evaluating Business Operations: Data Envelopment Analysis with Spreadsheets* (Kluwer Academic Publishers, 2002). He developed the DEAFrontier software which is a DEA add-in for Microsoft Excel. For information on his research, please visit www.deafrontier.com.

Author Index

A

Abdullah · 425
Adolphson · 407, 561
Afriat · 7
Aggarwall · 370, 371, 373, 376
Ahn · 31, 77
Aigner · 243, 313, 410
Akhavein · 352
Al-Faraj · 356, 395
Ali · 105, 119, 167, 447, 562
Alidi · 395
Allen · 104, 105, 121, 122, 123, 124, 125,
   126, 128, 129, 130, 132, 134, 406
Almond · 411
Anderson · 106, 310, 443, 444, 447, 452,
   482
Arnold · 9
Asenova · 426
Athanassopoulos · 109, 356, 357, 395,
   396, 408, 455, 456, 459, 461, 473, 552
Avkiran · 366

B

Balakrishnan · 182

Balk · 215, 217, 218
Banker · 13, 20, 21, 42, 43, 46, 47, 49, 54,
   55, 58, 61, 62, 64, 68, 72, 76, 77, 159,
   165, 204, 215, 232, 239, 260, 269,
   273, 299, 300, 301, 302, 304, 305,
   306, 307, 309, 310, 311, 312, 313,
   314, 315, 316, 317, 318, 319, 328,
   330, 331, 332, 336, 337, 364, 410,
   426, 459, 484, 493, 526
Bardhan · 28, 43, 46, 47
Barr · 539, 542, 545, 558, 559
Battese · 375, 410, 557
Bauer · 352
Baumol · 62
Beasley · 23, 105, 114, 117
Ben-Aderet · 425
Beran · 277
Berg · 355
Berger · 352, 353, 354, 366, 380, 473
Berndt · 307
Bessent · 108, 329, 347
Bhattacharyya · 354
Bickel · 277
Bjurek · 225
Blanchard · 342, 347
Boile · 426
Boussofiane · 104
Bowlin · 230, 231

Bradford · 325
Bradley · 347
Braglia · 426
Bridge · 324
Brockett · 24, 116, 117, 128, 183, 189,
    191, 198
Bu-Bshait · 395
Bukh · 355
Bulla · 402, 404, 422

C

Camanho · 357, 396, 406
Carotenuto · 426
Caves · 203
Chalos · 347
Chambers · 142, 143
Chang · 42, 43, 49, 72, 77, 309, 353
Charnes · 2, 3, 4, 5, 6, 7, 8, 9, 10, 12, 13,
    23, 25, 30, 31, 42, 43, 45, 61, 62, 77,
    78, 80, 81, 82, 89, 91, 93, 95, 102,
    106, 116, 118, 141, 159, 193, 204,
    230, 232, 233, 237, 238, 239, 244,
    249, 253, 254, 258, 260, 269, 270,
    299, 301, 303, 304, 307, 310, 315,
    324, 328, 347, 372, 403, 406, 410,
    426, 456, 459, 483, 493, 509, 510,
    524, 529
Chen · 209, 212, 224
Cherchye · 183
Cherian · 347
Chilingerian · 481, 484, 485, 488, 489,
    491, 494, 495, 496, 499, 500, 501,
    503, 506, 507, 509, 512, 515, 516,
    517, 521, 523, 524, 525, 526, 529
Christensen · 203
Chu · 425
Chubb · 324
Chung · 142, 143
Clark · 353
Clarke · 237
Clotfelter · 323

Coelli · 102, 375, 556, 557
Cohn · 324
Coleman · 418
Colwell · 380
Conover · 318
Cook · 23, 24, 106, 107, 109, 153, 154,
    157, 165, 169, 170, 171, 172, 173,
    356, 396, 404, 405, 422, 426
Cooper · 2, 3, 4, 5, 6, 7, 8, 9, 10, 12, 13,
    23, 29, 30, 31, 32, 33, 35, 42, 43, 45,
    46, 47, 49, 53, 56, 59, 61, 66, 68, 72,
    77, 82, 95, 117, 121, 128, 141, 157,
    159, 160, 165, 174, 183, 184, 188,
    189, 190, 193, 196, 198, 204, 230,
    232, 233, 234, 235, 237, 238, 239,
    240, 241, 243, 244, 245, 248, 249,
    250, 251, 252, 253, 254, 255, 259,
    260, 261, 299, 301, 303, 304, 307,
    310, 315, 324, 328, 347, 359, 383,
    384, 402, 403, 404, 405, 406, 410,
    411, 422, 426, 455, 484, 493, 502,
    509, 510, 512, 519, 529, 546, 547, 554
Cornia · 407
Corra · 410
Cowie · 426
Coyle · 416
Criswell · 404, 423

D

Das · 310
Datar · 310
Datta · 347
Davis · 380
Debreu · 7, 139, 141, 402
Delsi · 215, 216, 217, 218
Deng · 182, 189, 196, 198, 250, 251
Deprins · 269, 410, 425
Dervaux · 426
Desli · 215, 309
DeYoung · 352, 353, 366, 380

Dharmapala · 81, 82, 83, 84, 85, 86, 88, 89, 91, 92, 95
Diewert · 145, 203
Doyle · 173, 426, 456, 459
Drake · 356, 397
Ducharme · 277
Duncombe · 323, 347
Dunstan · 348
Dyson · 23, 107, 109, 117, 347, 357, 396, 406, 408, 552

E

Efron · 275, 277

F

Färe · 29, 32, 42, 62, 71, 72, 140, 141, 142, 143, 182, 183, 184, 185, 186, 187, 203, 207, 215, 225, 267, 268, 309, 347, 376, 403, 407, 414, 415, 416, 425, 550, 552
Farrell · 4, 5, 6, 7, 10, 27, 28, 139, 141, 146, 148, 268, 328, 338, 403, 405, 532
Farren · 347
Favero · 354
Firsch · 417
Fixler · 364, 380
Forrester · 415
Førsund · 55, 61, 309, 314
Fried · 403, 411
Friedman · 277
Frisch · 55, 416
Fukuyama · 62, 347, 354

G

Gang · 411
Geske · 324
Gijbels · 274, 285
Giokas · 356, 396, 397, 400

Girod · 411, 412, 418
Golany · 23, 44, 104, 105, 107, 116, 119, 121, 357, 397, 459, 493, 522, 524
Gold · 353, 355, 399
Grabowski · 366
Green · 106, 108, 173, 348, 426
Greene · 317, 516
Grifell-Tatje · 215
Grinman · 425
Grosskopf · 29, 32, 42, 62, 71, 72, 142, 182, 183, 184, 185, 186, 187, 203, 207, 347, 403, 407, 414, 415, 416, 550, 552
Gsatch · 312
Gu · 183, 188, 189, 196

H

Haag · 249, 356, 397
Hababou · 153, 396
Hall · 293, 449, 451
Halme · 112
Hancock · 352
Hanushek · 324, 325, 338, 342
Hao · 352
Hayes · 425
Haynes · 366
Hebner · 172
Hoopes · 404, 405, 407, 409, 423
Hougaard · 412
Howard · 444
Howcroft · 356, 397
Huang · 198, 230, 240, 241, 243, 250, 251, 252, 259, 260, 529
Humphrey · 352, 353, 354, 380
Husain · 425

I

Ijiri · 255
Iman · 318

J

Jagannathan · 261
Janakiraman · 316
Jaska · 249, 356, 397
Johnston · 169, 172
Jones · 459, 549
Judd · 324

K

Kabnurkar · 412
Kantor · 357, 398
Kao · 411, 412
Kaparakis · 352
Karsak · 426
Kazakov · 404, 405, 422, 426
Kemeny · 148
Kennington · 347
Kerstens · 426
Kibler · 418
Kim · 154, 157
Kinsella · 358
Kittelsen · 309
Kneip · 272, 273
Koopmans · 3, 5, 6, 10, 234, 260, 402,
    403, 410, 509
Korhonen · 112
Korostelev · 273
Kress · 170, 171, 173
Kuchta · 410
Kuman · 425
Kumar · 426
Kuosmanen · 183

L

Ladd · 323
Ladino · 356, 399
Lahman · 448
Land · 238, 244, 257, 260, 367, 411
Lang · 108

Laviolette · 411
Leibenstein · 183, 257, 324, 483
Lelas · 240, 241, 243
Lewin · 506, 522
Lewis · 282
Li · 75, 95, 106, 183, 188, 189, 196, 198,
    230, 240, 241, 243, 250, 251, 252,
    259, 260
Lindgren · 203, 207
Lindsey · 444
Lins · 106
Liu · 411, 412
Löthgren · 278, 407
Lotov · 425
Lovell · 29, 32, 42, 62, 71, 140, 182, 183,
    184, 185, 186, 187, 215, 238, 243,
    244, 257, 260, 327, 357, 398, 403,
    410, 411, 414
Luenberger · 140, 142

M

Mahler · 140, 146
Maindiratta · 43, 62
Maital · 357, 398, 483
Malmquist · 203, 204, 206, 207, 212, 215,
    219, 222, 223, 224, 225, 282, 309,
    370, 375, 376, 426, 546, 550, 554,
    556, 558, 563
Malt · 325
Manksi · 275
Mayston · 348
Mazur · 444
McCarty · 333, 347
McCutcheon · 169
McDermott · 427
McMillan · 347
Meeusen · 410
Mehrez · 347
Meza · 106
Miller · 444
Millington · 347

Miner · 342, 347

Miyashita · 427

Moe · 324

Moock · 324

Morey · 20, 21, 58, 68, 330, 331, 332, 336, 337, 364, 459

Morgenstern · 148

Moutray · 347

Muñiz · 329, 347

N

Nakanishi · 426

Nash · 357, 398

Natarajan · 309, 312, 313, 314, 316

Neralic · 77, 249

Nijkamp · 407, 408

Niskanen · 342

Nolan · 426

Norsworthy · 426

O

Oates · 325

Odden · 323

Odeck · 426

Olesen · 107, 110, 240, 241, 243, 244, 260, 411, 562

Ondrich · 338

Oral · 356, 398

Otis · 404, 410, 424, 425

P

Palmer · 444

Panzar · 62

Papi · 354

Paradi · 329, 355, 361, 363, 365, 366, 367, 368, 370, 386, 387, 395, 398, 399, 400, 427, 461

Park · 31, 32, 35, 59, 188, 273, 274, 402, 404, 411, 422, 502, 512

Parkan · 355, 399

Partangel · 407, 409

Pastor · 31, 32, 35, 59, 188, 355, 357, 398, 502, 512

Pedraja-Chaparro · 115, 406

Perez · 486

Peristiani · 366

Persaud · 426

Petersen · 106, 107, 110, 124, 208, 244, 260, 310, 411, 542, 562

Petroni · 426

Podinovski · 109, 127

Portela · 112

Post · 183

Press · 282

Primont · 141, 142

R

Ramanathan · 425

Ray · 45, 110, 114, 131, 215, 216, 217, 218, 309, 314, 332, 333, 334, 347

Reagan · 347

Reeves · 106

Reschovsky · 323, 341

Resti · 366

Rhodes · 2, 3, 4, 5, 6, 7, 8, 13, 42, 45, 141, 204, 238, 244, 253, 303, 324, 347, 403, 493, 509

Richmond · 427, 430

Rietveld · 407, 408

Roll · 23, 104, 107, 116, 121, 404, 405, 422, 459, 493, 522, 524

Roos · 203, 207

Rosen · 399

Rouatt · 377, 378, 379, 381, 383

Rousseau · 91, 249

Ruggiero · 21, 323, 330, 331, 332, 333, 334, 337, 338, 339, 342, 347

Russell · 203

S

Sahoo · 29
Salinas-Jiménez · 406
Samuelson · 178, 413
Sarangi · 410
Sarkis · 425, 427
Sarrico · 115, 347, 406
Sawada · 426
Schaffnit · 110, 111, 356, 365, 379, 385, 386, 398, 399, 427
Schaffnit-Chatterjee · 427
Schmidt · 243, 403, 410, 411
Scott · 280, 532
Seaman · 411
Seaver · 417
Seiford · 2, 7, 12, 29, 30, 31, 33, 42, 43, 53, 68, 71, 72, 77, 89, 90, 91, 92, 93, 94, 95, 107, 167, 183, 184, 190, 193, 209, 249, 261, 404, 447, 531, 542, 546, 547, 554, 559
Semple · 91, 249
Sengupta · 260, 411, 412
Sexton · 76, 95, 347, 486, 497, 498, 526
Shale · 406
Sharp · 444, 447, 452
Shen · 42, 43, 71, 72
Shephard · 7, 139, 140, 141, 146, 150, 267, 268, 402
Sherman · 353, 355, 356, 399, 481, 484, 485, 488, 489, 495, 496, 500, 506, 507, 509, 512, 515, 523, 524, 525, 526, 527, 529
Sheth · 412, 418
Shewhart · 423
Shin · 183
Silva Portela · 338, 347
Silverman · 278, 279, 280
Simar · 76, 265, 270, 275, 276, 277, 278, 279, 280, 282, 410
Simon · 230, 255, 256, 257, 260
Sinha · 237, 406

Sinuany-Stern · 347
Sleeper · 347
Smith · 348, 351, 406, 427
Soloveitchik · 425
Soteriou · 357, 399
Sowlati · 361
Spanos · 269
Stavrinides · 357
Stedry · 256
Sterman · 415, 417, 427, 430
Sterna-Karwat · 357, 398
Stigler · 183, 257
Storbeck · 357, 397
Sueyoshi · 68, 444
Sullivan · 406
Sun · 402, 426, 529
Svensson · 183
Swanepoel · 277
Symonds · 254, 260

T

Taggart · 347
Takamura · 2
Talluri · 427
Tambour · 278, 407
Tavares · 8
Taylor · 114, 116
Thanassoulis · 23, 104, 105, 107, 109, 110, 112, 114, 117, 121, 122, 123, 124, 125, 126, 128, 130, 132, 133, 324, 338, 347, 348, 408, 459, 552, 554
Theil · 318
Thompson · 23, 56, 66, 68, 81, 82, 83, 84, 85, 86, 87, 88, 89, 91, 92, 95, 106, 110, 112, 113, 114, 116, 117, 131, 148, 183, 189, 354, 366, 404, 423
Thore · 238, 244, 257, 260, 411
Thorn · 444
Thorogood · 347
Thrall · 42, 43, 46, 49, 56, 59, 60, 61, 64, 66, 68, 81, 82, 83, 84, 85, 86, 88, 89,

91, 92, 95, 112, 161, 183, 189, 196,
203, 224, 233, 446
Tibshirani · 277
Tone · 2, 7, 12, 29, 30, 31, 33, 42, 53, 68,
95, 132, 184, 203, 209, 214, 215, 261,
426, 546, 547
Triantis · 401, 404, 405, 407, 409, 410,
411, 412, 413, 417, 418, 423, 424
Tuenter · 153, 396
Tulkens · 353, 356, 400, 410
Tyteca · 425

V

Van Den Broeck · 410
Vander Vennet · 366
Vaneman · 412, 413, 415, 416
Varian · 55, 66, 112, 178
Vassiloglou · 356, 400
Vela · 329, 366, 367, 368, 370, 461
Vitaliano · 347

W

Walters · 407, 561
Wang · 183, 425
Weber · 347
Weinrach · 425
Wheelock · 269
Willig · 62

Wilson · 76, 77, 265, 269, 270, 275, 276,
277, 278, 279, 280, 282, 402, 404,
417, 422, 486, 497, 504, 525, 526
Wolstenholme · 416
Wong · 23, 114
Wooldridge · 318
Wyckoff · 324

Y

Yaisawarng · 333, 347
Yamakawa · 427
Yang · 329, 355, 363, 365, 387, 395, 400,
461
Yinger · 323, 347
Yolalan · 356, 398
Yu · 44
Yue · 372

Z

Zadeh · 411
Zenios · 357, 399
Zhang · 425
Zhu · 2, 17, 18, 23, 24, 42, 43, 45, 50, 71,
72, 89, 90, 91, 92, 93, 94, 95, 107,
117, 154, 159, 183, 190, 193, 209,
213, 249, 447, 494, 554, 555, 556
Zieschang · 364, 380

Subject Index

A

Additive model · 30, 31, 32, 37, 41, 44,
  58, 59, 60, 81, 92, 93, 96, 193, 196,
  200, 224, 231, 356, 502, 503, 519,
  520, 523, 534
Adequacy · 299, 300, 316, 317, 318,
  320, 341, 351, 458
Allocative efficiency · 28, 32, 34, 36,
  137, 143, 145, 146, 151, 200, 268, 307
Assurance Regions · 110, 112, 117,
  119, 120, 122, 128, 131, 132, 134
Assurance Regions type I · 110, 111,
  115, 121, 127, 129, 133
Assurance Regions type II · 112, 113,
  132, 133

B

Banker-Morey Model · 20, 21, 58, 68,
  330, 331, 332, 336, 337, 364, 459
Banking institution · 354, 355
Baseball · 443, 444, 447, 448, 449, 451,
  454
BCC · 13, 17, 29, 41, 43, 44, 45, 46, 48,
  49, 51, 52, 53, 54, 56, 57, 60, 61, 71,
  72, 73, 113, 204, 209, 232, 233, 239,
  245, 247, 269, 301, 303, 307, 310,
  315, 355, 356, 359, 361, 364, 366,
  382, 383, 384, 385, 386, 395, 396,
  397, 398, 400, 426, 493, 494, 523,
  540, 545, 550, 556, 561
Best-practice · 351, 352
Bootstrap · 265, 286, 287, 288, 289, 290,
  291, 295, 296
Bureaucracy · 344, 346

C

Catch-up · 204, 205, 206, 219, 224
Categorical variable · 6, 8, 22, 36, 514,
  533, 546, 550, 555, 561
Categories · 540, 541, 543, 550
CCR · 3, 4, 8, 9, 13, 14, 15, 16, 17, 19,
  20, 21, 22, 24, 29, 41, 42, 43, 45, 49,
  50, 51, 52, 53, 54, 56, 57, 59, 60, 71,
  72, 73, 82, 90, 91, 113, 204, 209, 239,
  244, 245, 253, 256, 269, 303, 355,
  356, 357, 361, 366, 382, 383, 384,
  395, 396, 397, 398, 399, 400, 445,
  446, 450, 484, 493, 523, 540, 545,
  546, 550, 552, 553, 556, 561, 563
Chance constrained DEA · 229

Competitive index · 362
Complementary slackness · 15, 85, 86, 92
Cone Ratios · 116, 119, 120, 128, 129, 130, 131, 133, 134
Confidence intervals · 265, 266, 274, 276, 277, 281, 282, 286, 291, 294
Congestion · 19, 36, 37, 177, 179, 180, 181, 182, 183, 184, 185, 186, 187, 188, 189, 190, 191, 192, 194, 195, 197, 198, 199, 200, 245, 249, 260, 426, 552
Contextual variables · 299, 300, 314, 315
Control Charts · 423, 435
Cost function · 62, 112, 139, 140, 145, 146, 537
Coverage · 19, 24, 265, 282, 286, 292, 293, 295, 404, 462, 488
CRS · 14, 45, 46, 52, 100, 108, 110, 112, 118, 123, 131, 132, 133, 168, 174, 204, 209, 215, 216, 218, 220, 221, 222, 224, 351, 427, 540, 550, 552
Culture · 362, 363, 367, 369, 370, 488

D

DEA · 1, 2, 3, 4, 6, 7, 8, 10, 11, 13, 17, 19, 22, 24, 25, 27, 31, 35, 36, 37, 38, 39, 41, 42, 43, 44, 51, 53, 55, 61, 62, 69, 70, 71, 73, 75, 76, 77, 81, 83, 84, 93, 95, 96, 97, 99, 100, 104, 105, 106, 107, 108, 109, 110, 111, 112, 113, 114, 115, 116, 118, 119, 120, 121, 122, 123, 126, 127, 128, 129, 130, 131, 134, 135, 136, 137, 138, 139, 140, 148, 150, 151, 153, 154, 155, 156, 157, 159, 160, 165, 167, 168, 169, 170, 171, 173, 174, 175, 177, 179, 180, 183, 184, 197, 198, 199, 200, 203, 204, 207, 208, 218, 224, 225, 226, 227, 229, 230, 232, 233,

237, 238, 239, 243, 244, 245, 248, 249, 252, 253, 254, 256, 260, 261, 262, 263, 264, 265, 266, 269, 271, 273, 274, 275, 295, 296, 297, 299, 300, 301, 302, 303, 305, 307, 308, 312, 313, 314, 315, 316, 317, 318, 319, 320, 321, 323, 324, 328, 330, 332, 333, 334, 335, 336, 337, 338, 342, 343, 344, 345, 346, 347, 349, 350, 352, 353, 354, 355, 356, 357, 358, 359, 360, 361, 362, 363, 364, 365, 366, 369, 370, 371, 372, 374, 376, 381, 382, 383, 384, 385, 387, 388, 390, 391, 392, 393, 394, 395, 401, 403, 404, 405, 406, 407, 408, 409, 410, 411, 412, 414, 417, 418, 419, 420, 421, 422, 423, 425, 426, 427, 428, 429, 430, 431, 432, 433, 434, 435, 436, 437, 439, 440, 441, 443, 444, 445, 447, 448, 449, 454, 455, 460, 461, 462, 466, 467, 470, 472, 478, 479, 481, 482, 483, 484, 485, 486, 487, 488, 489, 490, 491, 492, 493, 494, 496, 498, 499, 500, 503, 504, 506, 507, 508, 509, 511, 512, 513, 514, 515, 516, 517, 518, 519, 520, 522, 523, 524, 525, 526, 527, 528, 529, 530, 531, 532, 533, 534, 535, 536, 537, 539, 540, 542, 544, 545, 546, 547, 550, 551, 552, 553, 554, 555, 556, 558, 559, 560, 561, 562, 563, 564, 565, 566
DEA estimator · 275, 296, 299, 300, 301, 302, 317, 318, 319
DFA · 352
DMU · 1, 3, 4, 6, 7, 8, 9, 10, 15, 17, 19, 20, 21, 22, 24, 25, 27, 33, 45, 53, 72, 75, 76, 78, 79, 80, 81, 82, 83, 84, 87, 88, 89, 90, 91, 94, 95, 100, 101, 102, 103, 104, 107, 108, 109, 111, 112, 114, 115, 117, 118, 120, 122, 123, 124, 125, 126, 127, 128, 129, 130,

133, 140, 148, 149, 154, 155, 158,
159, 160, 161, 168, 170, 173, 174,
186, 187, 203, 204, 206, 209, 216,
218, 219, 220, 221, 222, 229, 231,
232, 234, 235, 236, 240, 241, 243,
249, 253, 254, 255, 256, 257, 299,
300, 301, 306, 307, 309, 312, 313,
327, 328, 329, 330, 331, 332, 359,
372, 375, 376, 384, 407, 409, 419,
425, 444, 445, 446, 447, 449, 483,
492, 493, 495, 506, 515, 531, 533,
542, 546, 547, 548, 550, 552, 553,
554, 555, 556, 557, 558, 559

E

Education · 4, 238, 261, 323, 325, 329,
343, 344, 345, 346, 347, 348, 565
Efficiency · 1, 2, 3, 4, 5, 6, 7, 8, 10, 11,
13, 14, 15, 17, 18, 19, 20, 22, 24, 25,
27, 28, 29, 30, 32, 35, 36, 37, 38, 41,
44, 45, 46, 47, 51, 52, 53, 54, 55, 59,
60, 61, 62, 63, 65, 69, 70, 73, 75, 76,
77, 92, 93, 96, 97, 99, 100, 101, 103,
104, 105, 106, 107, 109, 110, 111,
112, 114, 116, 120, 121, 126, 127,
128, 129, 130, 131, 133, 134, 135,
136, 137, 138, 139, 140, 141, 143,
145, 146, 147, 148, 152, 153, 156,
157, 158, 167, 168, 170, 173, 174,
177, 181, 183, 184, 185, 187, 190,
200, 201, 203, 204, 205, 206, 207,
209, 215, 216, 217, 218, 224, 225,
226, 227, 230, 231, 232, 234, 235,
236, 237, 238, 239, 241, 245, 250,
252, 253, 255, 257, 260, 261, 262,
263, 264, 265, 266, 268, 271, 276,
281, 285, 292, 296, 297, 298, 299,
300, 302, 305, 306, 308, 309, 310,
311, 312, 313, 314, 320, 321, 323,
324, 327, 328, 329, 330, 332, 333,
334, 335, 336, 337, 338, 340, 341,

342, 343, 344, 345, 346, 349, 350,
351, 352, 353, 354, 355, 356, 357,
358, 359, 360, 366, 367, 368, 370,
371, 373, 374, 375, 376, 377, 378,
382, 383, 387, 388, 389, 390, 391,
392, 393, 394, 402, 403, 404, 405,
406, 407, 408, 409, 410, 412, 413,
414, 415, 416, 417, 418, 422, 423,
424, 425, 426, 427, 429, 431, 432,
433, 434, 435, 436, 437, 438, 439,
440, 441, 443, 444, 445, 446, 448,
450, 452, 453, 454, 455, 456, 457,
458, 459, 460, 461, 463, 464, 465,
466, 467, 468, 469, 470, 471, 472,
473, 474, 475, 476, 477, 478, 479,
481, 482, 483, 484, 486, 487, 488,
489, 491, 492, 493, 494, 495, 497,
498, 499, 500, 501, 503, 506, 507,
508, 509, 512, 513, 514, 515, 516,
517, 520, 521, 522, 523, 524, 526,
528, 529, 531, 532, 533, 534, 535,
536, 537, 542, 546, 548, 552, 553,
556, 557, 558, 559, 564, 565, 566
E-model · 229, 230, 244, 245, 252
Engineering Applications · 421, 426,
439
Environmental Controls · 424, 434
Environmental factor · 169, 337, 362,
386
Equity · 341, 344, 345, 372, 397
Excel · 39, 71, 95, 97, 542, 544, 545,
546, 547, 548, 551, 554, 555, 556,
557, 558
Exclusive scheme · 209
Exogenous · 327, 328, 329, 334

F

FDH · 269, 273, 274, 275, 282, 285, 297,
352, 353, 355, 356, 393, 400, 410
Frontier-shift · 204, 205, 206, 219

H

Handicapping function · 329, 363, 364, 366, 387
Health · 199, 297, 380, 481, 482, 483, 484, 485, 486, 487, 488, 489, 490, 491, 492, 493, 494, 495, 502, 503, 504, 505, 506, 507, 508, 512, 513, 514, 515, 516, 517, 520, 522, 523, 524, 525, 526, 528, 529, 530, 531, 532, 533, 534, 535, 536, 537
Highway Maintenance Patrols · 422, 433
Hospital · 2, 8, 110, 136, 226, 482, 484, 485, 486, 487, 488, 489, 490, 491, 493, 494, 495, 497, 500, 501, 503, 504, 505, 506, 507, 508, 509, 511, 512, 513, 515, 524, 525, 526, 527, 528, 529, 531, 532, 533, 534, 535, 536, 537

I

Identification · 25, 100, 118, 188, 198, 224, 379, 417, 418, 420, 424
Index number · 5, 226, 364, 556
Inference · 266, 271, 275, 298, 300
Influential Observations · 441
Input separability · 300, 307, 308
Input-oriented · 13, 219
Intermediation approach · 355
Intertemporal · 208, 224, 225, 434
Iterated bootstrap · 277, 293, 294, 295

K

KPI · 351

L

Likert scale · 153, 157, 169, 170, 172, 174, 362

M

Malmquist index · 203, 212, 224, 226, 297
Malmquist Index · 207, 370, 376, 390
Management · 3, 6, 19, 23, 68, 69, 153, 154, 156, 170, 182, 198, 199, 200, 237, 254, 339, 349, 350, 351, 356, 357, 358, 360, 361, 362, 363, 367, 368, 370, 371, 375, 377, 380, 381, 383, 384, 385, 388, 403, 407, 409, 419, 422, 426, 427, 443, 444, 456, 458, 459, 460, 462, 466, 468, 471, 474, 476, 477, 478, 488, 491, 494, 495, 497, 522, 528, 531, 534, 536, 537
Monte Carlo simulation · 276, 387

N

Naive bootstrap · 265, 277, 278, 284, 285, 286, 292
Nation at Risk · 324, 344
Nerlovian indicator of profit efficiency · 143
Network DEA · 436
Niskanen · 342, 344
Non-discretionary · 327, 330, 336, 345, 346
Non-oriented · 222, 223
Non-parametric · 203, 207, 224, 260, 297, 303, 304, 305, 307, 308, 351, 352, 355, 388, 403, 483, 535

O

Operational thinking · 428

Ordinal data · 153
Ordinal weight restrictions · 447, 449, 451
Outlier · 76, 116, 418
Output-oriented · 13, 223

P

Panel data · 373
Parametric · 243, 266, 293, 295, 297, 299, 300, 302, 307, 309, 313, 316, 317, 318, 319, 351, 352, 353, 403, 535
Parametric functional · 299, 300, 309, 316, 317, 318
Performance analysis · 350, 363
Performance Measurement Science · 428
P-model · 230, 252
Production approach · 355
Productivity change · 204, 205, 224, 225, 299, 300, 309, 310, 311, 376
Profitability · 2, 112, 137, 138, 351, 356, 357, 379, 382, 455, 456, 457, 459, 465, 470, 471, 472, 473, 474, 478, 502, 503
Progress · 197, 320, 321, 391

Q

Qualitative data · 153, 169

R

R&D · 154, 156, 167, 173, 433
Radius · 79
Rank position · 153
Ratio analysis · 351
Resampling · 284, 285, 292
Returns to scale · 13, 29, 35, 43, 45, 46, 47, 51, 52, 55, 59, 204, 286, 287, 288, 289, 290, 291, 303, 345, 356, 512, 561

Returns to Scale · 41, 42, 43, 45, 46, 49, 54, 57, 61, 64, 65, 66, 71, 72, 73, 131, 132, 133, 215, 224, 395

S

SBM · 209, 210, 211, 212, 213, 214, 383, 384
Scale efficiency · 203, 215, 216, 225
Sensitivity · 75, 77, 83, 96, 97, 262, 297, 393
Serrano v. Priest · 341
Service quality · 173, 357, 367, 369, 481
SFA · 352, 353, 392, 489, 490
Software · 94, 95, 120, 131, 135, 137, 149, 156, 157, 420, 539, 540, 544, 545, 554, 555, 556, 557, 558, 559, 560, 561, 562, 563
Solver · 37, 39, 70, 71, 96, 97, 135, 200, 218, 226, 263, 390, 434, 534, 544, 545, 546, 547, 554, 555, 556, 557, 558, 562, 565, 566
Sports · 443
Super-efficiency · 97, 106, 208, 209, 223, 225, 226, 227, 310, 444, 445, 446, 448, 452, 453, 454, 542, 545, 550, 553, 554, 556, 558, 559
Super-SBM · 209, 210, 211, 212, 213, 214, 215

T

Terrestrially-Based Large Scale Commercial Power Systems · 423
TFA · 352, 353
Tobit · 333, 508, 514, 515, 516
Transit Systems · 425, 432
Translog function · 490

U

Unobserved DMUs · 99, 121, 122, 123, 125, 126, 130, 131, 133, 134

V

Value Judgments · 105
VRS · 14, 108, 111, 118, 122, 130, 131, 132, 159, 160, 163, 165, 168, 174, 204, 209, 215, 216, 223, 224, 225, 427, 540, 550

W

Weights restrictions · 99, 108, 109, 110, 111, 114, 115, 116, 118, 122, 129, 130, 131, 132, 133, 134
Window analysis · 25, 27, 77, 373, 376, 542, 546, 550, 552, 558

X

X-inefficiency · 324, 483

Early Titles in the
INTERNATIONAL SERIES IN
OPERATIONS RESEARCH & MANAGEMENT SCIENCE
Frederick S. Hillier, Series Editor, *Stanford University*

Saigal/ *A MODERN APPROACH TO LINEAR PROGRAMMING*
Nagurney/ *PROJECTED DYNAMICAL SYSTEMS & VARIATIONAL INEQUALITIES WITH*
   APPLICATIONS
Padberg & Rijal/ *LOCATION, SCHEDULING, DESIGN AND INTEGER PROGRAMMING*
Vanderbei/ *LINEAR PROGRAMMING*
Jaiswal/ *MILITARY OPERATIONS RESEARCH*
Gal & Greenberg/ *ADVANCES IN SENSITIVITY ANALYSIS & PARAMETRIC PROGRAMMING*
Prabhu/ *FOUNDATIONS OF QUEUEING THEORY*
Fang, Rajasekera & Tsao/ *ENTROPY OPTIMIZATION & MATHEMATICAL PROGRAMMING*
Yu/ *OR IN THE AIRLINE INDUSTRY*
Ho & Tang/ *PRODUCT VARIETY MANAGEMENT*
El-Taha & Stidham/ *SAMPLE-PATH ANALYSIS OF QUEUEING SYSTEMS*
Miettinen/ *NONLINEAR MULTIOBJECTIVE OPTIMIZATION*
Chao & Huntington/ *DESIGNING COMPETITIVE ELECTRICITY MARKETS*
Weglarz/ *PROJECT SCHEDULING: RECENT TRENDS & RESULTS*
Sahin & Polatòglu/ *QUALITY, WARRANTY AND PREVENTIVE MAINTENANCE*
Tavares/ *ADVANCES MODELS FOR PROJECT MANAGEMENT*
Tayur, Ganeshan & Magazine/ *QUANTITATIVE MODELS FOR SUPPLY CHAIN MANAGEMENT*
Weyant, J./ *ENERGY AND ENVIRONMENTAL POLICY MODELING*
Shanthikumar, J.G. & Sumita, U./ *APPLIED PROBABILITY AND STOCHASTIC PROCESSES*
Liu, B. & Esogbue, A.O./ *DECISION CRITERIA AND OPTIMAL INVENTORY PROCESSES*
Gal, T., Stewart, T.J., Hanne, T. / *MULTICRITERIA DECISION MAKING: Advances in*
   MCDM Models, Algorithms, Theory, and Applications
Fox, B.L. / *STRATEGIES FOR QUASI-MONTE CARLO*
Hall, R.W. / *HANDBOOK OF TRANSPORTATION SCIENCE*
Grassman, W.K. / *COMPUTATIONAL PROBABILITY*
Pomerol, J-C. & Barba-Romero, S. / *MULTICRITERION DECISION IN MANAGEMENT*
Axsäter, S. / *INVENTORY CONTROL*
Wolkowicz, H., Saigal, R., & Vandenberghe, L. / *HANDBOOK OF SEMI-DEFINITE*
   PROGRAMMING: Theory, Algorithms, and Applications
Hobbs, B.F. & Meier, P. / *ENERGY DECISIONS AND THE ENVIRONMENT: A Guide*
   to the Use of Multicriteria Methods
Dar-El, E. / *HUMAN LEARNING: From Learning Curves to Learning Organizations*
Armstrong, J.S. / *PRINCIPLES OF FORECASTING: A Handbook for Researchers and*
   Practitioners
Balsamo, S., Personé, V., & Onvural, R./ *ANALYSIS OF QUEUEING NETWORKS WITH*
   BLOCKING
Bouyssou, D. et al. / *EVALUATION AND DECISION MODELS: A Critical Perspective*
Hanne, T. / *INTELLIGENT STRATEGIES FOR META MULTIPLE CRITERIA DECISION MAKING*
Saaty, T. & Vargas, L. / *MODELS, METHODS, CONCEPTS and APPLICATIONS OF THE*
   ANALYTIC HIERARCHY PROCESS
Chatterjee, K. & Samuelson, W. / *GAME THEORY AND BUSINESS APPLICATIONS*
Hobbs, B. et al. / *THE NEXT GENERATION OF ELECTRIC POWER UNIT COMMITMENT*
   MODELS
Vanderbei, R.J. / *LINEAR PROGRAMMING: Foundations and Extensions, 2nd Ed.*
Kimms, A. / *MATHEMATICAL PROGRAMMING AND FINANCIAL OBJECTIVES FOR*
   SCHEDULING PROJECTS
Baptiste, P., Le Pape, C. & Nuijten, W. / *CONSTRAINT-BASED SCHEDULING*
Feinberg, E. & Shwartz, A. / *HANDBOOK OF MARKOV DECISION PROCESSES: Methods*
   and Applications

* A list of the more recent publications in the series is at the front of the book *